J. Breuste, H. Feldmann, O. Uhlmann (Eds.) **Urban Ecology**

Springer-Verlag Berlin Heidelberg GmbH

J. Breuste H. Feldmann O. Uhlmann (Eds.)

Urban Ecology

With 80 Figures and 51 Tables

Springer

Prof. Dr. Jürgen Breuste
Dr. Hildegard Feldmann
Ogarit Uhlmann

UFZ Centre for Environmental Research
Leipzig-Halle GmbH
Permoserst. 15
D-04318 Leipzig
Germany

Cover illustration:
Street map of Leipzig. © Falk-Verlag AG, Kartographie: GeoData
Photographs taken by J. Breuste

Library of Congress Cataloging-in-Publication Data
Urban ecology / J. Breuste, H. Feldmann, O. Uhlmann (eds.).
p. cm.
Results of a international conference held in Leipzig, Germany, June 25–29, 1997.
Includes bibliographical references.

ISBN 978-3-642-88585-3 ISBN 978-3-642-88583-9 (eBook)
DOI 10.1007/978-3-642-88583-9

1. Urban ecology – Congresses. 2. City planning – Environmental aspects – Congresses.
3. Sustainable development – Congresses. I. Breuste, Jürgen. II. Feldmann, H. (Hildegard), 1957–
III. Ohlmann, O. (Ogarit), 1971– .
HT241.U724 1998
307.76 – dc21 98-34103
CIP

Originally published by Springer-Verlag Berlin Heidelberg New York in 1998.

Cover design: design & production GmbH, Heideberg
Data conversion: Best-set Typesetter Ltd., Hong Kong

SPIN: 10761543 31/3111 – 5 4 3 2 1 – Printed on acid-free paper

Preface

In the last 3 decades urban ecology has developed increasingly from a scientific branch of Biology towards an interdisciplinary research field with applications in local and regional planning. Not only natural science disciplines, such as Biology, Soil Science, Hydrology, Climatology and Landscape Ecology contribute to the complex field of urban ecology, but also the humanities. Today it is clear that only a connection between the economic, cultural and social fields can lead to success in coping with the serious environmental problems of many cities world-wide. With the shift of the focal point from natural science based research to problem-oriented research on the improvement of the living environment of people in cities, Social Sciences, Cultural Sciences, Geography and Economics gain a stronger meaning in the research field of urban ecology than was the case 20 years ago.

Social Sciences and environmental problems are tightly bound together in many cities. In order to find solutions, strategies must include many different fields of expertise into account, but such inter-disciplinary measures naturally involve a high degree of complexity. Clearly, a theoretical basis is required which is oriented not only towards the national specifications of environmental and social problems and their investigation and management in the urban area, but which also takes a global view of the problems.

Exchange of information and experience is required, not only between disciplines, but also between nations and cultures since many problems are caused by the inter-twining of people and the environment in urban areas. Many of these problems are no longer confined to the boundaries of the cities, but can be viewed on a regional, inter-regional and even on an international scale; some prove themselves to be comparable world-wide, and others have a clear, regional, specific meaning.

Urban ecology, as a young, human-ecology oriented scientific discipline is of use with respect to world-wide problems, but it must find its own specific subject methods and theories to help solve these problems.

The research department for Urban Landscapes at the UFZ-Umweltforschungszentrum Leipzig-Halle GmbH (UFZ Centre for Environmental Research Leipzig-Halle) was set-up in 1992 and carried out research on urban ecology and urban development.

The department's initiative in 1996 to provide an international exchange of experience on the important focal points of urban ecological research was emphatically supported and subsidised through a scientific advisory board for urban ecology of the Federal Ministry for Education, Research and Technology (BMBF). The international conference on Urban Ecology took place in Leipzig from 25th to 29th June, 1997 under the patronage of UNESCO and with support from the European Commission and the Saxonian State Ministry for Science and Arts (SMWK).

The conference was seen as providing the start of a deep international discussion over the theoretical consolidation of urban ecology and the integration of differing scientific disciplines into this complex field. Scientists, planners and politicians were approached so that their views and contributions could be integrated into the discussion.

Several points were highlighted during the conference: one of these was the postulation of good practice examples as models, another was the transferral and generalisation of practical experience.

This led to the definition of the 5 main themes of the conference:

- the ecological city- model for quality goals and standards,
- integration of ecological, economic and social cultural bases for the improvement of the living environment,
- land use as a tool to direct the city in an ecological direction,
- ecologically acceptable mobility,
- integration of nature and landscape cover – the balance between nature and culture.

These themes show the range of the aspects dealt with in the conference. These extend from the discussion of political goals on the experience with integrated work, the problems with area and population growth, the environmental burdens through specific area uses and traffic, to the balance between nature and town development in view of growing urban agglomerations.

380 participants from 48 countries contributed their experience in 150 lectures in 15 workshops, in poster sessions and in many discussions. The participants agreed that the results of the conference should be made available in a general publication of the topics covered; this follows on from the existing volume. Heterogeneity in the means and ways of the portrayal, and also the differentiated views, on the same subject through different authors are not only made clear, but should also provoke further discussion. This book provides a cross section of the topics of urban ecology (oriented to the above main themes) with an international outlook; this provides a thorough overview of the subjects covered. At the end of the event the various working groups brought together the results of their discussions so that they directly emphasised the main focus of the conference and could point out ways for future work.

I would like to express my thanks to the international evaluation committee who were kind enough to take the time to proof-read the contributions and provided the authors with valuable advice for the revisions.

The editor hopes that this publication will provide a contribution to the development of the research field of urban ecology and initiate international discussion in the addressed themes.

Leipzig, Germany
August 1998

Jürgen Breuste

Contents

Introduction

1 Ecological Cities – Models, Environmental Aims, Standards

1.1 Workshop: Environmental Quality Aims and Ecological Aims

1.2 Workshop: Urban Climate

4.2 Workshop: Traffic Behaviour and its Possible Changes

4.3 Workshop: Traffic Emissions and their Ecological Impact

4.4 Poster

5 The Integration of Nature and Landscape into Urban Development

5.1 Workshop: Analysis and Evaluation of Urban Nature

List of Contributors

ABS, MICHAEL; Elßholzstr. 8, D-10781 Berlin, Germany

ADOMSSENT, MAIK; Lüneburg University, Institute for Environmental Communication, Rotenbleicher Weg 42, D-21335 Lüneburg, phone: +49 4131/78-29 24, fax: +49 4131/78-28 19, e-mail: adomssent@uni-lueneburg.de

AGRAWAL, MADHOOLIKA; Banaras Hindu University, Department of Botany, Varanasi – 221 005, India, phone: +91 542/317799, fax: +91 542/317074, e-mail: Madhoo@banaras.ernet.in

APEL, DIETER; HENCKEL, DIETRICH; German Institute of Urban Affairs (Difu), Straße des 17. Juni 112, D-10623 Berlin, Germany, phone: +49 30/39001-292, fax: +49 30/39001-100

ASAEDA, TAKASHI; Saitama University, Urawa, Saitama 338, Japan, e-mail: asaeda@env.civil.saitama-u.ac.jp

BALLACH, HANS-JOACHIM; Johann Wolfgang Goethe University, Department of Geobotany and Plant Ecology, Siesmayertraße 70, D-60323 Frankfurt/Main, Germany, phone: +49 69/79824757, fax: +49 69/79824822

BAUER, BERNHARD; UFZ Centre for Environmental Research Leipzig-Halle, Department of Urban Landscapes, Permoserstraße 15, D-04318 Leipzig, Germany, phone: +49 341/235-2342, fax: +49 341/235-2534, e-mail: bauer@pro.ufz.de

BAUMANN, FRAUKE; City of Bremen, c/o Senator for Environmental Conservation, Section Sustainable Development of the Urban Environment, Hanseatenhof 5, D-28195 Bremen, Germany

BECKER, UDO J.; Dresden Technical University, Chair of Transport Ecology, D-01062 Dresden, Germany, phone: +49 351/463 6504, fax: +49 351/463 7718, e-mail: becker@rcs.urz.tu-dresden.de

BEDÊ, LÚCIO CADAVAL; Brandt Meio Ambiente, Rua Timbiras 270, 30.140-060 – Belo Horizonte – MG, Brazil, phone: +55 31/2812258, fax: +55 31/225 6929

BEEKMAN, PIET; Eindhoven University of Technology, P.O. Box 513/5600 MB Eindhoven, The Netherlands, phone: +31 40/247 3691, fax: +31 40/245 2432

BERG, P.; Latvian Hydrometeorological Agency, 165 Maskavas str., LV-1019, Riga, Latvia, fax: +371 7/145154, e-mail: EPOC@meteo.lv

BERGEN, FRANK; Ruhr University Bochum, Chair for General Biology & Neurobiology, Faculty for Biology, D-44780 Bochum, Germany, fax: +49 234/7094-278, e-mail: axel.schwerk@ruhr-uni-bochum.de

BERGMANN, AXEL; Institute for Ecological Space Development Dresden (IÖR), Weberplatz 1, D-01217 Dresden, Germany, phone: +49 351/4679-0, fax: +49 351/4679-212

BERGMANN, MATTHIAS; Öko-Institut e.V. (Eco-Institute), Friedrichstraße 165, D-10117 Berlin, Germany, fax: +49 30/20 16 50 88

DE BOER, JOOP; Socio-Economic and Natural Science of the Environment, p/a Instituut voor Milieuvraagstukken, Vrije Universiteit De Boelelaan 1115, NL-1081 HV Amsterdam, The Netherlands, phone: +31 20/4449514, fax: +31 20/44495531, e-mail: JOOP.DE.BOER@IVM.VU.NL

VAN BOHEMEN, HEIN D.; Ministry of Transport, Public Works and Water Management, Road and Hydraulic Engineering Division, P.O. Box 5044, 2600 GA Delft, The Netherlands, fax: +31 15/2611361, e-mail: h.d.vbohemen@dww.rws.minvenw.nl

BÖCKER, REINHARD; University of Hohenheim, Institute for Landscape and Plant Ecology, D-70593 Stuttgart, Germany
BÖHM, PETER; UFZ Centre for Environmental Research Leipzig-Halle, Department of Urban Landscapes, Permoserstraße 15, D-04318 Leipzig, Germany, fax: +49 341/235-2534, e-mail: pboehm@pro.ufz.de
BÖTTCHER, ELLEN; Frauenstraße 11, D-04668 Grimma, Germany, phone/fax: +49 3437/918 577
BRACHER, TILMAN; IVU – Gesellschaft für Informatik, Verkehrs- und Umweltplanung mbH, (Society for Informatics, Traffic and Environmental Planning), Bundesallee 88, D-12161 Berlin, Germany, fax: +49 30/859 06-111, e-mail: tb@ivu-berlin.de,
BREUSTE, JÜRGEN; UFZ Centre for Environmental Research Leipzig-Halle, Department of Urban Landscapes, Permoserstraße 15, D-04318 Leipzig, Germany, phone: +49 341/235-2843, fax +49 341/235-2534, e-mail: breuste@pro.ufz.de
BURKERT, FRIED M; Egen 30 D-45549 Sprockhövel phone: +49 202/524585 Germany
CECCATO, VANIA; Royal Institute of Technology, Fiskartorpsvägen 15 A, S-10044 Stockholm, Sweden, fax: +46 8/7906761, e-mail: vania@infra.kth.se
CUNHA, RODRIGO C.; CETESB – Avenida Prof. Frederico Hermann Jr., 345, 05489-900 São Paulo, SP, Brazil
DEELSTRA, TJEERD; The International Institute for the Urban Environment, Nickersteeg 5, 2611 EK Delft, The Netherlands, fax: +31 15/2624873, e-mail: urban@theoffice.net
DIAS DOS SANTOS, ANTONIO CARLOS; Fundação Estadual de Engenharia do Meio Ambiente (FEEMA), Rio de Janeiro, Brazil, [3]German Society for Technical Cooperation (Deutsche Gesellschaft für Technische Zusammenarbeit (GTZ) GmbH), Eschborn
DIJST, MARTIN; Netherlands Graduate School of Housing and Urban Research, P.O. Box 80115, NL-3508 TC Utrecht, The Netherlands, phone: +31 30/2534442, fax: +31 30/2540604, e-mail: M.DIJST@FRW.RUU.NL
DOWNTON, PAUL FRANCIS; Centre for Urban Ecology, 84 Halifax Street, Adelaide, Tandanya Bioregion, SA 5000, Australia, fax: +61 8/8232 4866, e-mail: urbanec@dove.mtx.net.au, web site: http://www.eastend.com.au/~ecology/index.shtml
DUHME, FRIEDRICH; Chair for Landscape Ecology, Munich Technical University, D-85350 Freising, Germany
DZIKUS, ANDRÉ; United Nations Centre for Human Settlements (Habitat), P.O. Box 30030, Nairobi, Kenya, phone: +254 2/62 3060, fax: +254 2/624265, e-mail: andre.dzikus@unchs.org
ENGELBACH, WOLF; RWTH Aachen, Department of Environmental Engineering (ISA), D-52056 Aachen, Germany, phone: +49 241/805278, fax: +49 241/8888285; e-mail: engelb@isa.rwth-aachen.de
FAGGI DE LENZ, ANA MARIA; Cefybo-Conicet, Serrano 669, 1414 Buenos Aires, Argentina, phone: +54 1/6 33 3053, fax: +54 1/8 56 2751, e-mail: parodiana@cefybo.edu.ar
FARAT, RYSZARD; Institute of Meteorology and Water Management, ul. D'browskiego 174/176, 60-594 Poznañ, Poland, fax: +48 61/84 75-440, e-mail: dyr_imgw@rose.man.poznan.pl
FRANCO GOMES, MARIA RITA; Avenida Contorno, 8.000/608, Santo Agostinho 30.110-120, Belo Horizonte, Minas Gerais, Brazil
FREEMAN, CLAIRE; Resource and Environmental Planning, Massey University, P. Bag 11222, Palmerston North, New Zealand
FREY, JOCHEN; FÖJ Rheinland-Pfalz, PF 1951, D-55009 Mainz, Germany, phone: +49 6131/279463, fax: +49 6131/231971
FREYER, KLAUS; UFZ Centre for Environmental Research Leipzig-Halle, Department for Analytical Chemistry, Permoser Str. 15, D-04318 Leipzig, Germany, phone: +49 341/235-2297 or -2408 or 2922, fax: +49 341/235-2625, e-mail: freyer@ana.ufz.de

FRITSCHE UWE; Institut for Applied Ecology E.V., Darmstadt Office, Bunsenstr. 14, D-64293, Germany, phone: +49 6151/819124, fax: +49 6151/8191, e-mail: fritsche@oeko.de

GANTZ, HELGA; SAS Regional, Transport and Environmental Research, Cunnersdorfer Straße 25, D-01189 Dresden, Germany

GÉCZI, ROBERT; JATE University, Department of Physical Geography, P.O. Box 653, H-6701 Szeged, Hungary, phone/fax: +36 62/454158, e-mail: robi@earth.geo.u-szeged.hu

GEISLER, ERHARD; Leipzig University, Interdisziplinary Institute for Nature Conservation and Environmental Protection (IINU), Augustusplatz 10/11, D-04109 Leipzig, Germany, phone: +49 341/97-36345, fax: +49 341/97-36348

GHOSH, SANTOSH; The President, Centre for the Built Environment, 2/5 Sarat Bose Road, Calcutta-700 020, India, phone: +91 33/476 1495, fax : +91 33/466 0625

GLASAUER, HERBERT; Kassel University (GhK), Gottschalkstraße 28, D-34109 Kassel, Germany, phone: +49 561/804-3786, fax: +49 561/804-2485

GLOTZ-RICHTER, MICHAEL; City of Bremen, c/o Senator for Environmental Conservation, Section Sustainable Development of the Urban Environment, Hanseatenhof 5, D-28195 Bremen, Germany, phone: +49 421/361-6703, fax: +49 421/361-10875

GOODE, DAVID A.; The London Ecology Unit, Bedford House, 125 Camden High Street, London NW1 7JR, Great Britain, phone: +44 171/267 7944, fax: +44 171/267 9334, e-mail: sl@london-ecol-unit.demon.co.uk

GOEVERT, JAN; Johann Wolfgang Goethe University, Department of Geobotany and Plant Ecology, Siesmayertraße 70, D-60323 Frankfurt/Main, Germany

GORKHALY, GIRIJA PRASAD; Ministry of Housing and Physical Planning, Department of Housing and Urban Development, Barbar Mahal, Kathmandu, Nepal, phone: +977 1 427 143 or 413 621, fax: +977 1/224 139, e-mail: wtc@eee.mos.com.np

GÖTZ, KONRAD; ISOE – Institut für sozial-ökologische Forschung GmbH (Institute for Socio-Ecological Research), Hamburger Allee 45, D-60486 Frankfurt, Germany, phone: +49 69/70 00 12, fax: +49 69/777 341, e-mail: isoegmbh@t-online.de

GRAUTHOFF, MANFRED; Dr. Grauthoff Unternehmensberatung für Energie und Umwelt, Auf dem Jäger 36, D-44892 Bochum, Germany, phone: +49 234/29 54 78, fax: +49 234/29 47 47, e-mail: dr.grauthoff@cww.de

GROSSMANN, WOLF DIETER; UFZ Centre for Environmental Research Leipzig-Halle, Group Regional Future Modelling, Permoserstraße 15, D-04318 Leipzig, Germany, phone: +49 341/235-2282, fax: +49 341/235-2796, e-mail: wdgross@alok.ufz.de

GU, CHAOLIN; Chinese Academy of Sciences, Institute of Geography, Beijing 100101, China, fax: +86-10/64911844, e-mail: Gucl@iog.ac.cn

GUEVARA VELAZCO, A. VLADIMIR; National Climate Centre, Meteorological Institute, Havana, Cuba, e-mail: meteoro@ceniai.inf.cu

GULDAGER, SUSANNE; Ministry of Environment and Energy, Danish Forest and Landscape Research Institute, Hoersholm Kongevej 11, DK-2970 Hoersholm, Denmark, phone: +45 45/76 32 00, fax: +45 45/76 32 33, e-mail: sgu@fsl.dk or cnu@fsl.dk

GUNTHER, MARCO ANTONIO; CETESB – Avenida Prof. Frederico Hermann Jr., 345, 05489-900 São Paulo, SP, Brazil

HABBEL, WERNER; Dr. Krätzig Ingenieurgesellschaft mbH, Aachen, Süsterfeldstraße 81, D-52072 Aachen, German Society for Technical Cooperation (Deutsche Gesellschaft für Technische Zusammenarbeit (GTZ) GmbH), Eschborn, Germany, phone: +49 241/8 89 73-0, fax: +49 241/8 89 73-28, e-mail: dr.kraetzig-ing-gmbh@t-online.de

HAHN, EKHART; Paul Lincke Ufer 30, D-10999 Berlin, Germany, phone: +49 30/6117531, fax: +49 30/611 2320, e-mail: OEKOCITY@aol.com

HALME, EERO; Helsinki University, Department of Ecology and Systematics, P.O. Box 17, FIN-00014, Finland

HARTWIG, CHRISTIAN; City of Cologne, Department of Environment, Gürzenichstr. 6-16, D-50677 Köln, Germany, phone: +49 221/221-2770, fax: +49 221/221-4612, e-mail: christian.hartwig(betina.kuechenhoff)@×400.umweltamt.koeln.lion.de

HELD, THOMAS; Ruhr University Bochum, Institute of Geography, D-44780 Bochum, Germany, phone: +49 234/700-4790, fax: +49 234/7094-180, e-mail: held@geographie.ruhr-uni-bochum.de

HESSE, MARKUS; for Ecological Development of Economy IÖW (Institut für ökologische Wirtschaftsforschung (IÖW) GmbH), Giesebrechtstraße 13, D-10629 Berlin, Germany, phone: +49 30/884 59 4-24, fax: +49 30/882 5439, e-mail: mailbox@ioew.b.eunet.de

HILDMANN, CHRISTIAN; Berlin Technical University, Institute of Ecology, Department of Limnology, Hellriegelstraße 6, D-10629 Berlin, Germany, e-mail: hildoabi@mailszrz.zrz.TU-Berlin.de

HÖFER, WOLFRAM; Munich Technical University, Chair of Landscape Architecture and Planning, D-85350 Freising-Weihenstephan, Germany, phone: +49 8161/72 3250, fax: +49 8161/71 4158, e-mail: wolfram@pollux.edv.agrar.tu-muenchen.de

HOFFMANN, JOACHIM; Leipzig University, Interdisziplinary Institute for Nature Conservation and Environmental Protection (IINU), Augustusplatz 10/11, D-04109 Leipzig, Germany, fax: +49 341/97 36348; e-mail: hoffmann@rz.uni-leipzig.de

HOMICH, VALERY S.; Institute for Problems of Use of Natural Resources & Ecology, 10 Staroborysovski tract, 220114 Minsk, Belarus, phone: +375 17/264 23 12, fax: +375 17/264 24 13

HU, DAN; Chinese Academy of Sciences, Research Centre for Eco-Environmental Sciences, 19 Zhongguancun Road, Beijing 100080, China, fax: +86 10 62562775

HUCK, KARIN; Karlsruhe University, Institute of Petrography and Geochemistry, Kaiserstraße 12, D-76128 Karlsruhe, Germany, phone: +49 721/608-6155, fax: +49 721/697328, e-mail: karin.huck@bio-geo.uni-karlsruhe.de

HUCKSTEP, SCOTT; University of Missouri-Columbia, School of Natural Resources, 112 Stephens Hall, Columbia, MO 65211 U.S.A.

HÜMMELER, CHRISTOPH; Leipzig City Council, Department of Urban Planning, P.O. Box 780, D-04007 Leipzig, Germany, phone: +49 341/123-4923, fax: +49 341/123-4825, e-mail: chummele@leipzig.de

HÜTTER, ULRIKE; Institute for Water Research GmbH Dortmund, Zum Kellerbach 46, D-58239 Schwerte, Germany

ICHINOSE, TOMOHIRO; Chair of Landscape Ecology, Munich Technical University, D-85350 Freising-Weihenstephan, Germany, fax: +49 8161/714427, e-mail: tomohiro@dec.loek.agrar.tu-muenchen.de
Present address: Lab. of Landscape Ecology and Planning, Graduate School of Agricultural and Life Sciences, The University of Tokyo, Yajoi 1-1-1, Sunkyo-ku, Tokyo 113-8657 Japan phone: +81-3-3812-2111 ext. 5050 Fax: +81-3-5684-0623

ICHINOSE, TOSHIAKI; Centre for Global Environmental Research, National Institute for Environmental Studies, 16-2, Onogawa, Tsukuba 305, Japan, phone: +81 298/50-2598, fax: +81 298/58-2645, e-mail: ichinose@uesgc5.rcast.u-tokyo.ac.jp

INGEGNOLI, VITTORIO; University of Milan, Faculty of Natural Sciences, Milan, Italy, Professional Studio: 45 via Senato, 20121 Milano, Italy, phone: +39 2/797343, e-mail: vingegnoli@ iol.it

IPSEN, DETLEV; Kassel University, AEP (Working Group for Empirical Planning Research), Mönchebergstraße 17, D-34109 Kassel, Germany, phone: +49 561/804-3554, fax: +49 561/804-2485, e-mail: aep@hrz.uni-kassel.de

JAHN, THOMAS; Institute for Social-Ecological Research (ISOE), Hamburger Allee 45, D-60489 Frankfurt/Main, Germany, phone: +49 69/70 00 12, fax: +49 69/777 341, e-mail: isoegmbh@t-online.de

KABISCH, KLAUS; Leipzig University, Interdisziplinary Institute for Nature Conservation and Environmental Protection (IINU), Augustusplatz 10/11, D-04109 Leipzig, Germany, fax +49 341/97 36348; e-mail: kabisch@rz.uni-leipzig.de

KABISCH, SIGRUN; UFZ Centre for Environmental Research Leipzig-Halle, ÖKUS – Department of Ecological Economics and Environmental Sociology, Permoserstaße 15, D-04318 Leipzig, Germany, phone: +49 341/235-2366, fax: +49 341/235-2511, e-mail: kabisch@alok.ufz.de

KAGERMEIER, ANDREAS; Munich Technical University, Department of Geography, Arcisstraße 21, D-80290 Munich, Germany, phone: +49 89/2892-5095, fax: +49 89/2892-5380, e-mail: Kagermeier@lrz.tu-muenchen.de

KAISER, MATHIAS; Dortmund University, Chair of Landscape Ecology and Landscape Planning, Department of Spatial Planning, D-44227 Dortmund, Germany, phone: +49 231/755-2380, fax: +49 231/755-4856

KAKAREKA, SERGEY V.; Institute for Problems of Use of Natural Resources & Ecology, 10 Staroborysovski tract, 220114 Minsk, Belarus, phone: +375 17/264 23 12, fax: +375 17/264 24 13, e-mail: ikonicov@ns.ecology.ac.by

KASEK, LEONHARD; SAS Regional, Transport and Environmental Research, Cunnersdorfer Straße 25, D-01189 Dresden, Germany, phone: +49 351/4083209, fax: +49 351/4083290, e-mail: SAS-ISB@t-online.de

KATOH, KAZUHIRO; Experimental Station for Landscape Plants, Tokyo University, Hata-machi 1051, Hanamigawa-ku, Chiba 262, Japan, fax. +81 43/276-6330, e-mail: aster@hongo.ecc.u-tokyo.ac.jp

KAUSCH, STEFFEN; Dresden Technical University, Institute of Medical Microbiology and Hygiene, Löscherstraße 18, D-01309 Dresden, Germany

KEIDEL, THOMAS; UFZ Centre for Environmental Research Leipzig-Halle, Department of Urban Landscapes, Permoserstraße 15, D-04318 Leipzig, Germany, phone: +49 341/235-2390, fax: +49 341/235-2534, e-mail: keidel@pro.ufz.de

KÊPIŃSKA-KASPRZAK, MAGORZATA; Institute of Meteorology and Water Management, ul. D'browskiego 174/176, 60-594 Poznañ, Poland

KESTELOOT, CHRISTIAN; Catholic University Leuven, Institute of Social and Economic Geography, Leuven 3001, Belgium, fax: +32 16/322980, e-mail: C.kesteloot@geo.kuleuven.ac.be

KIM, KWI-GON; Seoul National University, College of Agriculture and Life Science, Department of Landscape Architecture, Suwon, Korea 441 744, phone: +82 331/290 2621 2628, fax: +82 331/292 4089 or 296 0777, e-mail: atmaranta@plaza.snu.ac.kr

KINDLER, ANNEGRET; UFZ Centre for Environmental Research Leipzig-Halle, ÖKUS – Department of Ecological Economics and Environmental Sociology, Permoserstaße 15, D-04318 Leipzig, Germany, phone: +49 341/235-2366, fax: +49 341/235-2511, e-mail: kindler@alok.ufz.de

KIVIKKOKANGAS-SANDGREN, RITVA; University of Helsinki, Department of Geography, PL4 (Yliopistokatu 3), 00014 University of Helsinki, Finland, fax: +358 09/191 22641, e-mail: ritva.kivikkokangas-sandgren@helsinki.fi

KLEINSCHMIDT, VOLKER; Sachverständigenbüro, c/o AIRAIL KG, Ollenhauerstraße 97, D-13403 Berlin, Germany, phone: +49 30/4135854, fax: +49 30/4137116 and Sachverständigenbüro c/o AIRAIL KG, Schleefstraße 4, D-44287 Dortmund, Germany, phone: +49 231/442294-0, fax: +49 231/442294-19

KOHLMANN, SABINE; Johann Wolfgang Goethe University, Department of Geobotany and Plant Ecology, Siesmayertraße 70, D-60323 Frankfurt/Main, Germany

KRÄMER-BADONI, THOMAS; Bremen University, KUA, ZWE "Work and Region ", P.O. Box 33 04 40, D-28334 Bremen, Germany, phone: +49 421/2 18-22 95, fax: +49 421/2 18-26 80, e-mail: tkb@kua.uni-bremen.de

KRAVCHUK, LUDMILA A.; Institute for Problems of Use of Natural Resources & Ecology, 10 Staroborysovski tract, 220114 Minsk, Belarus, phone: +375 17/264 23 12, fax: +375 17/264 24 13

KRESSLER, FLORIAN P.; Austrian Research Centers, Department for Environmental Planning, A-2444 Seibersdorf, Austria, phone: +43 2254/780-3876, fax: +43 2254/780-3888, e-mail: kressler@arcs.ac.at

KRÖGER, GABRIELE; Technical University Hamburg-Harburg, Institute for Urban Ecology, Kasernenstraße 10, D-21073 Hamburg, Germany, phone: +49 40/7718-3589, fax: +49 40/7718-2580
KUCHARCHIC, TAMARA I.; Institute for Problems of Use of Natural Resources & Ecology, 10 Staroborysovski tract, Minsk, 220114 Belarus, fax: +375 17/264 24 13, e-mail: ikonicov@ns.ecology.ac.by
KÜCHENHOFF, BETINA, City of Cologne, Department of Environment, Gürzenichstr. 6-16, D-50677 Köln, Germany
KUHN, ACHIM; Johann Wolfgang Goethe University Frankfurt/Main, Department of Geobotany and Plant Ecology, Siesmayerstraße 70, D-60054 Frankfurt/Main, Germany, phone: +49 69/798-24802, fax: +49 69/798-24702, e-mail: ac.kuhn@em.uni-frankfurt.de
KÜHN, MANFRED; Institute for Regional Development and Structural Planning (IRS), Flakenstr. 28-31, D-15537 Erkner, Germany, phone: +49 3362/793 238, fax: +49 3362/793 111
KUHPFAHL, ILONA; FHS Merseburg, Project Group Technology Assessment, Geusaer Straße, D-06217 Merseburg, Germany
KUHLKE, OLAF; Kent State University, Department of Geography, Kent, Ohio 44242-001, U.S.A., e-mail: okuhlke@kent.edu
KUNDEL, HELMUT; Institute of Ecology, North-East Estonian Department, 15 Pargi Str., EE2045, Jõhvi, Estonia
LANG, MARKUS; Office for Urban Renewal (ASW), City of Leipzig, D-04092 Leipzig, Germany, phone: +49 341/123-55 04, fax: +49 341/123-55 16, e-mail: mlang@leipzig.de
LEHMANN, ANDREAS E. H.; Friedrich-Schiller-University of Jena, Department of Geological Sciences, Burgweg 11, D-07749 Jena, Germany, phone: +49 3641/9-48721, fax: +49 3641/9-48602, e-mail: as@geo.uni-jena.de
LEHMANN, HARRY; Wuppertal Institute for Climate, Environment, Energy, Döppersberg 19, D-42103 Wuppertal, Germany, phone: +49 202/2492-124, fax: +49 202/2492-138, e-mail: harry_lehmann@wupperinst.org
LEONHARDT, CLAUDIA; Institute for Ecological Space Development Dresden (IÖR), Weberplatz 1, D-01217 Dresden, Germany
Liblik, Valdo; Institute of Ecology, North-East Estonian Department, 15 Pargi Str., EE2045, Jõhvi, Estonia, phone: +372 33 61 912, e-mail: valdo@ecoviro.johvi.ee
LOERINCZ, KAROLY; JATE University, Department of Experimental Physics, Dom ter 9., H-6720 Szeged, Hungary, phone: +36 62/454382, fax: +36 63/311154, e-mail: lorincz@physx.u-szeged.hu
LUSTIG, SANDRA; Berlin Technical University, Institute for Management in Environmental Planning, Franklinstrasse 28/29, FR 2-7, D-10587 Berlin, Germany, phone: +49 30/314-73325, or -27692, fax: +49 30/314-73517, e-mail: lustig@imup.tu-berlin.de
MARKER, ANDREAS; Dr. Krätzig Eng. Ltd, Süsterfeldstr.81, 52072 Aachen, Germany, phone/fax: +55 11/30306577 (CETESB); e-mail: andreasm@cetesb.br
MATOSHKO, NATALIYA; Hydrometeorological Agency, 165 Maskavas str., LV-1019, Riga, Latvia, fax: +371 7/145154, e-mail: EPOC@meteo.lv
MAYER, HANS-NORBERT; Carl von Ossietzky University Oldenburg, Institute for Sociology, Town Research, P.O. Box 2503, D-26111 Oldenburg, Germany, phone: +49 441/9706-367, fax: +49 441/9706-366, e-mail: mayer@psychologie.uni-oldenburg.de
MEINEL, GOTTHARD; Institute for Ecological Space Development IÖR (Institut für ökologische Raumentwicklung e.V.), Weberplatz 1, D-01217 Dresden
MENEZES DA COSTA CROCE, CRISTINA; Avenida Contorno, 8.000/608, Santo Agostinho 30. 110-120, Belo Horizonte, Minas Gerais, Brazil, phone: +55 31/2910309, fax: +55 31/2915714, e-mail: mcris@net.em.com.br
MEURER, MANFRED; Institute of Geography and Geoecology I, Karlsruhe University, Kaiserstraße 12, D-76128 Karlsruhe, Germany, phone: +49 721/6084367, fax: +49 721/696761, e-mail: manfred.meurer@bio-geo.uni-karlsruhe.de

Mihalca, Doina; West University of Timisoara, Faculty of Chemistry-Biology-Geography, Geography Department, Pestalozzi 16, 1900 Timisoara, Romania, phone: +40 56/190377, fax: +40 56/190009 or +40 56/190333

Millard, Andy; Metropolitan University, Faculty of Health and Environment, Brunswick Terrace, Leeds LS2 8BU, UK, phone: +44 113/283 2600, fax: +44 113/283 3190, e-mail: a.millard@lmu.ac.uk; c.freeman@lmu.ac.uk

Mladjenovic, Ivica; Architect, Bulevar Avnoja 113, ulaz 1, 11070 Beograd, Yugoslavia, phone/fax: +381 11/141-426

Mook, Veronika; Environment – Planungsgemeinschaft Stadt und Umwelt (Environment – Planning Co-operation Town and Environment), Ruschenstraße 8, D-45133 Essen-Bredeney, Germany, phone: +49 201/42 55 48, fax: +49 201/42 55 52

Moriyama, Masakazu; Research Centre for Urban Safety and Security, Kobe University, Nada, Kobe 657, Japan, phone/fax: +81 78/803-1045, e-mail: moriyama@kobe-u.ac.jp

Moser, Peter; UFZ Centre for Environmental Research Leipzig-Halle Dept. Urban Landscapes Permoserstr 15 D-04318 Leipzig phone: +49-341-235-2369 fax: +49-341-235-2534 e-mail: moser@pro.ufz.de

Mosha, Aloysius Clemence; University of Botswana, P.O. Box 00 22, Gabarone, Botswana, phone: +267 3552511 (0) 303132, fax: +267 3552511 (0) 356591, e-mail:moshaac@noka.ub.bw

Moss, Timothy; Institute for Regional Development and Structural Planning (IRS), Flakenstraße 28-31, D-15537 Erkner (Berlin), Germany, phone: +49 3362/793-185, fax: +49 3362/793-111, e-mail: mosst@irs.los.shuttle.de

Müller, Norbert; Berlin Technical University, Institute of Biology and Ecology, and Office for Nature Conservation Augsburg, Dr. Ziegenspeckweg 10, D-86161 Augsburg, Germany, phone: +49 821/3246015, fax: +49 821/3246050, e-mail: mueller-norbert@t-online.de

Münchow, Birgit; UFZ Centre for Environmental Research Leipzig-Halle, Department of Urban Landscapes, Permoserstraße 15, D-04318 Leipzig, Germany, phone: +49 341/235-2181, fax: +49 341/235-2534, e-mail: muenchow@pro.ufz.de

Musinszki, Anna; Dortmund University, Faculty for Regional Planning, August-Schmidt-Straße 10, D-44221 Dortmund, Germany

Netzband, Maik; Institute for Ecological Space Development IÖR (Institut für ökologische Raumentwicklung e.V.), Weberplatz 1, D-01217 Dresden, Germany, phone: +49 351/4679 257, fax: +49 351/4679 202, e-mail: maik.netzband@pop3.tu-dresden.de

Nguyen Huu Dung; Hanoi Architectural University, Director of the Research Centre for Architectural Indoor Climatology and Environment (RCAICE), Ministry of Construction, Dal hoc klen truc Hanoi-Tranh Xuan, Hanoi, Vietnam, phone: +84 4/835 6775, fax: +84 4/854 4288 or 844 825 2153

Niemelä, Jari; Helsinki University, Department of Ecology and Systematics, P.O. Box 17, FIN-00014, Finland, phone: +358 9/1917391, fax: +358 9/1917492, e-mail: jari.niemela@helsinki.fi

Nieves Poveda, María E.; National Climate Centre, Meteorological Institute, Havana, Cuba, e-mail: meteoro@ceniai.inf.cu

Niimitsu, B.; Research Centre for Urban Safety and Security, Kobe University, Nada, Kobe 657, Japan

Nilon, Charles; University of Missouri-Columbia, School of Natural Resources, 112 Stephens Hall, Columbia, MO 65211 U.S.A., phone: +1 573/882-3738, fax: +1 573/884-5070, e-mail: snrnilon@showme.missouri

Nohl, Werner; Workshop for Landscape and Free Space Development (Werkstatt für Landschafts- und Freiraumentwicklung), Stockäckerring 17, D-85551 Kirchheim (Munich), Germany, phone: +49 89/903 83 46, fax: +49 89/904 58 05

Norra, Stefan; Karlsruhe University, Institute of Petrography and Geochemistry, Kaiserstraße 12, D-76128 Karlsruhe, Germany, phone: +49 721/608-6155 or -3323 or -3325, fax: +49 721/69 73 28, e-mail: stefan.norra@bio-geo.uni-karlsruhe.de

NUPPENAU, CHRISTINE; Ministry of Environment and Energy, Danish Forest and Landscape Research Institute, Hoersholm Kongevej 11, DK-2970 Hoersholm, Denmark

OLARU, MARTIN; West University of Timisoara, Faculty of Chemistry-Biology-Geography, Geography Department, Pestalozzi 16, 1900 Timisoara, Romania, phone: +40 56/190377, fax: +40 56/190009 or +40 56/190333

ORTIZ BULTÓ, PAULO L.; National Climate Centre, Meteorological Institute, Havana, Cuba, e-mail: meteoro@ceniai.inf.cu

PASENAU, HORST; Dr. Krätzig Ingenieurgesellschaft mbH, Aachen, Süsterfeldstraße 81, D-52072 Aachen, Germany, phone: +49 241/8 89 73-0, fax: +49 241/8 89 73-28, e-mail: dr.kraetzig-ing-gmbh@t-online.de, German Society for Technical Cooperation (Deutsche Gesellschaft für Technische Zusammenarbeit (GTZ) GmbH), Eschborn

PAULEIT, STEPHAN; Chair for Landscape Ecology, Munich Technical University, D-85350 Freising, Germany, phone: +49 8161/714219, fax: +49-8161-714427 e-mail: paul@dec.loek.agrar.tu-muenchen.de

PATZ, RENATE; Merseburg, Project Group Technology Assessment, Geusaer Straße, D-06217 Merseburg, Germany, phone: +49 3461/46 2903, fax: +49 3461/46 2919, e-mail: renate.patz@ltg.fh-merseburg.de

PEITHMANN, ORTWIN; Vechta University, Faculty for Space Planning, P.O. Box 15 53, D-49364 Vechta, Germany, phone: +49 4441/15-419, fax: +49 4441/15-445, e-mail: opeithmann@ispa.uni-vechta.de

PEIU, NICOLAE; 11a/3, Macazului Street, 6600 Iasi, Romania, phone: +40 32/213 920, fax: +40 32/214357, e-mail: apmiasi@cccis.ro

PETERSEN, RUDOLF; Wuppertal Institute for Climate, Environment and Energy, POB 10 04 80, D-42004 Wuppertal, Germany, phone: +49 202/4292 117, fax: +49 202/4292 108, e-mail: Rudolf_Petersen@wupperinst.or

PERKS, WILLIAM T.: University of Calgary, Faculty of Environmental Design, Calgary, Alberta T2N 1N4, Canada, fax: +1 403/284-4399, e-mail: perksw@cadvision.com

PFEFFERKORN, WOLFGANG; Regional Consulting ZT GmbH, Schloßgasse 11, A-1050 Vienna, Austria, phone: +43 1-544 07 80-37, fax: ++43 1-548 49 56, e-mail: rc@regcon.co.at or pfefferkorn@rosinak.co.at

PIETSCH, JÜRGEN; Technical University Hamburg-Harburg, Institute for Urban Ecology, Kasernenstraße 10, D-21073 Hamburg, Germany, phone: [1]+49 40/7718-3111, [2]+49 40/7718-3589, [3]+49 40/7718-3588, fax: +49 40/7718-2580

POLIMENI, CLAUDIA MARCELA; Universidad Nacional de Cuyo, Instituto de Geografía, Facultad de Filosofía y Letras, Corrientes 122. 2° piso, Dpto A, 5500, Mendoza, Argentina, fax: +54 61/200112, e-mail: claudiap@logos.edu.ar

POPP, PETER; UFZ Centre for Environmental Research Leipzig-Halle, Department for Analytical Chemistry, Permoser Str. 15, D-04318 Leipzig, Germany, phone: +49 341/235-2297 or -2408 or 2922, fax: +49 341/235-2625, e-mail: popp@ana.ufz.de

PUKY, MIKLÓS; Hungarian Danube Research Station of the Institute of Ecology and Botany of the Hungarian Academy of Sciences, 2131 Göd Jávorka S. u. 14, Hungary, phone: +36 27/345023, fax: +36 27/345023, e-mail: pm@alfa.elte.hu

PULIAFITO, CARLOS; Mendoza University, Instituto para el Estudio del Medio Ambiente (IEMA), Arístides Villanueva 773, 5500 Mendoza, Argentina, phone/fax: +54 61/392939

PULIAFITO, ENRIQUE; Mendoza University, Instituto para el Estudio del Medio Ambiente (IEMA), Arístides Villanueva 773, 5500 Mendoza, Argentina, phone/fax: +54 61/392939, e-mail: epuliafi@umdz.edu.ar

QUERO, JOSÉ; Guerreiro, Pablo; Mendoza University, Instituto para el Estudio del Medio Ambiente (IEMA), Arístides Villanueva 773, 5500 Mendoza, Argentina, phone/fax: +54 61/392939

RAMOS DE CARVALHO, MARIA CRISTINA; Avenida Contorno, 8.000/608, Santo Agostinho 30.110-120, Belo Horizonte, Minas Gerais, Brazil

RAPAPORT, ERIC; Royal Institute of Technology, Department of Infrastructure and Planning, S-100 44 Stockholm, Sweden, phone: +46 8/190 9247, fax: +46 8/190 6761, e-mail: ericrapa@infra.kth.se

REMMLER, FRANK; Institute for Water Research GmbH Dortmund, Zum Kellerbach 46, D-58239 Schwerte, Germany, phone: +49/231-544-1300, fax: +49 231/544-1220, e-mail: 101651.3326@compuserve.com

RICHTER, MATTHIAS; University of Hohenheim, Institute for Landscape and Plant Ecology, D-70593 Stuttgart, Germany, fax: +49 711/459 2831, e-mail: richterm@uni-hohenheim.de

RICHTER, URSULA; IMU-Institute GmbH, Pohlandstr.19, D-01309 Dresden, Germany, phone: +49 351/3119107, fax: +49 351/3119118, e-mail: imu-dresden@t-online.de

RINK, DIETER; UFZ Centre for Environmental Research Leipzig-Halle, ÖKUS - Department of Ecological Economics and Environmental Sociology, Permoserstaße 15, D-04318 Leipzig, Germany, phone: +49 341/235-2015, fax: +49 341/235-2511, e-mail: rink@alok.ufz.de

RIPL, WILHELM; Berlin Technical University, Institute of Ecology, Department of Limnology, Hellriegelstraße 6, D-10629 Berlin, Germany, e-mail: riplggdb@mailszrz.zrz.TÜ-Berlin.de

RÖSCH, ANDREAS; UFZ Centre for Environmental Research Leipzig-Halle, Group Regional Future Modelling, Permoserstraße 15, D-04318 Leipzig, Germany

RUNGE, KARSTEN; Hamburg Technical University, Kasernenstr. 10, D-21073 Hamburg, Germany, phone: +49 40/7718 3589, fax: +49 40/7718 2580, e-mail: runge@tu-harburg.d400.de

SALES, JOSÉ ARNALDO; Fundação Estadual de Engenharia do Meio Ambiente (FEEMA), Rio de Janeiro, Brazil, German Society for Technical Cooperation (Deutsche Gesellschaft für Technische Zusammenarbeit (GTZ) GmbH), Eschborn

SAUERWEIN, MARTIN; Martin Luther University Halle-Wittenberg, Department of Geography, Domstraße 5, D-06108 Halle (Saale), Germany, phone +49 345/55-26033, fax +49 345/55-27169, e-mail: sauerwein@geographie.uni-halle.de

SCHÖTTLER, UWE; Institute for Water Research GmbH Dortmund, Zum Kellerbach 46, D-58239 Schwerte, Germany

SCHRAMM, MICHAEL; IWAB - Consultants for Water, Soil, and Waste Management, Marie-Alexandra-Straße 27, D-76135 Karlsruhe, Germany, phone: +49 721/32599, fax: +49 721/32502, e-mail: iwab@compuserve.com

SCHWEITZER, HANNELORE; Dresden Technical University, Institute of Medical Microbiology and Hygiene, Löscherstraße 18, D-01309 Dresden, Germany, phone: +49 351/3177432, fax: +49 351/3360313, e-mail: schweitzer@imib.med.med.tu-dresden.de

SCHWERK, AXEL; Ruhr University Bochum, Chair for General Biology & Neurobiology, Faculty for Biology, D-44780 Bochum, Germany

SEIDLING, WALTER; Technical University of Berlin, Institute for Ecology and Biology, Schmidt-Ott-Straße 1, D-12165 Berlin, Germany, phone: +49 30/314-71194 or +49 30/781 55 02, fax: +49 30/781 55 02

SEKHAR, MATLI CHANDRA; Water & Environment Division, Regional Engineering College, Warangal - 506 004, India, fax: +91 8712/76547, e-mail : mcs@recw.ernet.in

SEUL, HEINRICH; CREAM Consultants, Gut Fahrenbach, D-37216 Witzenhausen, Germany

SIMON, KARL-HEINZ; Kassel University, Centre for Environmental Systems Research, Kurt-Wolters-Straße 3, D-34109 Kassel, Germany, phone: +49 561/804 2273, fax +49 561/804 3176, e-mail: simon@usf.uni-kassel.de

SNICKARS, FOLKE; Royal Institute of Technology, Fiskartorpsvägen 15 A, S-10044 Stockholm, Sweden

SÖKER, ROLAND; *anstiftung* ggmbh, Daiserstraße 15 Rgb., D-81371 München, Germany phone: +49 89/747460-14, fax: +49 89/747460-30

STAHR, KARL; University of Hohenheim, Department of Soil Science and Ecology, D-70593 Stuttgart, Germany, e-mail: kstahr@uni-hohenheim.de

Stauch, Carola; University of Stuttgart, Institute for Landscape Planning and Ecology, Keplerstr. 11, D-70174 Stuttgart, Germany, phone: +49 711/121-4140, fax: +49 711/121-3381, e-mail: Carola.Stauch@ilpoe.uni-stuttgart.de

Steinnocher, Klaus T.; Austrian Research Centers, Department for Environmental Planning, A-2444 Seibersdorf, Austria

Stephan, Bernd; Zwischenhausen 3 1/2, D-35037 Marburg, Germany, phone: +49 6421/62871, e-mail: Stephan@stud-mailer.uni-marburg.de

Steusloff, Sabine; Institute for Geography and Geoecology I, Karlsruhe University(TH), Kaiserstraße 12, D-76128 Karlsruhe, Germany, phone: +49 721/608-4722, fax: +49 721/696761, e-mail: Sabine.Steusloff@bio-geo.uni-karlsruhe.de

Struk, Mikhail I.; Institute for Problems of Use of Natural Resources & Ecology, Belorussian Academy of Sciences, 10 Staroborisovski trakt, 220114 Minsk, Belarus, phone: +375 17/264 23 12, fax: +375 17/264 24 13, e-mail: ipipre@bas18.basnet.minsk.by

Stüben, Doris; Karlsruhe University, Institute of Petrography and Geochemistry, Kaiserstraße 12, D-76128 Karlsruhe, Germany, phone: +49 721/6083322, fax: +49 721/697328, e-mail: doris.stueben@bio-geo.uni-karlsruhe.de

Sukopp, Herbert; Berlin Technical University, Institute for Ecology, Schmidt-Ott-Straße 1, D-12165 Berlin, Germany, phone: +49 30/314 71190, fax: +49 30/314 71324

Surjadi, Charles; Atma Jaya Catholic University, Center for Health Research, 14440 Jakarta, Indonesia, phone/fax: +62 21/6682512, e-mail: puslit@link.net.id

Takabayashi, H.; Research Centre for Urban Safety and Security, Kobe University, Nada, Kobe 657, Japan

Tent, Ludwig; District Office (Bezirksamt) Wandsbek/GU 40, Robert-Schuman-Brücke 8, D-22041 Hamburg, Germany, phone: +49 40/6889-2658, fax: +49 40/6889-2023, e-mail: bawva42@t-online.de

Ufermann, Kay; Technical University Hamburg-Harburg, Institute for Urban Ecology, Kasernenstraße 10, D-21073 Hamburg, Germany, phone:, +49 40/7718-3588

Uricher, Angelika, Öko-Institut e.V. (Eco-Institute), Friedrichstraße 165, D-10117 Berlin, Germany, fax: +49 30/20 16 50 88

Vu Thanh Ca, Saitama University, Urawa, Saitama 338, Japan

Vu Thi Vinh; Hanoi Architectural University, Dal hoc Klen truc Hanoi, Thanh Xuan, Hanoi, Vietnam, fax: +84 4/8541616

Tourbier, Joachim T.; Dresden Technical University, Institute for Landscape Architecture, Helmholtzstr. 10, D-01062 Dresden, Germany, phone: +49 351/463 3453, fax: +49 351/463 7081, e-mail: tourbier@msn.com

Treutler, Hanns-Christian; UFZ Centre for Environmental Research Leipzig-Halle, Department for Analytical Chemistry, Permoser Str. 15, D-04318 Leipzig, Germany, phone: +49 341/235-2297 or -2408 or 2922, fax: +49 341/235-2625, e-mail: treutler@ana.ufz.de

Tschense, Holger; Leipzig City Council, New City Hall (Neues Rathaus), Environmental Authority, D-04229 Leipzig, Germany, phone: +49 341/123-3500

Tyler, Mary-Ellen; University of Calgary, Dean, Faculty of Environmental Design 2500 University Drive N.W., Calgarg, Alberta T2N 1N4, Canada

Velguth, Peter H.; University of Wisconsin, Department of Horticulture, 1575 Linden Dr., Madison, U.S.A., fax: +1 608/262-4743, e-mail: pvelguth@facstaff.wisc.edu

Veretennikov, D.; Latvian Hydrometeorological Agency, 165 Maskavas str., LV-1019, Riga, Latvia, fax: +371 7/145154, e-mail: EPOC@meteo.lv

Votsmeier, Thomas; Deutsche Gesellschaft für Qualität e.V. (German Society for Quality), August-Schanz-Straße 21 A, D-60433 Frankfurt/Main, Germany, phone: +49 69/9 54 24-0

Wagler, Dietrich; Leipzig University, Interdisciplinary Institute for Nature Conservation and Environmental Protection (IINU), Augustusplatz 10/11, D-04109 Leipzig, Germany, phone: +49 341/97 36346, fax: +49 341/97 36348, e-mail: dwagler@rz.uni-leipzig.de

WANG, RUSONG; Chinese Academy of Sciences, Research Centre for Eco-Environmental Sciences, 19 Zhongguancun Road, Beijing 100080, China, fax: +86 10 62562775, e-mail: wangrs@sun.ihep.ac.cn

WANG, YUANFEI; Department of Geography, ECNU, 200062, Shanghai, P.R. China

WEBER, MARKUS; Brandt Meio Ambiente, Rua Timbiras 270, 30.140-060 – Belo Horizonte – MG, Brazil, phone: +55 31/2812258, fax: +55 31/225 6929, e-mail: bma@qnet. com.br

WEHLING, PETER; Institute for Social-Ecological Research (ISOE), Hamburger Allee 45, D-60489 Frankfurt/Main, Germany

WEILAND, ULRIKE; Berlin Technical University, Institute for Management in Environmental Planning (FR 2-7), Franklinstraße 28/29, D-10587 Berlin, Germany, phone: +49 30/314 27692, fax: +49 30/314 73517, e-mail: Weiland@IMUP.TU-Berlin.de

WELLENSIEK, CHRISTIANE; CREAM Consultants, Gut Fahrenbach, D-37216 Witzenhausen, Germany, phone: +49 5542/8876, fax: +49 5542/5767, e-mail: well@cre-a-m.com

WERHEIT, MARTINA; Martin Luther University of Halle-Wittenberg, Institute of Geography, Heinrich-und-Thomas-Mann-Straße 26, D-06099 Halle (Saale), Germany, phone: +49 345/55-26 059, fax: +49 345/55-27 144, e-mail: werheit@geographie.uni-halle.de

WICKOP, EVELYNE; UFZ Centre for Environmental Research Leipzig-Halle, Department of Urban Landscapes, Permoser Straße 15, D-04318 Leipzig, Germany, fax: +49 341/235-2534, e-mail: wickop@pro.ufz.de
Present address: Wiesbaden City Council Department of Environmental Protection, Luisenstr. 23 D-65185 Wiesbaden, Germany

WITTIG, RÜDIGER; Johann Wolfgang Goethe University, Institute of Botany, Department of Geobotany and Plant Ecology, Siesmayerstraße 72, D-60323 Frankfurt/Main, Germany, phone: +49 69/798-24747, fax: +49 69/798-24702, e-mail: R.Wittig@em.uni-frankfurt.de

ZHENG, JINGCHUN; Department of Geography, ECNU, 200062, Shanghai, P.R. China

ZIDEK, LADISLAV; Rajec City Council, Námestie SNP 2/2, 01501 Rajec, Slovak Republic, phone: +421 823/422030, fax: +421 823/422224

ZISENIS, MARCUS; Eisenacher Straße 54, D-10823 Berlin, Germany, phone:+49 30/784 1788

Advisory Board

Introduction

Urban Ecology – Scientific and Practical Aspects

Herbert Sukopp

1
Introduction

We are living in a century of rapid urbanisation. The United Nations forecasts that by the year 2025, 60% of the world's population will be living in urban areas, compared to 29% in 1950. The 50% mark will be reached between the years 2000 and 2010. In 2025, more than a dozen cities will have over 20 million inhabitants, and some will have over 30 million. 23 of the 25 biggest urban conglomerations on the planet will be in Africa, Asia and Latin America, rather than in Europe or North America.

Today, 12.7% of the land area of the Federal Republic of Germany is used for settlements or transport: this represents an increase of almost three percentage points over the last 30 years. This expansion seems set to continue because of the increasing demand for space, in particular for leisure uses and rubbish deposits.

The word 'urban ecology', in German *Stadtökologie*, is used in two different ways. In the normative usage, it describes urban design programmes at the political and planning level (see Deelstra, this volume). Within the natural sciences, however, urban ecology is used to refer to that area of biology which is concerned with urban areas. 'Ecology' is understood here as the science of the relationships amongst living creatures and communities as well as their relationship to their environment (Sukopp, Trepl, 1995). My task here is to present urban ecology as a natural science and illustrate the relationship of this discipline to politics, environmental policy-making and urban development. In the scientific sense, 'ecology' describes and analyses without forming value judgements. In planning, the same word means 'urban design' in the normative sense.

Urban ecology as a natural science is a young discipline. For a long time, it was thought that urban areas were not worth studying with regard to ecology. Cities were seen as anti-life. It was assumed that few plants or animals could survive in an urban setting and that urban animal and plant commu-

Berlin Technical University, Institute for Ecology, Schmidt-Ott-Straße 1, D-12165 Berlin, Germany

J. Breuste, H. Feldmann, O. Uhlmann (Eds.)
Urban Ecology

nities were products of coincidence. Attempts to discover patterns or the reasons for such patterns were regarded as futile.

This view began to change over 25 years ago. Since then, ecological aspects of cities have been intensively researched. Soon, it became obvious that urban communities were not merely coincidental.

Early studies of ecology in cities shared the astonished realisation that environments created by humans provide habitats for characteristic species and that these species reoccur under similar conditions. Analyses have shown that urban areas include a wide variety of habitats, organisms and communities. The variety of species found in urban areas often exceeds that in the urban hinterland.

With increasing suburbanisation, it becomes more and more difficult to separate analyses of the urban area from those of the hinterland. The word 'hinterland' is used here to refer to areas which are functionally connected to the city although they cannot be precisely defined in space. Relationships between towns and their hinterlands involve economic, cultural and ecological aspects. The immediate hinterland, or suburban area, must also be differentiated from the distant hinterland. Urban areas are, today, provided with foods, raw materials and natural resources from a distant hinterland which stretches across the globe. A dense transport network involving land, water and air makes it possible to transport huge volumes of goods and resources.

The import of agricultural products into Germany and the EU has taken place for centuries and is still having ecological effects. The growing disparity between urban and rural areas therefore creates not only social and economic problems, but also ecological ones. We are living a global lifestyle, and our cities are connected, in an ecological sense, to ecological systems worldwide, not just to systems in the immediate hinterland.

1.1 The City as a 'New Type of Environment'

Urban ecosystems differ from non-urban ones in a number of ways. Although most of the factors which affect urban ecosystems also operate in non-urban areas, the combination of these factors means that unique ecosystems develop with species combinations peculiar to urban areas.

Cities have particular climatic characteristics (Landsberg 1981, Kuttler 1993, Schirmer et al. 1993). The combined effects of the greenhouse effect, the high heat capacity of buildings and of urban ground and actual heating of buildings leads to an annual average air temperature 0.5–1.5°C warmer than normal for the region. This means that the growth season for plants in the inner-city area is extended, blossoming and leaf formation occurs earlier, and heat-loving plants and animals from more southerly latitudes migrate into the city. The city heat island usually covers the same area as the built-up area, but changes in wind direction can mean other areas are temporarily heated.

Air pollution is closely linked to the temperature increase. Trace gases, excepting low-level ozone, occur at concentrations five to fifty times higher than usual.

Material inputs into cities have exceeded outputs for centuries (Mumford 1961). This has meant that the ground level in older districts has risen by several metres. Surface sealing and the resulting reduced infiltration capacity, as well as the extraction of groundwater for human usage, further reduced the groundwater level. This means that urban soils have become considerably dryer over the centuries - except in gardens and parks, which have often become moister through extensive watering. Urban soils are generally eutrophicated and over-compacted. Soils in industrial areas, under waste deposits, on Leaching fields (land onto which waste water was directed as a form of purification) and on roadsides are in some cases heavily polluted.

Urban buildings do not only affect the urban topography - they also have ecological effects. For many animals in urban habitats, building surfaces resemble natural rock faces. Microclimatic effects are also of ecological importance: depending on their exposure, external building walls can be subject to high temperatures which can be maintained over larger periods.

Thus, urban areas cannot be seen as ecologically homogenous with regard to climate, soil and buildings. Rather, a mosaic of different biotope types, dependent on the small-scale distribution of land uses, can be distinguished. These biotopes can usually be clearly separated, and are themselves relatively homogenous. 'Natural' ecological conditions are outweighed and modified by the type of land use in an area. For this reason, ecological mapping is often based on land use types (Arbeitsgruppe "Methodik der Biotopkartierung im besiedelten Bereich" 1993). In addition to the mosaic-like biotope pattern, concentric zones can be distinguished for many European cities (intensive and extensive built-up areas and inner and outer suburban zones; see Fig. 1). Urban building and economic development divides the city into more and less intensively built zones. In the inner suburbs, allotment gardens, rubbish deposits, waste and slag heaps are often found. Forests and large parks are typical for the outer suburbs.

Urban development has thus had extensive effects in terms of air pollution and warming, changes in the ground water level, and a rise in ground level. The extent of the 'cultural layer' is related to the eutrophication of many areas, and the compacting or loosening of soils within the settlement. Eutrophication through urban waste affects not only rubbish deposits and Leaching fields, but also all the water courses in the areas, and the composition of terrestrial and aquatic communities.

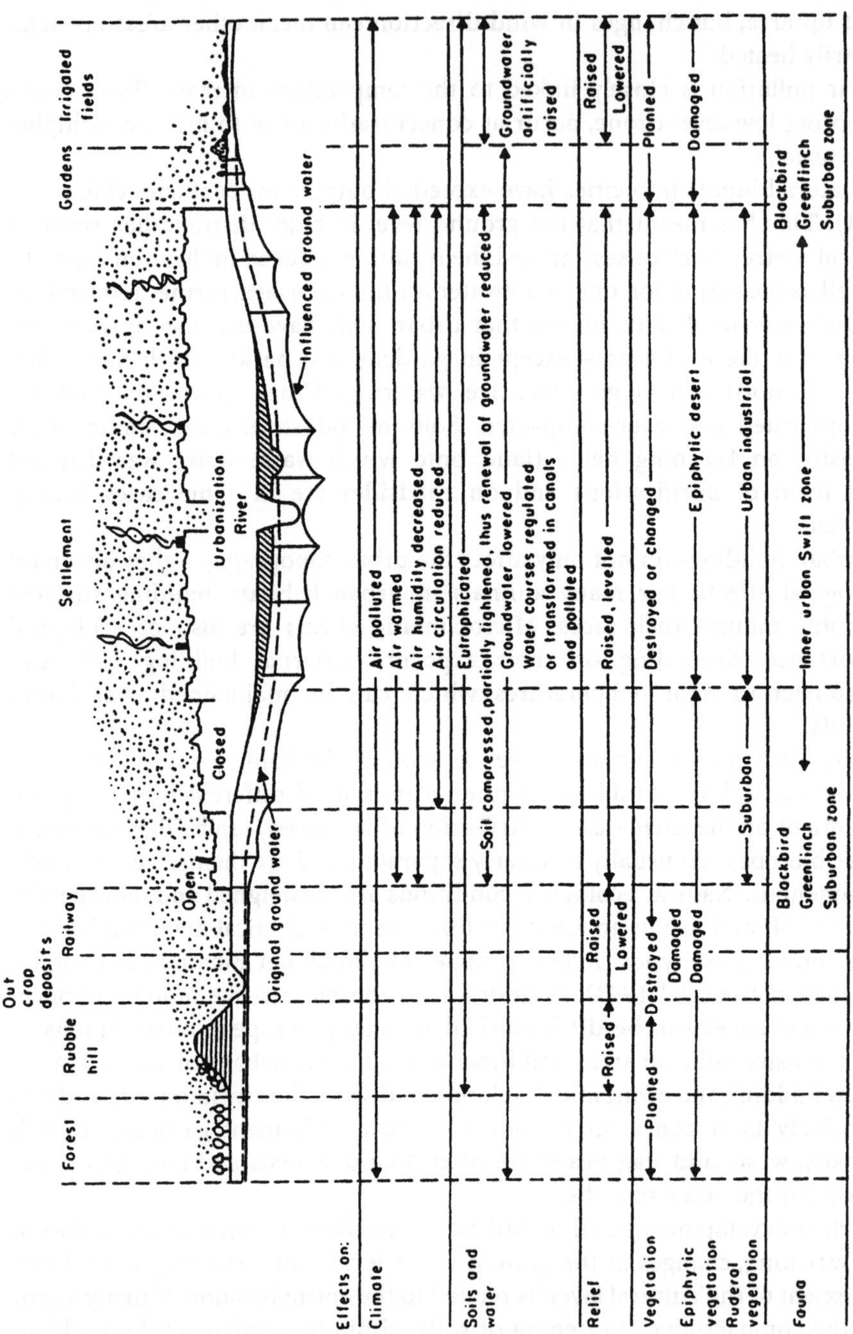

Fig.1. Intensive and extensive built-up areas and inner and outer suburban zones

1.2
Species Variety and Composition of Urban Flora and Fauna

The altered soil, climate and water conditions have a corresponding impact on the composition of species occurring in urban areas (Wittig 1991, Klausnitzer 1993). These changes are greater the larger the urban area (Falinski 1971) and the closer one gets to the city centre. Comparisons between urban and rural areas and between different cities and city edge - inner city gradient studies have been established as methods of analysing these differences (Sukopp 1968, Kunick 1974, 1981).

The number of ferns and flowering plant species per unit area is higher in cities with over 50,000 inhabitants than in the surrounding area (Haeupler 1974). In Central Europe, the number of ferns and flower plant species correlates closely with population size and/or density (Klotz 1990, Brandes, Zacharias 1990, Pysek 1993). In small and medium-sized towns, between 530 and 560 species are usually found; 650–730 species are found in cities with 100,000 to 200,000 inhabitants, 900–1000 species in older cities with populations of between 250,000 and 400,000, and in cities with over a million inhabitants, the number of species usually exceeds 1,300.

There are various reasons for this.

- Urban areas are heterogeneous, consisting of a variety of settlement structures, land uses and small-scale habitats. This creates many unusual ecological conditions.
- Historically, species which have been introduced into an area through human activity, directly or indirectly, begin their dispersal in urban areas and therefore occur there most frequently. With increasing settlement size, trade and traffic in and out of the city increases, and thus the proportion of non-native species in the flora increases.

The number of non-native flowering plants and diasporas which have been introduced into Central European urban areas has been estimated. At least 10,000 species have been introduced (Sukopp 1976, for trees and shrubs Kowarik 1992). However, when the number of hemerochores (species and subspecies which have only reached the area through direct or indirect human assistance) who have become established, 385 in total, is taken into account, it can be seen that less than 5% of the species and sub-species introduced have become naturalised.

The method by which hemerochores immigrate and their actual extent is subject to continual change. Both reflect the economic development and changes in the region (Sukopp 1976). The most important changes in the range of species immigrating have been the result of changes in transportation technology and in the method of immigration. Habitat and biozone changes affecting immigration may have been of lesser importance (Trepl and Sukopp, 1993).

This is shown by the relationships between certain species and certain methods of transportation. For example, the use of rye straw and *Strandwiesenheu* from Sicily and southern Italy as packing material facilitated the immigration of a limited number of 'fruit companions' such as *Reseda lutea, Chenopodium vulvaria,* and *Plantago coronopus* for a limited period (for Leipzig: Fiedler 1937; Gutte 1992).

Apart from numerous non-native species, many native organisms also profit from urban habitats, especially if the conditions are similar to their original habitats. In the natural Central European landscape, these urban apophytes may originally have occurred on tree-free sites such as on gravel or mud banks along rivers and streams, in open areas within forests, on the *Lagerfluren* of large mammals or on ground disturbed by fire or earthquakes. As most large cities are located on large rivers, erosion and sedimentation areas along the river act as important initial habitats for urban apophytes (Falinski 1971, Schmitz 1996). However, environmental changes and species transport cannot fully explain the changes in the composition of urban flora and fauna. Often, genetic changes in individual organisms occur which are only possible when the organism has become widespread in cities and in the immediate hinterland. *Xanthium albinum* is one example. Today, this plant is widespread in Central Europe, but only a few decades ago it was unknown. It was first identified in eastern Brandenburg in 1830. Its closest relatives are found in North America, but not *Xanthium albinum* itself. This leads us to the conclusion that one of these species was imported from North America and established itself as a separate species in a relatively short period (Wagenitz 1964). The spread of the evening primrose (*Oenothera* spp.) in Europe seems to have followed a similar pattern. Today, more than 15 species have been identified, few of which are identical to their North American ancestors. These European species developed over a period of 350 years after the first importation of the American parent species to Europe.

Both of these examples show that when conditions permit, new species can develop over a short period of a few centuries. New species and sub-species have developed over historical time-scales, not just over geological ones.

The number of flowering plant species introduced to Central Europe over the last 300 years has exceeded the number lost (Landolt 1992, for Zürich).

A high number of fern and flowering plants can be found at the city edge. In city centres and large new housing areas, species plants are low, however. The same applies to fauna, at least to many invertebrate groups, birds, and mammals, excepting large predators. However, a relatively large number of species can still be found in inner cities.

Animals vary in their dependence on human settlements. In Central Europe, some species have never been found outside city areas. Others are found in such high concentrations that they can be described as typical urban species. This includes street pigeons, house sparrows, crested larks and swifts. Species which usually live on rocks and boulders in mountain areas often occur in urban areas, where buildings act as 'rock faces'. Species which usually live in caves are also found in cellars and other rooms.

Behavioural changes take place as a response to the changed urban conditions: for example, the mating season of house sparrows takes place earlier as a result of artificial lighting. These types of ecological issues with respect to animals have been the subject of many contributions to the Leipziger Symposium on Urban Ecology between 1981 and 1989 and in 1994 and 1996.

1.3 Species Communities and Ecosystems

Urban ecosystems differ completely from those of their hinterlands with regard to resource and energy budgets. In the middle latitudes, urban energy use in oil, coal, gas, electricity and food usually amounts to 25–50% of the natural energy use. In addition, the use of insolation energy for the biosynthesis of organic compounds is very limited, as the biomass of the primary producers is low in urban areas. Green plants do not form the 'energy base' as they do in natural ecosystems. The consumers – humans and many individuals of a small number of domestic and wild animal species – must be provided for the import of energy in the form of resources from the hinterland. 'Destructors' play a very limited role. Flows of materials rarely form complete circles, so that both urban ecosystems and those in the hinterland are burdened with high levels of waste.

Despite the rapid changes in flora and fauna, habitat-related combinations of organisms are formed, comprising of species which originated in differing habitats. An outline of urban plant communities is given by Wittig 1991 and by Brandes 1990.

The close relationship between certain species and the urban environment means that these can be used as bioindicators for specific environmental factors. Sticky goosefoot, *Chenopodium botrys*, is an example of a temperature-dependent plant. It originated in southern Eurasia and in the Mediterranean region, and has been resident in Berlin since 1889, where it is typically found on ruderal areas in the inner city. Usually, this plant is found on sandy and stony soils on river banks and on talus slopes. Thus, roadsides, wine-growing areas and derelict sites became secondary habitats. Human activity in Central Europe has increased the number of these open, chalky, sandy to gravelly locations. Despite this, large long-term groups of this species can only be found north of the Alps in areas with warm summers, such as in the upper Rhine floodplain, in Berlin and the Ruhr area and on the burning spoil heaps in Lille (Sukopp 1971).

Specific urban-industrial environmental effects are often investigated using lichen on tree bark (epiphytes). The distribution pattern of various lichen species is examined because they differ in their sensitivity to air pollution. A lichen-free-zone in the city centre is usually followed by a zone of transition, where lichen growth markedly increases towards the city edge. The 'normal zone' begins where the number of species, their combination, and their extent, corresponds with the conditions in the hinterland. Lichen growth has been mapped in Leipzig since 1976 at intervals of approximately ten years, thus documenting changes in air quality in this period.

1.4
The History of Urban Research

Pfeil (1972) asserted that research on urban areas always meant research on large cities. This applies to all the disciplines which investigate specifically urban topics. She saw societal problems and health concerns as a main reason for research being carried out.

The beginnings of urban ecology can be found in investigations into the flora (e.g. Deakin 1855, Nylander 1866) and fauna of urban areas, into their occurrence and extent, as part of the 'natural history' tradition. In 1823, Schoŭw used the expression 'plantae urbanae' for plants which occurred near villages and towns and added: 'In most cases, these plants are of foreign origin, which explains why they are only found near towns and villages'.

Weidner (1939), Rudder, Linke (1940) and Peters (1954) made the first attempts at general syntheses and summaries. The extensive destruction in urban areas during the Second World War led to investigations of the flora of ruined sites (e.g. Scholz 1956). Gilbert (1989, translated into German 1994), Wittig (1991) and Klausnitzer (1993) have written up-to-date summaries.

Urban ecology has developed from landscape ecology through intensive research in settled areas down to the biotope level (Sukopp 1990, Sukopp and Wittig 1993) and has become established as a 'landscape ecology specialisation par excellence' (Leser 1991).

2
Science and Environmental Policy

In this section, the current discussion about sustainable and environmentally sensitive urban development is picked up and existing as well as new challenges for the science of ecology are presented. Renate Mayntz (1997) and Hubert Markl (1997) have written general papers on this subject. They will form the basis for discussions at this conference on the results of the Habitat

II conference in Istanbul and on the current 'European Sustainable Cities and Towns Campaign' based on the Charta of Aalborg in 1994.

These initiatives are an attempt to link global phenomena and local action more closely (Local Agenda 21, City Climate Alliance). Or is Elisabeth Lichtenberger right when she says that urban ecology as political practice can only be achieved as part of an international environment policy? National solutions to urban problems are not specific to individual urban areas, she believes, but involve cities internationally.

Environmental policy and environmentally-concerned politics need reliable information about the environment and how it can be preserved. Environmental policy is concerned both with environmental protection and with making use of the environment: it should seek to balance the use of environmental resources with their conservation for future generations.

An ideal relationship between environmental research and environmental policy is often envisaged as follows (Mayntz 1997): Science and politics are two clearly differentiated social systems with a clear division of labour between them. Politics formulates the goals for state action and carries out the measures necessary, while the role of science is to provide information about how these goals can best be achieved. This normative model is based on systems theory. The clear functional division between science and politics in intended to guarantee the autonomy of both systems in fulfilling their tasks. The American social scientist Adam Wildawsky titled his well-known book on scientific advice to politicians 'Speaking Truth to Power'.

Many areas of environmental policy are indeed concerned with objective problem-solving. The overall goals, to protect human health and to guarantee human life on earth, are not in dispute. However, when it is unclear whether a certain situation is, in fact, a problem, how problems have been caused, or how they can be solved, then - according to this model - politicians cannot act. For this reason, scientific knowledge plays a central role in areas such as environmental protection.

Although, as stated, the main issues involve objective problem-solving and not, or not principally, value-related issues or conflicts on distribution, the division of labour between science and politics described above is often disrupted. One reason for this, although not the only one, is the uncertainty of scientific 'knowledge'. In the dual model of scientific advice, it is assumed that scientists can indeed 'speak the truth' to those in power. The scientific establishment wants its statements and its opinion to go unchallenged - until they are disproven. Its status and its right to be listened to depend on this acceptance. However, science cannot always deliver certainties. Politicians need reliable facts in order to be able to act, and if they are justifying their actions with scientific information, they want to be sure that this information is not disputed. This definite knowledge cannot, however, always be achieved. It is a useful fiction which is very important for both parties involved. This unfortunately prevents us recognising that cognitive uncertainty has real implications for political processes.

Assessment principles have been developed in many areas of human behaviour to judge how incomplete data can be interpreted in the case of cognitive uncertainty. These principles are characteristic of or have been laid down for certain groups and institutions. Often, they appear as alternative rules for behaviour, only one of which may be chosen. For example, either something – a material or an action – can be regarded as harmless until proven harmful, or it can be regarded as harmful until proven harmless. When cognitive uncertainty arises, interests as well as assessment principles decide the political conclusions to be taken from incomplete data in need of scientific analysis. In this situation, scientists themselves are often in disagreement.

When varying scientific interpretations of an uncertain subject meet competing political interests, scientific controversies can easily become political controversies, and the competing 'expert opinions' make the debate more heated rather than, as political advisors would like to believe, making it more objective. As long as scientific statements about the level of danger posed by a certain material or type of behaviour are uncertain and in need of interpretation, government and opposition can point to opposing expert opinions and, armed with scientific legitimation, can insist even more dogmatically that their point of view is the 'correct' one.

The environmental risks associated with our technological civilisation only become clear gradually, and it is usually scientists who make us aware of them. This is because the risks and dangers which may exist are typically aggregate effects with indirect, delayed results. Science cannot yet provide definite answers on environmental issues in particular; neither with regard to the existence of danger and the level of risk, nor on its causes and possibilities for treatment. In the area of environmental protection, therefore, political advice by scientists in accordance with the 'dual model' is more the exception than the rule. Time is needed, sometimes – as in the case of the rain forests – a long time, before a complete scientific consensus is reached about the nature and causes of the problem, if this ever happens. Until this point is reached, science creates and intensifies conflicts rather than making political debate more objective. The sooner this phenomenon is recognised by both sides, the better scientists and politicians will be able to co-operate (Mayntz 1997).

What, then, can environmental research achieve? Environmental research creates better scientific understanding about the relationships between humans and their abiotic and biotic environment and thus helps solve environmental problems. However, it is unable to restore a supposedly lost 'natural' balance between humans and nature. With given population levels and with given types of production and consumer behaviour, environmental research can help to create new biosphere conditions which would be considerably preferable, taking all factors into account, to the present situation. But in attempting this, new conditions will be created, some aspects of which will themselves be regarded as problematic. These will lead to new challenges for research and to new attempts to put this knowledge into use. As an applied science, environmental research is per se one of the factors changing environmental conditions.

To take an example (Markl 1997): Ecological and ethological research attempts to find out how to allow as many plants and animal species as possible to survive in our cultural landscapes, and then to implement these findings. By doing this, however, we are creating conditions of selection and evolution in which the species which we wish to preserve actually change continually by adapting to these conditions (see the examples outlined above involving *Xanthium* and *Oenothera*). We observe that well-known animal species have changed their behaviour completely, for example having lost all fear of humans and abandoning their original way of life, collecting food from rubbish heaps or artificial feeding. It is a nonsense in evolutionary biology terms, however, to speak of the degeneration of noble wild creatures by humans, because these species have merely successfully adapted to changed conditions. Our attempts to preserve nature in our landscapes and way of life merely create a new form of nature in which humans play a very important role.

Just as the tarpan and field pigeon have not actually died out, although they no longer exist in their original form, having changed their genetic pool into numerous horse and domestic pigeon species, it will be impossible, within a few millennia, to determine what actually makes up the 'natural' environment. Our research and our actions, building up from generation to generation, will mean that nature will continually change and develop. The relationships between humans and the environment, and environmental problems, will also change. Environmental research, as an applied science, is unavoidably part of the mechanism by which humans change their environment according to their own goals. It is part of nature itself, as is everything which humans do.

An assessment of the present-day situation and of probable future developments in relations between humans and the environment can help to reach some conclusions about the results of our actions. These are based on 'if-then' statements derived from scientific causal analysis, and on non-scientific value judgements. The following are examples of such statements:

- If we judge it to be important that at least 50% of the species currently existing can survive long-term, we must ensure that at least 10% of the current habitat areas are preserved for the species communities without any human interference. If we wish to protect more than 50%, even more biotope areas are needed and must also be left untouched.

- If we want a rich variety of birds, plants or insects within our cultural landscapes, we must use these landscapes in such a way – with regard to land use, pesticides, landscape design, etc. – that these species can find a home there. Scientific environment research is capable of providing detailed information on how this can be done.

The method of argumentation by which scientific environmental research is combined with non-scientifically based value judgements is always as follows:

Scientific environmental research first discovers the various factors causing the loss of biodiversity; in fact, research is often the reason for such a loss being noticed in the first place, such as a slow secular dying-out process or a gradual poisoning through environmental chemicals or elements in food which can only be statistically established by means of large-scale epidemiological investigations. Then, a value judgement must be placed on these findings. These values require their own normative justification – it cannot be scientifically proven whether the loss of a species or the heating-up of the planet is good or bad. This requires a value judgement whose arguments are not based on fallacious 'naturalistic' conclusions regarding the current situation.

Once a consensus or near-consensus about the judgement to be given to a certain phenomenon has been established – which is a societal and not a scientific process – scientific research again plays a role, as technological research accompanying and enabling the measures to be carried out. The relevant measures must be available. As this always involves influencing and directing human behaviour, it is obvious that it is not merely a scientific or technological process. The entire repertoire of psychological, education, sociological, economic and legal functions is required in order to change behaviour such that it meets the goals set.

These goals can only be successfully met by means of worldwide co-operation between all the societal forces involved – scientific, economic, political, administrative and judicial. Applied science is essential for this process, but cannot achieve it alone. As strange as it may sound, environmental research and environmental management can only develop in this way if it is more than applied science and technology – if it is prepared to play its role in the joint effort made by the human, natural and social sciences which will be required to play a part in managing cities as living spaces (Markl 1997).

References

Arbeitsgruppe "Methodik der Biotopkartierung im besiedelten Bereich" (1993) Flächendeckende Biotopkartierung im besiedelten Bereich als Grundlage einer am Naturschutz orientierten Planung. Natur und Landschaft 68: 491–526

Brandes D, Zacharias D (1990) Korrelation zwischen Artenzahlen und Flächengrößen von isolierten Habitaten, dargestellt aus dem Bereich der Regionalstelle 10 B. Florist Rundbr 23, 2: 141–149

Deakin R (1855) Flora of the Colosseum of Rome. London, VIII: 237

Deelstra T (1998) Towards Ecological, Sustainable Cities: Strategies, Models and Tools. The European Experience In: Breuste J, Feldmann H, Uhlmann O (eds) Proceedings International Conference "Urban Ecology" Leipzig 1997. Springer, Berlin Heidelberg New York

Falinski JB (ed) (1971) Synanthropisation of plant cover. II. Synanthropic flora and vegetation of towns connected with their natural conditions, history and function. Mater Zakl Fitosoc Stos U W Warszawa-Bialowieza 27: 1–317 (Polish, English summary)

Fiedler O (1937) Die Fremdpflanzen der Mitteldeutschen Großmarkthalle zu Leipzig 1932–1936 und ihre Einschleppung durch Südfruchttransporte. Hercynia 1: 124–148

Gilbert OL (1989) The ecology of urban habitats. London, New York (German 1994: Städtische Ökosysteme. Radebeul)

Gutte P (1992) Fremdpflanzen in Großstädten – dargestellt am Beispiel Leipzigs. Acta Acad Scient 1: 90–93

Haeupler H (1974) Statistische Auswertung von Punktrasterkarten der Gefäßpflanzenflora Süd-Niedersachsens. Scripta Geobotanica 8

Klausnitzer B (1993) Ökologie der Großstadtfauna. 2nd edn. G Fischer, Jena Stuttgart

Klotz S (1990) Species/area and species/inhabitants relations in European cities. In: Sukopp H, Hejny S, Kowarik I (eds) Urban Ecology. The Hague pp 99–103

Kowarik I (1992) Floren- und Vegetationsveränderungen infolge der Einführung und Ausbreitung nichteinheimischer Gehölzarten in Berlin und Brandenburg. Verh Bot Ver Beih 3: 1–188

Kunick W (1974) Veränderungen von Flora und Vegetation einer Großstadt, dargestellt am Beispiel von Berlin (West) Diss Techn Univers Berlin

Kunick W (1981) Comparision of the flora of some cities of the Central European Lowlands. In: Bornkamm R, Lee JA, Seaward MRD (eds) Urban ecology. 2nd European Ecological Symposium, Berlin 1980. Oxford Blackwell, pp 13–22

Kuttler W (1993) Stadtklima. In: Sukopp H, Wittig R (eds) Stadtökologie. Stuttgart pp 113–153

Landolt E (1992) Veränderungen der Flora der Stadt Zürich in den letzten 150 Jahren. Bauhinia 10: 149–164

Landsberg H (1981) The urban climate. International Geophysics Series 28. New York

Leser H (1991) Landschaftsökologie. 3rd edn. UTB 521 Ulmer, Stuttgart

Markl H (1997) Naturwissenschaftliche Forschung und Umweltpolitik. Lecture Bonn March 12, 1997. 25 Jahre Sachverständigenrat für Umweltfragen

Mayntz R (1997) Möglichkeiten und Grenzen der Umweltpolitik – Beratung aus sozialwissenschaftlicher Perspektive. Lecture Bonn March 12, 1997. 25 Jahre Sachverständigenrat für Umweltfragen

Mumford L (1961) Die Stadt. Geschichte und Ausblick. Köln-Berlin

Nylander W (1866) Les lichens du Jardin du Luxembourg. Bull Soc Bot France 13: 364–372

Peters H (1954) Biologie einer Großstadt. Heidelberg

Pfeil E (1972) Großstadtforschung. Entwicklung und gegenwärtiger Stand. Hannover.

Pysek P (1993) Factors affecting the diversity of flora and vegetation in central European settlements. Vegetatio 106: 89–100

Rudder B de, Linke F (eds) (1940) Biologie der Großstadt. Steinkopff, Dresden Leipzig

Schirmer H, Kuttler W, Löbel J, Weber K (eds) (1993) Lufthygiene und Klima. Ein Handbuch zur Stadt- und Regionalplanung. Düsseldorf

Schmitz G (1996) Phytophagenkomplexe von *Artemisia vulgaris* L. und *Tanacetum vulgare* L. (Asteraceae) und deren Beeinflussung durch zunehmende Urbanität der Standorte. Univ Diss Mikrofiche-Ausgabe 3, Mikrofiches: 24x (unpublished)

Scholz H (1956) Die Ruderalvegetation Berlins. Diss Freie Univers Berlin

Sukopp H (1968) Der Einfluß des Menschen auf die Vegetation. Lecture to qualify as a university teacher (Habilitationsvortrag)

Sukopp H (1971) Beiträge zur Ökologie von *Chenopodium botrys*. L. Verh Bot Ver Prov Brandenburg 108: 3–74

Sukopp H (1976) Dynamik und Konstanz in der Flora der Bundesrepublik Deutschland. Schriftenreihe Vegetationsk 10: 9–26

Sukopp H (ed) (1990) Stadtökologie Berlin. P 455

Sukopp H, Trepl L (1995) Stadtökologie. In: Kuttler W (ed) Handbuch zur Ökologie. Berlin pp 391–396

Sukopp H, Wittig R (eds) (1993) Stadtökologie. G Fischer, Stuttgart p 402

Trepl L, Sukopp H (1993) Zur Bedeutung der Introduktion und Naturalisation von Pflanzen und Tieren für die Zukunft der Artenvielfalt. Rundgespräche der Kommission für Ökologie 6: 127–142

Wagenitz G (1964) Compositen (Korbblütler) II. Teil. In: HEGI, Illustrierte Flora von Mitteleuropa Vol VI, 3. Munich

Weidner H (1939) Die Großstadt als Lebensraum der Insekten, ihre Biotope und ihre Besiedlung. Verh VII. Intern Kongr Entomologie 2, pp 1347–1361

Wittig R (1991) Ökologie der Großstadtflora. G Fischer, Stuttgart p 261

Towards Ecological Sustainable Cities: Strategies, Models and Tools

Tjeerd Deelstra

1 Defining Sustainable Urban Development

Sustainability has become an important political issue since the World Commission on Environment and Development launched *Our Common Future*. The Commission brought two strands of thought together: the pressure for socio-economic development and the need to limit its harmful impacts on the environment. (World Commission on Environment and Development 1987).

Our Common Future is by nature general and global in its aims. If they are to be implemented, the recommendations of such an international policy outline must be translated into action plans for local communities. Initiatives with respect to this have been taken by associations of municipalities nationally as well as world-wide. At the European level the *Sustainable Cities and Towns Project* helped formulate a set of guidelines for sustainable local development based on comprehensive urban ecological research. The main principles were adopted by a large number of cities through their signature of a charter when they convened in Aalborg, Denmark, in 1994 (European Commission 1996).

Citizens may have very different concepts of local sustainability or may stress particular aspects of development and environment when it comes to setting their priorities for cities and towns. This can originate from their position as consumers, businessmen, government officials or experts. Concepts will also be embedded in a cultural context.

A quick survey among conference participants demonstrates how different approaches can be.

Professor Mary-Ellen Tyler (Canada) focuses on *urban form*: "When I think about sustainable cities I compare European cities with North American cities and towns. The loosely organised American city requires a lot of energy and land. The compact European city allows for a more sustainable lifestyle with less consumption of resources."

The International Institute for the Urban Environment, Nickersteeg 5, 2611 EK Delft, The Netherlands

J. Breuste, H. Feldmann, O. Uhlmann (Eds.)
Urban Ecology

Professor Kwi-Gon Kim (Korea) concentrates on *equity*: "Sustainability means equity between people: between people living now, and between generations".

Professor Herbert Sukopp (Germany) stresses *the resilience of nature*: "For me, sustainability in urban development is compatibility with the environment".

Three experts, three reference points.

In the UNESCO MAB (Man and Biosphere) programme, which included the first international research project on urban ecology, three types of science were brought together: natural sciences, engineering/planning and the humanities (Celecia 1991). Fundamentally different methods are used in these three fields of knowledge, each of the different scientific cultures dealing with different facts and values.

Despite a large number of contrasting disciplines within the natural sciences, integrated methods have been developed for the analysis of ecosystems (including those influenced by urbanisation). Consequently, there has been a development in the acquisition of knowledge about nature in and around cities. Contemporary practice in green space management and design is heavily influenced by this know-how and citizens as well as wildlife and plants profit from it.

Environmentally sound urban planning and design is not yet so well developed but serious efforts are currently being made to overcome this deficiency (Deelstra 1993). Management of flows of energy, water and materials (or waste), food and products, people and services (information) is the starting point for new concepts in urban planning.

More efficient and effective infrastructure is required along with more integrated forms of land use; the analysis of the life-cycle of the urban fabric determines its design.

The human ecological dimension of urban development is the least developed, although first attempts can be observed in action oriented social research, participatory approaches to urban planning and in the work of the *healthy cities movement.* (Deelstra, Yanitsky 1991; WHO Healthy Cities Project Office 1994).

2
Different Viewpoints, Unified Strategies

When developing strategies for sustainability, different viewpoints can provide inspiration. These might provide a means of enriching conceptions and making them more varied; contrasting views can be appreciated as being complementary to each other (The International Institute for the Urban Environment n. d.).

Conflicts of interest can arise during discussions which are intended to prepare concrete actions in cities. However, since stakeholders need each other, these conflicts must be solved. Assessing possible scenarios for the future can help to find compromises and new pathways towards co-operation (The International Institute for the Urban Environment 1996).

The *European Awareness Scenario Workshop* method is a proven tool for solving conflict situations. Discussions between stakeholders are supported with the help of extreme scenarios (Fig. 1). One extreme favours mere technological solutions whilst the other stresses more organisational approaches.

For instance, a high-tech strategy for the saving of drinking water, using complex household fittings, will perhaps stimulate fine mechanical industries and informatics. A choice for the more organisational extreme will have another impact on the economy and urban way of life. An example of an organisational solution could be to make a neighbourhood pond which is to be cleaned and controlled by, say, school children or elderly people. Both extremes can equally save costly drinking water, the first by considerable reduction in end use and the second by using rainwater stored in the pond instead of drinking water for toilet flushing.

Extremes are also determined by the level of initiative i.e. top down from public offices or through individual choices in the market. Examples in the area of water management are municipal double-piping systems versus a compost toilet or water saving shower.

The extremes in scenarios stimulate creative thinking about possible futures and make people realise that choices can be made. The interaction of the stakeholders produces feasible options and draws attention to bottlenecks that must be overcome. Scenario workshops can help construct a spirit of mutual trust and co-operation by breaking down barriers and reaching a consensus.

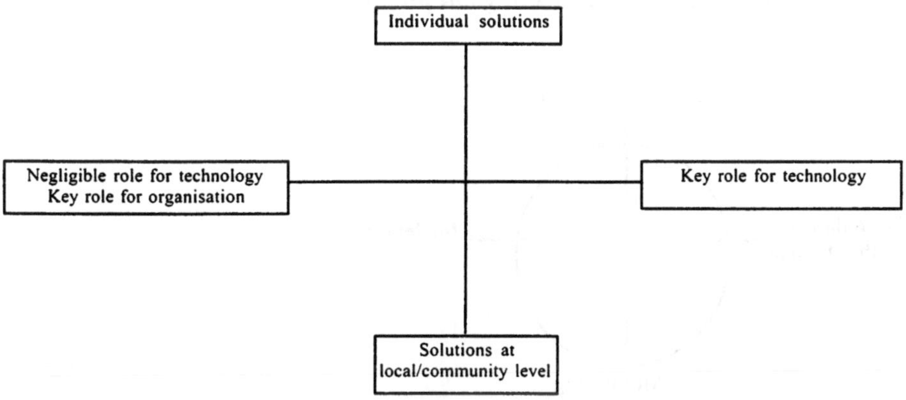

Fig. 1. Four extremes make four possible scenarios

Some cities in which scenario workshops have been held have installed semi-permanent platforms or *Urban Forums for Sustainable Development* (European Commission 1994). These Forums help to establish policies for local sustainability, conceived as a common effort with shared responsibility.

3
Policy Life Cycle

Policies will start when an agenda has been formulated. Such an agenda should be derived from commonly recognised local problems and issues. The local agenda can be inspired by Agenda 21, the *Rio 92* convention on the earth's development in the twenty-first century. Most nations have adopted the sustainability principles of Agenda 21 which imply a large number of local measures.

Awareness raising actions and public information campaigns are essential tools for arriving at an agenda for action based on the wide support of the local community (The International Institute for the Urban Environment 1998).

The next step is to formulate a plan of approach (Fig. 2). The content of the plan is highly dependent on local circumstances and the commitment of stakeholders. Sustainable development requires extreme reductions in consumption and resource use in cities. Main fields of action should be determined and an outline should be given of the necessary measures to be taken, for instance in the fields of transport and industry. Such a plan must be based on a local view of a sustainable economy, and can also incorporate *design principles for urban planning* (Deelstra 1993).

When considering implementation, learning from experiences elsewhere will be useful. Therefore it is advised that cities look for (international) co-operation and exchange lessons in *good practice*.

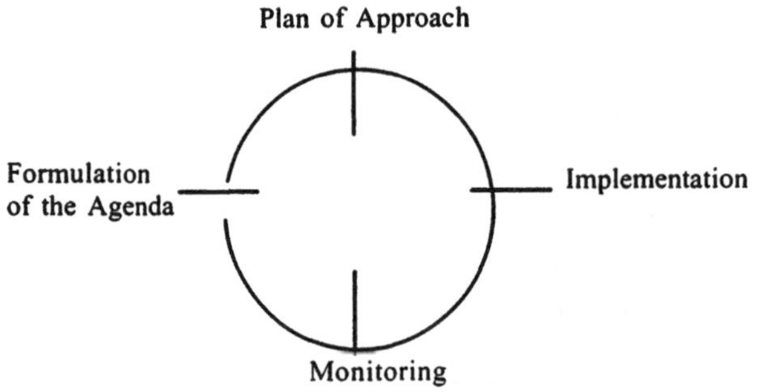

Fig. 2. Policy life-cycle

A key question is: can conservation stimulate community values, better architecture and improved environments for people to live in? Will this in turn support more environmentally sound behaviour by citizens? Well thought out projects can stimulate public discussion on the experiential value of sustainable development of cities.

Monitoring is crucial: to measure is to know. The performance of the city must be assessed. How can the performance of the city (as well as local actors) be improved?

It is of course not physically possible to monitor and measure everything and it would be too expensive, therefore *indicators* are used. Indicators are selected representative and relevant elements that help us visualise sustainability. A comparison between this picture and the situation in previous years will show a process, hopefully towards sustainability. The first measurement establishes a baseline against which future progress can be measured.

The choice of indicators to be used depends on three factors (The International Institute for the Urban Environment 1994):

- in what way sustainability is defined,
- which environmental, social and economic problems and features in the area are to be measured,
- which measures are to be evaluated (consumption patterns to be changed, transport and traffic systems to be improved, etc.).

Indicators can also be used in the discussion on how to reallocate resources and to reorganise policies to favour those areas of the city which the indicators have demonstrated to have the greatest need.

Disseminating the results of monitoring is very important in generating local support for the paths to be followed.

References

Celecia J (1991) UNESCO and Science – From Ideals and to Actions. In: Deelstra T, van Emden HM, de Hoop WH, Jongman RHG (eds) The Resourceful City – Management approaches to efficient cities fit to live in. Royal Netherlands Academy of Arts and Sciences Amsterdam

Deelstra T (1993) The Quest for Sustainability in Urban Development and Design. In: Rautsi J (ed) The European City Today – The Helsinki Round Table on Urban Improvement Strategies. Ministry of Environment Helsinki

Deelstra T, Yanitsky O (1991) Cities of Europe – The public's role in shaping the urban environment. Mezhdunarodnye otnoshenia Publishers Moscow

European Commission (1994) City and Environment. Office for Official Publications of the European Communities. Luxembourg

European Commission (1996) European Sustainable Cities. Office for Official Publications of the European Communities. Luxembourg

The International Institute for the Urban Environment (1994) The European Sustainability Index Project – Project Report. Delft

The International Institute for the Urban Environment (1996) Local Scenario Workshops in the LEC Network – Final Report. Delft

The International Institute for the Urban Environment (1998) The Ecological Footprint of Cities. Delft

The International Institute for the Urban Environment (nd) Our City in the Picture. Delft

WHO Healthy Cities Project Office (1994) Action for health in cities. WHO Regional Office for Europe Copenhagen

World Commission on Environment and Development (1987) Our Common Future. Oxford University Press Oxford, New York

1 Ecological Cities – Models, Environmental Aims, Standards

Perspectives on the Environment: New Options

Santosh Ghosh

We are now at the turn of the century and we should look back to what took place one hundred years ago. At the end of the nineteenth century, within a span of fifteen years, various inventions took place – the automobile, the telephone, domestic use of electricity (although electricity itself was developed earlier), the lift or elevator, cinema etc. – and a few years later the development of the aeroplane. These inventions caused discussions about future life styles in the twentieth century. Speed, mobility and everything new and innovative were the orders of the new life styles.

During the twentieth century unprecedented development took place in every field of science and technology. The extent and pace of development that has taken place in this century was not seen in the five thousand years of human civilisation, but this type of development brought disasters to human beings and damage to the environment, again unparalleled in human history. Two World Wars were fought as well as various other wars killing millions of people and resulting in millions of refugees. Disparity of living standards widened. According to the World Bank estimate the ratio of rich and poor countries was 1:50 in 1950 and is now 1:80 and this ratio will become greater by the turn of century; there is also wide disparity within nations and cities.

There are five concepts in the area of human settlement. At the beginning of the century the concept of garden cities and industrial cities arose, then came the modern architecture, pioneered partly by Germany, attached to which was a short lived post modernism period (although it came much later). Just before and after the second World War, statutory town planning practice called for the preparation of plans based on landuse; town planning codes and building rules were then developed. After this came the conservation movement and 1975 was declared the European year of architectural renaissance. The conservation of nature, historic buildings, monuments and areas brought about a new awareness and, with it, the environmental movement which started in the seventies after the decade of unrest and new consciousness of the youth in the sixties.

The President, Centre for the Built Environment, 2/5 Sarat Bose Road, Calcutta-700 020, India

J. Breuste, H. Feldmann, O. Uhlmann (Eds.)
Urban Ecology

The Untied Nations Conference on the Environment took place in Stockholm in 1972 and the UN Habitat I conference on human settlement in Vancouver in 1976. The environmental movement was officially recognised through a wider perception of the environment and development at the conference in Rio de Janeiro in 1992. This century's last mega conference was held in Istanbul – the UN Habitat II in 1996; after the debates, dialogues and concourses, came a final declaration. This consisted only of a few pages of statements after all the discussion, but it provides a broader agenda for the future.

There are several options for the 21st century. Sustainable development, a sustainable environment and similar words appear many times but the methods of integration of the environment and ecology are still unclear. Many programmes have been initiated which aim to improve the environment: – the Sustainable Cities programme adopted by the United Nations with the help of the World Bank; the Healthy Cities Plan initiated by the World Health Organisation; the Environmental Management Plan adopted by the United Nations again with the help of the World Bank and a multitude of other plans and programmes. The United Nations declared 1997–2006 as the international decade for eradication of poverty.

These plans and programmes often have overlapping jurisdiction and overlapping concepts. There is a lack of coordination in every city and there is gap between the macro concept and micro programmes and between the promises and actual performances. Good principles remain on paper as their implementation is difficult. There is no interaction between the Government departments, universities and the non-Governmental organisations. People's participation is advocated but not desired in many cities.

It reminds me of the satirical statement of the poet Goethe who was once a student in Leipzig. "If our children grow up with all the ideas and principles they are taught then everyone will be a genius."

While at international conferences green agenda issues such as climatic change, the ozone layer etc. are discussed, the cities concentrate on the brown agenda of air, water and land pollution and solid waste disposal. The slogan "think globally and act locally" has little meaning, local action plans often lack vision and have no concern for regional environmental issues. One example is the prevention of pollution in urban rivers, land erosion and flooding; this is counteracted by the construction of dams in many areas which is responsible for ecological disturbance. There is now a movement in India, the USA and some other countries against the construction of large dams which disturb the ecology and cause long term suffering of human beings.

I am an architect and urban planner and had the opportunity to work under a totalitarian administration, under a Kingdom, under a democratic government institution and under a Communist-coalition Government, but the attitude is the same everywhere. Until recently architects were taught to design buildings on vacant or new land. They are happy to see some of the old buildings demolished in prime locations so that new architectural masterpieces can be built. They are not concerned about the immediate surroundings and are limited by municipal regulations. The urban planner has always advocated city expansion by building satellite towns or low density communities. Like leapfrog urbanisation there is leapfrog planning. Building in the historic central area is avoided in most city plans, except perhaps for traffic improvements.

In this process of planning and development the historic buildings and areas have been destroyed. High-rise buildings or skyscrapers are being constructed in cities. Skyscrapers were developed in the USA, then came to Europe and are now present in third world cities. Every city wants to have the Manhattan or New York "look". Most office buildings consist of about 40% storage space but due to computer, internet, electronic mail, fax and the development of cybernatics even corporate business culture is changing and in the USA lower high-rise buildings are now being constructed.

Some third world cities such as Kualalumpur, Shanghai and others claim to have taller buildings than Sydney. These high-rise buildings are built with pile foundations and many of these buildings are weakened in the event of an earthquake and may collapse, as occurred in Mexico and Manila. In some cities the water level has dropped, causing a scarcity of water. Besides increasing traffic congestion, excessive use of energy and the methods of construction of high-rise buildings mean that such buildings are unsustainable.

Culture is an important aspect of every city, especially of older cities. The destruction of historic buildings and areas affects this culture, as well as the ecology of the city. It should be remembered that ancient cities like Mohenjadaro or Harappa in the Indus valley civilisation and many other cities in other civilisations were abandoned due to environmental degradation. The beautifully designed and constructed Fatehpur Sikri, the Mughal city in Agra was abandoned immediately after completion due to scarcity of water and other problems. Last year a planning meeting took place in Seoul, Korea for the organisation of an international conference on "the disappearance of Asian cities", as the character and flavour of Asian cities is being lost.

The rapid urbanisation connected with cities has destroyed agricultural land, forest cover and water bodies. In this process many plant, animal and bird species have become extinct. Ecological disasters occur, birds have disappeared from many cities, and flash floods are common due to the silting up of rivers. Many third world cities are subjected to such disasters, although European cities are not immune as it can be seen by the flooding of Dutch towns and Venice.

The Indian poet Rabindranath Tagore prayed to God to take away the cities and bring back the countryside. The poet's dreams can be realised if greenery and water – the green and the blue – are brought to cities. Architects and planners have created many dream cities but did not think about the carrying capacity of our earth.

Environmental problems are serious in third world cities where many ecological disasters occur. Third world cities have become very large; out of sixteen cities with a population of more than 10 million, 11 are in the third world and of these 7 are in Asia. In some countries about 30 or 40% of the urban population live in one city. The cities have become an oasis in the desert of rural poverty, but within cities one can see poverty and plenty side by side. The list of environmental problems ranges from pollution to poverty, from loss of wetlands to a reduction of biodiversity. Many programmes and projects have been undertaken to improve the environment of cities e.g. on traffic and transportation, water supply, sewerage, drainage and solid waste management, sanitary and infrastructure development and economic incentives and social welfare. The World Bank emphasises the provision of basic services to the poor, improving human health and strengthening municipal finance. Such developments were undertaken with the help of the World Bank and other agencies.

The late Indian prime minister Indira Gandhi said at the UN Conference on the Environment in Stockholm in 1972 that the environment could not be improved in the condition of poverty. To the poor sustainability means survival. In many cases local issues overshadow the metropolitan objectives. Projects with a shorter gestation period could not be co-ordinated within long term strategic plans -lack of co-ordination became a headache, co-ordination of authorities under the same political power did not result in co-ordinated projects. O. Henry, an American writer, when asked once about the future of New York said that, "it will be a beautiful city if they ever finish the projects."

Urban development projects are generally not oriented towards sustainability. Environmental Impact statements have became more of a ritual rather than a document which can be implemented. In Bangkok canals are being filled up for new roads. In Calcutta and Dhaka wetlands are being filled up to build townships. Improved roads, water supply, drainage etc. in urban areas attract real estate developers who commence private building projects, and this scenario will surely become more widespread.

At the UN Habitat II Conference in Istanbul UNESCO circulated a document calling for the humanisation of cities. The social dimension of urban development has been neglected but in cities with a high concentration of people, human ecology is just as important as physical ecology.

Western societies, especially in urban areas, have suffered from a breakdown of family life (an effect of technological civilisation) with a high rate of divorce, violence and insecurity together with rising unemployment and homelessness.

Twenty five years ago any European woman would think twice about visiting an Asian city because of insecurity, but today Asian women will think twice about visiting many American or European cities. The gender issue is grossly neglected in cities. In the recent Annual report of the World Health Organisation it is said that western diseases like cancer and heart disease are becoming more widespread, whilst Aids has no boundary.

There is conflict in cities between the environment and development. A German writer once said, and was echoed by many German architects, that Berlin was not a city. It was a cold war issue for so many years, a city of conflict in political thinking and now it is a city of conflict between the environment and development, like so many cities.

To the poor of the third world, sustainable development is survival while to others it means the ability to co-exist with the environment. The main issue in ecology is humankind's relationship with nature; cities are closed system, whereas ecology demands an open system. The third world environmental issues in cities are related to the provision of health, hygiene and housing and control of population, poverty and pollution.

Religion developed with the relationship of nature to the cosmos. All the religious prophets-Moses, Jesus Christ, Buddha, Confucious, Mohammed and others - were born in Asia and showed us the light. The dictionary meaning of ecology (interrelationship of organisms) is important here as cities are no longer communities, but commodities.

In countries like India or China ancient literature, and religious texts provided ample references and discourses on ecology. The prohibition on the cutting of trees, filling up water bodies and regulation on where to build and where not to build form the basic ecological principles. Even today a tribal woman in India's remote area knows how to collect rain water, how to recycle the waste or how to use solar energy to dry food.

We are now using solar energy again. The second solar civilisation has arisen and considerable research has been carried out already in India. Urban agriculture (including horticulture, fishery etc.) is another developing field with the inclusion of "ecofarming" which helps to preserve the environment. In Calcutta and other cities people are producing vegetables using natural compost from solid waste and fishes from recycling the sewage effluent of the city, without the help of the Government or experts.

In Manila, there is network of solid waste dealers. Resource recovery is extensive in Shanghai and the sale of recycled products is widespread. In Curitiba, Brazil 70% of households separate garbage and in slum areas food, bus fares and exercise books are exchanged for garbage.

Experimental eco-cities and eco-villages are now being built in many countries emphasising sustainable development, but unlike fashions in clothing or new types of music, these eco-city ideas do not spread very quickly.

It is important to make cities an attractive place in which to live in the following ways: by reducing pollution, introducing green areas and agriculture, recycling solid waste and waste water, reusing resources, increasing the use of public transport, increasing the number of pedestrian and bicycle paths, conserving the cultural heritage, miniaturisation of eco-friendly businesses and industry. Above all, the long-term ecological situation should be integrated into the development and implementation of the city's plans which should also take into account the socio-economic framework of the city.

Workshop: Environmental Quality Aims and Ecological Aims 1.1

Environmental Accounting on a Communal Level

Jürgen Pietsch and Gabriele Kröger

1 Introduction

The guidelines for sustainable development demand the global reduction of resource consumption and the orientation to the carrying capacity of the environment. The project "Environmental Accounting on a Communal Level" (Fig. 1) is concerned with the problem of how this sustainable concept can be transformed from a global to a community level.

The paradigm of sustainable development requires integrated, systematic modes of perception. Instead of the traditional divisions (environmental i.e. water, soil and air; or society based divisions i.e. ecology, economy and society), one has to find a new model for the reality. One way to leave the traditional, sectoral and emission oriented approach is to take the whole life cycle into account.

In recent years methods of material and energy accounting were developed and used. In this way the ecological impacts of products or companies can be compared and assessed. A corresponding method is lacking for the communal and regional level, thus the central idea is to develop a methodical framework for environmental accounting of communal actions. The approach is based on a comprehensive, integrated perception and analysis of city structures, in which environmental impacts can be described by drawing up a balance between material and energy input, quality of areas and quality of the environmental fields.

The methodical framework is derived from the approach for Life Cycle Analysis of the German Federal environmental agency (Umweltbundesamt).

Starting from an ecological perception of urban-industrial areas, the method will be used to analyse and evaluate which environmental impacts are the result of communal activities (e.g. the results of different kinds of water supply systems).

Technical University Hamburg-Harburg, Institute for Urban Ecology, Kasernenstr. 10, D-21073 Hamburg, Germany

J. Breuste, H. Feldmann, O. Uhlmann (Eds.)
Urban Ecology

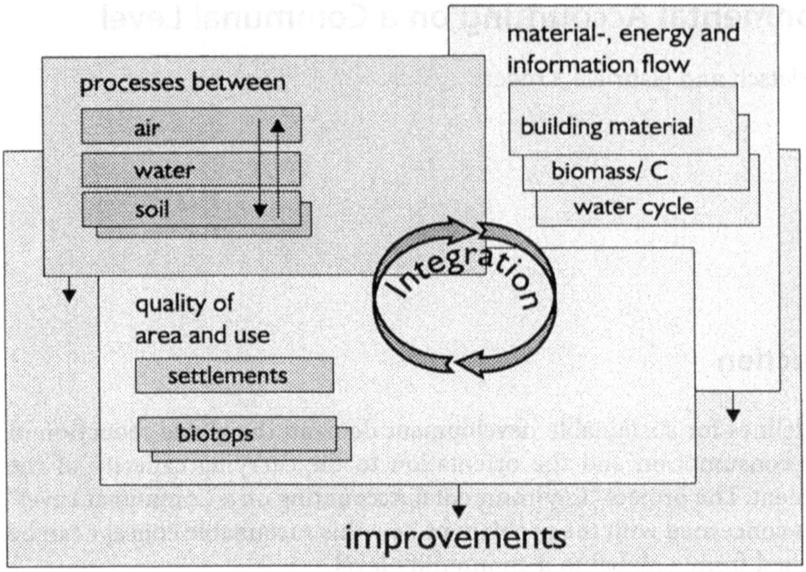

Fig. 1. Approaches for environmental accounting on a communal level

2 Objectives

With this approach we want to support the perception of processes such as:

- material and energy flow,
- the change in quality of areas and their uses,
- the impacts of pollutants on environmental fields and their interactions,
- cumulative effects,
- the postponement of environmental problems.

This approach opens new horizons for environmental planning within the community. Greater efficiency and a decrease in costs can be arrived at by leaving "end-of-the-pipe" strategies; opportunities for an improved planning process and measures for different individuals and organisations can be drawn up.

3 The Methodical Framework

The methodical framework comprises five steps: objectives, system balancing, impact balancing, forecast frame and possible improvements (Fig. 2). Instead of following a sequential method, we pursue the idea of integrated work.

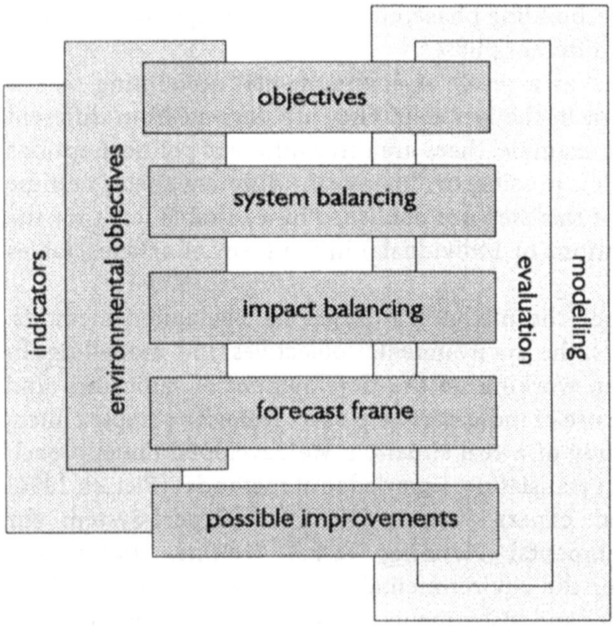

Fig. 2. Methodical framework

The Life-Cycle analysis shows that the central figure of the "system balancing" is the model. It shows the whole system and its interactions between the single parts and the environmental impacts. It illustrates the energy and material flow as well as the financial flow. In contrast to simple black-box-models, we need the illustration of the relationship to space and time, and the connections of involved organisations and individuals.

The central point of the "impact balancing" is also a model which shows the understanding of the single and cumulative impacts. The method of evaluation has to be drawn up at this step. A good method is the 'critical input', 'critical level' and 'critical structure' created by the Wissenschaftlicher Beirat der Bundesregierung Globale Umweltveränderungen (WBGU).

There is a need to look at the whole development of the analysed system. One has to consider not only the present, but also the past and future development in order to realise the relevant processes in which the system or parts of it are involved. Phases of the life cycle with their different rhythms must be characterised and identified. For example the life-time of a building is different from the life-time of its roof, its windows, its heating, its air-condition or its refrigerators.

With the additional step 'forecast frame' we try to assess especially the future development. Using scenario techniques and other methods, the system-behaviour over the span of time is analysed. The whole life-cycle consists of the planning phase, building phase, etc. The use and application phase is often sub-divided into different phases.

'Possible improvements' as a result of environmental accounting on the communal level are shown in the last step. They are derived from different views and analyses. If, for example, there are three different political options to renew a housing estate it is possible to illustrate the different effects of these alternatives. The results of this step are published in a suitable form for the relevant bodies (organisations or individuals) in the form of articles, tables and figures.

In Figure 2 one can see the methodical basics i.e. evaluation-methods, development of indicators, the environmental objectives and modelling. In recent years we have been working on the 'development of indicators' and 'evaluation methods'. The use of indicators help us to deal with complex interactions in a balancing study of a real situation. We developed the approach "MethodenWissen" (direct translation: knowledge of methods) (Pietsch 1994) and the computer-aided expert-system "EXCEPT" (Expert-System for Computer-Aided Environmental Planning Tasks) (Hübner 1992). The methodical framework for the environmental accounting is based on these already existing approaches and instruments which contain two central ideas. The first aspect is the elaboration of a structured base of the existing specialised knowledge (for example traffic planning or water supply) of the analysed problem. Different ways of comprehension and perception are illustrated. Starting from this basis the correct evaluation methods are chosen and developed, combining and expanding methods such as ABC-method, ecological risk analysis, use-value-analysis or cost-benefit-analysis.

A modular structure is obtained through structuring the indicators according to the relevant main emphasis. One module or indicator-set could be focused on, e.g. water supply, the flow of different materials or energy flow. It is also possible to draw up a balance between these modules.

'Environmental objectives' show the specific environmental aims of the community. These are important for orientation, especially if there is detailed spatial and process-oriented information available.

'Modelling' helps to show the perception and understanding of the analysed system. By illustrating the different kinds of interactions, the relevant actors etc., a transparent basis for communication and evaluation are built up. The model was created on different levels in the form of pictures, descriptions, diagrams and tables.

For computer modelling 'Petri-Netze' (petri-network) and the software 'umberto' are used. They consist of two kinds of locations (storages and transformation points) and illustrate the material and energy flow between the two.

4
Co-operation

The studies carried out by Prof. Baccini (ETH Zürich, Switzerland) and Prof. Brunner (TU Vienna, Austria) play a very important role as a basis for the material and energy flow analyses in communities. They started together with the study 'Metabolism of the Anthroposphere' (Baccini and Brunner 1991) at the end of the 1980's. Prof. Baccini is currently developing a regional material management system for biomass (Baccini und Bader 1996). Prof. Brunner is analysing the anthropogenic material fluxes of cities like Vienna (Brunner 1996). These studies are used as background information for particular developments, methods etc.

5
The Range of Application

To develop the methodical framework through examples, an intensive practical co-operation with cities and regional organisations is set up. We want to analyse fields which are very different in topic and scale.

At the moment we are starting with three accounts. One deals with an harbour area in Hamburg-Harburg, where different planning options in an area of mixed uses are analysed. The second is a comparative study of two different building alternatives in a new residential area of Jena which will be accounted. Last but not least, we are using these methods to develop a new basis for spatial planning by characterising the material input and output of different areas and uses in a community.

References

Baccini P, Bader H-P (1996) Regionaler Stoffhaushalt. Erfassung, Bewertung und Steuerung. Spektrum Akademischer Verlag Heidelberg, Berlin, Oxford

Baccini P, Brunner PH (1991) Methabolism of the Anthroposphere, Springer Verlag Berlin, Heidelberg, New York

Brunner PH (1996) Der anthropogene Stoffhaushalt der Stadt Wien. Projekt PILOT. Manuskript des Instituts für Wassergüte und Abfallwirtschaft. Abteilung Afallwirtschaft

Hübner M (1992) EXCEPT - ein System zur Unterstützung und Dokumentation von Bewertungsvorgängen in der Umweltverträglichkeitsprüfung. In: Günther O (Hrsg) Konzeption und Einsatz von Umweltinformationssystemen, Informatik Fachberichte 301, Springer, 1992

Pietsch J (1994) MethodenWissen - Generierung von Wahrnehmungszusammenhängen und partizipativ-evolutionäre Wissensaquisition. In: Hilty L.M, Jaschke A, Page B, Schwabl A: Informatik für den Umweltschutz, 8. Symposium Hamburg, S. 233–240.

Wissenschaftlicher Beirat der Bundesregierung Globale Umweltveränderungen (1994) Welt im Wandel: Die Gefährdung der Böden. Jahresgutachten 1994

Urban Development and Environmental Policy: Objectives Outline of a Multi-disciplinary Research Programme

Joop de Boer[1] and Martin Dijst[2]

1 Introduction

In recent years there has been a growing awareness that environmental policy has much ground in common with spatial planning. This awareness should be seen against the background of several conditions:

1. It has become clear that the achievement of both environmental and spatial policy objectives might seriously be hampered by the current economic growth unless more care is taken to steer this development in the right direction.
2. It is a well known fact that many urban areas which were important economic centres in the past are currently in need of major revitalisation to prevent further decline of their social and ecological living conditions.
3. It has appeared that cities and the urban way of life can potentially make important contributions to a more sustainable future for society as a whole.

In this context the concept of "sustainable cities" has become a kind of catch-phrase that inspired quite a number of authors to emphasise the potential synergy between spatial planning and environmental planning. In considering the contribution of scientific research to a sustainable future it is important to set the stage for a more multi-disciplinary approach. In this chapter we will give an outline of a multi-disciplinary research programme focused on the relationship between urban development and environmental policy objectives. This project was commissioned by the Dutch Ministry for Housing, Spatial Development and the Environment (De Boer & Dijst, 1997).

[1] Socio-Economic and Natural Science of the Environment, p/a Instituut voor Milieuvraagstukken, Vrije Universiteit, De Boelelaan 1115, NL-1081 HV Amsterdam, The Netherlands
[2] Netherlands Graduate School of Housing and Urban Research, P.O. Box 80115, NL-3508 TC Utrecht, The Netherlands

J. Breuste, H. Feldmann, O. Uhlmann (Eds.)
Urban Ecology

2
Analytical Framework

As a first step in addressing the issue of the integration of urban development policies with environmental policies, an analytical framework has to be designed that covers comprehensively the relationship between urban development and its environmental effects. This framework should integrate the analytical concepts that are used in urban and environmental research. Two perspectives are important in this connection: an *area related perspective* and a *chain related perspective*. Each perspective can be described in terms of functions, activities and physical processes, such as cycles or fluxes of substances and materials.

For example, the *functional characterisation of an area* is an important aspect of zoning plans. These plans specify which parts of an area can be used for housing, industry, recreation, agriculture, or nature development, while taking into account that these functions will interfere with one another if they are located too close together. Alternatively, if these functions are located too far from each other, they will cause effects that are also unwanted, such as an increase in traffic. Therefore, the design of a zoning plan requires a careful and transparent trade-off between social, economic, spatial and environmental effects, based on a coherent functional analysis.

The *functional characterisation of a product or service* is the starting point for the method of Life Cycle Assessment (LCA), which deals with the environmental effects of a product covering its full life cycle, from resource extraction through production and use, to the processing into final wastes. LCA links environmental effects to social and economic functions.

Functions are achieved by entities such as spatial and technological structures and behaviour patterns. These, in turn, are linked with physical processes causing flows of substances and materials which are diffused into the environment where they accumulate and disturb life-supporting systems.

Combining the functional approach of areas with that of products and services might contribute to a better understanding of the causes of these environmental problems as well as to the design of more efficient and effective solutions. An elaborated functional analysis is particularly useful as it will add a spatial dimension to LCA-methods and a substance-oriented dimension to spatial planning.

The common ground of spatial and environmental policy is also related to the pressures that are caused by the often diverging functions of a certain area. Obviously, the effects of the physical processes are heavily dependent on the vulnerability of the ecosystems and the people who live in the affected area. The most vulnerable effects will show themselves in a loss of biodiversity, health and well-being.

One way of alleviating these effects is to take measures to improve the environmental and spatial quality of the area. This improved environmental and spatial quality is not only important from the viewpoint of sustainability and the protection of life-supporting systems, but can also have a value of its own.

In summary, this preliminary framework can contribute to a better understanding of the environmental dimension of urban activities and the urban way of life by taking into account not only ecological processes, but also those social, economic, technological and spatial processes that might either hamper or promote environmental policy objectives.

3 Research and Policy Themes

On the basis of information received from a limited survey of recent literature, personal interviews and a workshop with representatives from both research schools and the ministry, we formulated five broad research themes for the research programme.

3.1 Urban Development as a Function of Regional and (Inter)national Processes

Urban development is embedded in economic, social, cultural, technological, demographic and institutional processes at regional and (inter)national scales. These functional linkages influence spatial processes such as the migration of households and firms, the use of locations and the usefulness of existing buildings. These linkages will also change the use of natural resources, energy and space and, in the end, biodiversity and human well-being. To provide local and national governments with the instruments to anticipate these developments and their consequences, analytical methods have to be designed which can identify them in an early phase.

These functional relationships at different scales should also be considered in the process of policy making. Various policy makers within different government institutions are involved in strategic choices that might have direct or indirect consequences for urban development and the environment. A more integrated approach requires that these strategic choices are analysed and contrasted with a view to their compatibility and combined impacts.

3.2 Urban Areas as Framework for Eco-efficient Facilities

Urban areas can offer important opportunities to realise eco-efficient facilities. These include technological and organisational arrangements for energy chains, water management, transport systems, green areas and industrial parks. There are many ideas about these topics, but also many questions that have to be answered, especially about the degree of sustainability that might be achieved by various systems, their financial costs, their optimal level of scale, as well as the best way to initiate and organise their implementation. One specific area for further research is the design of analytical methods to weigh-up the different concepts and initiatives.

3.3 Function of Urban Areas for Citizens and Firms

In many current policy plans a direct relationship is assumed between urban form and environmental pressure. Positive environmental effects are expected for, e.g. degree of compact, mixture of functions, building in the proximity of public transport. However, reviews of evaluation studies show little consensus about the environmental effects of urban forms. An important explanation for this lack of consensus refers to the diverging life-styles among the urban population.

Although we can expect some positive effect from compact urban forms in comparison with more spread-out forms, we still need to know in what circumstances and to what degree this effect occurs, and in what way we can most effectively exploit this potential of compact urban forms.

In order to understand the complex relationships between urban form, transportation systems and (mobility) behaviour, we need more research starting from an *activity-based perspective*. This perspective is derived from the view that observed mobility patterns are the result of a complex decision-making process where individuals and firms try to achieve particular goals in the pursuit of their activities within certain spatio-temporal and institutional constraints.

With this approach we can acquire more knowledge about the practicable strategies to exploit the potentials of environmentally sound urban forms. These strategies can be focused on (re)constructing activity places, such as dwellings and offices. Other types of policies can also influence spatial processes: institutional rules, pricing policies and information services. The activity based approach makes it possible to weigh-up different strategies and policies in an integral way.

Extension and intensification of built-up areas dominate in spatial planning. However, the foreseeable extension of built-up areas in the Netherlands in the next few decades is about 1% of the overall area. Most of the building stock and infrastructure in the Netherlands in the year 2030 already exists! This means we have to pay far more attention to the management of the existing urban areas than to the extension of these areas. It appears, however, that most urban research in the Netherlands is focused on new building sites and not on the existing areas. Alternatively, management policies have to be developed aimed at matching people, firms and the built environment in an environmentally sound way.

It is generally agreed that the urban way of life might significantly contribute to the development of consumption and production patterns that use natural resources and energy in a more efficient way. A good starting point is to distinguish products from functions. For example, environmental pressures caused by the use of private cars can be reduced by making a distinction between 'owning transport modes' and 'communication'. The need for communication can be fulfilled by renting a car instead of owning one. More research is needed into such possibilities to decouple products and their function.

3.4 Institutional Dimensions

The main question to be answered in this programme is: which strategies of urban development are feasible to improve the sustainability of urban activities and the urban way of life? In addition, much attention has to be paid to the study of efficient methods of implementing the relevant strategies in the context of government policy-making at local, regional and national levels.

Reference

De Boer J, Dijst MJ (1997) Stedelijke ontwikkeling en milieu: contouren van een onderzoekprogramma voor het Ministerie van VROM. Amsterdam/Utrecht: SENSE/NETHUR.

Environmental Quality Goals and Standards as a Basis and Way to an Agenda 21 for Leipzig

Holger Tschense

1 Leipzig's Environmental Quality Goals

With the resolution of the UN- Conference for the Environment and Development in Rio de Janeiro in 1992 and passing of the closing document "Agenda 21", local communities especially were called upon to carry out actions leading to environmentally sound, forward-looking development.

The city of Leipzig is aware of this answerability in the field of environmental politics and for many years has been traveling along paths towards a sustainable improvement of the environmental conditions on a local level.

An essential contribution was made in this context with the establishment of environmental quality goals for Leipzig, drawn up by the city council. They supplement and support a series of others in the last years, partly through political resolutions, accepted instruments, concepts, measures and adopted commitments such as:

- A subsidy programme of the city of Leipzig for the reduction of air pollution (including a CO_2 reduction programme and installation of renewable energy)
- Concepts for traffic politics including „traffic reduction in the inner city"
- Energy concept oriented towards climate protection,
- Membership of "Climate alliance of the European States with the indigenous people of the rainforests/Climate alliance e.V",
- Signing of the "Heidelberg Declaration

to name just a few examples.

In their complexity the environmental quality goals represent, in special ways, an essential step to the putting into place of a Local Agenda in Leipzig; whose whole process began only a short time ago. The goals were outlined in 1992 and in the following years were continually further developed.

Leipzig City Council, New City Hall (Neues Rathaus), Environmental Authority, D-04229 Leipzig, Germany

J. Breuste, H. Feldmann, O. Uhlmann (Eds.)
Urban Ecology

In the framework of a long, intensive voting and consensus finding process within the city administration and also with external (scientific) specialist institutions, the aims fall by the wayside. The aims are in a close sense ecological aspects comprising social, economic and development politics – basically corresponding to the original setting of the prepared outline of Agenda 21. Social, economic and development political achievement of goals remain close to the 'ecological' aspects.

After taking part in a wider publicity campaign and a closing public hearing in Autumn 1995, the last revisions from the numerous useful pieces of advice received were made and the final version was put through the city administration on the 20th of June 1996.

With this, the environmental quality goals stand as politically professed intentions for the provision and maintenance respectively of healthier, acceptable living conditions as well as for the sustainable development of the city of Leipzig. They will be put into place in accordance with the principles of environmental concepts and testing instruments.

The handling of the environmental quality goals deals, however, primarily not with the putting in place of the usual rules, rather more on the precautionary and sustainable protection of people and nature in the next 10 to 15 years, to be structured in 5 year parts. An important instrument for environmental protection is thus revealed – it is neither a legal rule nor a reason for restrictive environmental politics, rather it should make a balanced and forward-looking way for city development possible.

An extrapolation of the environmental quality goals must come about through the changing political and legal framework conditions, as well as the further development of scientific knowledge and practical experience.

Good arguments and sound knowledge are necessary, thus I hope for the corresponding impulses through your work and exchanges at this conference.

The contents of Leipzig's concept for environmental quality goals and standards differentiates itself from similar instruments in other cities, essentially through a comprehensive consideration of all essential complexes of the provisional environmental protection – such as health provisions and nature and resource protection. Goals and standards have been formulated; they have also been established for the area of emission protection (for the separate media of air, water, noise, soil), local climate, relaxation, landscape/open area protection, waste and energy.

With the help of numerous other instruments (mentioned at the beginning), measures and concepts for the development of environmentally friendly traffic concepts will be pursued. These will be closely related to the energy concept of the city of Leipzig with the recommendation to further change to environmentally friendly energy producers with the highest possible degree of effectiveness and smallest possible use of primary energy (to reduce the air pollutants).

As a further point of importance, in the last years there have been increasing demands for open space for industry, settlements and transport. Towards the end of the financial year 1996 the increase in the number of new registrations for planning permission on open spaces was 5%; a reduction to 2% will be attempted by the year 2005. This should be done through giving absolute priority to protection and reactivation of already built-on and settled old city areas before any new land is used for building.

I am sure that we have drawn up good goals but I realise that putting these into place is not without problems due to several reasons:-

- The environmental quality goals and standards of the city of Leipzig do not have any legal binding outside of the city, only internally. This is especially relevant when the standards and goals go outside of the established valid, legal restrictions. In the framework of the provision of environmental protection they are, however, to be strived for in mid- to long-term (2000–2010),
- In numerous cases important conflicts of goals arise,
- The achievement of some goals does not lie directly within the city's sphere of influence; many problems are 'imported' from the surroundings and the environmental quality goals will only be agreed on within the boundaries of the city of Leipzig,
- The cited problem of the exclusion of social, economic and politically oriented development goals (through which the entire, sustainable development should be attained), should be solved step-by-step through the recently started process to the formation of a local Agenda 21; but with this we stand at the beginning – but it is at least done.

2 Leipzig's Agenda 21

After making its resolution to the 'Environmental quality goals' the city parliament engaged the administration to facilitate the organisation of the recommended communication process (Chapter 28 of Agenda 21) inside the city structure of Leipzig.

The administration found helpful partners in an active city parliament, in institutions and persons who have initiated a process for 'Leipzig's Agenda 21' starting from the ground level.

The combining of the named partners led to a public call (in June 1996) to begin the process of the development of Leipzig's Agenda 21 in the city.

As a result of this action a working group was set up, made up of groups/associations, researchers, economists and local administrations. This group (predominantly in an honorary occupation) has proposed a model for the organisation of the communication process (Fig. 1).

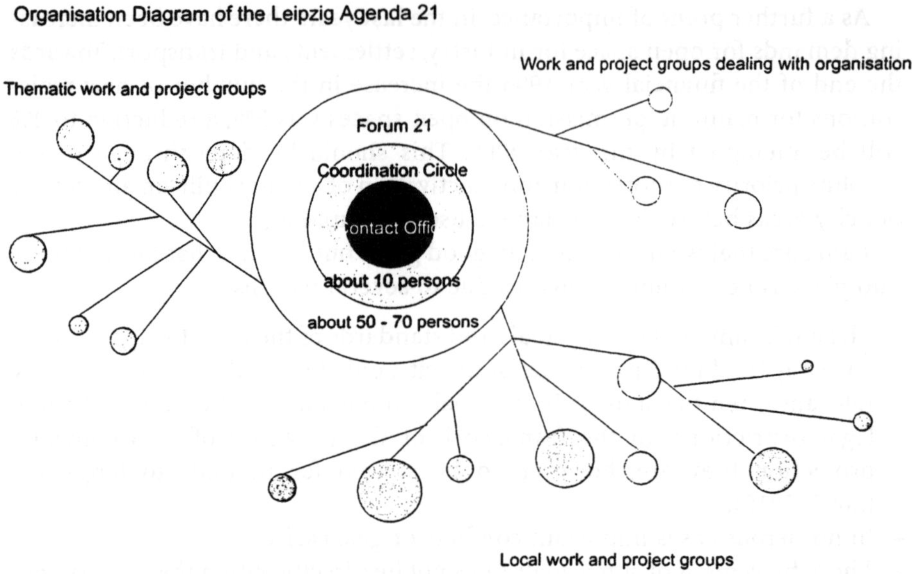

Fig. 1. Organisation scheme for Leipzig's Agenda 21

After the drawing up of the proposed organisational scheme, the way to a consensus building about the work in the project group was found. In the meantime the following points about certain subjects have been raised:

- Theme areas,
- Affiliation of places (for example residential areas),
- Affiliation of fields/departments,
- or simply through the desire to work together to form a specific project.

The following work groups were formed:

- AG Values- Activity-Lifestyle
- AG Regional economic cycles
- AG Citizen participation
- AG Mobility
- AG Women and Agenda 21
- AG Energy
- AG Education for tomorrow
- AG Allocation of existing work
- AG Environmentally friendly production and marketing
- AG Subsidising investments

They are open to all those who wish to contribute and work towards the general goal.

These activities should be grasped by a contact office and expanded. Thus the work would take on a service function, especially for the mediation of information within and between the project groups as well as with regards to the public relations and political bodies. The office would then serve as a contact and administrative body.

The work of the office should be steered and advised through a co-ordination group made up of representatives of various social institutions from the fields of business, groups/ associations, administrations and research/education. These should also, along with others, serve to provide focus points for the stimulation of specific activities. The session rota is to be decided on by the co-ordination group itself.

A quarterly, day-long Forum 21, made up of about 50 to 100 representatives from various organisations, should reinforce and accompany the agenda process and work as a multiplier for the whole city.

One of the essential supports for the success of the agenda process is the balanced participation of the various social groups and institutions.

In an ideal case an infinite number of small 'Cells' come into being from this structure. These 'cells' search for a consensus on the theme '*future-sustainable* Leipzig' and the corresponding political decisions could be derived from their results.

However it is known to us that such an ever-changing way of doing things inevitably provokes a change in the thinking process and requires a new understanding of the relationship between citizens, administration and politics.

Conflicts – also with regard to established competences- are natural and their solution requires the definite readiness of all participants to respond exactly to the requested dialogue.

We are all agreed that the putting into place of the 'environmental quality goals' builds a good basis for the communication process.

On the other hand starting out from the ideal case, on Day X each business, employer, institution, even each family would know what it means for them to deal with a future-looking, sustainable Leipzig, in correspondence with the planned local goals. I am not revealing any secrets to you when I state that many difficult steps are necessary to achieve this.

In the next months the task will be to network the activities of existing groups, initiatives and institutions and the newly founded working groups with the work of various specialist departments of the city administration and the political structure. With respect to all new ways for the participation of citizens, the integration of commerce, education and research – (through which we, as administrators and politicians, have high hopes that our work will be made easier) – the political decision making will remain reserved for the mayor of the city and the city parliament. Each specialist department will be given an 'Agenda Commission' (for the initiatives they are responsible for) in order to guarantee the working together of people and so that at any time separate clients can be included in the process and present their own proposals.

We are clear that the preservation of the basics of life and the avoidance of social conflict require the increase of certain solid economic developmental factors in order to come to a deciding positional factor.

We must search for an integration of ecological, economic, social and cultural aspects if we wish that our citizens are able to find healthy, acceptable and sustainable living conditions in cities.

It is the precisely the contents of Agenda 21 and these goals that the city of Leipzig wants to put in place.

Environmental Quality Targets for Urban Structural Units in Leipzig with a View to Sustainable Urban Development

Evelyne Wickop

1 Introduction

The redevelopment process taking place in east German cities since the reunification of Germany presents a unique chance for the integration of aspects of sustainable urban development into the transformation process from the beginning. Unfortunately, while cities are being radically reshaped, development patterns from western cities are being copied and ecologically comprehensive approaches are often not being taken into account.

2 Starting Points for Sustainable Urban Development in Leipzig

There are four basic starting points for cities wanting to tread the path of sustainable urban development: the way of using resources, the processes of exchange with the hinterland, the urban spatial structures, and the integration of economic and social components. When considering these four starting points for Leipzig, the development is often contradictory to the principles of sustainable urban development. Since reunification, Leipzig has experienced a high decrease in population and a massive loss of workplaces, as well as a shift in employment structure. Diverging and competing processes between the city and the surrounding area are typical for the transition process. Leipzig is changing from a former compact, industrialised, low-traffic urban region in need of extensive renewal to a decentralised, low-density region using up large areas of space and producing a large amount of traffic (Breuste 1996).

UFZ Centre for Environmental Research Leipzig-Halle, Dept. Urban Landscapes, Permoser Str. 15, D-04318 Leipzig, Germany, * present address: Landeshauptstadt Wiesbeden Umweltamt, Luisenstr. 23, D-65185 Wiesbeden, Germany

J. Breuste, H. Feldmann, O. Uhlmann (Eds.)
Urban Ecology

3 Environmental Quality Targets as a Method for Implementing Sustainability Targets in Urban Development

It is necessary to formulate principles and targets to deal with the natural resources and emissions of harmful substances in cities regarding the sustainability model for the development trends described. These environmental quality targets can only be implemented effectively if a clearly defined land area is designated. Conceptual plans exist only at a city-wide planning level and there is a lack of all-encompassing concepts based on the integration of ecological, social and economic development components. The aim of this project, which is being carried out by the Department of Urban Landscapes in the Centre for Environmental Research together with the City Planning Department of Leipzig, is to draw up a quality-target concept for urban structural units.

3.1 Urban Structural Units as a Basis for a Transferable Urban Typology

The orientation of urban development towards sustainability requires suitable methods for the analysis of land use patterns and urban spatial structures. Urban structural units are defined as '*areas with physignomically homogenous character, which are marked in the built-up area by a characteristic formation of buildings and open spaces*'. Thus, they integrate areas with similar environmental conditions and can provide a collective impression of the morphological situation and the characteristics which can be expected in different areas. In contrast to the method of structuring the city into administrative regions or districts, urban structural units are based on the key ecological features, 'use' and 'structure' (Duhme, Pauleit 1992). Urban structural units are located between the micro and macro-scale of a city and can therefore act as a decision-base for planning as well as research, taking both local characteristics and the potential for generalisation into account (see Fig. 1). Through this approach the evaluation of specific features in various areas and the ability to transfer these features to similar areas should be guaranteed.

Altogether 10 main urban structural units with 26 sub-units have been identified for the entire city of Leipzig (Böhm et al. 1996). The features which were applied to evaluate the urban structural units are *land use, building age, density and structure, green area structure and degree of surface sealing.* To cover the typical urban structures of the built-up area of Leipzig four heterogeneous districts have been chosen using specific criteria. The typical urban structural units of the study areas are: 'terraced and detached curb-close apartment buildings with built-up or open courtyard (late 19th and early

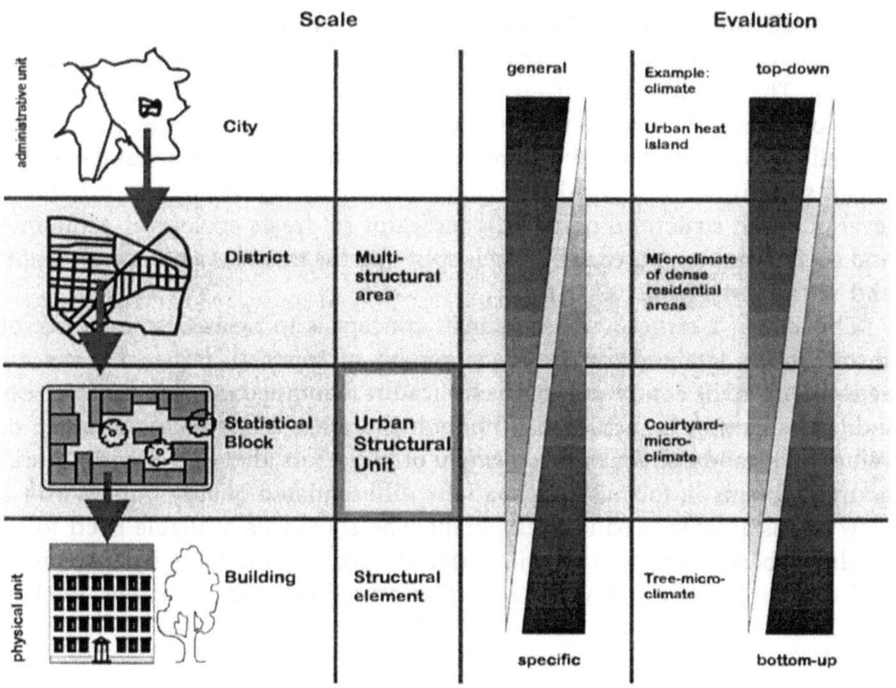

Fig. 1. Scale related assessment and evaluation of urban structures (own illustration)

20th century)'; 'large new prefabricated housing estates (since 1960)'; and 'old, high density industrial and commercial areas'. Some study areas contain 'detached and semi-detached houses' as well as 'villas', others are too heterogeneous to be usefully considered.

3.2 Development of Quality Targets with Regard to Sustainability on the Basis of Indicators

Once formulated principles, guidelines and starting points for sustainable urban development are available (which can be used to describe the urban system qualitatively) indicators to measure the degree of sustainable urban development attained are needed. In this context, the integration of ecological, economic and social components, which are usually treated separately, is important for realisation of the sustainability concept on a local level, something not often done at present. Sustainable urban development evolves from the overall concept of sustainability and is concerned with resource use, processes of exchange, urban spatial structures and social and economic integration. The development of city-wide indicators and targets are based on these principles in order to take into account the specific situation in Leipzig as well as to derive locally differentiated quality targets and standards for urban structural units. For this reason an indicator set was developed on the level of urban structural units. This indicator set treats ecological, economic and social components equally and is split into the three areas: 'natural', 'built' and 'social' environment.

The aim of a target and assessment concept is to measure the degree of sustainability attained for the development of concrete, targeted plans for action. Due to the complexity of the indicators a homogenous system to assess indicators cannot be achieved. Although there are a significant number of values and standards for the assessment of indicators, there are obviously still many problems in formulating spatially differentiated quality targets with a view to sustainable urban development. The standards available need to be modified, based both on a re-orientation towards the model of sustainability, and on the planning relevance of urban structural units (Werheit, Wickop 1997).

3.3 Investigations of Selected Urban Structural Units

First, theoretical considerations of the characteristics, effects and causal relationships operating in urban structural units were carried out. The perceived strengths and weaknesses of several selected urban structural units were considered from an ecological point of view. While 'villas', for instance, are highly attractive and have high ecological properties in terms of soil, water and air quality, they are inefficient in terms of use of space and resources. 'Terraced apartment blocks' suffer environmentally in terms of pollution, but are highly efficient and usually enjoy a wide variety of functions within the area. The 'large new housing estates' are often regarded as being unattractive; however, their micro-climate is relatively good and, in resource-use terms, they are highly efficient.

After this theoretical consideration of urban structural units, investigations in the study areas with regard to ecological, housing and population aspects were carried out. Population change is a measure of people's quality of life in an area, coupled with their ability to move. The population decrease in Leipzig is particularly acute in older areas with extensive and dilapidated 19th century housing, whereas the population in areas with the 'detached and semi-detached houses' even increased by 15%. Up to 25% of flats in the older structural units are empty. Poor conditions and restoration work are by far the most important reasons.

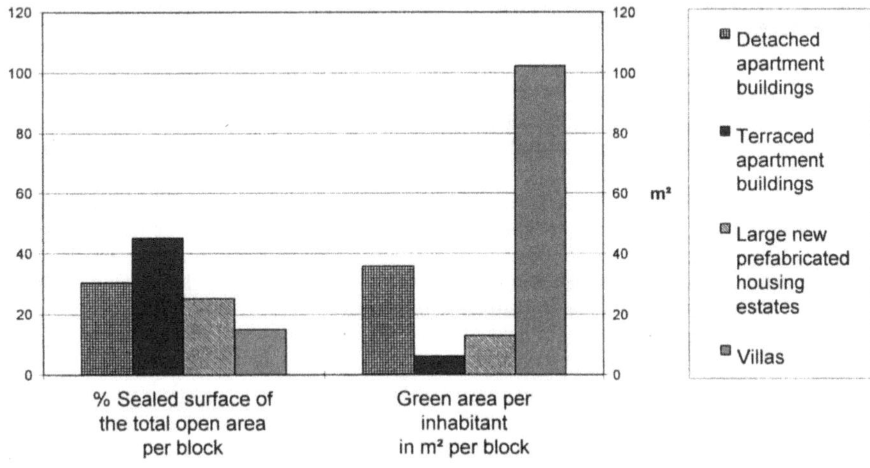

Fig. 2. Degree of surface sealing in relation to the amount of the green area per inhabitant in selected urban structural units (source: Erhebung der Gewerbe-, Handels-, Büro- und Wohngebäudestruktur in Leipzig, Stadt Leipzig, Stadtplanungsamt)

Considering the percentage of surface sealing of the total open area per block, the analysis clearly shows the differences between the urban structural units (see Fig. 2). The percentage of surface sealing in the urban structural units is mainly influenced by the building structure. The supply of green space area is strongly related to the population density of the structural units: the 'large, new prefabricated housing estates' with a very high population density have a very low amount of green space per inhabitant, although the total amount of green space is actually reasonably high.

4
Prospects

With reference to the model of sustainable urban development, the current situation in Leipzig is very mixed. Present processes in Leipzig are developing in contradiction to the concept of sustainability in many cases. Nevertheless, due to its recent history, Leipzig has the opportunity to integrate the aims of sustainable urban development into the current transition process. To make local decisions more effective, concrete targets must be developed for the present trends. Spatially differentiated quality targets and standards for urban structural units could make a valuable contribution by orientating the transition process in Leipzig towards sustainability.

References

Böhm P, Breuste J, Heyn C, Wickop E (1996) Map: Urban Structural Units of Leipzig. UFZ

Breuste J (1996) Grundzüge des Wandels von Umwelt und Stadtstruktur in Leipzig, ein Überblick. In: Breuste J (ed): Stadtökologie und Stadtentwicklung: Das Beispiel Leipzig. Berlin pp 11–32

Duhme F, Pauleit S (1992) Strukturtypenkartierung als Instrument der räumlich-integrativen Analyse und Bewertung der Umweltbedingungen in München. Teil 1: Ziele und Methodik. Freising

Werheit M, Wickop E (1997) Nachhaltige Stadtentwicklung in Ostdeutschland-Qualitätszielkonzepte für Stadtstrukturtypen in Halle und Leipzig. UVP-report 2/97

The Role of „Process Chains Analysis" in Investigating Environmental Problems of Cities

Karl-Heinz Simon[1] and Uwe Fritsche[2]

1 Introduction

The so-called 'process-chains approach' used here can be regarded as part of life-cycle assessment in a broader framework. A computer tool called GEMIS/TEMIS ('total emission model of integrated systems') is available to support the type of analysis mainly required in complex evaluation tasks. Various databases comprise part of the computer tools; these consist of several hundreds of pre-defined components of technical infrastructure, representing stages or important devices within the system. The whole system is used for satisfying demands for energy carriers or materials and products. Last year's analysis was mainly a cover analysis of energy systems, whereas more recent efforts deal with material fluxes and complete infrastructure systems.

In this chapter some aspects of the starting point for the analysis 'demand structures' are discussed, followed by information on the method 'process chains analysis'. In order to be more specific and to show what kind of results can be gained by using the approach, some examples are given, namely of certain supply structures and related emissions which would occur if these demands were to be satisfied.

2 Demand Structures as a Starting Point

We take demand patterns as a starting point for our analysis. These patterns can be composed of demands for energy, food, mobility, or other requirements which are likely to induce energy and material fluxes and cause the release of air-borne emissions and residues. Demand patterns could be specified for household levels, functional subsystems (for example local authori-

[1] University of Kassel, Centre for Environmental Systems Research, Kurt-Wolters-Str. 3, D-34109 Kassel, Germany
[2] Institute for Applied Ecology e.V., Darmstadt Office, Bunsenstr. 14, D-64293 Darmstadt, Germany

J. Breuste, H. Feldmann, O. Uhlmann (Eds.)
Urban Ecology

ties), or the whole city with its huge machinery to satisfy the demands of citizens, enterprises and service organisations on the city's territory.

From our point of view, upstream (and downstream) processes have to be integrated into the analysis to produce a complete picture of the environmental burdens resulting from the demands (Fig. 1). In most cases, these environmental impacts occur outside the city's own territory, for example mining processes or food production, or with respect to transportation processes as an essential part of the whole supply system. Besides these upstream processes, additional sources of environmental impacts have to be taken into consideration, i.e. amounts of energy to keep systems running as well as energy and materials incorporated into the devices (steel, concrete etc.) (Enquête-Kommission Schutz der Erdatmosphäre 1994).

Emissions of climatic relevance are selected as an example of results produced by the tool, as they are, at present, of extraordinary importance. Other calculation results show the amount of air-borne emissions, residues, costs, area influenced, etc. The tool offers additional assistance in evaluating which components of the supply system are responsible for which types of emissions and how much they contribute to certain environmental impacts.

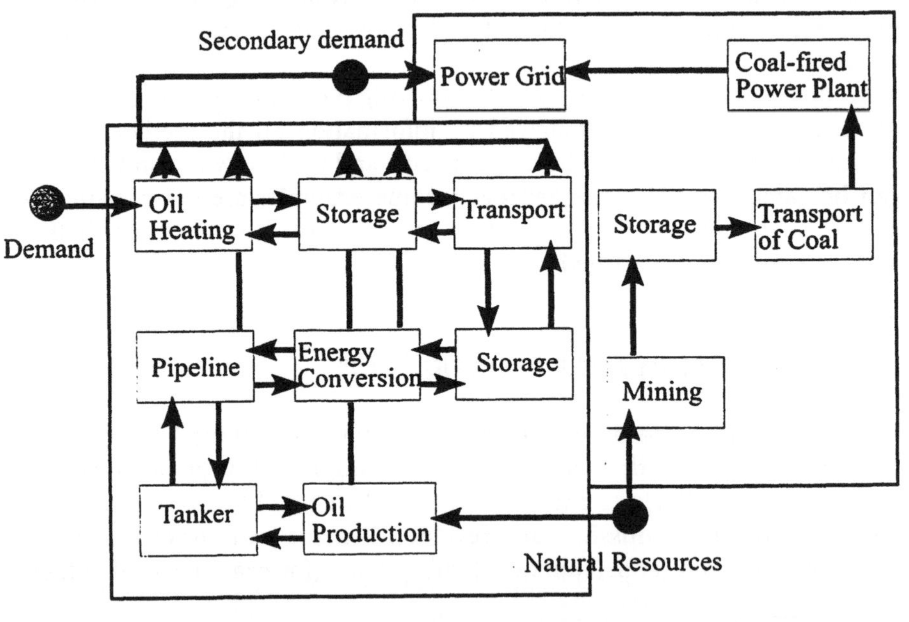

Fig 1. The structure of process-chains or networks – energy supply with oil as an example. A demand is sent across the chain until it reaches natural resources as the starting point of the supply activities. Branching occurs due to auxiliary energy demands etc.

3
The Process-Chains Approach

A large set of single input-output-models represents important process stages of the supply structures for energy, materials, food, transportation, and other technical or organisational subsystems.

The machinery works as follows: a certain demand is specified (using the scenario option) and is fed into the supply model (using interfaces for predefined or user-defined process chains). All those subsystems which participate in fulfilling the specified demand are integrated into the evaluation process step by step, and the chains are searched through until sources for primary energy or materials are reached. During searching the processes network at all stages where sources for emissions (or other environmental impacts) exist, a calculation routine is activated to sum up the amount of emissions occurring at that point. After having examined the whole network the total amount of emissions can be specified and presented as the result of the evaluation.

4
Examples of Demand Patterns and Related Results

Tables 1 and 2 show a prototypical demand structure for three types of households and the evaluation results using the GEMIS-approach.

The example only shows some of the capabilities of our approach and should not be misunderstood as a model of a real problem situation. Further

Table 1. Prototypical households – dispatch of demands (Pkm means „person-kilometres") (Source: Fritsche et al. 1997)

		Type 1 „Oil"	Type 2 „Gas"	Type 3	„Eco"
Room heating		18.000	18.000	9.000	wlood
Water heating	in	2.400	2.400	2.400	solar
Cooking	kWh	1.000	1.000	1.000	gas
Refrigerating		350	350	350	
Lighting		250	250	250	high eff.
Car-based mobility		20.000	20.000	17.500	
Trams and buses	in	2.000	2.000	2.500	
Railway	Pkm	2.000	2.000	10.000	
Flights		10.000	10.000	-	
Bicycle, walking		6.000	6.000	10.000	
Paper products	in kg	50	50	50	
Waste disposal		450	450	450	

Table 2. Calculation results for prototypical households (Source: Fritsche et al. 1997)

	Type 1 „Oil“	Type 2 „Gas“	Type 3 „Eco“
Sulphur dioxide	13,8	9,4	5,8
Nitrogen	35,2	33,4	18,6
Carbon dioxide	14'100	11'900	4'900
Methane	7,8	30,4	11,5
Nitrous oxide	1,2	1,2	0,7
Carbon dioxide equiv.*)	14'700	13'000	5'400

*) The equivalence values sum up the contributions of all those emissions which are relevant to climate change

steps towards integration of additional turnovers representing further sources of environmental impacts would be necessary to gain a more realistic picture of what is going on in real problem areas. This would be combined, for example, with an analysis of life-style patterns or alterations as a basis for empirically sound data about consumption of energy and products.

It is certainly not surprising that the calculation results show that the type 3 'eco' option seems to be favoured compared with the other two types in the example. Somewhat surprising is the extent of the differences, i.e. the grade of reduction one can achieve in restructuring supply structures. In our example, only a reduction factor of 2 or 3 was realised despite manifest interventions into the systems. In the context of discussions about ways towards sustainability it is obvious that such a reduction factor will not be sufficient.

5
Résumé

The process-chains approach can contribute to the assessment of environmental impacts of certain demand patterns and promises several benefits. First, it suggests a generally applicable tool which could provide a kind of standardisation of assessment procedures, as already tested in the field of energy planning. Second, with this common tool a comparison of assessment results could be obtained, i.e. results from different projects and approaches could be related one to another.

In general, the discussed approach has the advantage that the analysis could start at different stages of complexity; it could be applied to individual demand patterns as well as to consumption patterns of larger units, such as a certain supply sector or a whole city. In addition to the well-known Baccini-approach (Baccini 1993), the process-chains approach would assure integration of upstream processes which are of great importance to the overall evaluation.

References

Baccini P, Daxbeck H, Glenck E, Henseler G (1993) Metapolis - Güterumsatz und Stoffwechselprozesse einer Stadt. Nationales Forschungsprogramm Stadt und Verkehr, 34A, 34B, Zürich

Fritsche U et al. (1997) Gesamt-Emissionsmodell Integrierter Systeme (Softwarepaket) http://www.usf.uni-kassel.de/service/gemis/

Enquête-Kommission Schutz der Erdatmosphäre (1994) Studienprogramm Band 1/II, Teilbericht J: Landwirtschaft und Ernährung. Economica Verlag, Bonn

Environmental Quality Targets in the City of Wiesbaden

Thomas Votsmeier

1 Definition

Environmental Quality Targets define the envisaged qualities of resources, capacities or functions under specific material, spatial and temporal aspects. They should differentiate and list the valued environmental components, they should have a time-table, be action and project oriented and should consider local predispositions. Environmental Quality Targets put the general objectives of environmental policy into concrete terms. They can be used as a part of a system including environmental guidelines, standards and indicators. They should be specific enough to guide action and help to decide between alternative actions. They are put into operation through environmental standards.

The aims must be based on the principle of sustainability (ensuring long term protection of the environment) and must therefore be preventative in nature. Environmental Quality Targets are an instrument that furthers the precautionary principle. The targets should be designed across media, to allow for interaction, interconnection of problems and the functioning of ecosystems. They connect scientific knowledge with political aims inside the community.

Environmental Quality Targets have no legal requirements as long as they are non-binding guiding principles, but they can fill legal gaps. However, Environmental Quality Targets should always be more stringent than legislated limits or standards – otherwise one need not bother to establish them but just follow the law.

2 Functions of Environmental Quality Targets

Environmental Quality Targets can be used as control mechanisms in the context of local Environmental Impact Assessment. There they build the back-

Environmental Department of Wiesbaden, Luisenstraße 23, D-65185 Wiesbaden, Germany

J. Breuste, H. Feldmann, O. Uhlmann (Eds.)
Urban Ecology

ground against which the environmental strengths or weaknesses of a (local) project or plan can be evaluated. They can also be used as *guidelines for policy development and as a management instrument.*

The discussions on sustainability (Local Agenda 21) have spurred the need to develop appropriate indicators for monitoring whether or not local communities are moving in the right direction. The frameworks for selecting indicators often ask for the definition of broad visions, narrowing them down to general targets and only then selecting the measurable indicators.

The goals and targets that have been defined within an Environmental Quality Target concept can be reached by using different kinds of actions. To obtain a realistic chance for reaching the targets it is useful to integrate the necessary measures into an Environmental Quality Target concept.

The great number of possible actions, measures and instruments that can generally be taken by different actors (private persons, local administration, regional administration, business, . . .) is to be reduced to the main ones that should really be used by the entire local administration.

For example the target to reduce air pollution concentration caused by traffic to a level 50% below the legal standard can result in decisions to introduce important changes in the local traffic management scheme. Traffic is one area where the city (in Germany) has a high level of planning discretion. Even where the city has no or little authority (for instance with regards to regulating the business sector or with influencing the German railway service), negotiations based on voluntary co-operation can result in improvements.

Financial and political priorities should be set for realisation of the necessary measures by the different departments inside the administration (planning, environmental, traffic, waste management, purification plants, green plot management, . . .).

If priorities have been defined and endorsed by the council and one department is responsible for controlling and co-ordinating their implementation, the Environmental Quality Target concept obtains the function of an internal management instrument.

3
Procedure for Setting up the Environmental Quality Targets

Targets should be goal oriented and should be applied in a systematic way to all departments of a local authority. Pragmatism, rather than scientific rigour, should be applied in the process. The entire administration should assist the environmental department in the process of creating the catalogue of targets. In this way conflicting targets will be recognised and minimised or even marked in the concept.

The establishment of a system of Environmental Quality Targets cannot be seen as a one time task; the catalogue must be reviewed and modified constantly as new knowledge is gathered. 'New knowledge' refers both to the knowledge gained through the practice of environmental protection and the leading political opinions in the local community, as well as to the growing base of knowledge about ecology and environmental problems world-wide.

In the city of Wiesbaden, for example, the EQTs endorsed in 1990 were modified in 1995. This was caused by actual environmental problems (ozone levels, climate protection) which have now been supplemented in the catalogue of targets, and by new financial developments (decreasing of available money for realisation of expensive measures) that lead to a change in several time-oriented targets (consequence: the same target is to be reached some years later).

Procedure for setting up the targets-proposals based on the experience in the city of Wiesbaden:

1. a mandate from the city council or the mayor/chief executive is useful;
2. establish committee with representatives from each department and each political group in the council;
3. acquire different kinds of environmental targets (local, regional, medial, ecosystem-oriented, ...);
4. analyse focal points of pollution inside the city;
5. develop general goals;
6. formulate hierarchy of philosophy, vision, and targets;
7. define measures necessary for reaching the targets;
8. circulate the findings within the administration with the public and discuss within each group;
9. change targets if necessary;
10. discuss results with the committee;
11. repeat the internal and external discussions until consensus is reached on a preliminary catalogue of Environmental Quality Targets;
12. Council group representatives jointly introduce Environmental Quality Targets' catalogue for discussion and decision into the full Council;
13. Council approves or disapproves;
14. administration is to be instructed to realise measures and assess local activities against the Environmental Quality Targets;
15. evaluate the results and the targets after some years.

4 Position of Environmental Quality Targets in the Interconnection of Instruments in the City of Wiesbaden

The Environmental Quality Target concept contains the guidelines for environmental policy of the council. It concentrates on the main statements of

Table 1. Target Levels

Target Levels general	Examples	
Target Level 1: Define overall objectives, philosophy, vision.	The features of nature and their potential in the city of Wiesbaden are to be saved and developed for a long period as a basis for existence of the people living in the city. The state of the local environment must be saved, it is not allowed to change for the worse.	
Target Level 2: Detail vision and philosophy by separating 'environment' into various segments (different environmental media, environmental potentials, human health).	The following media targets and standards shall be reached as far as possible, they are to be made spatially concrete by the landscape plan (environmental plan that is to be made law for nature protection).	
Target Level 3: Define quality targets and standards for the various environmental segments.	By 1995 the disposal of waste in the disposal pit of Wiesbaden is to the reduced to 50% of the value of 1989; by 30% by 2000.	The use of drinking water is to be reduced by 2000 to 200 l/ inhabitant/day or absolutely to 20.2 million m^3 per year.
Target Level 4: Establish avoidance and minimisation targets for causal chains that lead to pollution and strain.	The use of landscape and soil for built-up areas (housing, industrial, traffic) is to be minimised.	The concentration of toxic substances in the soil is not allowed to rise, based on values of 1990.
Target Level 5: Derive measures and instruments for the municipality to reach the targets or to implement strategies, plans, projects aimed at reaching the target levels.	In all planning proceedings inside the city of Wiesbaden the environmental planning (as a part of the environmental department) assesses the contents of the plan/map and develops alternative intentions and measures to minimise the negative impacts on the environment.	To reduce the concentration of air-pollution inside the streets following measures have to be derived: - improve the public bus system - put into place the parking management system - reduce the number of parking spaces for cars - reduce the speed limit down to 30–50–70 km/hour in the different types of streets. . . .

already existing environmental endorsements and has, to a small extent, the function of a local environmental programme. The Environmental Quality Targets contain the most important measures and instruments that should be carried out to achieve the goals. They contain no priorities for realisation of the different measures and no immediate directives to the administration. They have to be made spatially concrete in connection to the local environmental (landscape) plan. They could be used in an auditing process, especially for an external environmental audit as they do not contain targets concerning the internal environmental administration management. The Environmental Quality Targets are appropriate for use as guidelines/framework in assessments (local EIA). In every single EIA they have to be made concrete.

5
Example of the Structure of a Local Environmental Quality Target Concept

Table 1 exhibits the five step hierarchical system used in defining targets and subsequent measures in the city of Wiesbaden.

In 1996 a report containing the concept and the results of the evaluation was published by the City of Wiesbaden.

Further Reading

Umweltdezernat der Landeshauptstadt Wiesbaden (1996) Umweltqualitätsziele Kontrolle der Umsetzung - Novellierung 1995. In: Umweltbericht Nr. 10, Wiesbaden

Assessing the Metabolism of Urban Systems for Urban Planning

Stephan Pauleit and Friedrich Duhme

1 Goals

Concepts and targets for sustainable urban development and scientific support of the adequate decision-making process are urgently needed. This goal can only be achieved on reproducible data on the relationship between the socio-economic efficiency and use of natural resources. Furthermore, this relation has to be established for urban working units, both structural and functional.

2 The Structure Type Approach

Urban structural units are physiognomically homogeneous and delineate distinct configurations of built-up and open spaces in the city e.g. single-family housing, multi-storey blocks, industrial areas, parks, agricultural lands etc.

The structural units were grouped into 29 structure types. Features of land cover i.e. the percentages of built-up and of other sealed surfaces, vegetation cover etc. were quantified for every single unit by interpreting monochromatic aerial photographs, scale 1: 5000 and occasionally 1: 1000. The data base was further processed by means of a geographic information system (GIS).

The survey was coupled with statistical standard databases such as socio-demographic data and detailed information on the built-up area. Environmental data were available from several sources i.e. thermal imagery and information on the energy demand by space heating (Pauleit and Duhme 1995).

Chair for Landscape Ecology, Technical University of Munich, D-85350 Freising, Germany

J. Breuste, H. Feldmann, O. Uhlmann (Eds.)
Urban Ecology

3 Case Study: The Hydrology of the Cityscape

3.1 Goals and Methods

This case study investigates the hydrology of the structure types. The main emphasis is placed on establishing the relationship between water fluxes and the physical and the socio-demographic framework conditions. The goals are:

- to assess the annual hydrological balance of the structure types and to estimate stormwater runoff rates,
- to quantify the area required for trenches for on site storage and infiltration of the runoff from a rainstorm of 40 mm per hour,
- to explore the options for rainwater use as a substitute for drinking water.

Rainwater infiltration and stormwater runoff from sealed surfaces into the sewage system were computed for each unit by means of published coefficients. To compute the area required for stormwater infiltration we used a simple model of a trench (3 m wide at the top, 2 m at the base, 2 m deep) filled with coarse gravel (or brick rubble). The storage capacity was assumed to be one third of the overall volume of the trench.

We assumed that half of the total annual precipitation can be collected from the roofs and be stored in tanks to substitute for drinking water in some cases e.g. for toilet flushing. We investigated the feasibility of substituting the current water demand for garden irrigation, toilet flushing, laundry and house cleaning, as well as a water saving standard.

3.2 Results

Figure 1 shows the impact of the settlement pattern on the hydrology of the test area. Complementary to this, the mean infiltration and runoff rates of the structure types were calculated.

Infiltration rates are much reduced, particularly in densely built-up districts in the inner city, industrial and commercial areas. Railways, on the other hand, are especially important for the overall hydrological balance of the test area. While railways cover little more than 5% of the test area, it is here that 17% of total annual precipitation infiltrates.

More than half of the stormwater is converted into runoff in the built-up areas and some 80% of the total runoff of 720,000 m^3 is produced in the built-up areas. Thus, the mean runoff harvest of multi-storey blocks approaches 30 l per m^2/hr. Low density housing areas, on the other hand, have mean runoff rates slightly above 10 l m^2/hr.

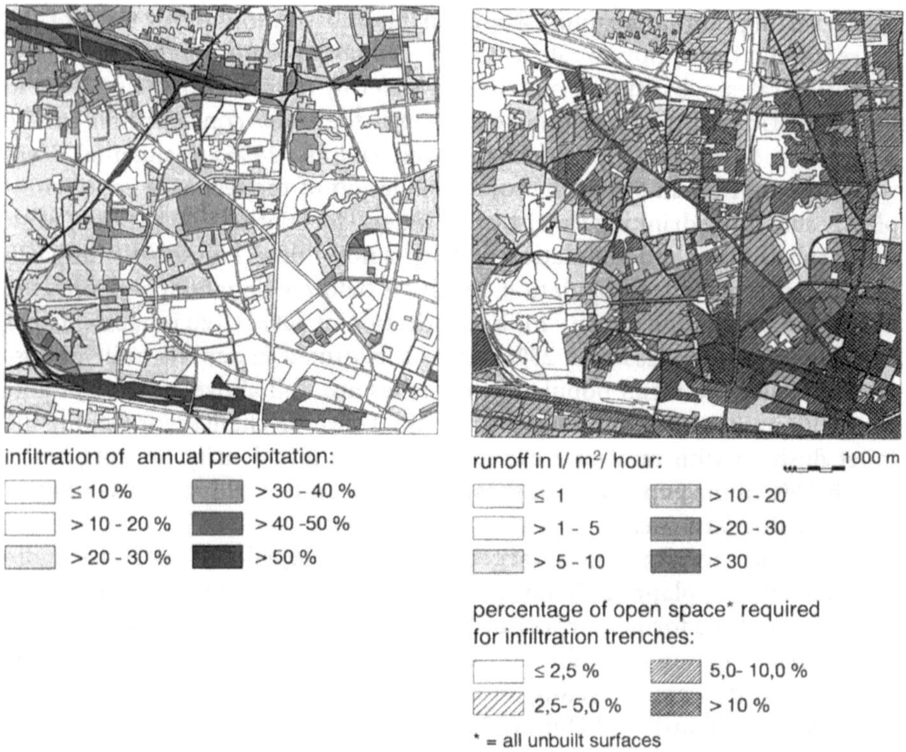

Fig. 1. Rainwater infiltration, stormwater runoff and areas required for infiltration trenches

3.3 Quantifying Hydrological Targets and Comparing Options for City Planning

Infiltrating the stormwater runoff on site will help to reduce both the strain on the sewage system and to enhance groundwater recharge. This can be achieved by constructing infiltration trenches in the structural units that collect the runoff from stormwater events. For complete infiltration of stormwater runoff 434 km of infiltration trenches are necessary in the test area which covers 130 ha. More than 60% of the infiltration trenches are required in densely built-up residential areas.

Grass covered areas only can be converted into infiltration trenches available for this purpose as those areas covered by trees and shrubs have to be conserved. Low density housing areas have enough open spaces for infiltration trenches, while at least 13 ha of sealed surfaces will have to be removed in densely built-up quarters. This has to be considered as a minimum figure, as in reality only parts of the existing grasslands can be used for construction of infiltration trenches due to open space uses, necessary distances of the trenches from house walls etc.

Harvesting rainwater runoff would enable substitution of one third of the current water consumption rate of 85 l per capita/day. Thus, given present consumption rates, substitution of drinking water by rainwater may not seem to be a realistic goal for these quarters. Even if the water consumption can be reduced to 54 l/capita/day (Burkhardt and Duhme, 1996), residential areas with little more than 6% of the total population of the test area could completely substitute drinking water by rainwater. The relationship between rainwater substitution and population density is non-linear. Full substitution can only be achieved when population density is below 35 inhabitants per ha, the maximum population density for a substitution of 50% is 120 inhabitants per ha. Therefore, often less than 25% can be substituted in densely built-up districts because the mean population densities are well above 200 inhabitants per ha for multi-storey buildings and blocks. However, chances for feeding the toilet flushing with rainwater are fairly good in low density housing areas. Nearly 90% of the population of these structure units could make use of rainwater for this purpose.

In a similar way, ecological targets can be developed and tested for more detailed levels of planning. Thus, for the development of a new town quarter on the area of the former airport of Munich, several options for the substitution of drinking water by rainwater can be offered. Roof greening on the one hand and waste water recycling on site were considered as further factors in this study (Burkhardt and Duhme 1996).

4 Conclusions

Planning and decision-making for sustainable urban development have to be based on reliable information on the relationship between the socio-economic and the ecological performance of the cityscape and its sub-units. Case studies can show how the structure types and structural units may serve as a common geographic reference to analyse the ecological implications of the city's physical pattern and its different functions e.g. with respect to the hydrological behaviour.

The distinct structure types have different sets of physical land use properties and hence also differ with respect to their ecological performance. The typology is based on established categories of zoning and development plans which may be further modified for the purpose of urban ecosystem analysis. Therefore, the approach will facilitate the transfer of environmental data into the planning process.

Particular emphasis has to be placed on exploring the relationship between ecological features such as stormwater runoff or energy demand for space heating, attributes such as building types, configurations and densities, and socio-demographic features such as population density, space demand per inhabitant etc. Indicators such as energy demand per unit floor area and year serve as a basis for comparing the different structure types and units and for linking the results with other studies. Furthermore, it will be a main challenge for the social sciences to apply appropriate methods and indicators for linking the metabolic behaviour of the structure types with the activities of the inhabitants. Investigating these relationships for the spatial compartments of the urban system will be the task of our further studies.

References

Burkhardt I, Duhme F (1996) Ökologische Bausteine Messestadt-Riem. Teil I Stadtplanung. Dept. for Town Planning (ed) Munich

Pauleit S, Duhme F (1995) Developing Quantitative Targets for Urban Environmental Planning. Land Contamination & Reclamation 3(2): 64–66

Workshop: Urban Climate 1.2

The Urban Climate and Municipal Climate Protection – Different Sides of the Same Coin?

Manfred Grauthoff

1
The Situation, Problems and Questions

With respect to the greenhouse effect and anthropogenic emissions of greenhouse gases, climate protection is an important component of concepts for sustainable development.

Urban climatology nowadays mostly has only minor importance in planning processes and will perhaps in the future be out of the picture totally when considering climate protection.

Resulting questions:

- Are climate protection and urban climatology totally different aspects except for the word 'climate' in both of them?
- Is the way of dealing with urban climate nowadays superseded or even wrong?
- Are there connections between municipal climate protection and urban climatology which have been out of the picture until now?

2
Aims of Sustainable Development for Germany

The greenhouse effect threatens to cause irreversible changes globally to climate and nature as well as a major disruption to society in the next fifty years. To prevent this development in Germany the following aims of environmental policy have to be achieved within the next half century (BUND, MISEREOR 1995):

- a reduction of the consumption of non-renewable energy to less than half of that currently used;

Dr. Grauthoff Unternehmensberatung für Energie und Umwelt (Dr. Grauthoff Management Consultation for Energy and Environment), Auf dem Jäger 36, D-44892 Bochum, Germany

J. Breuste, H. Feldmann, O. Uhlmann (Eds.)
Urban Ecology

- a reduction of the consumption of non-renewable raw materials to less than 20% of that today;
- no further encroachment by traffic and urban areas: limits set to the areas utilised today;
- a complete change in agriculture and forestry to ecological and near-natural working methods (organic farming . . .);
- a major reduction of polluting emissions.

3 Relations between Municipal Climate Protection and Urban Climatology

Climatic requirements on local and regional scales must be compatible with the needs of climate protection and sustainable development.

A broadly accepted system of principles for urban climatology has been unavailable up to now (Grauthoff 1994)!

When aiming for a healthy urban climate the needs of sustainable development must be taken into account.

4 The Climatic Function of Air Exchange

Nowadays the function of air exchange in urban areas by cold air or country breezes is optimised to equalise given climatic loadings.

Using the climatic function of air exchange for the amelioration of the urban climate is an *'end-of-the-pipe-technology'*, comparable to the policy of 'high chimneys' in the sixties and seventies!

Realising sustainable development will alter the climatic and air-hygiene situation of built-up areas, for example by:

- the reconstruction of the urban building structure to minimise energy needs;
- changing the structure of urban surfaces to minimise sealing and maximise natural surfaces;
- a reduction of traffic loadings by reducing traffic and minimising its emissions.

Consequently the aim of urban climatology today (named 'conservation and re-creation of areas for air exchange to reduce pollutant loads') will be out of date in the near future (Grauthoff, Mook 1997a).

5
The Climatic Function of Urban Climatic Oases

With respect to the needs of human biometeorology, climatologists currently demand the conservation or creation of urban climatic oases (i.e. local areas with good bioclimatic conditions).

Changing the urban building structure and the traffic infrastructure with regard to sustainable development will result in healthy bioclimatic conditions over most of the city. With this fact in mind, efforts to conserve isolated climatic oases are antiquated!

More interest should be taken in developing new concepts for town planning, for instance (Mook, Grauthoff 1997):

- concepts for mixed development with high, but not too high, density; priority given to inner-city development and land recycling;
- concepts for open spaces with high bioclimatic value by reducing the area needed by traffic;
- taking into account interior rooms of buildings and the need for healthy living and working conditions;
- building and living using solar energy in new and reconstructed houses and settlements (i.e. Ministerium für Bauen und Wohnen NRW 1998).

6
The Interrelation between Urban Climate and the Natural Balance

Nowadays urban climate is evaluated only with respect to human needs; the urban natural balance is not considered.

For sustainable development it is also necessary to evaluate the urban climate with respect to the natural balance (Grauthoff 1994): The guideline is to achieve an extensive adaptation between climate and ecosystem through:

- reducing the climatic effects of human interference with the environment to a minimum;
- trying to compensate for unavoidable damage to the ecosystem.

For urban climatology this means that:

- the ideal of garden cities with dispersed building structure is antiquated;
- concentrated building structures with optimised bioclimatic conditions have to be achieved; urban encroachment has to be halted.

7 The Interrelation between the Local/Regional and Global Scale with Respect to the Climate

The protection of the local and regional climate is a part of the global climate protection.

- All measures resulting in a reduction of the emission of anthropogenic trace gases with climatic relevance reduce the global consequences of the anthropogenic greenhouse effect for the environment and for humans.
- These measures also reduce the consequences of the anthropogenic greenhouse effect on regional climatic conditions, for instance in a country such as Germany.
- The consequences of the anthropogenic greenhouse effect will probably exceed the natural rates of change and therefore have negative results for our ecosystems. Wetland habitats or even forests might serve as examples to show these effects.

There is an interactive system between the cumulated local effects of human activities and the global climatic consequences and their feedback to the local and regional climatic conditions and to ecosystems.

8 Results, Consequences and Requirements for Urban Climatology

The emissions of trace gases with climatic relevance, especially CO_2, have to be balanced within the climatic analysis of a project or area (e.g. a town).

The aims for environmental and climatic quality of the area under consideration have to be defined in detail in an early and permanent coordination between planner and climatologist.

Avoiding and minimising climatic loadings at their origin shall have priority. Trying to ameliorate the urban climate through air exchange is not enough.

The climatologist has to contribute actively to detailed concepts for sustainability of a project or area, for instance through concepts for:

- reducing the emissions of traffic, industry, commerce and housing;
- unsealing and re-vegetating urban surfaces;
- minimising energy needs (Lehmann, Reetz 1995);
- housing and living 'with the sun', which means using solar energy in passive and active ways (Grauthoff, Mook 1997b).

An important step in this direction is currently being carried out in Nordrhein-Westfalen (NRW) in Germany. The government made a proclamation for 'building with the sun – 50 solar settlements in NRW'. For this project a scientific guide has been developed by a study group, co-ordinated by the author. This planning guide contains requirements and criteria for solar settlements which integrate energy with ecological and social aspects of life (Ministerium für Bauen und Wohnen NRW 1998).

References

BUND, MISEREOR (ed) (1995) Zukunftsfähiges Deutschland. Birkhäuser, Berlin Basel Boston

Grauthoff M (1994) Das Schutzgut "Klima" in der UVP: Nicht bewertbar und damit nicht von Wert? UVP-report 8, H 5: 253–257

Grauthoff M, Mook V (1997a) Das Schutzgut Klima – ein alter Hut angesichts von Klimaschutz und nachhaltiger Entwicklung? UVP-report 11, H 2: 91–92

Grauthoff M, Mook V (1997b) Energetische und ökologische Konzepte für Solarsiedlungen in Nordrhein - Westfalen. In: Tagungsband renergie '97 – Messe mit Kongreß für regenerative Energie. Hannover: 131–134

Lehmann H, Reetz T (1995) Zukunftsenergien. Strategien einer neuen Energiepolitik. Birkhäuser, Berlin Basel Boston

Ministerium für Bauen und Wohnen NRW (ed) (1998) Mit der Sonne bauen – 50 Solarsiedlungen in Nordrhein - Westfalen, Planungsleitfaden. Düsseldorf

Mook V, Grauthoff M (1997) Nachhaltige Stadtentwicklung – Worthülse oder eine realisierbare Zukunftsvision? Naturschutz und Landschaftsplanung 29, H 8: 247–253

A Case Study on the Effects of Vegetation on the Climate in the Urban Area

Takashi Asaeda and Vu Thanh Ca

1 Purpose of the Study

The purpose of this study is to give a detailed investigation into the effects of vegetated areas on the thermal climate of a new city in the Tokyo Metropolitan area - the Tama New Town - through field observations of the ground surface temperature, atmospheric temperature and other atmospheric conditions during hot summer days.

2 Observational Procedure and Equipment

The study area is located within the Tama New Town, a new, well-planned city in the west of the Tokyo Metropolitan area. At present the town covers an area of 30.2 km^2 and has a population of about 410,000. The study area, which is near the Tama Central Station, has an area of about 1.2 km^2. A large proportion of the ground surface in the study region is covered by commercial and office buildings, together with asphalt paved parking lots. There are several green areas in the region, the largest one being at the Tama Central Park.

Observations were carried out during the summer season from August to September 1994. During the observations, hourly values of air temperature, relative humidity and wind velocity were measured at many locations at 1.2 m above the ground level; foliage temperature and temperatures of the grass surface, wall and pavement surfaces were also measured at the same locations. Solar radiation, downward longwave radiation, net radiation, albedo, together with soil temperature at the ground surface and 5 cm and 10 cm depths were measured in the centre of a wide grass field in the centre of the park. Although observations were carried out several times during the summer of 1994, all observation periods, except August 26–27, were rejected because of variable weather since the purpose of this study is to investigate

Saitama University, Urawa, Saitama 338, Japan

J. Breuste, H. Feldmann, O. Uhlmann (Eds.)
Urban Ecology

the effect of vegetation on the climate of an urban area during hot summer days.

3
Effects of Grass and Wooded Areas on Heating and Cooling

During the observation day the weather was fine with the maximum solar radiation flux reaching 780 W/m^2 at 12.00. The highest air temperature on the grass field (32.4°C) was recorded at 14.00, while the lowest air temperature (23.5°C) was recorded at 06.00. Relative humidity reached 96% in the early morning and 46% at noon, corresponding to the partial vapour pressure of 27.9 and 23 mb, respectively. It was relatively calm at night but the wind became strong from 09.00 until 19.00, with a maximum wind velocity of 7.3 m/s recorded at 15.00, predominantly coming from the south.

Data on surface temperature at many locations in the study area (not shown) show that the surface temperature of the asphalt road and concrete pavement rose to more than 55°C at noon, almost 23 K higher than that of the air. At the same time, grass surface temperature reached its maximum of 40.3°C, only about 8 K higher than the air temperature. Foliage temperature above the wood reached 37.8°C, 5.4°C higher than that of the air above the grass surface. This shows that at noon the grass surface and trees heat the air to much smaller extents than do the asphalt road and concrete pavement. The measured vertical structure of temperature of the soil under the grass surface (not shown) shows that a small heat conductivity of the soil prevents the conduction of heat to deep soil layers, i.e. the below surface heat storage is small. The heat stored under the grass surface is rapidly released to the atmosphere in the evening. Even before sunset, at 17.00, the temperature of the grass surface decreased to less than that of the air and the grass surface began to cool the atmosphere.

Other studies (Asaeda et al. 1996, Doll et al. 1985) revealed that in contrast to the grass surface, paved surfaces store a large amount of heat during the day and continue to heat the atmosphere throughout the day and night.

At 14.00, the air temperature difference between the coolest and hottest places in the study area amounted to more than 2.5 K. The lowest air temperature of 31°C is observed in the centre of the park. The highest air temperature near the Tama Central Station, a commercial area near the entrance of the park, and a parking lot in the south-west of the park was more than 33°C. With a hot asphalt surface, the most likely candidate contributing to the high air temperature is the sensible heat released from the surface. Since this site is located in the commercial area, anthropogenic heat released by air conditioning, traffic and other commercial activities can contribute significantly to the heating of air. The northern entrance to the park, with an open pavement surface, is located next to a street with heavy traffic, surrounded by offices and commercial buildings and has an air temperature of 33.7°C. The excessive anthropogenic heat released by air conditioning and traffic at this site are the most likely contributors to the high air temperature, except for the sensible heat exchange between the surface and the air. Not far from here, in the Partenol Street (also located in the centre of the commercial area) the air temperature is less than 32°C, despite an expected high anthropogenic heat flux. This is because in the morning the street is in the shade due to high rise buildings located on both sides of the street. The cool ground surface results in a cooler air temperature, compared with that of a sunlit area, despite the high value of anthropogenic heat. At noon, the newly sunlit pavement surface has a lower temperature compared to that of other open spaces. Despite the lack of accurate data on anthropogenic heat at the time, it can be remarked that the energy released from heated pavement surfaces is much more important than anthropogenic heat for heating of the air.

A cool place is observed at the Tama Central Park, where air temperature reached less than 31°C. Even though the park was exposed to solar radiation throughout the morning, evaporation from the grass surface and trees caused low ground surface temperature and consequently low air temperature.

At 20.00 (not shown) the hottest place is observed near the Tama Central station with an air temperature of 28°C. The coolest air extends over the park and the surrounding areas. The air temperature difference between the two locations amounts to more than 2 K. At this time, the heating of the atmosphere is due to the release of heat from the pavement surface in the form of sensible heat, upward longwave radiation and anthropogenic heat. Ground surface in the park is even cooler than the overlying air, due to evaporation, and there is no anthropogenic heat release; the air in the park is cooled by the ground surface. On the other hand, near the Tama Central Station, the sensible heat released from the heated pavement surface, together with extensive release of anthropogenic heat, makes the air much hotter than that in the park.

4
Conclusion

From the observational results, it is found that even though it is small, the Tama Central Park is significantly cooler than the surrounding area during both the day and night. With strong local southern winds during the day, it is to be expected that the cool air from the park can significantly contribute to the reduction of the heat intensity in the town.

Kanda et al. (1991) found that a park of 1.2 km × 1.0 km can produce an air temperature difference of 4 K between the park and the surrounding city. Considering that a much smaller park was selected for this study, the results can be taken to be acceptable.

Acknowledgements. The contribution of Dr. Murakami of Nihon Koei Co., Dr. Fujino, Saitama University, many people from Nihon Koei Co. and graduate and undergraduate students of Saitama University in the observation and processing of data are greatly appreciated.

References

Asaeda T, Vu TC, Wake A (1996) Heat storage of pavement and the effects on the near surface atmosphere. Atmospheric Environment, Part B, Urban Environment, 30 pp 413–427

Doll D, Ching JKS, Kaneshiro J (1985) Parameterization of subsurface heating for soil and concrete using net radiation data Boundary - Layer Meteorol, 31 pp 351–372

Kanda M, Inagaki S, Hino M (1991) Field observation and numerical experiments on the effect of plant and water surface on urban meteorological environment. Annual J. of Hydraulics, JSCE, 35, 585-590.

Assessment of Human-Biometeorological Conditions in Different Kinds of Urban Structures

Bernhard Bauer

1 Introduction

Creating an 'ecological city' means that we have to search for strategies in order to develop urban structures. One important aspect in planning urban areas is the consideration of human-biometeorological issues. Investigations of urban climate show that the urban space consists of a variety of microclimates which are characterised by the individual structure of buildings and the arrangement of urban open areas. A high degree of sealing, the inauspicious arrangement of buildings and a lack of vegetation can further intensify thermophysiologically unfavourable meteorological conditions. Every urban structure has climatic potentials which can be developed through taking measures to improve microclimatic conditions. Ecological urban planning thus requires detailed information about microclimatic differences and their human-biometeorological impacts.

2 Human-Biometeorological Assessment

This study focuses mainly on the complex of thermal effects in human-biometeorology. The basis for the assessment of thermal conditions is the description and mapping of surface and building structures, including vegetation. The graphic implementation of mapping is performed using the ARC INFO Geographical Information System. Combining maps and databases allows spatial features to be linked to experimental investigations.

Mobile measurements are carried out to record the meteorological parameters relevant for the heat balance of a human being, such as air temperature, vapour pressure, wind velocity and mean radiation temperature. The

UFZ Centre for Environmental Research Leipzig-Halle, Department of Urban Landscapes, Permoserstr. 15, D-04318 Leipzig, Germany

J. Breuste, H. Feldmann, O. Uhlmann (Eds.)
Urban Ecology

Table 1. Assessment Index PET (Physiological Equivalent Temperature) according to Matzarakis and Mayer 1996

PET (°C)	Thermal Sensitivity	Grade of Physiological Stress
	very cold	extreme cold stress
4		
	cold	strong cold stress
8		
	cool	moderate cold stress
13		
	slightly cool	slight cold stress
18		
	neutral	no thermal stress
23		
	slightly warm	slight heat stress
29		
	warm	moderate heat stress
35		
	hot	strong heat stress
41		
	very hot	extreme heat stress

established technique (originally described by Höppe 1992) is used to determine the mean radiation temperature. The mean radiant flux density absorbed by a person is established by means of integral radiation measurements taken with a slewable pyranometer (CM21) and pyrgeometer (CG1). This allows the mean radiation temperature to be calculated using the Stefan-Boltzmann law.

Human-biometeorological assessment of the complex of thermal effects is then performed on the basis of MEMI (Munich Energy Balance Model for Individuals) (Höppe and Mayer 1987) using the meteorological parameters measured as input values during the experiments. The benchmark parameter used is the PET (physiologically equivalent temperature) derived from MEMI (VDI Guideline 3787 Part 2 (1997). Table 1 shows the assessment index of the physiological equivalent temperature correlated with the thermal sensitivity and the grade of physiological stress.

3 Human-Biometeorological Conditions in Different Kinds of Urban Structures

The low orographic structure of the City of Leipzig means that microclimatic differences are primarily attributable to the arrangement of the various urban structures. The areas of investigation chosen are located in representative urban structures which are typical of Leipzig's townscape. The examples presented here are situated in a courtyard of a large prefabricated housing estate and a park with adjacent apartment buildings.

3.1
Influence of Building Structures

Owing to the relatively simple, homogeneous design of the courtyard, the microclimatic differences are chiefly determined by the building structure and the shade it offers. Whereas the daily patterns of air temperature (max. 28°C) and vapour pressure (a mean of 14 hPa) exhibit a uniform path at all measuring points, large differences are observed in the course of the radiation balance throughout the day. In this courtyard the short-wave radiation from the upper-half-space reached values between 50 and 800 W/m^2 depending on the influence of the building. The differences are lower in the long-wave radiation directed upwards, although differences of up to 70 W/m^2 were still noted depending on the degree of exposure. This resulted in the mean radiation temperature in the shade of the building (e.g. a north-exposed measuring point) being about 30 K lower than in the areas with direct solar radiation, where mean radiation temperatures exceeding 60°C were attained (e.g. south-exposed measuring points).

In Figure 1 the diagram of diurnal variation of physiologically equivalent temperature (PET) clarifies the spatial and temporal variability of human-biometeorological conditions in an inner courtyard on a cloudless summer day. The highest grade of physiological stress (extreme heat stress) is found

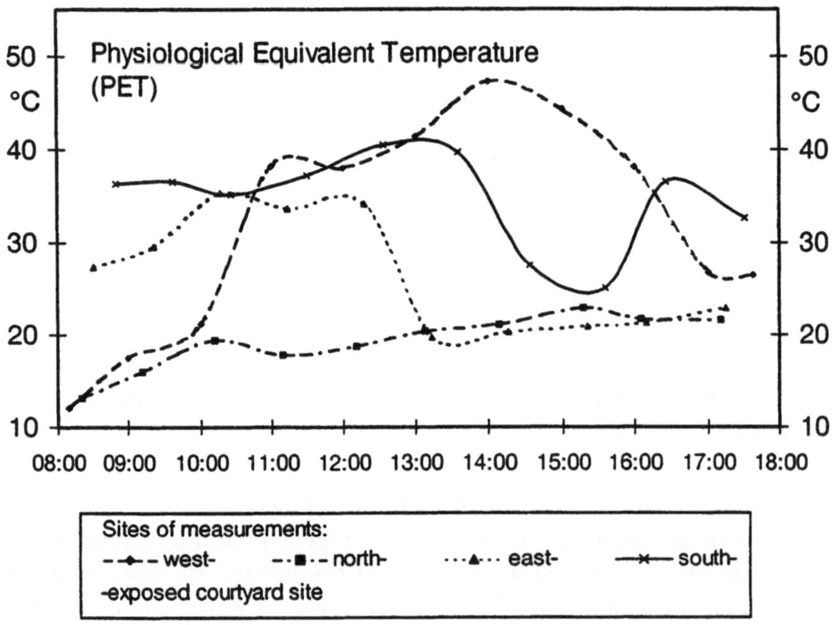

Fig. 1. Diurnal variation of PET (Physiological Equivalent Temperature) in a courtyard in Leipzig-Grünau; Mobile measurements at 19.08.96; all time information in CET; Calculation after VDI 3787 Part 2 (1997)

at the south-exposed courtyard site. These conditions are caused by high short-wave radiation and high long-wave radiant flux densities from the ground and the vertical building fronts. This demonstrates the close correlation between the PET and the mean radiation temperature. On sunny days a high proportion of direct solar radiation becomes a dominating factor in human-biometeorological assessment using MEMI (Höppe and Mayer 1987).

The arrangement of a few single deciduous trees could be enough to shade the south exposed courtyard site during summer months with the effect of reducing thermal heat stress peaks. The following examples show how large the influence of vegetation structures can be.

3.2 Influence of Vegetation Structures

Examination of the short-wave and long-wave radiant fluxes at various measuring points along a line leading from a park to the adjacent building structures clarifies the impact of vegetation structures on the radiation components at around noon on a cloudless summer's day (air temperature: max. 28°C). Owing to the low horizon limitation, the largest short-wave radiant fluxes are measured on the open space with meadow. At this measuring point, reflection from the lower-half-space is also greater then at the other measuring points. In the long-wave radiant fluxes it is found that in a largely sealed street canyon, terrestrial radiation from the lower half-space is about 70 W/m^2 greater than on the similarly sunny meadow. At the measuring point in the meadow, the lowest values for long-wave radiation are measured from the upper half-space owing to the low horizon limitation, as the effect of buildings on either side is absent and thus virtually only atmospheric counterradiation is recorded. The lowest long-wave radiant flux density, in comparison to the other sunny measuring points, is observed at the measuring point of open space with meadow.

In the area of tree shade, mean radiation temperatures are observed which are about 20 K below the values of sunny areas around noon. This results in human-biometeorological differences between strong heat stress (PET 38°C) in the street canyon without trees and slight heat stress (PET 28°C) in the street with trees. Similar results of this tree shade effects are also described by Höppe and Mayer 1987.

4 Final Remarks

VDI Guideline 3787 Part 2 (1997), provides a tool for taking human-biometeorological concerns into account in town planning and urban structuring. The results show that there is potential for improvement within built-up structures; this can reduce the frequency of thermophysiological stress situations. Above all, increasing vegetation and reducing the degree of sealing contributes to the improvement of the microclimate. However, architectural design measures also enable improvements in the neighbourhood's microclimate to be realised.

References

Höppe P (1992) Ein neues Verfahren zur Bestimmung der mittleren Strahlungstemperatur im Freien; Wetter und Leben vol. 44, no. 1–3: 147–151

Höppe P, Mayer H (1987) Planungsrelevante Bewertung der thermischen Komponente des Stadtklimas; Landschaft + Stadt, 19: 22–29

VDI-Richtlinie 3787 Blatt 2 (Gründruck) (1997) Methoden zur human-biometeorologischen Bewertung von Klima und Lufthygiene für die Stadt- und Regionalplanung; Teil 1: Klima

Matzarakis A, Mayer H (1996) Another Kind of Environmental Stress: Thermal stress; Newsletter No. 18; WHO Collaborating Centre For Air Quality Management and Air Pollution Control at the Institute for Water, Soil and Air Hygiene – Federal Environmental Agency

Models for Setting up a Biometeorological Warning System over a Populated Area in Havana

Paulo L. Ortiz Bultó, María E. Nieves Poveda, and
A. Vladimir Guevara Velazco

1 Introduction

Bronchial Asthma (BA) is a disease affecting a large proportion of the population in the tropics, including Cuba. BA is caused by a multiplicity of factors, weather variations being one of these. The weather influences the thermoregulatory mechanism of patients, causing a significant increase in the number of affected people during periods of the year, when greater climatic variations are present.

The forecasting of the degree to which the number of sick people will increase with climatic variations is of great interest to both the meteorological and medical community.

The objective of this work is to propose a model which will help to predict the number of BA cases on two temporal scales (monthly and weekly) for a locality in Havana.

2 Description of Series

In this study two time series are used, corresponding to monthly and weekly temporal periods.

2.1 Epidemiological Series

Data from the emergency services was obtained to produce weekly and monthly series of BA cases from January 1993 to December 1996.

National Climate Centre, Meteorological Institute, Havana, Cuba

J. Breuste, H. Feldmann, O. Uhlmann (Eds.)
Urban Ecology

2.2 Climatic Series

These were obtained from the Casablanca station which is located at the Meteorological Institute in Havana. The following variables were involved in this study:

Maximum and Minimum Temperatures, in °C (XT, NT); Mean Temperature, (XT + NT)/2, in °C (MT); Atmospheric Pressure, in hPa (AP); Vapour Pressure, in mm of Hg (VP); and Relative Humidity, in % (RH). All these values were corrected to describe the behaviour around Casino Deportivo, using a correction factor calculated using the difference between the Casablanca station and an observation point.

3 Methodology

Many authors have tried to predict and alert the coming of asthma and respiratory diseases cases and the way in which the weather can influence human health. These relationships have often concentrated on the association of isolated elements with the disease using linear regression; results so far have been unsatisfactory.

In this study, two complex indexes are used to evaluate the action of climatic variables on the disease. The first describes the weekly climatic conditions and the second describes the climatic anomalies of the month, with the objective of using them in the auto-regressive models with exogenous variables, where the climatic index is taken as the exogenous variable. This index could be written as:

$$I_m = .743(XT) + .902(NT) + .882(MT) - .562(AP) + .892(VP) + .488(RH) \tag{1}$$

$$I_w = -.9123(XT) - .9778(NT) - .9748(MT) + .6643(AP) - .9674(VP) \tag{2}$$

These indexes are the result of Principal-Components Analysis (PCA) which describes the combined action of climatic variables in term of anomalies, where the first component explains 78% of the variation of the monthly conditions; and 87% in the case of the weekly conditions (Ortíz and Cabanas (1995)).

It is observed that the monthly and weekly behaviour of the incidence of BA might be explained by the anomalies or climatic variations. This was shown by carrying out a test for exogenous variables (Ender (1995)) to determine the status of the index in the model.

We could show that in the monthly behaviour pattern in winter (November to March) the index clearly describes the effect of the variations and their influence on the incidence of BA. The BA rate increases when there is a significant anomaly present. In Cuba these anomalies are common in winter, when there are high contrast values due to the unstable weather conditions. In summer (weeks 70–95) the weather anomalies are positive thus the rate of BA decreases. However, if extreme conditions arise in summer, the number of BA cases will increase.

These indexes enable us to use models to postulate the possible incidences of BA according to the current weather conditions.

4
Monthly Forecast Model of BA to the Locality

The application of multiples series techniques is relatively new in Human Biometeorology, especially with regard to the evaluation of the influence of weather on human health. These techniques can be used to predict the behavioural pattern of BA.

It was shown previously that BA cases are related to the behaviour of the climatic conditions within the health field. We now consider a seasonal autoregressive process with s observations perseason, using exogenous variables to describe the observed behaviour, given by this equation:

$$(\lambda_1 X_{t-s} + \lambda_2 X_{t-2s} + \ldots + \lambda_p X_{t-ps}) - \sum K_j Y_{ik} = c + a_t \tag{3}$$

In terms of the backs hift operator

$$\Lambda_P(B^S)X_t - \sum K_j Y_{jk} = c + a_t \tag{4}$$

where: $\Lambda_P(B^S) = 1 - \lambda_1 B^S - \lambda_2 B^{2s} - \ldots - \lambda_p B^{ps}$

This is a seasonal AR process of order p, where the {λp} are the seasonal autoregressive parameters and Yt the series of exogenous variables, the parameters resulting from this equation.

Box and Jenkins (1994) propose that this is unrealistic and introduce a correlative structure between observations within a season, by allowing the noise input {at} in equation 3, to be represented by a second model $\Phi_P(B)\,X_t = a_t$ (5) where {at} is a white noise process, and $\Phi_P(B)$ are polynomials in B of order p.

The proposed model can be used to describe the relationships we found. The exogenous variables in the model do not correspond exactly with the measurements of the meteorological station of Casablanca, but necessary corrections were introduced to rectify this matter.

4.1 Pattern of the Monthly Forecast

The BA cases were predicted over a 6 month period using the above function; the forecast was found to be correct in 83% of cases.

The results obtained enable the medical authorities to be warned about possible future incidences of BA. Necessary measures can then be taken to mitigate the influence of the climate on human health and to plan the allocation of resources.

4.2 Model for the Weekly Forecast

In the case of the weekly series seasonality was not presented due to the length of the data, thus the model was reduced to the expression:

$$\Phi_P(B)X_t - \sum K_j Y_{jt} = a_t \tag{6}$$

The forecast function was adjusted using a skill of .80 for the independent samples, with values of highly significant parameters and an MSE of 60 cases.

After obtaining the results, the prediction was used in a real life situation.

4.3 Quality of Fulfilled Forecasts

The application of the model to predict the behaviour of BA in urban climatic conditions produces satisfactory results.

When the anomalies described by the index were matched, the response of the model to the causes of increases and decreases of BA cases improved.

Another important fact is that the forecast had a success rate in over 90% of cases, thus it might be possible to offer this application to the entire health service in Havana City, thus contributing to an improved quality of life for the population.

The stability of forecasts is controlled constantly through the test of CUSUM on the accumulated residues, to verify that there are no residues in the model which could be considered as maladjusted, or that incorrect information has been entered in the model.

5 Conclusions

The models developed to predict the incidence of BA cases to assist the health service in Casino Deportivo can be used to control asthma in the locality, and foresee possible "rush" weeks. The results of this study prove that the models are general, enough to be able to be used in other municipalities or health areas within Havana. Models of this nature enable the medical authorities to take measures to plan the allocation of available resources, with the objective of improving their services, as well as the quality of life for the patients, and thus preventing an increase in the incidence of this disease.

References

Box GEP, Jenkins GM (1994) Time Serie Analysis, Forecasting and Control. Holden, San Francisco London Amsterdam

Ender W (1995) Applied Econometric Time Series. John Wiley and Sons, Inc. New York

Ortíz P, Cabanas G (1995) Modelos de pronóstico de asma bronquial con variables exógenas. Boletín de la Sociedad Meteorológica de Cuba. Vol 1, #2. (In Spanish)

Wei William WS (1990) Time Serie Analysis. Univariate and Multivariate Methods. Addison-Wesley. New York

Workshop: Air Pollution 1.3

Air Quality in a Tropical Mega-City – Rio de Janeiro, Brazil

Werner Habbel[1,3], Antonio Carlos Dias dos Santos[2,3], Horst Pasenau[1,3], and José Arnaldo Sales[2,3]

In 1986 an environmental control project was set up under the technical cooperation between the State Foundation of Environmental Engineering, the Environmental Protection Agency of the state of Rio de Janeiro (FEEMA), and the German Agency for Technical Cooperation (GTZ). This project was revitalised in 1993 to include further investigations in air quality monitoring. One of the objectives of these monitoring activities was to analyse the effects of meteorological conditions on air pollution. These activities were implemented in collaboration with the Meteorological Department of the Federal University of Rio de Janeiro (UFRJ) and we are presenting the first preliminary results.

The southern part of Brazil is situated in the semi-humid/humid outer tropical zone of the southern hemisphere. According to the classification scheme of Köppen/Geiger, Rio de Janeiro is located in a transition zone between the Aw/Af-climate and the Cw/Cf-climate, but with extreme annual alterations. In general temperatures never fall below 18°C, and rain falls during the entire year with high precipitation during the southern summer (December to March).

The urban agglomeration of Rio de Janeiro, the so-called Metropolitan Area of Rio de Janeiro (RMRJ), covers an area of 6500 km^2 that includes 17 municipalities. After São Paulo, Rio de Janeiro represents the second most important economic centre of Brazil; 12 million inhabitants earn about 15% of the Brazilian income.

The Metropolitan Area is morphologically characterised by distinct features: a mountain chain extends over all of the western and northern parts of the suburban area; some isolated partial inselberg-type elevations (with altitudes of up to 900 m) divide the southern city into distinct basins; and the northern city is part of a large alluvial plain. Apart from these features, Guanabara Bay and variable land use determine the major meteorological events. The climatic conditions can be described as follows: low values of pre-

[1] Dr. Krätzig Ingenieurgesellschaft mbH, Aachen, Süsterfeldstr. 81, D-52072 Aachen, Germany
[2] Fundação Estadual de Engenharia do Meio Ambiente (FEEMA), Rio de Janeiro
[3] German Agency for Technical Cooperation (Deutsche Gesellschaft für Technische Zusammenarbeit (GTZ) GmbH), Eschborn

J. Breuste, H. Feldmann, O. Uhlmann (Eds.)
Urban Ecology

cipitation in winter combined with high air pressure; and lower winter temperatures accompanied by low values of relative humidity.

The annual distribution of days with typical thermal inversion layers results from these climatic conditions and the topography of the region (Fig. 1). The high frequency and minor heights of the layers during winter periods cause a significant deterioration in air quality.

Air quality in the Metropolitan Area of Rio de Janeiro is affected by the emissions of industry and road traffic. Heating and power stations, which contribute significantly to air pollution in temperate regions, are negligible. Due to the climate, heating is unknown and the major part of electric energy (more than 90%) is produced by hydroelectric power stations. These two types of emission sources can be allocated to separate regions. Most of the industries, especially the large ones, are situated in the northern and north-eastern part of the Metropolitan Area, whereas the traffic emissions dominate the residential and commercial areas in the south and south-west. Other sources of emissions are dust from unpaved roads and uncontrolled open burning, such as burning of waste.

Whereas industrial emissions exhibit high amounts of sulphur oxides, vehicles contribute more to CO emissions. The increasing number of vehicles will provoke an increase in atmospheric contamination.

In order to assess air quality, FEEMA started a monitoring programme in 1968. During the past 30 years, the initial network has been adapted to make use of modern technologies. The air monitoring network consists of High-Volume-Samplers which measure particle concentrations, as well as fixed and mobile monitoring stations which measure gaseous pollutants.

One of the objectives of the technical co-operation between FEEMA and the GTZ is to investigate the influence of meteorological conditions on the variability of air pollution. As a first approach, a short-term monitoring campaign was carried out for a period of six months in 1994. Monitoring comprised measurements of particle concentrations and meteorological parameters; 11 sampling sites were installed in the Metropolitan Area of Rio de Janeiro.

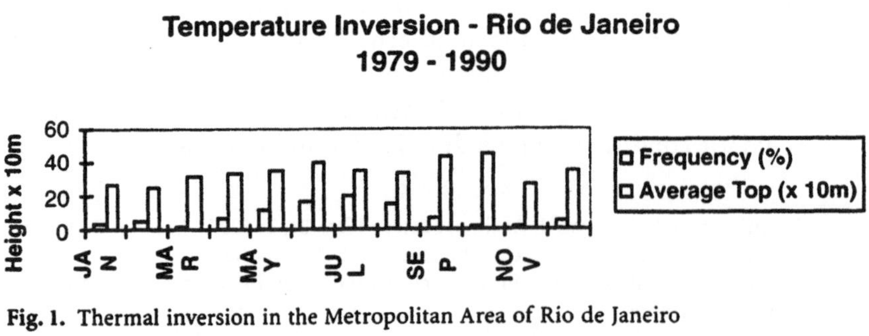

Fig. 1. Thermal inversion in the Metropolitan Area of Rio de Janeiro

The lowest concentrations were measured at the sampling stations Itaguaí and Sumaré (Fig. 2), i.e. the background stations, with average values of 43 μg/m³ and 40 μg/m³, respectively. The highest concentration was measured in São João de Meriti at 262 μg/m³.

Among the most typical weather conditions is the approach of cold fronts from the southwest, which is related to moderate gales from constant directions. This general weather condition causes specific wind fields close to the ground and influences the formation of thermal inversion layers.

These wind and thermal conditions have distinct effects on the particle concentrations in the ambient air. An approaching cold front causes very strong atmospheric stability, which is often accompanied by temperature inversion. The north-east wind distributes the pollutants, emitted mainly in the north-eastern part of Rio de Janeiro, over the central and southern region. During a passing cold front, the unstable atmospheric conditions lead to rainfall and an ensuing decrease in particle concentrations.

In summary, these first results indicate that air pollution may be associated with general weather conditions. In order to extend the base of data and to consolidate our first results, we are continuing these special investigations. We are presently improving the network with respect to quality and quantity and are installing additional monitoring stations in regions which have not yet been monitored. Gaseous pollutants will also be measured in the programme.

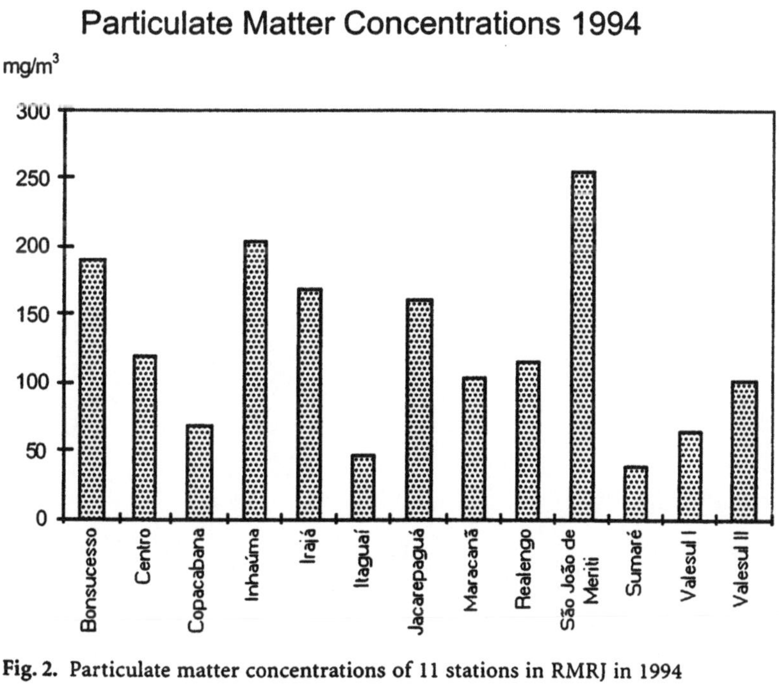

Fig. 2. Particulate matter concentrations of 11 stations in RMRJ in 1994

Since there are no tools to influence meteorological conditions which cause high pollution levels, processes to reduce or avoid emissions have to be developed and applied in order to avoid critical immission situations. These may include low-emission processes in industries as well as catalytic converters, exhaust gas emission control and traffic regulation. We hope that the results of our work will enforce the development and adequate application of such measures.

The aim of future investigations is to develop a model for the prediction of critical immission situations which is linked to the usual weather forecast. We hope this will contribute to constant clean air in and a blue sky above our beautiful city of Rio de Janeiro.

Airborne Pollutants from Mobile Sources for the City of Mendoza, Argentina

Enrique Puliafito, Carlos Puliafito, José Quero, and Pablo Guerreiro

1 Introduction

Mendoza (33°S, 68°W) is a city of 885,000 inhabitants, located on the east side of the Andes mountains (750 m.a.n.s.l.). Due to its orographic and meteorological conditions this city faces increasing air pollution.

Some important meteorological facts are: precipitation of 180 to 220 mm/annum, most of it during the summer months (November to March), low relative humidity (50%), low incidences of fog, few days with covered skies (65–75 days/annum), most days with clear skies (more than 125 days/annum); predominant winds are: S 13%, NE and W 12%, SW 9.8%, SE 8.2% and calm days 35%, the mean wind intensity value is not greater than 2 m/s.

The climate may be classified according to the Köeppen scale as BWakw: dry desert, with mean temperature of the hottest month >22°C, cold and dry in winter with mean temperature below 18°C.

Zonda winds – warm and dry NW winds similar to Föhn winds-, prevail in the higher layer most of the year and anticyclonic situations in the winter months (May to Oct.) with a high probability of frost, which contributes to a higher degree of pollution due to the generation of remarkable thermal inversion layers, (Norte 1995).

2 Sources of Contaminants

The most important source of contamination is private and public transport, which reaches more than 100,000 vehicles per day in the metropolitan area of the city.

The main fixed point sources are located on the periphery of the city. Energy production, oil refineries and petrochemical industries are located south west of the city, two cement industries are located on the north edge of

Universidad de Mendoza, Instituto para el Estudio del Medio Ambiente (IEMA), Arístides Villanueva 773, 5500 Mendoza, Argentina

J. Breuste, H. Feldmann, O. Uhlmann (Eds.)
Urban Ecology

the city and on the east side are several agro-industries. Depending on the wind direction and its intensity, these industries may have an increasing effect on the surroundings.

During the winter and spring months, two types of frost may affect the agriculture area of Mendoza: irradiation (or convective) and advective (cold air masses). High air pollution may occur when using waste oils to prevent freezing; the local authorities are promoting the use of gas to replace waste oil as fuel.

3 Systematic Monitoring and Immission Values

Through its 'Direction for Environmental Control', the Ministry of Environment and Public Works has carried out monitoring since 1970 (in some stations) in around 20 sites. It monitors daily: Total Suspended Particles (TSP) (24 h mean values) – by filter capture and reflectometry; NO_x (24 h) – by colorimetric Griess-Saltzmann; and weekly: 24 h mean values lead (Pb) – by colorimetric ditozone and sulphur dioxide SO_2 – by colorimetric West and Gacke, modified by Pate. The actual mean annual values are 80 $\mu g/m^3$ and 50 ppb for particles and nitrogen respectively.

The Institute for Environmental Research (IEMA) at the University of Mendoza continuously monitors black carbon and PAH poliaromatic compounds (Aethalometer GIV reflectometry), surface ozone (Horiba ozone monitor) and meteorological parameters. The annual mean value of these measurements reaches 40 $\mu g/m^3$ for ozone, 20 $\mu g/m^3$ for black carbon and 53 $\mu g/m^3$ for PAH. These and other measurements are carried out in cooperation with the UFZ-Centre for Environmental Research Leipzig-Halle.

Table 1 summarises the annual mean values for three different monitoring sites in the City. (Puliafito et al. 1995). (a) San Juan and Lavalle streets represent the conditions downtown, (b) Godoy Cruz represents a residential area in the southern part, (c) San Martin Park, in the west, is the main recreation area of the city. From these tables it is possible to see that the downtown area is the most severely affected, especially due to vehicle exhaust gases. Nevertheless a significant improvement was seen during 1991. In this year the public transport routes were modified, before then, 80% of the public transport routes went through these streets. After this date almost all through traffic in these streets consisted of private cars and electric-powered public transport. In recent years an increase in nitrogen levels has been observed, produced by a growth in the number of private cars (source of values: Katz, Saenz 1991, Tables 3 and 4 pp 3–4).

Table 1. Annual mean air pollutant values for selected sites in Mendoza (1987–1995)

Site	a) Downtown Mendoza (San Juan and Lavalle Streets)								
Contaminant	87	88	89	90	91	92	93	94	95
TSP ($\mu g/m^3$)	302	331	290	225	138	86	83	71	80
NO_x (ppb)	92	45	50	47	26	20	21	28	30
Lead ($\mu g/m^3$)	0.5	0.5	0.7	0.3	0.2	0.2	0.3	0.3	0.4
SO_2 (ppb)	32	4.3	9.5	8.3	2.3	1.6	1.6	4.4	1.7
Site	b) Godoy Cruz				c) Park San Martín				
Contaminant	92	93	94	95	91	92	93	94	95
TSP ($\mu g/m^3$)	64	65	55	45	15	13	11	12	13
NO_x (ppb)	11	16	18	18	8	8	11	12	13
Lead ($\mu g/m^3$)	0.3	0.2	0.2	0.2	0.1	0.1	0.2	0.2	0.2
SO_2 (ppb)	1.6	1.1	1.5	1.6	2.0	3.4	1.6	1.9	1.9

Table 2. Annual travelled kilometres and carried passengers

Fuel	km × 10^6	pass. × 10^6
Diesel buses	77	145,7
CNG buses	2,5	6,8
Gasoline care	700	56

Table 3. Vehicle emissions according to fuel type in g/km

CO	HC	NOx	TSP
9	2	5	3.8
28	2.4	7.9	0.1
18	2.5	2.4	0.2

4
Study of the Emissions of Public and Private Transport

Table 2 shows the composition of public transport. Four groups can be recognised, according to fuel type: (a) diesel, (b) natural gas (CNG), (c) electric powered buses and (d) gasoline fuelled private transport without catalytic converter. The estimated emissions of contaminants per g/km is given in Table 3. (WHO 1992, Katz and Saenz 1991). The emissions were calculated using Table 2 and Table 3 and the following equation: $E_t\ (g) = N\ (veh).\ e\ (g/km).\ d\ (km/veh)$

Table 4. a) Annual emissions in tn/a, b) Annual emissions (g/passenger)

a) Annual emissions in tn/a					b) Annual emissions (g/passenger)			
Vehicle	CO	HC	NO_x	TSP	CO	HC	NO_x	TSP
Diesel	693	154	385	293	4,75	1,05	2,64	2,01
CNG	70	6	20	0,25	10,25	0,88	2,94	0,03
Gasoline	12,600	1750	1680	140	225	31,25	30	2,5

where E_t is the total emission for the time considered, N amount of vehicles for the same time, e specific emission for each fuel and d is the mean distance travelled.

The calculated emissions reach 300 tn/annum of particles from diesel combustion and 140 tn/annum from gasoline vehicles. For nitrogen compounds the emissions are 385 tn/annum from diesel and 1680 tn/annum from gasoline consumption. Table 4 shows the annual emissions in tn/annum. and in g/passenger These values show that personal vehicles are the principal contaminators in the city. However, the particle (TSP) level is the most worrying value, more than 60% of which originates from diesel combustion, mainly from the public diesel buses and almost 40% from gasoline vehicles and other sources. NO_x values are the second highest emission values, which originate from gasoline fuelled private transport. Compressed natural gas (CNG) has been proposed as an alternative fuel for public transport, as well as increasing the number of electric buses (trolley-type) to reduce drastically the particle emissions.

5 Conclusions

The most important source of contamination is public and private transport, which reaches monthly mean immission values over 100 µg/m³ for TSP and 50 ppb for NO_x in winter. The current trend shows an increase in these parameters caused by an important growth in the number of automobiles.

The annual emission in grams per passenger for TSP are 2 g/pas using the public transport, and 2.5 g/pas for private transport; for nitrogen the values are 5 g/pas and 30 g/pas for public and private transport respectively.

The emission calculations confirm the immissions records and suggests that more effort should be done to reduce emissions from mobile sources, by encouraging the use of public transport, reducing the travelled distances and incorporating new technologies.

References

Katz RA, Saenz AA (1991) Diesel vehicles emissions control in Santiago, Chile: Itinerary and Technical Developments, SAE International, Rep 912423

Puliafito E, Puliafito JL, Behler JC, Alonso P (1995) La calidad del aire en Mendoza: Contaminación y efectos sobre la salud. In: Martínez Carretero E, Dalmasso AD (eds) Mendoza Ambiental, Mendoza, Argentina, pp 207–242

Norte F (1995) Condiciones meteorológicas favorables a la contaminación atmosférica en Mendoza In: Martinez E et al. (eds) Mendoza Ambiental, pp 207–242, Mendoza, Argentina

WHO (1992) World Health Organization: Urban pollution in megacities of the world, UNO (UNEP), Blackwell Publishers

Suburban Commercial Estates and their Impact on the Situation of Immissions in the City of Leipzig

Dietrich Wagler[1], Klaus Freyer[2], Peter Popp[2], and Hanns-Christian Treutler[2]

1 Introduction

Over the past five years, a process of structural transformation has been underway in eastern Germany. One element of this is the development of large shopping centres and commercial estates in the suburban areas of cities. The construction of these greenfield commercial estates is accompanied by large-scale land usage and the sealing of large areas, making them impermeable, as well as effects on the micro-climate, the generation of additional motorised traffic, increased vehicle emissions, and changes to the immissions' situation with respect to atmospheric pollutants typical of exhaust emissions (Enquete-Commission 1994).

This paper describes investigations carried out using the example of Wachau Commercial Estate on the south-eastern outskirts of Leipzig in the immediate vicinity of two housing estates, an allotment compound where people spend their free time, and two hospital complexes.

Our study addressed both the causes (the additional transport links) and their effects, namely the impact on the immissions' situation.

2 Methods

Customer surveys recorded customers' shopping frequency at the shopping centre (the number of trips there per week or month), their place of residence, and how often they use the petrol station. These data were used to establish the shopping centre's catchment area, the distances travelled by shoppers, and the frequency distribution of the distance categories.

Traffic counts were carried out on several days by people counting vehicles both at the commercial estate and on the access routes at 15-minute intervals.

[1] University of Leipzig, Interdisziplinäres Institut für Natur- und Umweltschutz (IINU), Augustusplatz 10/11, D-04109 Leipzig, Germany

[2] UFZ Centre for Environmental Research, Department for Analytical Chemistry, Permoser Str. 15, D-04318 Leipzig, Germany

J. Breuste, H. Feldmann, O. Uhlmann (Eds.)
Urban Ecology

For the production of an accurate microscalar model of the *traffic-emission–immission* effect chain with respect to time, *traffic models* (speed on uncongested stretches and tailback speed ahead of cross-roads with and without traffic lights) and models for emission factors of the various vehicle types are required. These will depend on the influencing factors acting on emissions and their mutual dependencies (traffic volumes, distance, speed, acceleration, specific emission factors and vehicle types). On the roads most frequently used in the commercial estate, a total of 55 different emissions sources must be taken into account: 5 nearby cross-roads, 19 tailback zones with varying lengths preceding cross-roads, and 31 sections with free-flowing traffic. A key role is played in emission models by the emission factors, which depend on average speed and vehicle type for each vehicle pollutant (carbon monoxide, nitrogen oxides and hydrocarbons). The German Department of the Environment has the emissions factors updated at regular intervals (Hassel et al. 1994, 1995).

For *dispersion calculations* the DASIM programme developed by Darmstadt Technical College (a three-dimensional current, turbulence and diffusion model) was used (Blinda and Hopf 1993).

Measurements of the concentration and subsequent analyses of the *BTEX components* (benzene, toluene, ethyl benzene, m/p-xylene and o-xylene) were carried out using the PN8000 sampling unit, the DA8000KA desorption unit and autosampler (produced by AMA) and the Hewlett Packard's HP5890II gas-phase chromatograph. Later these measurements were augmented by field analysis using an HC1000 mobile gas-phase chromatograph produced by Airmotec.

3 Results and Discussion

Customers survey: taking into account shopping frequency, the average distance was 5.2 km and 8.2 km for shoppers from Leipzig's surroundings. The mean travelling distance for all shoppers was found to be 6.3 km.

The frequency distributions of the distance categories, the frequency of shopping trips per week and the numbers of customers per week (approximately 40,000, estimated from traffic counts) produced the following total distances travelled (round trips):

Shoppers from Leipzig	292,000 km/week
Shoppers from the surroundings	251,000 km/week
Total	543,000 km/week

Traffic counts: the daily curves of traffic flows have a typical shape, and feature a slight maximum between 10.00. and 12.00, and a clear afternoon maximum between 15.30 and 17.30. On the basis of the analysis of counted and calculated traffic flows, the proportion of traffic additionally generated by the erection of the commercial estate can be calculated to be about 44% of the total traffic. Calculating the exhaust emissions to form a basis for calculations of immissions and model verification requires all traffic flows (in our case 76) and all vehicles types represented in the study area to be recorded over time as precisely as possible. During traffic counts, vehicles were classified into the following optically distinguishable types: 2-stroke cars, lorries, and other cars. Licensing statistics were used to subdivide lorries into light and heavy goods vehicles.

Modelling: The emissions factors' dependence on speed can be modelled by means of an exponential function (for hydrocarbons and carbon monoxide and both two-stroke and four-stroke vehicles) and a quadratic function (for nitrogen oxides emitted by two-stroke vehicles, four-stroke vehicles and light goods vehicles). In the case of benzene, emission factors are generally linked to the fuel's benzene content (Moussiopoulos et al. 1992). We performed our calculations with an emission factor of 2.5% of the emission factor for hydrocarbons, which was determined in a few random tests of fuels from different suppliers.

Figure 1 shows a typical emissions profile along a cross-roads. The values shown are total emissions from 5 m intervals plotted against the distance from

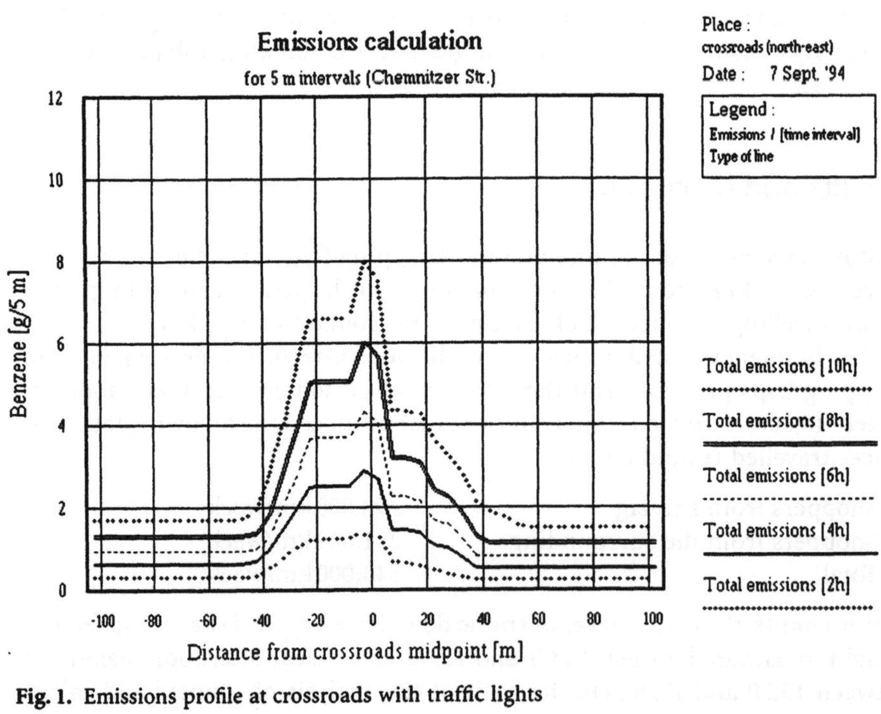

Fig. 1. Emissions profile at crossroads with traffic lights

the centre of the cross-roads. A peak in emissions occurs in the immediate vicinity of the cross-roads. Before and after the immediate cross-roads area increased emissions due to tailbacks and the effect of acceleration result in an increased emissions' function compared to emissions in the uncongested areas. The emissions on the roads and cross-roads of the commercial estate are augmented by two other sources – emissions at the car-park and vaporisation at the petrol station. The information gathered provides an accurate representation of emissions at the commercial estate over space and time. The survey area was divided into a three-dimensional grid to calculate immissions using the DASIM programme.

The benzene concentrations were initially calculated for concrete immission situations. Annual means of benzene concentration were calculated from the mean emissions (calculated from averaged daily curves) and wind direction statistics recorded in Leipzig. Around the hospitals, housing estates and allotments concentrations are at least 1–2 μg/m³. Large parts of the survey area are polluted by levels of 2–5 μg/m³ or 5–10 μg/m³. Pollution is highest at the car-park, the cross-roads and the petrol station; in some areas benzene pollution exceeds 50 μg/m³.

BTEX measurements: in Table 1, a number of measurements in the survey area are compared with the calculated immission values of concrete situations. The table is structured so that the calculated immission intervals are shown in the first line and the relevant measurements are shown in the second line. The correspondence of the measurements and the calculated values at the locations Dösener Weg, the car-park and the north-east cross-roads can be described as good. In contrast, the measurements obtained for Meusdorf estate were always higher than the dispersion calculations. The reason for this is not yet known.

To summarise, in the example described here, the construction of the commercial estate has led to a rise in the volume of traffic and additional emmissions of air pollutants. This results in the following additional emissions per

Table 1. Comparison of model values and actual measurements

Date, time	Dösener Weg	Car-Park	North-east crossroads	Meusdorf
10 May, 10:30	2–5	5–10	5–10	0–1
	3.4	**12.9**	**8.4**	**3.7**
15 June, 16:15	2–5	2–5	10–15	1–2
	1.9	**5.2**	**13.9**	**2.4**
7 Sept., 14:15	2–5	10–15	5–10	0
	4.8	**8.3**	**8.5**	**4.0**
7 Sept., 16:30	2–5	10–15	5–10	0
	6.2	**10.2**	**10.5**	**5.6**

year: 200 t of CO, 62 t of NO_x, 92 t of NMHC and 4.6 t of benzene. The actual measurements of benzene pollution coincide with simulated values.

References

Blinda S und Hopf A (1993) Handbuch zur Anwendung des Ausbreitungsmodells DASIM. Technische Hochschule Darmstadt, Institut für Meteorologie

Enquete-Kommission "Schutz der Erdatmosphäre" des Deutschen Bundestages (Hrsg) (1994) Mobilität und Klima, Economica Verlag Bonn

Hassel D, Jost P, Weber F-J, Dursbeck F, Sonnborn K-S, Plettau D (1994) Abgasemissionsfaktoren von Pkw in der Bundesrepublik Deutschland. Erich Schmidt, Berlin

Hassel D, Jost P, Weber F-J, Dursbeck F, Sonnborn K-S, Plettau D (1995) Abgasemissionsfaktoren von Nutzfahrzeugen in der Bundesrepublik Deutschland für das Bezugsjahr 1990. Erich Schmidt, Berlin

Moussiopoulos N, Oehler W, Zellner K (1992) Kraftfahrzeugemissionen und Ozonbelastung. Springer, Berlin

Assessment of Ambient Air Quality in an Urban Ecosystem

Matli Chandra Sekhar

1 Introduction

Assessment of air quality has become very important for several reasons. Particulates are a major reason for complaints concerning air pollution as they are visible and offensive. Gaseous pollutants, especially from industries and auto-exhausts, also pose considerable discomfort especially in the urban areas (Kenneth, Warner 1976). Several methods of monitoring air quality are being used all over the world to assess the levels of pollutants in the ambient air (Sherwood 1970). In view of the problems associated with conventional monitoring, it is essential to locate suitable sampling sites for monitoring pollution and to evaluate suitable identifying parameters to indicate the sources of pollution and control the emissions (Liptak 1974). The southern states of India have been fortunate with respect to air pollution but there are several pockets where air pollution problems have already become critical. This is the appropriate time to take adequate measures to ensure that future urban industrial and vehicular growth does not result in serious environmental degradation. This calls for assessment of air quality in the urban ecosystem using the available air quality data. The present study aims to study the air quality of Hyderabad city based on the landuse pattern.

Hyderabad is an ancient Muslim city with lakes, massive boulders strewn about on rocky escarpments, palaces, mosques, mausolea and fortresses. It is an inland city in the south central part of India lying on longitude 78° 28′ East and latitude 17° 27′ North. The city spreads over an area of 200 sq km and its elevation is around 545 m above mean sea level. The city is situated on the banks of the river Musi, a tributary of the river Krishna. The climate is dry with a brief rainy reason during the south-west monsoons. The mean annual rainfall is about 850 mm. The temperature varies between 9°C in winter to 45°C in peak summer. In the long run such studies will help to generate information to aid in planning pollution control strategies so as to keep the pollutants within safe limits.

Water & Environment Division, Regional Engineering College, Warangal - 506 004, India

J. Breuste, H. Feldmann, O. Uhlmann (Eds.)
Urban Ecology

2
Materials and Methods

Air quality data in Hyderabad was collected to assess the existing ambient air quality. Sufficient care is taken to collect the data based on the landuse pattern of the city and, at the same time, prominent landuses which are certain to result in air pollution are selected for the purpose of the study. For the sake of comparison, residential landuse where pollution is expected to be minimum is included in the study.

3
Results and Discussion

The nuisance value of the dust makes it an important parameter and dust fall rate gives an overall idea of the general air quality of a particular place. Dust fall generally refers to particles of sufficient size and density that are separated from air by gravity and trapped in the dust fall jars. The dust fall rates and the nature of dust with respect to pH are presented in Table 1. The nature of dust varied around neutral pH conditions. Another important observation is that the major part of the dust is water insoluble, indicating that it is chemically inert and it is airborne from badly maintained road surfaces and pavements. Dust fall rates are higher in the commercial areas followed by industrial and residential zones. This is mainly because of the heavy traffic volumes on badly maintained, congested road surfaces. Various industrial operations in the industrial area account for the dust in this zone. Construction activity is the major contributor to dust in the residential zone.

Table 1. Dust Fall Rate & Sulphation Rate

Mont h	Sampling Station - Landuse	pH	Total Dust (gm)	DFR M tones/ sq.km/month	Sulphation Rate (mean of 4 Samples)
Nov'9 3	Industrial	7.0	0.07	3.7	0.12
	Commercial	7.2	0.31	14.4	0.09
	Residential	7.6	0.05	2.2	0.16
Dec'9 3	Industrial	6.9	0.17	6.2	0.08
	Commercial	7.4	0.30	10.3	0.06
	Residential	7.2	0.10	4.0	0.16
Jan'9 4	Industrial	6.4	0.09	3.7	0.14
	Commercial	7.4	0.30	13.1	0.14
	Residential	7.1	0.03	1.2	0.20
Feb'9 4	Industrial	7.1	0.16	6.5	0.12
	Commercial	7.6	0.17	6.9	0.04
	Residential	7.4	0.12	5.2	0.16

Sulphation rate, which is a cumulative measure of sulphur dioxide, is a better parameter to cover longer periods continuously. The values of the sulphation rate are presented in Table 1. Sulphation rates are high in industrial zones as combustion invariably features in most industrial processes. Higher sulphation rates in the residential zones is a peculiar observation which could not be explained by the limited data available for the present study.

Results of the analyses of the data collected during the study are presented in Table 2. Close observation of the SPM levels indicates that the highest concentrations are recorded in the commercial zones. This is because of the badly maintained road surfaces and heavy traffic in these zones. Another possible reason for this is the concentration of various anthropogenic activities in the commercial zone. Comparison of the SPM concentrations in the industrial zones yields higher values at Balanagar, which is the major industrial area of the city. The concentrations observed in the area are slightly lower than the threshold values specified by the Central Pollution Control Board for industrial zones. Hence, it is high time that the emissions from the polluting industries are effectively controlled and further expansion in these zones is restricted.

The concentrations of sulphur dioxide presented in Table 2 do not indicate any particular trend in the different zones from which data is collected. Higher values of sulphur dioxide are anticipated in the industrial zone as it is an end product of the combustion process. The auto-exhaust emissions also contain considerable sulphur dioxide, especially during traffic delays at intersections. This is the possible explanation for all the stations recording similar levels of sulphur dioxide without much variation. However, long term air quality data is necessary to understand the trends more meaningfully.

The concentrations of oxides of nitrogen presented in Table 2 indicate higher values for commercial and industrial zones with lower levels in residential zones. Oxides of nitrogen are the end products of the combustion process, thus high values are obtained in industrial zones. Auto-exhaust emissions also result in oxides of nitrogen and hence high levels are certain to occur in commercial zones. However, the levels of oxides of nitrogen are far below the levels specified by Central Pollution Control Board.

4 Conclusions

Industrial emissions are, however, only one aspect of air pollution in the present study. Significant contributions from transportation activity especially in the existing road conditions is also leading to deterioration of air quality. The air pollution levels of the residential area are relatively high in comparison to many of the larger cities in India. Going by the results, only

Table 2. Air quality at different sampling locations (All concentrations in microgram per cubic meter)

Station Month	Uppal			Nacharam			Balanagar			Habsiguda			Kothi		
	No. of Samples	Daily Max.	Mean	No. of Samples	Daily Max.	Mean	No. of Samples	Daily Max.	Mean	No. of Samples	Daily Max.	Mean	No. of Samples	Daily Max.	Mean
SUSPENDED PARTICULATE MATTER															
Nov'93	15	239	189	-	-	-	8	430	420	2	218	171	1	329	-
Dec'93	6	309	192	5	182	137	22	525	355	3	281	169	3	367	275
Jan'94	-	-	-	27	310	235	6	600	485	3	146	131	3	311	233
Feb'94	-	-	-	10	280	240	-	-	-	3	192	162	3	289	261
SULPHUR DIOXIDE															
Nov'93	18	20	11	-	-	-	10	23	17	12	20	11	6	-	-
Dec'93	9	14	9	6	17	14	24	33	13	12	11	9	18	15	5
Jan'94	-	-	-	19	46	16	5	14	7	12	15	13	18	8	6
Feb'94	-	-	-	8	5	3	3	5	5	18	9	5	18	5	3
OXIDES OF NITROGEN															
Nov'93	22	37	26	-	-	-	13	37	30	12	9	8	6	-	-
Dec'93	10	54	41	7	23	20	37	43	33	12	10	9	18	23	22
Jan'94	-	-	-	44	43	24	9	33	27	12	10	6	18	39	24
Feb'94	-	-	-	19	54	38	4	32	32	18	10	9	18	13	9

SPM levels in ambient air exceed the standardised norms; this is mainly due to the dry ground conditions existing in and around Hyderabad.

Percentages of sulphur dioxide and oxides of nitrogen on an annual average have been found to be within the specified limits. However, it is widely reported that these levels increase periodically and the concentrations are spatially uneven, but on an annual basis work out to be lower than the limits laid down by the organisation concerned (A citizens Report 1985. The periodical shooting up of pollutant levels is only a pointer to the inadequacy of safety measures.

The polluting industries have to be identified and emission surveys carried out to control the pollution at the source for preventing deterioration of ambient air quality. The pollution resulting from transportation activity can be reduced by improving the road surfaces and by efficient traffic management, which allows smooth traffic movement and hence lower gaseous emissions. The data collected during the present study pertains to short term air quality survey, and hence clear pollution trends could not be established. However, study of long term air quality data can lead to more established findings and conclusions.

Acknowledgements. The author wishes to express his deep sense of gratitude to Prof. Breuste and the organisers of the International Conference on Urban Ecology for their financial assistance, without which the author's participation and presentation of the paper would not have been possible. The author is also thankful to Dr. Feldmann and Ms. Uhlmann MSc. for their continuous patience in answering his queries time to time.

References

Centre for Environmental Concerns (1985) A Citizens Report. Hyderabad: The State of Art of Physical Environment. Centre for Environmental Concerns, Hyderabad

Kenneth W, Warner CF (1976) Air Pollution Its origin and Control, IEP A Dun-Donnelley Publisher, New York, USA

Liptak BG (1974) Environmental Engineering Handbook, Vol.2, Chilton Book Company

NEERI Report (1983) Air Quality of Selected Indian Cities

Sherwood PT, Bower PH (1970) Air Pollution from Road Traffic: A Review of Present Position, RRL, Report No. LR 352, Birmingham

Estimation of Air Quality by Complex Index

Valdo Liblik and Helmut Kundel

1 Introduction

In the areas of oil shale mining and processing in north-eastern Estonia the simultaneous presence of 8–10 or more pollutants is typical for the general air pollution situation. In these conditions there is a real need to describe air quality in a simplified form. One way is to make use of special Air Quality Indices which permit the presentation of the pollution situation data in such a way that prevailing conditions are more easily understandable than when using concentration data as such for many components in the air (Ilvessalo 1995).

In Canada, Finland, Hong Kong and other countries, in the case of multi-component pollution fields a sub-index is calculated for each pollutant by a segmented linear function and the maximum of all sub-indices is selected as the complex index for the given site (Lee and Gervat 1995, Aarnio et al. 1995, Ott and Thom 1978). In some investigations (Guide for control of air pollution 1991, Kallaste 1991) the sub-indices were also calculated, but the fact that harmful effects of different groups of pollutants may differ in the case of the equivalent overcoming of limit values was accepted, and the complex index by summation of sub-indices or by root-mean-square method was established. A method based on a power equation was recommended for calculation of an air quality index (Ilvessalo 1995), but this method does not take into account the various biologically dangerous effects of pollutants.

This chapter deals with development of the Air Quality Complex Index (AQI) which takes into account the limit concentrations of pollutants in the air with respect to health and nature, different dangerous effects of pollutants etc.

Institute of Ecology, North-East Estonian Department, 15 Pargi Str., 41537, Jõhvi, Estonia

J. Breuste, H. Feldmann, O. Uhlmann (Eds.)
Urban Ecology

2
Methods: Calculation of Air Quality Complex Index

The estimated method for calculation of Air Quality Complex Index (AQI) combines the advantages of the methods described above. It is based on the comparison of the individual pollutants' concentrations (C) in the above ground air layer with their limit concentrations (LimC) in the ambient air. The principal is that the pollutants' concentrations below LimC have no harmful effects on inhabitants. The calculation of AQI is based on the following presumptions:

1. characterisation of pollution situation and effects on local inhabitants and nature is possible (Table 1);
2. value of the AQI for the multi-component pollution field should always exceed or equal the maximum value of sub-indices (AQI_{sub}) for individual pollutants;
3. the short-term (0.5-hr) maximum (MaxC), daily mean ($MeanC_d$) and annual mean concentrations ($MeanC_y$) of pollutants as well as corresponding limit values ($LimC_m$, $LimC_d$ and $LimC_{ny}$) should be taken into account,
4. the various biologically dangerous and synergetic effects of pollutants should be accepted.

Table 1. Characterisation of air quality by complex index

AQI	Air quality/ Pollution level	Effects on man	Effects on plants, ecosystems
0–50	Good/Very low	No effects	No or very slight effects
51–100	Satisfying/Low and moderate	No effects (the effects are improbable)	Effects on materials and natural objects
101–300	Insufficient (unhealthy)/ Relatively high	Influence on some sensitive individuals (respiratory illnesses, heart diseases)	Remarkable effects on plants (lichen, trees etc.), effects on materials
301–500	Poor (very unhealthy)/ High	Remarkable effects on general groups of public	Remarkable and strong effects on natural objects, materials
>500	Very poor (extraordinarily unhealthy)/ Very high	Strong effects on general groups of public	Strong effects on natural objects, materials

For every pollutant two sub-indices must be calculated: at maximum and average concentrations at which the air monitoring or calculated data have been used. The greatest value of sub-indices is the Air Quality Index for individual pollutant (AQI_{ind}) and AQI = MAX{$AQI_{i,ind}$} for multi-component pollution fields. The AQI_{sub} may be calculated by the following formulas:

I range ($AQI \Leftarrow 50$), where $C \Leftarrow 0.5 LimC_d$):

$$AQI_{sub} = 50C/0.5LimC_d \quad (1)$$

II range ($50 < AQI \Leftarrow 100$), where $C > 0.5LimC_d$

and $C \Leftarrow LimC_d$ or $C \Leftarrow LimC_m$):

$$AQI_{sub} = 50[(C - 0.5LimC_d)/(LimC - 0.5LimC_d) + 1] \quad (2)$$

III range ($AQI > 100$), where $C > LimC_d$ or $C > LimC_m$:

$$AQI_{sub} = 100(C/LimC)^n \quad (3)$$

where: if C = MaxC, then LimC = $LimC_m$; if C = $MeanC_d$ or C = $MeanC_y$, then LimC = $LimC_d$; n-coefficient which depends on the category of pollutant harmful effect (Guide 1991).

In Estonia the following categories of harmful effects of pollutants have been established: I (n = 1.7) – extraordinarily hazardous substances (ozone, Cl, Hg, etc); II (n = 1.3) – very hazardous substances (styrene, H_2S, HCHO, NO_x, etc.); III (n = 1) – hazardous substances (SO_2, oil shale fly ash. etc.) and IV (n = 0.9) – slightly harmful substances (NH_3, CO, aliphatic hydrocarbons, etc.).

This index system may be used for estimation of air quality with respect to nature. As a rule, the plants are more sensitive to chemicals and the ecosystems are affected if the pollutants' concentration C in the air during long period exceeds 50% from $LimC_d$ value (Aarnio et al. 1994). Critical loads and guiding concentrations ($LimC_{ny}$ as a yearly mean) that limit the pollutants' load in the air with regard to vegetation (trees, lichen etc.) lie usually in the ranges of $LimC_{ny} = (0.1–0.5)LimC_d$. This is because if the value of $LimC_{ny}$ for plants is known, then in the formulas (1–2) the value of 0.5 $LimC_d = LimC_{ny}$.

Depending upon AQI the air quality during the day, and pollution level in the different areas will be characterised by a special scale (Table 1). Estimation of air quality during longer periods occurs by counting the number of unhealthy days. Zonation of landscapes through calculated MaxC and $MeanC_y$ values can be carried out (Table 2).

Table 2. Values of AQI_{sub} for zonation of landscapes by air pollution levels

AQI_{sub}	Air pollution levels (Tab. 1)					
	very low	low	moderate	rel.high	high	very high
by MaxC	0–50	51–65	66–100	>100	>100	>100
by $MeanC_y$	≦5	≦25	≦50	≦50	51–100	>100

3
Results and Discussion

The index developed was used for the first time in 1996 for analysing the air monitoring data (SO_2, NO_2, H_2S, NH_3, HCHO, phenols, dust) and condition of air pollution in the towns of the north-east Estonian industrial area - Kohtla-Järve (in Järve region), Jõhvi and Kiviõli. The estimation of air quality for the public was carried out using AQI values. It was established that in Järve region and Kiviõli, which are located in the neighbourhood of oil shale processing plants, the total number of unhealthy days (AQI > 100) during the observation period was relatively high - 20–27% or 75–97 day yr^{-1}. In the town of Jõhvi, which is located about 12 km from Järve, the number of unhealthy days was 3.1 times smaller.

The index-method was also used for zonation of NE Estonian landscape (about 3370 km^2) by air quality with respect to the dangerous effect on nature. For that purpose the results of calculations of MaxC and $MeanC_y$ concentration fields of pollutants were used. It was shown that in the period of 1960–1995 the pollution situation was essentially changing: if in 1960–1990 the very high, high, relatively high and moderate pollution zones (levels) occupied about 56–67% (in 1990 – 35%) of the whole territory, then in 1995 the low and very low zones prevailed (90%). In many areas (Kohtla-Järve town, Saka, Kurtna Landscape Reserve) the role of sulphur dioxide has been increasing in the last few years, compared to that of alkaline fly ash.

4
Conclusion

The estimated air quality complex index is a new, effective and more easily understandable instrument for assessment of the air quality in the urban and natural areas of the NE Estonian regional system. The results and estimated air quality index can be useful in future investigations of effects of multi-component pollution fields, as well as for analysing the everyday air monitoring data.

Acknowledgements. This study is supported by Estonian Science Foundation (grant No. 2038).

References

Aarnio P, Hämekoski K, Koskentalo T, Virtanen T (1995) Air quality, monitoring and air quality index in the Helsinki Metropolitan area, Finland. In: Proc of the 10th World Clean Air Congress, Espoo V 2, Proc 201

Aarnio P, Koskentalo T, Hämekoski K (1994) Ilmanlaatu pääkaupunkiseudulla vuanna 1994, YTV. Helsinki

Guide for control of air pollution, RD 52.04.186-89 (1991) Moscow, pp 385–398 (in Russian)

Ilvessalo P (1995) A new method for calculation of an air quality index. In: Proc of the 10th World Clean Air Congress, Espoo Vol 3, Proc 527

Kallaste T (1991) Comprehensive assessments of air pollution in major towns. In: Estonian Environment 1991. Environmental Report 4. Helsinki, pp 34–35

Lee FYP, Gervat GP (1995) The air pollution index system in Hong Kong. In: Proc. of the 10th World Clean Air Congress, Espoo, Vol 3, Proc 524

Ott WR, Thom CC (1978) Environmental Indices – Theory and Practice. Ann Arbor Science Publishers, Michigan

Workshop: Water Management 1.4

Sustainable Berlin by Process Coupling between City and Catchment

Christian Hildmann and Wilhelm Ripl

1 Analysis

Present landscape and urban processes are uncoupled because of the high irreversible matter losses endangering the city's own existence and the existence of the surrounding area.

Solar energy provides the energetic potential for all landscape processes. Solar energy is dissipated by water with its three processor properties (Ripl 1992):

1. Most of the energy is attenuated by the physical process of evaporation and condensation.
2. The chemical cyclic process of solution and precipitation becomes an irreversible-loss process if there is a directed transport between solution and precipitation, e.g. the runoff from rainwater.
3. In the biological cyclic process of photosynthesis and respiration, energy is bound within a small area and then released after a time delay.

The efficiency of the landscape can be calculated as the ratio of the matter lost in irreversible-loss processes to matter directed in cyclic processes; both calculated as a charge flow in protons. The gross primary production of vegetation can be used as an estimation for the matter directed in cyclic processes.

Irreversible matter losses amount to approx. 1 t/ha/a (total mineral salt losses without NaCl) (Ripl et al. 1996). The main components of the matter losses are the base cations (Ca 263, K 26.5, Mg 18.9 kg/ha/a). The nutrients N and P constitute only a small fraction (N: 21.3 kg/ha/a). With such losses, the productivity of the ecosystem and the corresponding water cycle of a locality is broken down more and more as time goes on.

An undisturbed biogeocoenotic succession leads to minimised matter losses from a site and increases the sustainability of the area. Solar energy is

Berlin Technical University, Institute of Ecology, Department of Limnology, Hellriegelstr. 6, D-14195 Berlin, Germany

J. Breuste, H. Feldmann, O. Uhlmann (Eds.)
Urban Ecology

dissipated almost completely by cyclic material processes. Any remaining energy leads to the heating of the adjacent air layer and, via air mass exchange (wind), results in its dissipation at another location, or alternatively it can force the chemical-loss process; because of this, small temperature amplitudes (either daily or seasonally) indicate the high efficiency of an ecosystem at a particular place.

This efficiency concept is equally valid for the urban landscape because base cations and nutrients are similarly delivered to local rivers. Obviously urban areas show much higher matter losses compared to rural areas. For example, for the western part of Berlin total salt losses of 5600 kg/ha/a (without NaCl) have been calculated (Ripl and Koppelmeyer 1990). The town of Neumünster (Schleswig-Holstein) shows twice the matter losses of the surrounding landscape.

The following processes contribute to increases in matter losses from an urban area:

- Since the introduction of water-borne sewerage, running waters have been polluted with higher matter loads. The outflow of sewage plants in Berlin during 1995, in terms of the most significant base cation, amounted to ca 320 kg Ca/ha/a.
- Part of the high losses of base cations found in sewage water is as a result of the dynamic fluctuations in groundwater levels caused by its use for drinking water. Due to the lowered groundwater levels, each rainfall event results in highly-fluctuating „wetting" and „drying" phases in the upper soil horizons. These encourage the mineralisation of organic material by fungi and bacteria. In addition, organic acids are produced, which contribute to the dissolution of base cations. The dissolving of calcium carbonate in the deeper soil layers becomes important only when the existing acids are no longer buffered in the upper soil.
- The low amount of vegetation in the city, and the low retention of rain water, results in a decrease in evaporation and in more extreme amplitudes of air and surface temperatures. Remote thermal images show the poor thermal efficiency of the city. Sensible heat dissipation by air mass exchange will influence the surrounding area of the city because energy dissipation by evaporation will be spatially shifted further away from the city's surroundings. If the upper soil dries out more frequently, the probability of matter losses is increased.
- Examination of a city cannot be reduced to only the actual built-up town area. The daily supply of products to a central European city uses not only its surrounding vicinity, but reaches almost all of the world.

In this respect the practice of agriculture is highly relevant. Present-day production of arable crops need a well-drained soil. Along with each rainfall event, upper soil humidity alternates in wet/dry phases with frequently changing levels which raises the matter losses. These are far from being balanced by present levels of fertilisation, thus the agricultural production function of any locality is already put at risk in the medium-term.

Society is willing to shift the increasing environmental problems, temporally, into the future, since the full extent of these problems will only be apparent later. The problems are also shifted spatially, because the area used by the city is far away from the city itself and managed with high irreversible matter losses. Neither a temporal nor a spatial process-coupling of the matter flow of production, consumption and mineralisation exists in this situation.

This de-coupling of processes is only possible because of readily available cheap additional energy (coal, petroleum fuels, natural gas, nuclear energy). Before these energy sources were readily available, the minimisation of energy use and transport distances were determining criteria limiting the allocation of agricultural production close to the centres of high demand – the consuming cities. At the same time, the available energy for the supply of food and water and its disposal from the city, limited the further extension of cities.

Today, the uncontrolled growth of cities, as well as the distant location of supply areas, seems to depend only on political wishes, since transport costs for the supply of cities are only a small part of their total budget. In this respect, cities are as dependent on the cheap supplies of additional energy for their present size and expansion as if it were a drug. A breakdown of readily-available energy supplies, or a drastic increase in energy prices, would result inevitably in a reduction of the present extent of cities. A rapid jump in energy prices could perhaps result in totally uncontrollable social unrest which could even escalate to the level of civil war. The city has no choice, it has to become sustainable.

2
An Approach for a Sustainable Town Area

Processes have to be coupled again, (1) to minimise matter losses, (2) to reduce the dependency on additional energy, so that the city can become sustainable.

The city has to be re-integrated into its surrounding area to ensure its existence in the longer term. The first priority must be the maintenance of the basis for its physical existence, this is not negotiable. To achieve this, the irreversible matter losses have to be minimised. Other approaches, social, economic and cultural, can then build on this fundamental principle.

2.1 Modules and Elements for a Sustainable Town

Modules and elements of a sustainable city will be sketched out below. The capital city of Berlin can serve as an example. The main focus will be on the water cycle and water management, since water is the medium through which transport, reaction and cooling processes can control the rate of matter losses.

The catchment area of the River Spree contributes to the main part of Berlin's water, but the total run-off (discharge), with respect to the catchment area, is low, especially for the area between Lübben and Berlin. The disturbed water cycle of the landscape is seen as a fundamental reason for the low ratio of area precipitation to run-off.

In order that the run-off from this part of the drainage basin can be increased, the landscape has to once again become wetter by increasing the proportion of vegetation and organic matter in the soil. This offers the possibility to couple the process of re-establishing more permanently wet soils with the recycling flow of base cations and nutrients, as well as biomass production.

It is necessary to allow biologically-treated water (the treated water from sewage plants) to be passed through artificial wetlands such as reedbeds or polders. To retain the base cations and nutrients from this treated sewage water, an extensive evaporating water surface should be achieved. Considering the case of Berlin, an area of ca. 260 km^2 would be required for an annual loading of ca. 800 mm of biologically-treated water to artificial wetlands. Due to their high evaporation the temperature of the wetland area is cooler, particularly during the day. An increased condensation in the form of dew or light rain would be expected as a result. The high production of biomass from such artificial wetlands is a multi-use resource. Reeds can be used as a building material, for energy production, etc..

The middle part of the Spree catchment area contains the open brown coal pits in the Lausitz area which also delivers pumped ground water („Sümpfungswasser"). By increasing the area of lowered ground water table, an increase in the run-off of 20–30% is achieved. The expected decrease in pumped ground water caused by a lessened rate of extraction of brown coal is seen as critical for the water supply of Berlin (MUNR and SenStadtUm 1994). To offset this, additional storage basins would have to be built (first: Lohsa II), but critical water levels are expected to be only for up to two months in the year (maximum).

Due to the existence of the open coal-mining pits, there emerges what can be seen as a great groundwater depression funnel, which has, in parts, a depth of up to 70 m. The open pits and the groundwater depression funnel should be refilled as soon as possible. This should start with the open pits nearest to

Berlin, since this is the only way to increase the run-off area of Berlin quickly. If refilling were to start in the area higher up the drainage basin, then a part of the water which presently drains into the groundwater depression funnel during its passage, will have to be pumped out again. One open pit situated near to Lübben could be used as a storage basin to supplement low water flows.

For the extraction of drinking water within and in the immediate surroundings of Berlin, surface water should be used as much as possible. The pumping should be done in such a way that the groundwater is not in any way affected and is not made more dynamic. Even today, a major part of the drinking water of Berlin is extracted from bank filtration with deep wells; due to the associated processes, the littoral reedbeds have become severely damaged (Wöbbecke and Ripl 1990).

To improve the quality of surface water in general, activities throughout the entire catchment area are required. Here the focus is on the land owners and/or managers, as well as foresters, who manage the area. Besides the production of food and renewable resources they should be paid as „water landlords" (Ripl 1995) – payments with respect to a regular run-off in terms of both its quantitative and qualitative aspects – from their respective catchment areas. Restored, or artificial, constructed and managed wetlands along the riparian length of surface waters can be used as „retention systems" to retain base cations and nutrients leaching from the landscape before they reach the streams.

The hilltops and steeper slopes, both surrounding and within catchments, should not be managed, or only very extensively, because here the leaching of base cations will first result in the break down of the vegetation.

Also, in order to improve the climate within urban areas, a primary focus has to be on the water cycle and the overall water budget. Better temperature damping by means of higher amounts of evaporation has to be achieved. In this way, the almost semi-arid, even steppe-like, climate of cities can be developed more and more towards the climate found in woodlands and forests. To achieve this change in climate, rainwater within cities should be collected and evaporated by way of vegetated areas such as small wetlands situated among the various buildings, rooftops should be intensively vegetated („green roofs") and the fronts or sides of houses and buildings should be „greened". Houses should be greened more intensively and high buildings could be equipped with elements with a soil layer to be mounted in front of the building. Compared to the drainage of rain water, the advantage of evaporation is that the groundwater is not made any more dynamic and so the leaching of base cations and nutrients is very much retarded.

It is possible to imagine „high-rise" greenhouses on the outskirts of the city. In such buildings it will be possible to produce vegetables intensively and in close proximity to their main market. The high-rise greenhouses have the capacity to further increase process-coupling. For example, the required soil beds can be produced by the re-growth of low moor peat within the managed wetlands.

If matter cycles are to be closed within close spatial proximity, then the probability for matter losses also decreases, since total transportation distances are minimised. The closer the production of daily goods is to the town or city, the lower is the need for goods traffic, thus the less energy required for transportation and their associated costs. Many traffic problems can be solved in a pedestrian town by the town itself.

The better the matter cycles are closed, and the lower the material losses are, the more sustainable the city becomes. In a similar way, the city, along with its surrounding area, represents a biogeocoenosis (ecosystem) with all its required elements (Ripl and Hildmann 1995). All matter losses have to be compensated for -which is both very costly in terms of energy and financially. These compensations for matter losses must limit the growth of the city.

Only the sustainability of its physical basis can grant society the desired number of degrees of freedom it wishes for the long term: for a lasting social structure, to give sufficient scope for its creative ambitions, to preserve its culture and ensure its education.

Acknowledgements. We would like to thank Steve Ridgill, Trebon, for valuable suggestions and the English translation.

References

MUNR, SenStadtUm (1994) Wasserwirtschaftlicher Rahmenplan Berlin und Umland. Entwurf. - Berlin, Potsdam

Ripl W, Koppelmeyer B (1990) Stoffbilanzen für die Stadt Berlin. In: Nährstoffbelastung der Gewässer in Berlin und Umgebung: Bestandsaufnahme und Sanierungskonzepte, 1–16

Ripl W (1992) Management of Water Cycle: An Approach to Urban Ecology. In: Water Poll. Res. J. Canada 27(2): 221–237

Ripl W (1995) Nachhaltige Bewirtschaftung von Ökosystemen aus wasserwirtschaftlicher Sicht. In: Fritz P, Huber J, Levi HW (eds) Nachhaltigkeit in naturwissenschaftlicher und sozialwissenschaftlicher Perspektive, 69–80

Ripl W, Hildmann Ch (1995) Die Stadt als energiedissipative Struktur – ein systemökologischer Ansatz. In: Schr.-R. f. Vegetationskunde 27: 201–207

Ripl W, Janssen Th, Hildmann Ch, Otto I (1996) Entwicklung eines Land-Gewässer Bewirtschaftungskonzeptes zur Senkung von Stoffverlusten an Gewässer (Stör-Projekt I und II). Im Auftr. BMBF und LAWAKÜ - Berlin

Wöbbecke K, Ripl W (1990) Untersuchungen zum Röhrichtrückgang an der Berliner Havel.- In: Landschaftsentwicklung u. Umweltforschung, 71: 94–102

Infiltration of Storm Water Runoff with Respect to Soil- and Groundwater Protection

Frank Remmler, Ulrike Hütter, and Uwe Schöttler

1 Introduction

Nowadays, managed infiltration of runoff is often practised as an alternative approach to storm water drainage. This kind of infiltration management increases the input of water per infiltration area unit (m^2) as well as the volume and the spectrum of substances. When aiming to achieve protection of soil and water it is therefore necessary to question whether, and in which cases, storm water infiltration is sustainable and environmentally sound.

2 Potential Pollution of Storm Water Runoff

Many research projects have investigated the quality of precipitation and different types of surface runoff. The results give evidence for variations in pollution depending both on the local conditions and on the analytical conditions of the specific investigation. Due to their toxicity, persistence and accumulation in biological systems particular attention must be paid to organic and inorganic micro-pollutants.

An assessment of the quality of storm water runoff is difficult because there are, as yet, no special quality standards. The isolated cases of the investigations cannot be transferred to other kinds of catchment areas. Currently the different kinds of surface runoff components can only be classified using a simplistic approach. In Table 1, some typical surfaces are arranged in quality categories according to their expected potential degree of pollution.

The granting of permission to infiltrate depends on the quality of the storm water runoff and the kind of treatment facilities. The unpolluted and less polluted fractions can be reintroduced immediately into the natural water cycle, as long as some requirements concerning the location, structure and operation of the infiltration system are fulfilled. The runoff of the categories A and B can be infiltrated by procedures which allow runoff to pass through the

Institute for Water Research GmbH Dortmund, Zum Kellerbach 46, D-58239 Schwerte, Germany

J. Breuste, H. Feldmann, O. Uhlmann (Eds.)
Urban Ecology

Table 1. Classification of storm water runoff in different quality categories

Quality Category		Origin of surface runoff
A = harmless	A_1	- non metallic roof surfaces in residential areas and in industrial estates and industrial areas if their pollution is demonstrably comparable to residential areas
	A_2	- bicycle paths and footpaths with small crowds of people and animals - yard surfaces in residential areas with restricted use (e.g. carwash forbidden, no use of pesticides)
B = tolerable		- streets with parking lots and a low traffic density (average daily traffic <15,000 vehicles) - open air ground with large crowds of people (e.g. pedestrian precinct, markets, open air events) - metallic roof surfaces, yards and streets in industrial estates and roof surfaces in industrial areas if their pollution is demonstrably comparable to residential areas - runways of airports
C = not tolerable		- main roads and major roads with a high traffic density and car parks with frequent inflow and outflow of traffic - paved track systems and surfaces where planes and vehicles are refuelled and/or washed - surfaces with storage and handling of water-endangering substances as well as liquid manure - roof surfaces, yards and streets in industrial areas - open air ground with large crowds of animals (e.g. slaughterhouses) - waste plants and composting facilities

topsoil. Sub-surface infiltration is only permissible in special cases for runoff in category A_1. In general surface runoff of category C is not allowed to be infiltrated.

3
Requirements for the Infiltration

In accordance with comprehensive soil and water protection, several requirements concerning the location, structure and operation of infiltration systems must be fulfilled. Many aspects of the following criteria are based on the experiences at several infiltration sites within research projects in Germany in the last few years. Some of these practical aspects are already components of a German Standard and an additional report of a working party.

3.1
Location

A sufficient residence time of the seepage water in the unsaturated soil ensures cleaning effects by biological, physical and chemical processes. Thus, the *average* hydraulic conductivity of the unsaturated zone below the system should range between 1×10^{-3} and $5 \times 10^{-6}\,ms^{-1}$. The groundwater level should be more than one metre below the bottom of the infiltration system. To avoid the mobilisation of pollutants no residual pollution should be present beneath the infiltration system.

3.2
Structure

The basic criterion for groundwater protection is that infiltration of runoff should be via the topsoil. Materials for the infiltration systems should only be allowed when they cause no detrimental changes to the seepage water and groundwater in the long-term. There are specific requirements for the soil properties of the infiltration area to obtain sufficient cleaning effects for the infiltration systems:

- minimum thickness of 30 cm for the topsoil layer,
- a hydraulic conductivity of $1\times 10^{-3}\,m\,s^{-1}$ for the first 30 cm of the soil to guarantee a sufficient residence time for biological degradation processes,
- an adequate proportion of silt, clay and organic material in the topsoil to ensure a high buffer capacity,
- a soil-pH between 5.5 and 8 to minimise the mobility of heavy metals.

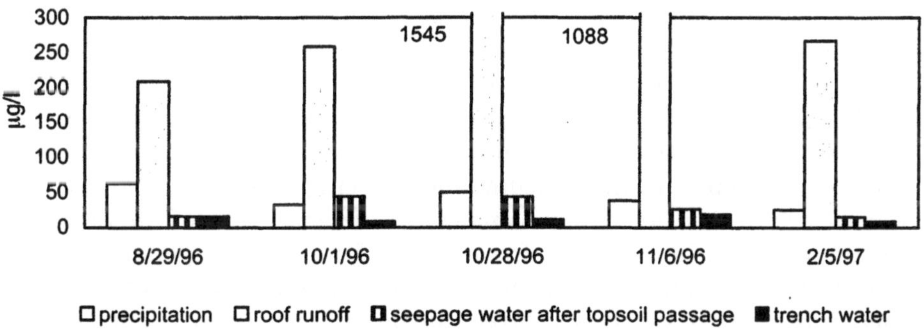

Fig. 1. Zinc concentrations in the precipitation, roof runoff and seepage water at the day care centre

Figure 1 illustrates the cleaning potential of the topsoil in an infiltration trough. Water samples were taken at a trough-trench-system at a day care centre in Dortmund. The network there provides the opportunity to investigate the quality of the precipitation and the storm water runoff on its flow-path to the groundwater. The zinc concentrations are described for five precipitation events. The airborne input varies between 30 and 60 µg/l but the runoff absorbs a lot of zinc from the roof drainage system, so that the concentrations of zinc sometimes exceed audit values for the investigation of groundwater (LAWA 1994). After the passage of the water through the topsoil the concentration of zinc is significantly reduced.

3.3 Operation

There are two major aims for the operation of infiltration systems. The cleaning potential of the infiltration system due to the adsorption of contaminants is of great importance as well as minimising the risk of remobilisation of these adsorbed pollutants and their release into the environment. Both aims can be achieved by appropriate maintenance and cleaning activities to guarantee proper functioning of storm water infiltration systems. To guarantee reliable functioning of the topsoil component, hydraulic weaknesses of the soil must be avoided. Plants must be renewed if vegetation damage or decrease occurs and such activities such as removal of leaves in autumn must be carried out. In the direct infiltration area the application of pesticides is forbidden. The infiltration systems must be inspected to see whether there are new unauthorised connections or a change of use of the connected areas which could negatively influence the storm water quality.

During the operating time chemical substances accumulate in the soil. To ensure durable adsorption capacity of the topsoil these accumulated substances must be removed, as the need arises, using suitable methods. There is a need for investigations of the topsoil and/or the seepage water and groundwater at regular intervals to assess the cleaning and adsorption capacity of the seepage area. All these maintenance, cleaning and inspection activities should be fixed in an established operational schedule.

The results of soil investigations in two different infiltration systems show the importance of the mentioned requirements (Fig. 2). The first investigation site is located in Dortmund. Soil samples at different depths of the infiltration troughs were investigated after 12 and 18 months of operation and compared with a control sample taken during the construction work. The second investigation site is a central trough in Berlin with a operating time of 80 years. Soil samples were taken in the middle of the trough and outside the infiltration area at various depths to obtain a comparative control.

In order to estimate the accumulation and mobilisation effects of substances, the soil samples were investigated with the pH_{stat}-procedure, which determines the ability of substances to be leached by water with pH 4 (Obermann and Cremer 1992). This procedure takes into consideration the long-term leaching behaviour in particular. Figure 2 shows the proportion of zinc which can be mobilised by the pH_{stat} leaching test with pH 4 in the topsoil of the infiltration trough (influx and margin area) in Dortmund after 12 and

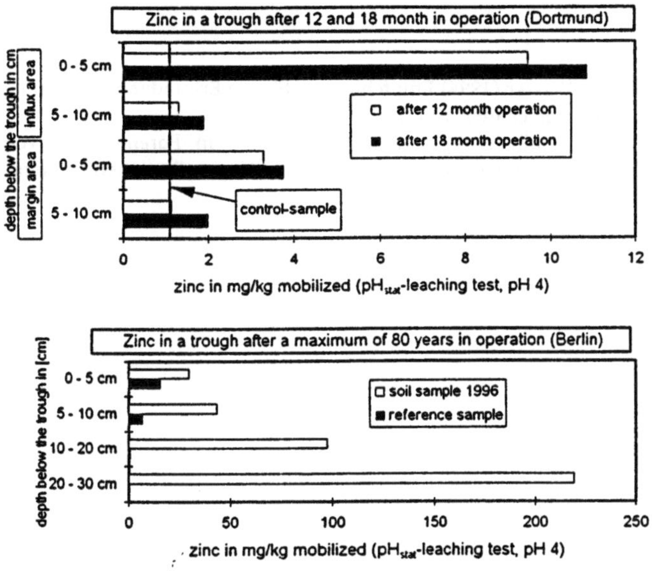

Fig. 2. Mobilisation of zinc in two infiltration troughs

18 months of operation. The investigations showed a high accumulation of zinc in the top 5 cm and, of course, a higher accumulation in the influx area. The results for zinc in the topsoil of the infiltration trough in Berlin showed, in contrast to the results in Dortmund, increasing, and a higher rate, of mobilised proportions of zinc with depth. The first immobilised zinc in the soil is apparently moving deeper towards the groundwater. The almost exhausted neutralisation buffer capacity for acids and the pH of the soil (4.3) make long-term immobilisation in the topsoil impossible.

4
Conclusions

The risk of groundwater pollution can only be minimised by using infiltration systems with a passage through the topsoil and by following specific requirements for the planning, construction and operation of infiltration systems. A long-term ecologically tenable storm water infiltration is only possible by developing and using suitable instruments to monitor the long-term effectiveness of such systems and by developing and using proper cleaning activities for these systems.

Acknowledgement. We thank the *German Environment Foundation* and the *Federal Ministry of Education, Science, Research and Technology* for their financial support of this work.

References

LAWA Länderarbeitsgemeinschaft Wasser (1994) Empfehlungen für die Erkundung, Bewertung und Behandlung von Grundwasserschäden. Stuttgart

Obermann P, Cremer S (1992) Mobilisierung von Schwermetallen in Porenwässern von belasteten Böden und Deponien. Entwicklung eines aussagekräftigen Elutionsverfahrens. LWA (ed) Essen

Requirements and Possibilities of Best Management Practices for Stormwater Run-off from the View of Ecological Town Planning

Mathias Kaiser

1
Goals for Restructuring Management Practices

Overburdening of the sewer system is occurring more frequently, especially after a heavy rain, e.g. a summer thunderstorm, as the system cannot handle the amounts of water arriving in such a short period of time, but generally flooding only occurs for a short time. In a combined system the flood water is always mixed with waste water.

Some of the sewers in need of repair could be redeveloped by installing another pipe inside the existing one or by being made watertight from the inside with localised patching. Both methods, however, reduce the cross section of the sewage pipe thus allow less water to flow through it. Apart from reasons of urban ecology, the financial aspects of repair makes it sensible to alter the normal urban water cycle (see Fig. 1, left side) so that less storm water will flow down the sewage pipes in the future.

Fig. 1 (right side) shows that it is necessary to increase considerably the amount of evaporation, surface run-off and ground water regeneration (Sukopp 1990). Thus the main task of restructuring the conventional urban water cycle is to create a new comprehensive system which is similar to a natural system. According to today's knowledge such a comprehensive system can consist of the following: tanks for processed water, infiltration depressions, troughs and trenches, moist habitats, expanses of open water, drain pipes with access shafts for maintenance and monitoring and – if necessary – conventional pipes.

2
The Method of Approach for this Research

In the two cities of the survey (Dortmund and Zwickau) a complete evaluation of the town is neither possible nor logical. Standardised types of settle-

Chair of Landscape Ecology and Landscape Planning, Department of Spatial Planning; University of Dortmund, D-44227 Dortmund, Germany

J. Breuste, H. Feldmann, O. Uhlmann (Eds.)
Urban Ecology

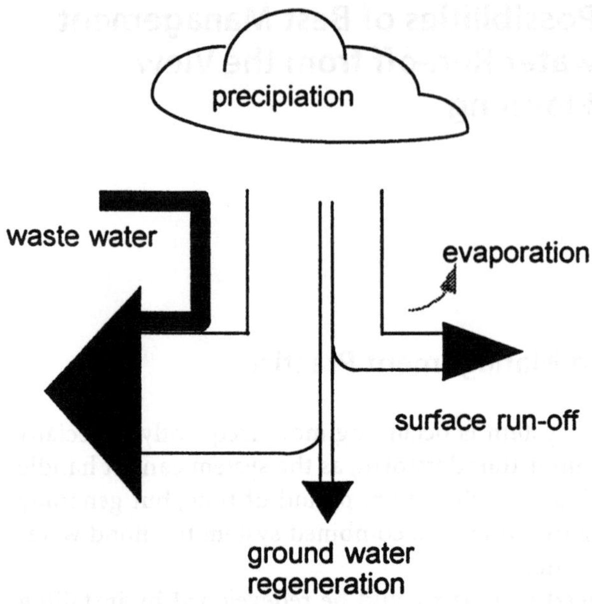

Fig. 1. Conventional urban water cycle

ment structures were developed to obtain transferable results (Sieker et al. 1992). These types portray the total spectrum of urban planning situations which occur; these depend on two variables: the amount of storm water disposal (dependent on the share of paved surfaces) and the possibilities for water infiltration (dependent on the amount of open space).

Subjects of the survey are residential areas as well as commercial districts. The residential areas were chosen according to the following criteria:

- type of housing,
- location in the town,
- type of infrastructure,
- type of property,
- proportion of paving,
- proportion of open space,
- spatial separation between developed space and open space,
- type of roof,
- proportion of infiltration space,
- infiltration system,
- infiltration network,
- connection between run-off ditches.

Both an existing and a newly planned building area were chosen as residential study sites. Areas of the same type were chosen in former East Germany

(Zwickau) as well as in former West Germany (Dortmund), where significant differences were obvious. The reason for choosing both new and old sites is that the development of measures to optimise ecologically the urban water cycle is both possible and necessary for new urban developments as well as for integration into existing building areas. For example, it can be combined with planned measures for improvement of residential areas or renewal measures.

The selection of the commercial districts was not dependent on density or structure (although these criteria were important for selection of the residential areas), instead, these areas were selected on the basis of whether or not soil contamination was suspected.

3 Ecological Valuation of the Alteration of the Water Cycle

The contribution of the surface runoff can be reduced nearly completely in favour of infiltration and evaporation with the aid of the solutions developed. The contribution of evaporation capacity should be observed as it can play an important role in cities from the point of view of the urban climate. Instead of reducing the infiltration capacity as the conventional solution does, it is increased to produce a contribution of 70–75%. This shows an overcompensation compared to the natural state before it was constructed (around 30 to 40%). From the point of view of urban ecology this is not seen to be critical as the contributions to infiltration of urban fields with canalised, sealed areas are already greatly reduced. When considering the household water in construction areas (including street areas in the immediate vicinity) it becomes obvious that the total contribution of infiltration is reduced again to under 40%. Compared to the situation before the construction, the increased infiltration rate does not mean a negative shift in urban ecological terms, but can be rated as compensation for the already existing displacements of the water balance (Kaiser 1996).

3.1 Extensive Integration of Natural Rainwater Management into the Planning Process

The practical planning development and scientific observation of a great number of settlement areas of new buildings and standing buildings have shown that often, theoretically, enough free areas are available for rainwater management. The share of very compact areas (cities and small parcelled „Gründerzeit" building) plays a secondary role concerning the entire expanse.

It also was shown that the reservation of areas obtained especially for natural rainwater management in practice results in stubborn resistance, thus the prospects for future natural rainwater management are very restricted for a widespread practical use. Currently the first steps of natural rainwater management are to try to defuse this problem by minimising the specific space demand (ratio of connected sealed area to provided management area 10:1). This, together with covering and reshaping of the areas, is found to be acceptable (troughs 30–50 cm deep with underlying trenches, exclusion of other uses e.g. for protection or development of a biotope etc.).

In contrast to the above mentioned optimisation path, we tried another method which did not even try to minimise the demand for space, but only to minimise the grade of function covering and thereby exclude other uses of rainwater management areas. We endeavoured to gather and manage the rainwater on as large an area as possible. Infiltration troughs are planned with a maximum dam up height of 10–15 cm instead of 30–50 cm.

With respect to planning, the natural rainwater management becomes a subordinate function of the area. As opposed to structural uses (buildings, parking lots, streets) it does not demand exclusive usage of the area. The degree of exclusion of other uses becomes higher the less natural the character of an area/system of rainwater management is chosen.

Similarly to the discussion about the arrangement and use of the covering of streets in the 1980's, the following principles are applicable here:

- separation principle space saving system the design and arrangement of rainwater managing areas/systems excluding other uses to a high degree;
- integration principle minimisation of the extent of disturbances of other planned uses on these free areas.

The integration principle refers to the new building and to the current standing buildings. It reveals advantages concerning the following ecological valuation factors:

- natural characteristics of the household water in the management areas (infiltration, evaporation, household groundwater);
- large areas for the succession of vegetation stands in the management areas;
- minimisation of diggings, banks of earth and deposits (ground protection);
- very limited use of resources such as gravel, lava.

From an economic point of view important advantages are:

- no further need of space explicitly for infiltration, rather extensive integration into free space used for other purposes;
- minimisation of the building costs by abstaining from underground constructions (drains, trenches, canals etc.);
- minimisation of the need for maintenance due to the landscape/free-space-integrated system of the management areas/systems;

- minimisation of the planning costs by avoiding the need for relatively high ranking official approvals in single cases (water law permission is not applied for, this is of special importance for an owner structure of small parcels e.g. for one-family-houses)

In practice wide (quantitatively important), and until now mostly unused, possibilities result from the developed strategy of natural rainwater management. For situations in very compact settlements with a very small amount of free areas we have to fall back upon less natural systems, such as trough-trench-systems.

References

Kaiser, M (1996) Auswirkungen der Versickerung von Niederschlagswasser auf den Naturhaushalt am Beispiel des Gewerbegebietes Flautweg in Dortmund. In: Kommunale Umweltaktion UAN (ed): Ökologischer Wasserhaushalt. Hannover, p 111 ff

Sieker, F, Erbguth W, Finke L, Heher B, Herlyn U, Rudolph KU, Schötter U (1992) Forschungsantrag "Möglichkeiten einer naturnahen Regenwasserbewirtschaftung in Siedlungsgebieten, untersucht und demonstriert am Beispiel der Städte Dortmund und Zwickau". Institut für Waserwirtschaft, Universität Hannover

Sukopp, H (ed) (1990) Stadtökologie. Das Beispiel Berlin. Berlin

Comprehensive Stormwater Management in Urbanising Townships in the US and its Implications for Germany

Joachim T. Tourbier

1 Introduction

The practice of stormwater management is evolving from what has been called the 'common enemy rule' into a 'Blue-Green Technology approach'. The common enemy approach has long been an engineering standard for stormwater, whereby water is shed away from potential 'damage areas' as fast as possible. This implies that flood problems caused by increases in runoff are being passed on to properties downstream, adversely affecting people, causing loss of life, damage to properties and problems to public health. Flood catastrophes along the Oder River, the Rhine and along other rivers in Europe illustrate the drama of these problems only too graphically, thus advances made in the US have implications for Germany and other countries.

The Blue-Green Technology approach is an extension of current practices in the US, internalising the responsibility for managing runoff from land surfaces. It is a way of implementing sustainable development objectives. This relates closely to the Agenda 21 as set forth at the Earth Summit in Rio de Janeiro.

Through application of a 'user pays principle', those who cause problems are being charged for solving them. In the US this concept has been adhered to through the use of so called 'Best Management Practices' that are implemented through a Stormwater Management Ordinance on the municipal level. The ordinance requires that each development maintains pre-development flood peaks. By expanding this approach, conservation associations (Tourbier 1996a) and municipalities (Tourbier 1994b) are now exploring a 'no net increase' approach as a standard for new development.

Through the 'Blue-Green Technology' approach the water-cycle is maintained through a self sufficient recharge and supply system that holds back stormwater and cleanses it through the actions of green plants. It replicates the processes of a woodland setting on a pre-development site where up to

Dresden Technical University, Institute for Landscape Architecture, Helmholtzstr. 10, D-01062 Dresden, Germany

J. Breuste, H. Feldmann, O. Uhlmann (Eds.)
Urban Ecology

40% of the precipitation is intercepted by tree surfaces and evapotranspired. On the uneven forest floor up to 50% of the precipitation is held back in depression storage and recharges the groundwater. Only 10% of the precipitation is turned into runoff. By utilising these processes on a development site we transfer the costs for large unsightly detention basins into a system of aesthetic landscape features with blooming fields of wildflowers, luscious groves of woody plants, and sparkling water features that provide for a functional landscape and a sensible use of open space.

2 Goals and Objectives

Goals and objectives are defined in a public process and then transferred into standards applied to the approval process for future uses:

1. Maintain pre-development runoff peaks - control systems for development sites shall be designed to assure that post-development runoff peaks are no greater than pre-development peaks.
2. Maintain the pre-development volume of groundwater recharge - annual post-development recharge rates for groundwater on development sites shall be maintained at pre-development levels to assure that there is no reduction in the annual recharge to the aquifer.
3. Control erosion and sedimentation of water resources - the total amount of sedimentation entering a water body during the development process shall be reduced to an acceptable level.
4. Abate runoff pollution-non-point source pollution associated with stormwater runoff from development sites shall cause no degradation of water quality to receiving waters and be kept as close as possible to pre-development conditions.
5. Integrate a comprehensive combination of measures - 'treatment trains' that combine measures for the linkage of riparian areas to form linear greenways providing multiple uses and a full range of benefits.

A Blue-Green stormwater management approach that implements the objectives stated above is achieved as stated below.

Measures that supplement each other and are arranged in sequence in the path of runoff concentration are linked in so called 'treatment trains'. In such a series the size of each measure will be reduced by the capacity and performance of the preceding one. The groups of measures (as described in this chapter), with examples of practices described here, should be practised sequentially:

The greening of structures is achieved through construction of one of two types of vegetated roof covers. Extensive roof gardens can be built with a thin growing medium applied to existing roofs. Intensive roof gardens use a thicker aggregate layer to store precipitation for irrigation; facades can be greened to help with the evapotranspiration of light rains.

Porous Pavement allows water to seep in and will not freeze up in the winter when a sufficient gravel base is supplied. Porous asphalt is made by simply reducing the amount of bitumen used in a bituminous-gravel mix. Modular pavements let water infiltrate through apertures that are seeded with grass.

The design of impervious surfaces to create small drainage areas – parking lots which are usually a continuous expanse of asphalt can be broken up in subsections through grading during construction, or through retrofitting by installing cobblestone 'dip swales' that intercept runoff on the surface and divert it to various types of treatment features. The length of slope is a component of the equation to calculate runoff; by reducing it peak flows are lowered.

Runoff can be considered to be contaminated (Harmfulness Class 1), or not contaminated (Harmfulness Classes 2 and 3), depending on the surfaces of origin. Non-polluted runoff from roofs, pedestrian ways and lightly travelled roadways will not require treatment before it is released into receiving bodies of water or infiltrated into the ground. Runoff from Harmfulness Class 1 surfaces should be filtered and treated through terrestrial or aquatic systems involving plants.

Grassy Vegetative Filters can take the form of wildflower meadows created to trap over 95% of suspended solids and related pollutants. The trapping efficiency will depend on the steepness of slope, filter length and on the type of vegetation. Previously concentrated runoff flows can be reconverted into a thin sheet flow through 'level spreaders' that provide a level, non-erosive edge. Lawns that are cut only four times a year will grow into meadows. Wildflower meadows should be seeded, or better still planted with small 'plugs' of pre-grown perennials to achieve the desired function and appearance.

Constructed Wetlands – nitrification and de-nitrification processes and uptake by plants in constructed wetlands remove nitrogen and phosphorus from stormwater. Constructed wetlands can be planted with attractive rushes, reeds, yellow water iris, violet blooming pickerelweed and other decorative plants.

Wet ponds with extended retention have been found to have pollution removal capabilities (for total suspended solids) of up to 98% and greater.

Pre-development groundwater recharge can be maintained on most sites when the first 1 to 2 inches of a storm event are infiltrated. This can be achieved through above-ground infiltration devices, or through below-ground measures. Above-ground measures are preferred because they can enhance a site. The upper soil horizons have a water renovating capacity and maintenance can be performed relatively easily.

Infiltration Berms are gently sloping landscaped depressions that follow the contour of the land and are constructed by excavating a channel and placing fill to form a ridge down the slope. Berms have been successfully used in London Grove Township Pa. (Tourbier 1996b) to blend unobtrusively into the landscape and to infiltrate storms with a two year frequency.

Vegetated Infiltration Swales with Checkdams function as a grassed surface conveyance device for stormwater that not only is less expensive than storm sewers, but also holds back water in small depressions for infiltration.

Gravel filled seepage areas provide temporary storage in the void spaces between the gravel. They may permit a secondary use, such as walkways, on their surface where paving patterns can create visual interest.

All the previously mentioned measures should be explored for location in sequence as a treatment train, prior to the final basin. This will detain and retain stormwater volumes and will make the final basin smaller in size than in a conventional design, inundating it less frequently and permitting secondary uses. A second option is a detention pond with a permanent pool, designed to offer all the amenities of a real estate lake, without the large and undesirable water level fluctuations of a customary design.

By combining and linking the explored measures into 'treatment trains', functional linear landscape systems can be created which form a composite with existing water features in stream valleys. Together they function as greenways that extend from upland development to the riparian zone. Fig. 1 shows components of such a combination, utilising the runoff originating from paved surfaces to provide a linkage to riparian lands. This offers an opportunity to create Blue-Green landscape features that should be connected to existing swales and stream valleys with their steep slopes and floodplains that can not be developed. Open space along existing streams can be replanted into a riparian forest buffer to filter runoff and to help to shade and cool water, thus increasing its oxygen carrying capacity and the health of streams. Treatment trains with their riparian elements will form functional open spaces that can structure urban growth in the suburbs (Tourbier 1994a), separate incompatible uses and provide an edge of natural areas to urban places.

3 Conclusion

The benefits of Blue Green technology consist of reductions in the runoff curve number (CN) and the lengthening of the time of concentration, which in turn lowers the peak runoff flows that are to be controlled. Sites can be designed so that there is no post-development net increase in runoff peaks, runoff pollution, or impairment of groundwater recharge. Blue-Green Technologies are functional, self maintaining, provide for multiple uses, recreation

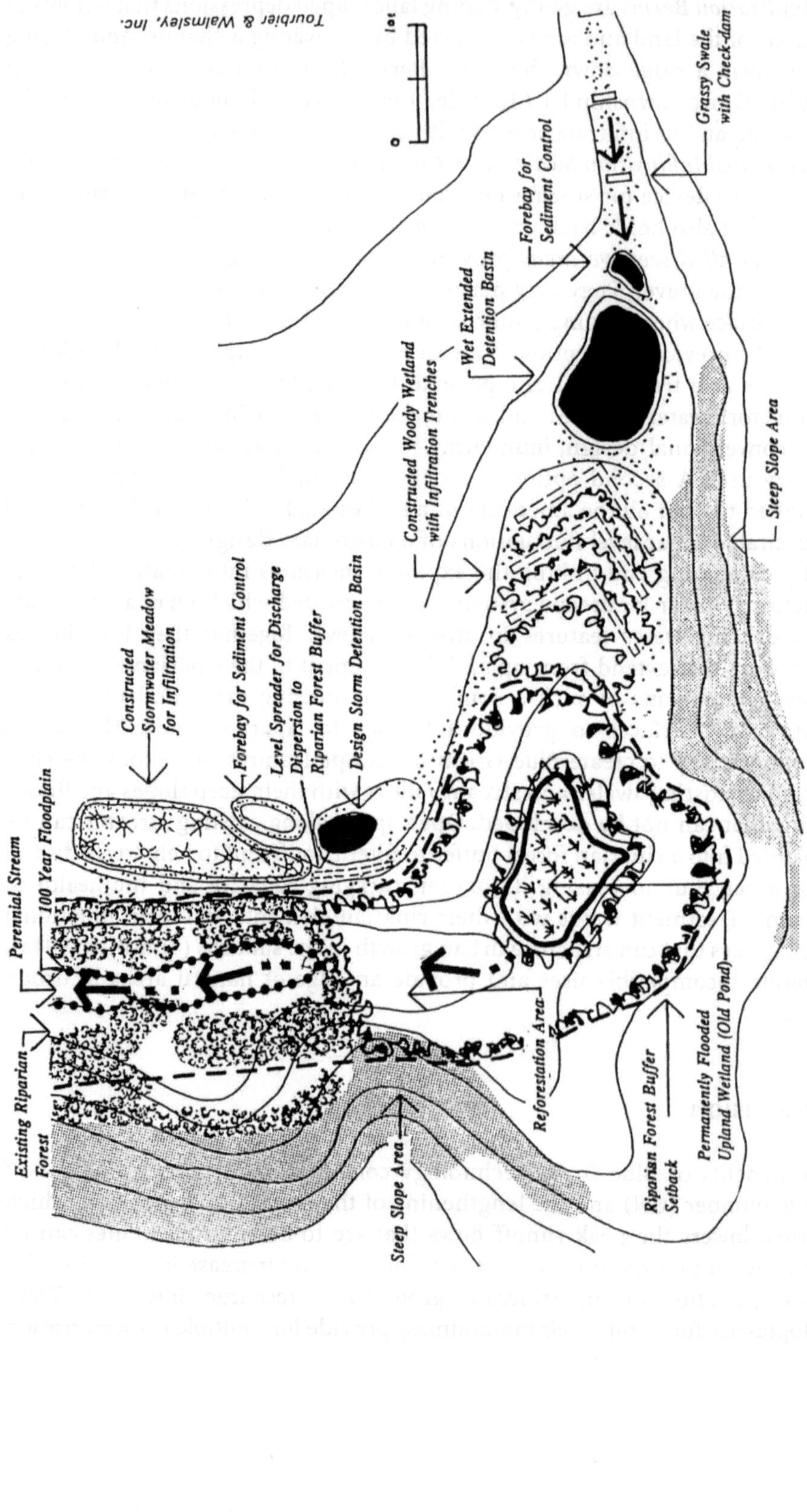

Fig. 1. A combination of measures are linked with existing riparian areas create a functional open space system

and wildlife observation and, last but not least, they add amenities to a site that are appreciated by urban residents (Bookout 1994).

Many would like to see Blue-Green technology practised more widely. The strategy is practical and examples of installations have already been made, implementing at least some, if not all, of the concepts given. The author and the Great Swamp Watershed Association in New Jersey, USA will be documenting examples of its application. Planners or designers who have made an installation that implements some of the concepts described here and would like to bring this to the attention of a wide audience ought to contact the author by writing to the Institut für Landschaftsarchitektur at the TU Dresden.

References

Bookout LW (1994) Value by Design: Landscape, Site Planning, and Amenities. Washington, D.C.: The Urban Land Institute

Tourbier JT (1994a) "Open Space Through Stormwater Management: Helping to Structure Growth on the Urban Fringe." Journal of Soil and Water Conservation, p 49

Tourbier JT (1994b) London Grove Township Stormwater Management Ordinance. Avondale, PA: London Grove Township

Tourbier JT (1996a) "Stormwater Management for Water Quality Improvement and Infiltration in the Great Swamp Watershed" New Vernon, NJ: Great Swamp Watershed Association

Tourbier JT (1996b) London Grove Township Stormwater Management Guidance Manual. Avondale, PA: London Grove Township, pp 27–29

Input and Output of Airborne Aggressive Substances on Green Roofs in Karlsruhe

Sabine Steusloff

1 Problem Outline

The run-off from roofs is one of the sources of diffuse input of airborne aggressive substances into water. In the event of precipitation, dust and aerosols are discharged together with the suspensoids of the atmosphere and can be transported by the sewer system into the receiving stream. However, suspensoids which are accumulated on vegetation or on the roof material as a dry deposit will be washed into the substrate, the filter layer and the drain layer of the roof when it rains. This causes a delay or even diminishes the suspensoid run-off. Considering these circumstances, a balance of the input and output of airborne heavy metals for green roofs was performed.

2 Methods

Two flat green roofs, an extensive one and a semi-intensive one with backwater irrigation, were compared with a flat gravel capped roof. Both of the green roofs are constructed in three layers: a substrate layer, a filter layer and a drain layer. The substrate layer of the extensive roof was typically composed of recycled material (broken tiles), mineral surcharge substances (pumice stone), clay and compost, its drain layer consists of broken tiles and pumice stone (thickness of construction: 10 cm). The substrate layer of the semi-intensive roof is a mixture of peat, bark compost, clay, pumice stone and inflated slate with mineral long-term fertiliser as an additive, its drain layer consists of inflated clay (thickness of construction: 22 cm). The storage capacity of water on the semi-intensive roof construction is decisively enlarged by a rain water retention of 4 cm. The models have a base of 1 m^2 and were constructed on the same roof-top study site in November 1995.

Institute for Geography and Geoecology, Lehrstuhl I, University of Karlsruhe (TH), Kaiserstraße 12, D-76128 Karlsruhe, Germany

J. Breuste, H. Feldmann, O. Uhlmann (Eds.)
Urban Ecology

The green construction systems were established twice. One of the models was left without plants, the other one was sowed with seeds of *Festuca rubra*, a sweet grass. This procedure allows the influence of the vegetation layer on the reservoir ability and retention capacity of aggressive substances to be looked at separately. The precipitation and the run-off were sampled after each event and mixed samples were produced monthly. The dynamics of each precipitation event were recorded by means of an ombrometer and the weather conditions were registered constantly using a climate station. The wet and the dry deposition were taken at the roof-top study site. The run-off and the green roof substrates as well as the roof vegetation were treated by voltametric analysis to determine their content of copper, zinc, cadmium and lead. In addition the pH-value and conductivity of the precipitation and run-off quantities were measured.

3 Results

3.1 Results of the Run-Off Measurements

The role of green roofs as a temporary reservoir for rain water is demonstrated in Table 1.

In particular, during the vegetation period, the green roofs with grass hold back more than 90% of the rain water. During the relatively low precipitation in September at the study site (21.8 mm), the storage of rain water is at about 100%, when regarding the semi-intensive system. The water demand of the plants is illustrated by considering the extensive roof in August: the extensive grass roof holds back rain water thirty times more than the one without plants. Furthermore during the autumn and winter months, the run-off coefficient is about 0.4 for the extensive system and close to 0.3 for the semi-intensive one. During these months there is no visible influence of the vege-

Table 1. Total amount of precipitation [mm] and percentage of run-off from the roofs (July 1996–December 1996)

	Precipitation [mm] = 100%	Gravel [%]	Extensive/ green [%]	Extensive [%]	Semi-inten./ green [%]	Semi-inten. [%]
July	39.5	57.2	5.6	8.3	0.7	2.0
August	23.3	64.4	0.4	12.9	0.4	0.8
September	21.8	60.5	3.2	5.0	0.0	0.0
October	124.8	70.7	42.8	46.6	19.2	25.9
November	40.6	78.0	48.0	48.0	27.8	27.8
December	31.9	81.5	35.4	35.4	30.7	30.7

tation layer anymore. The run-off coefficient of the gravel-capped roof lies between 0.6 and 0.8.

3.2
pH-Value and Conductivity

As the Table 2 shows, the conductivity of the run-off is enlarged to a great extent when compared to that of the precipitation.

The conductivity of the gravel capped roof run-off is 4 times higher than that of the precipitation, but it rises to 16 and 11 times higher in the run-off from the green roof models. For the semi-intensive roof this is due to the previously mentioned long-term fertiliser. Conductivity of the extensive roof run-off may be caused by plaster residues in the recycled building material (Appel 1994). The average conductivity of the grass model run-off is considerably lower than that of the run-off from the models without grass; this is induced by the nutrient demand of the plants. The pH-values of the run-off are nearly neutral, while the pH of the precipitation lies at 5.6, which is typical for urban areas (Mascher et al. 1997). Due to the buffer capacities of the roof substrates, the pH-value of the water is higher after its passage through the substrates.

Table 2. The average pH- and conductivity-values [µS/cm] in the precipitation and run-off from roofs (July 1996–December 1996)

	Precipitation	Gravel	Extensive/ green	Extensive	Semi-inten./ green	Semi-inten.
pH-values	5.6	7.0	7.7	7.7	6.5	6.5
conductiv.	31.7	122.3	526.5	781.2	356.9	429.6

3.3
Investigation of Heavy Metals

The run-off characteristics of the model roofs as described above decisively influence the concentrations of heavy metals in the run-off. The retention capability of the models for heavy metals seems to depend mainly on the retention capability for rain water. The evaluation of the heavy metal investigations is shown in Table 3. The amount of heavy metal deposited by the wet deposition was set to 100% for each type. The relative values for the run-off from the roof models represent the heavy metal fraction which was measured in the run-off.

Comparing the heavy metal contents in the run-off from the green roof models with that in the run-off from the gravel capped roof, the retention

Table 3. Contents of copper, zinc, cadmium and lead in the run-off from the roofs in percentage of the wet deposition (=100%)

Metal	Time period	Gravel [%]	Extensive/ green [%]	Extensive [%]	Semi-inten./ green [%]	Semi-inten.[%]
copper	Jul.–Sept.	19.8	2.7	6.2	0.2	1.0
	Oct.–Dec.	19.7	66.3	79.6	41.7	63.1
zinc	Jul.–Sept.	14.3	4.1	12.5	0.8	1.2
	Oct.–Dec.	19.6	27.5	39.5	8.4	9.7
cadmium	Jul.–Sept.	60.5	8.3	29.3	1.9	6.1
	Oct.–Dec.	58.7	38.0	49.9	11.7	59.1
lead	Jul.–Sept.	34.9	0.9	2.4	0.4	1.3
	Oct.–Dec.	26.6	9.1	17.1	6.1	7.9

capability of the green roof systems is obvious. The semi-intensive system with grass for example stores over 99% of the copper, zinc and lead input in the summer months, whereas cadmium seems to be more mobile and is kept back to 98%. Compared to this, the gravel capped roof holds back only 40% of the cadmium input over the whole time period. Looking at all roof models, cadmium is the most mobile of the investigated heavy metals. Copper, zinc and lead are stored away between 80% and over 99% during the summer months by the green roof systems. This corresponds in general to the findings of Köhler (1989). During the autumn and winter months the retention ability of the green roofs is reduced for all of the investigated heavy metals. During this time period, the retention capability for copper and zinc on the extensive green roofs is even lower than on the gravel capped roof, which shows nearly the same values during the summer and winter months. So far, evapotranspiration effects have not been investigated but will be considered in future research.

4 Conclusions

Among the positive ecological and economic attributes of a green roof is its ability to act as a temporary reservoir for rain water. The accumulation potentials for airborne heavy metals of modern green roofs appear to exceed those of a conventional capped gravel roof, with the exception of copper and zinc, during winter periods. Taking into account the pH-values of the run-off, some of the above mentioned findings can be related to the low solubility product of the investigated heavy metals which causes them to accumulate in the substrates. Further research will be necessary to explain some diverging results, in particular with respect to copper. Additional work will be dedicated to the influence of the changing pore structure of the substrates due to the growth of the roof vegetation. The measurements, presented in this chapter, provide quantitative material for underpinning the basic observation of the retention capabilities of green roofs for heavy metals. However, these measurements bring up new details which need to be further clarified when working towards the development of well understood rules of application.

References

Appel R (1994) Dachbegrünung mit Recycling-Baustoffen - Den Ökologiegedanken zu Ende gedacht. Dach und Grün 3: 16–17

Dürr A (1994) Dachbegrünung - ein ökologischer Ausgleich. Bauverlag GmbH, Wiesbaden Berlin

Köhler M (1989) Ökologische Untersuchungen an extensiven Dachbegrünungen. Verhandlungen der Gesellschaft für Ökologie (Essen 1988) 18: 249–255

Liesecke H-J (1989) Grundlagen der Dachbegrünung. Patzer, Berlin Hannover

Mascher F, Schaffler K, Kessler HH, Fischer G, Marth E (1997) Inhaltsstoffe nasser Deposition als Indikatoren für Luftqualität und Schadstoffdeposition im Raum Graz. Gefahrstoffe - Reinhaltung der Luft 2: 65–69

Poster 1.5

Development of the City of Poznañ and Changes in the Network of Water Courses and Climate

Ryszard Farat and Magorzata Kêpiñska-Kasprzak

Poznañ - a town with a population of six hundred thousand - is situated in mid-western Poland in the valley of the Warta river and neighbouring plateaus. The oldest settlement centre in Poznañ (tenth century) developed on the right hand side of the Warta river. Intensive spatial development of Poznañ started from the beginning of the twentieth century. The population of the town tended to grow throughout the entire period of the development of the city. Periodic drops in the population were associated with wars which were fought in Poland. Recent years (after 1990) brought about a negative natural growth, typical for agglomerations in highly developed countries.

Changes in the hydrologic conditions in the Poznañ area, connected with climatic changes, are difficult to account for due to human interference with the natural course of water circulation. However, while in the Middle Ages the river pattern of Poznañ played an important role in the development of the city area, of its economy and the system of defence, in the middle of the 17th century its significance began to decline gradually and finally to disappear (Kaniecki 1993).

Intensification of urbanisation processes also resulted in changes of local climatic conditions as evidenced by, among others:

- Higher air temperatures in comparison with the background (the strongest contrast between the town and its surroundings can be observed during windless and cloudless weather) (Fig. 1),
- Existence of urban islands of heat and cold, changes in the speed and direction of wind,
- Significant spatial diversification of annual totals of atmospheric precipitation and heavy rainfall in the period from May to September.

In the long-term plans (developed in recent years) and current plans for the Poznañ agglomeration, the main emphasis is placed on restoration of natural valley areas. These are characterised by exceptional natural scenery and the plan is to re-open the river network which existed in the past. Also,

Institute of Meteorology and Water Management, ul. D'browskiego 174/176, 60-594 Poznañ, Poland

J. Breuste, H. Feldmann, O. Uhlmann (Eds.)
Urban Ecology

Mean air temperature -

autumn, 7 a.m.

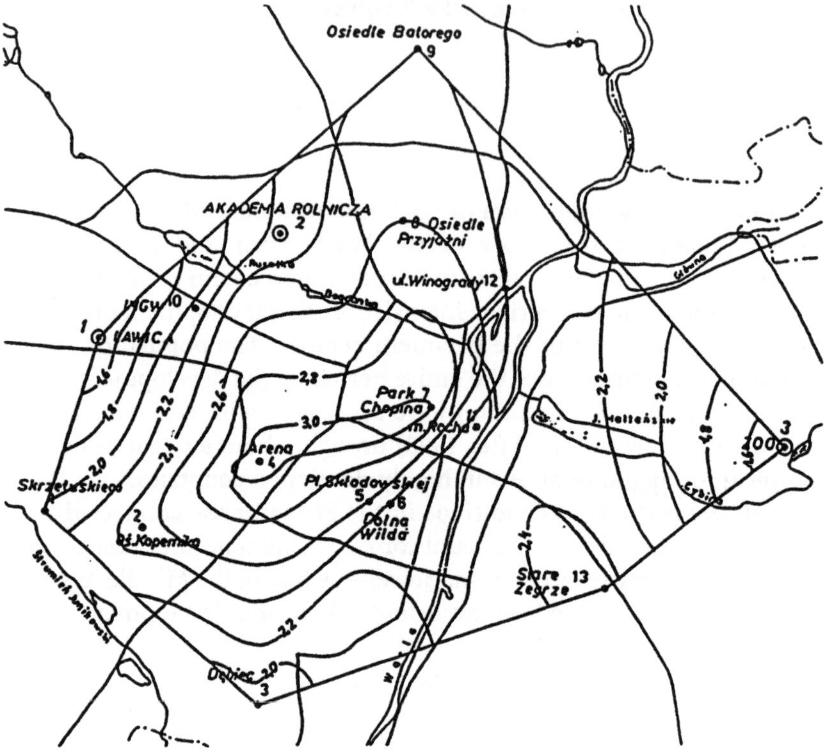

⊙ meteorological station

• point of the measurement

Fig.1. Mean air temperature - autumn, 7 a.m.

the effect of the observed changes in the climate are taken into account when planning future development of the city. Climatic conditions are now taken into consideration when building new sewage systems, constructing new housing estates as well as when modernising the old ones.

Reference

Kaniecki A (1993) The history of the city written with water. Aquarius, Poznan

Thermal Characteristics of Kolozsvár Urban Area

Robert Géczi[1] and Karoly Loerincz[2]

1 Introduction

The investigation of effects modifying climate in human settlements is a very important topic of urban ecology. During the recent decades the urban areas and the ratio of urban population have grown continuously so that the analysis of modifying effects and their physical explanation are required. On the other hand the city should be considered as a whole system, which includes not only buildings and streets but also infrastructure greenery and energy webs. This system functions on basic processes of energy flow. Our investigation is based on the modeling of the thermal characteristics of different built-up surfaces.

Kolozsvár is situated in the valley of the Szamos River, between 300 and 700 m above the sea level at the junction of three topographically different areas: the eastern part of Apuseni Mountains, the hillocks of Câmpia Transilvaniei and the hills of Podisul Somesan Plain.

2 Methods

The investigation supposes a common summer day with optimal meteorological conditions: this means a calm period which occurs after a cold front and cloudless sky. The steps are as follows:

1. Preparing the map. Removing the irrelevant details.
2. Defining the cities built-up environment index. Usually, two categories of urban areas are distinguished: open and built-up ones. According to these categories, a land-use classification of each 200 m square was adopted and scored: 2.5 for very densely built-up territory (concrete surfaces); 2.0 for districts without gardens; 1.5 for neighbourhoods with green patches; 1.0 for suburbs with gardens; and 0.0 if territories were not in use.

[1] JATE University, Department of Physical Geography, Po. Box 653, H-6701 Szeged, Hungary
[2] JATE University, Department of Experimental Physics, Dom ter 9., H-6720 Szeged, Hungary

J. Breuste, H. Feldmann, O. Uhlmann (Eds.)
Urban Ecology

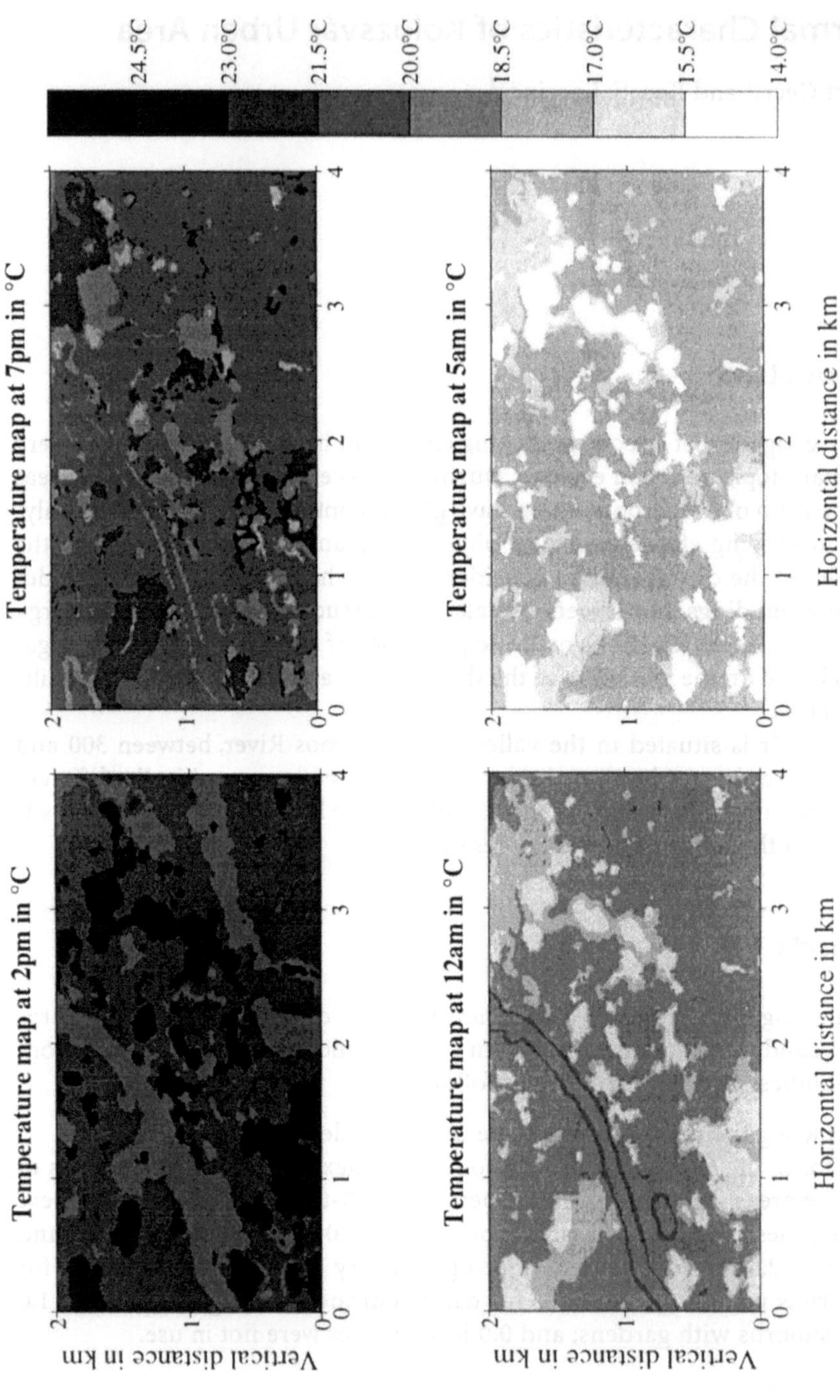

Fig.1. Spatial temperature distribution

3. Solving the differential equation for the thermal system:

$$\frac{d}{dt}(c\rho T(x,y)) + \frac{d}{dr}\left(\lambda \frac{d}{dr} T(x,y)\right) = f(x,y)$$

4. Plotting the results. Defining the coloured levels.

3 Results and Discussion

At 2 p.m. the spatial distribution shows the highest temperature in a part of the town centre and the territories which are built of concrete. The housing estates appear as the second warmest area of the town. Along the river the large mass of water moderates the temperature extremities, thus in afternoon it decreases the temperature. Cold areas appear where are mainly vegetated open spaces. Between the two extremities are situated the moderately warm territories like the suburbs, the neighbourhoods with green patches, open spaces with grass and the built-up areas with detached houses, especially those with vegetation. The map at 7 p.m. and 12 p.m. shows a relatively balanced situation. The warmest areas are the riverbank and around the esplanade. On the other hand the concrete and brick surfaces, where the vegetation is completely absent, turn cold. The temperature decreases step by step until 5 a.m. (Fig. 1)

Heavy Metals: Chemical Time Bombs in Urban Soils

Thomas Held

1 Introduction and Problem

Soils in conurbations are characterised by an accumulation of heavy metals. According to data given by Späte and Werner (1991) for North Rhine-Westphalia, allotments are a type of land use with an above average heavy metal load. Zn, Pb, Cu, As, and Cd are accumulated especially. Considerable amounts of the stored heavy metals can be released by a change in environmental factors which influence the Capacity Controlling Properties (CCPs) (Hekstra 1993), e.g. pH-value, organic matter content, redox-property and soil texture. Any change within this very complex system of CCPs, caused either by environmental factors and/or human impact, influences the sink function of soils in one or the other way. Pollutants which can be triggered by changing environmental factors are called Chemical Time Bombs (CTBs) (Stigliani et al. 1991). The extent of the human impact on the load and the CCPs of soils via land use must be investigated further.

In urban-industrial ecosystems the sink-function of soils is at risk due to an overload of different pollutants. The heavy metal load of allotments and town forests in the old-industrialised municipality of Witten, situated in the south-east of the Ruhr in North Rhine-Westphalia, Germany, was investigated. These two types of land use are established on loess soils, but are very different regarding the cultivation methods.

2 Results

The two land uses differ in amounts of important CCPs. Whereas urban forest topsoils accumulate higher amounts of organic matter than allotments, allotments have much better pH-values and CECs. The range of total soil-contents of heavy metals is shown in Table 1.

Institute of Geography, Ruhr University Bochum, D-44780 Bochum, Germany

J. Breuste, H. Feldmann, O. Uhlmann (Eds.)
Urban Ecology

Table 1. Minimum- and maximum values [mg/kg TS] of the background load (subsoils) and the topsoils

Land use/element	Cd	Zn	Pb	Cu
urban forest topsoils	0,1–0,2	56–78	98–127	14–20
urban forest subsoils	0,1	44–86	15–21	6–16
allotment topsoils ( ) = pasture	0,7–1,3 (3,6)	105–161 (719)	71–82 (188)	20–28 (185)
allotment subsoils ( ) = pasture	0,1–0,2 (0,5)	37–48 (132)	12–16 (23)	12–14 (20)

Table 2. The CTB-potential of topsoils in % of the total amounts

Land use/element	Cd	Zn	Pb	Cu
allotment topsoils	12–91	34–57	62–94	57–74
urban forest topsoils	<17–67	14–31	68–78	36–43

While the subsurface horizons of both land use systems indicate the local background load, the top soils of the allotments show an accumulation of Cd, Pb, Zn, and Cu. Despite the high accumulation factors, the level of contamination does not exceed the current critical limits or approximate values for soils (EWERS et al. 1994). Only the load of the river pasture is higher than the target values for soil cleaning, but is well within the existing threshold values.

To estimate the amount of reversible fixed heavy metals which can be triggered by changing environmental factors and/or land use, the soils were tested with a 0.05 m EDTA-solution. The results (shown in Table 2) show that considerable percentages of the total contents can be set free into other compartments of the environment.

3 Conclusions

Heavy metal extraction by using EDTA-solutions is an appropriate method for monitoring the CTB-potential of accumulated heavy metals in topsoils. Allotment top soils have excellent CCPs due to liming and the increased volume of the top layer. Even the comparatively highly contaminated meadow of the river Ruhr still has a sink function due to the pH-values and the good quality of the organic matter. The soils of urban forests, however, are completely acidified due to non sustainable land use and acid rain, and have lost the function to accumulate heavy metals, except for lead. Mobile heavy metals are bleeding into other compartments of the ecosystem.

The governing factor for the heavy metal inventory of soils and their sink function in urban ecosystems is land use as a control mechanism for other CCPs, which become land use related factors. The large accumulated amounts of heavy metals might be triggered by changing land use. For the purposes of urban development planning, regional planning and soil protection it is crucial to assess the CTB-risk caused by non-sustainable and/or changing land use.

References

Ewers U, Viereck L, Herget J (1994) Bestandsaufnahme der vorliegenden Richtwerte zur Beurteilung von Bodenverunreinigungen und synoptischen Darstellung der diesen Werten zugrundeliegenden Ableitungskriterien und -modelle. UBA-Text 35/94, Berlin

Hekstra GP (1993) Can Climate Change Trigger Non-Linear and Time-Delayed Responses to Pollutants Stored in Soils, Sediments and Ground Water? Mitt. Dt. Bodenkundl. Ges. 69: 267–276

Späte A, Werner W (1991) Erfassung und Auswertung der Hintergrundgehalte ausgewählter Schadstoffe in Böden Nordrhein-Westfalens. Landesamt für Wasser und Abfall Nordrhein-Westfalen, Materialien zur Ermittlung und Sanierung von Altlasten, Bd. 4, Düsseldorf

Stigliani WM, Doelman P, Salomons W, Schulin R, Smidt GRB & S.E.A.T., Zee VD (1991) Chemical Time Bombs: Predicting the Unpredictable. Environment 33(4)

On-Site Measurement Technique – a Useful Instrument for the Investigation of Environmental Pollution

Karin Huck

1 Introduction

The urban area is a very complex ecosystem with an immense, varying environmental impact on man, fauna and flora. Every organism has to endure different influences by numerous harmful substances that have may weaken individuals to greater or lesser extents. In cities there are many sources of such pollutants such as from industry, domestic emissions, traffic, waste water and also environmental pollution from hazardous waste sites.

The amount and type of pollution is crucial to the quality of life in urban areas. It is therefore important to analyse and to quantify this environmental influence to be able to provide answers about ecological conditions. The identification, quantification and modelling of contaminants and their distribution is a complex field of research because of the large number of possible parameters involved. This chapter reveals the Field Screening Method as an innovative way of capturing these environmental influences.

2 Discussion

The conventional procedure in assessing pollutants in the environment includes sampling, sample transportation to the laboratory, sample preparation and finally, the analysis and interpretation of analytical results. However, this method is time consuming and expensive to carry out (Gottlieb et al. 1997).

Field Screening, especially on-site measurement techniques, are an innovative way to avoid the disadvantages of the classical procedure. In Field Screening the sample does not have to be brought to the laboratory, rather the laboratory is brought to the area where sampling takes place.

University of Karlsruhe, Institute of Petrography and Geochemistry, Kaiserstrasse 12, D-76128 Karlsruhe, Germany

J. Breuste, H. Feldmann, O. Uhlmann (Eds.)
Urban Ecology

The instruments which are used are, on the one hand, conventional measurement techniques (e.g. gas chromatography, infrared spectroscopy, etc.) and, on the other hand, technical innovations such as sensor techniques, immunoassays, etc.. These instruments must, however, first be modified for use in the field.

The equipment has to be robust (insensitive to shock, humidity and variation in temperature, etc.), miniaturised, portable or mobile and independent of power supply. The instruments also should be easy to handle and should be able to detect a great number of different pollutants. Furthermore, on-site evaluation of the data must be possible.

On-site analysis avoids mistakes which may occur during the transportation of the sample to the laboratory (e.g. the loss of volatile organic compounds). Due to the rapid evaluation of results, sampling can be brought into line with the particular on-site conditions.

As mentioned above, for investigations of urban systems it is necessary to have a complete description of the quantity and distribution of the harmful substances. This is often only possible if a dense sampling grid is placed on the sector of interest, especially if a great variation of pollutants occurs in the area.

These investigations are generally laborious and often fail as they require much time and money. With an easy, economical and fast analysis, like on-site-analysis, it is possible to collect a larger number of samples and therefore obtain a more detailed grid of the field.

The chosen instrument depends on the objectives of the investigation and on the individual on-site conditions. In order to find an adequate description of the reality it is necessary to use an analytical method (or a combination of different methods) to obtain the 'right' number of data of the 'right' quality.

As a first step it is often sufficient to use an economical screening method (like Test Kits, etc.) to obtain a trustworthy statement about the presence or absence of environmental pollutants. Samples which are free of contamination need no further analysis, only contaminated samples require further investigations with costly measurement techniques. This strategy reduces costs because only the points of interest have to be investigated.

3
Case Studies

In the USA, the market of Field Screening techniques increased in the last few years (Hansen 1997). The Environmental Protection Agency (U.S.EPA) has run a biannual conference since 1989 on the topic, 'Field Analytical Methods for Hazardous Wastes and Toxic Chemicals'.

Also in Europe, Field Screening is a young and rapidly developing interdisciplinary subject and on-site measurement techniques are available for a number of applications (Huck 1996). In Germany, several field comparison test series were initiated by the Landesanstalt für Umweltschutz of Baden-Württemberg (State Environmental Department) to check the suitability of various on-site measurement techniques for use in the field (Landesanstalt für Umweltschutz Baden-Württemberg 1996 a, b and 1997, Maurer et al. 1997). Furthermore, a series of international conferences on the topic 'Field Screening in Europe' is planned to be held in Karlsruhe, Germany, to discuss the state of the art and to define future aims of European activities for Field Screening methods and strategies (commenced on September 29, 1997).

References

Gottlieb J, Huck K, Maurer A (1997) Field Screening - neue Methoden und Strategien bei der Umweltanalytik vor Ort. GIT Laborfachzeitschrift 1

Hansen P (1997) The U.S. EPAs Environmental Technology Verification Program. Proceedings: Field Analytical Methods for Hazardous Wastes and Toxic Chemicals, Las Vegas

Huck K (1996) Literaturstudie Vor-Ort-Analytik. Texte und Berichte, Landesanstalt für Umweltschutz, Baden-Württemberg 28/96

Landesanstalt für Umweltschutz Baden-Württemberg (1996a) Vergleichende Prüfung von Vor-Ort-Analytik-Geräten in Sinsheim am Modellvorhaben ehemalige Fa. Reinig. Texte und Berichte zur Altlastenanalytik 23/96

Landesanstalt für Umweltschutz Baden-Württemberg (1996b) Vergleichende Prüfung von Vor-Ort-Analytik-Geräten in Rastatt bei der MVG. Texte und Berichte zur Altlastenanalytik 27/96

Landesanstalt für Umweltschutz Baden-Württemberg (1997) Vergleichende Prüfung von Vor-Ort-Analytik-Geräten in der Neckartalaue Stuttgart. Texte und Berichte zur Altlastenanalytik 34/97

Maurer A, Eitel W, Kern F (1997) Results and conclusions of the study: applicability of on-site analysis and sensor technique. Proceedings of the first international conference "Field Screening Europe", Kluwer Academic Publishers

Feasibility Study on Effective Utilization of Sewage Heat in Urban Area with GIS

Toshiaki Ichinose, Hiromitsu Kawahara, Keisuke Hanaki, Takemi Ito, and Tomonori Matsuo

1 Introduction

In an urban area, materials such as products, water and energy are inputs which are emitted in the form of garbage, sewerage water, exhaust gas and exhaust heat as the results of urban activities (Hanaki and Ichinose 1995). This procedure could be called "urban metabolism", likened to the metabolism of a creature. The reduction of environmental load caused by urban activities is, therefore, optimal for the urban metabolic structure. Nowadays heat energy for space heating or hot water supply can be collected from sewerage water by heat pumps and from garbage incineration plants; the effective use of these kinds of heat energy closes the energy flow in urban areas and reduces the environmental load and the wastage of resources.

In the case of the heat supply from a garbage incineration plant, supply of high quality energy is possible through co-generation, as its heat source is very hot compared with the case of sewerage water. The area for heat supply is, however, limited, as is the location of the incineration plant itself. In contrast there are many opportunities for heat collection and supply in the case of sewage heat when the network of sewer lines is as dense as in Tokyo. Such utilization of sewage heat requires quantitative conformity between heat demand and waste heat supply. Geographic Information Systems (GIS) are an effective tool for searching for the suitable area on the map with the use of certain rules.

This chapter reports on the study on the location of the district heating systems for effective use of sewage heat in urban area. In this study GIS was developed as the analysing tool and the optimized location plans were discussed in Tokyo on the basis of the GIS analyses of spatial conformity between heat demand and waste heat supply. The author estimated the temporal and spatial structures of energy consumption in Tokyo on the basis of a very precise digital geographic land use data set containing the number of build-

Center for Global Environmental Research, National Institute for Environmental Studies, 16-2, Onogawa, Tsukuba 305, Japan

J. Breuste, H. Feldmann, O. Uhlmann (Eds.)
Urban Ecology

ing stories at each 25 m grid point; this is input as basic data into the GIS. The characteristics of the relationships between demand and supply in each sewage treatment district and geographical influences were also discussed.

2
"Simulation Model of Sewer System" as GIS

"Simulation Model of Sewer System (SMSS)" developed for this study possesses the raster data of land use and number of building stories, the polygon data of the co-ordinate system of sewer lines and the functions for calculations and mappings; thus it has the characteristics of a GIS. Land use (business) in each 250 m grid cell, areal energy consumption and areal heat demand for each business are input into the SMSS. Calculation of heat demand in each 250 m grid cell and flow of sewerage water and collectible sewage heat at the node point of sewer lines are possible using SMSS. In the following discussion heat demand is limited to household and commercial sectors.

With SMSS, collectible heat was calculated at several node points of sewer lines in several sewage treatment districts (Fig. 1) and the optimized location plans of district heating systems (heat pumps) in these districts were dis-

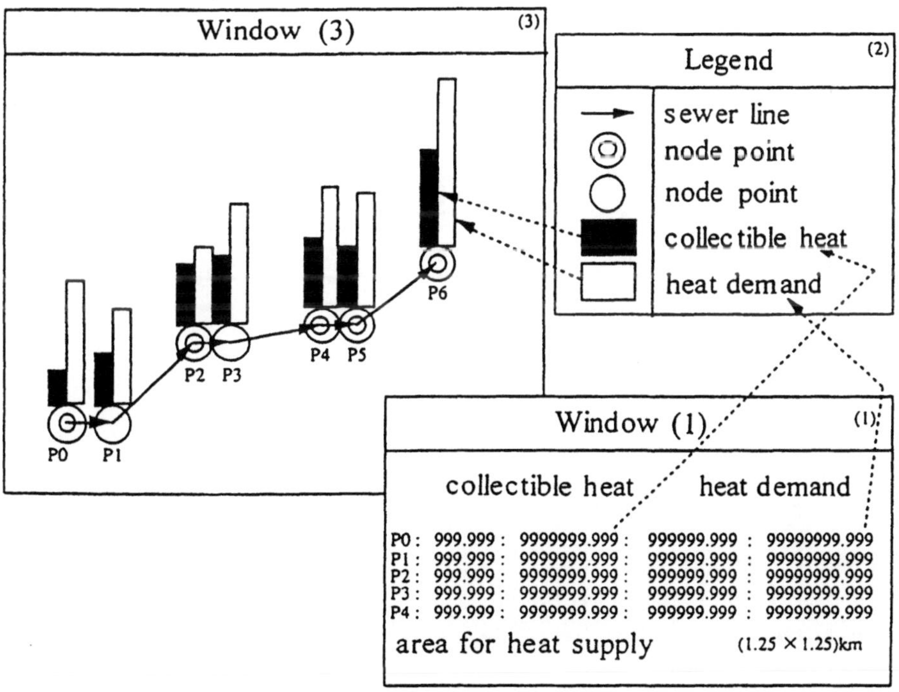

Fig. 1. Display of Simulation Model of Sewer System

cussed. The gross amount of usable heat along a sewer line is maximized in the optimal plan.

In the case of identifying one location on a sewer line, the site where the difference between collectible heat and heat demand is at a minimum could be chosen. In the case of identifying more than two places, the reduction of heat by heat collection upstream should be considered.

3 Concluding Remarks

The characteristics of the relationships between heat demand and waste heat supply in each sewage treatment district and geographic influences were discussed. Large amounts of usable heat are shown in sewage treatment districts with long sewer lines where there are many possibilities for heat collection and a large demand for heat. GIS can be used to analyse the optimal location of district heating systems.

Reference

Hanaki K, Ichinose T (1995) Relief of Impacts of Urban Activities on Global Environment. In: Murai S (eds.) Toward Global Planning of Sustainable Use of the Earth. Elsevier Science B.V.

Anomalies of Heavy Metals in Soils of Urban Landscapes

Sergey V. Kakareka and Valery S. Homich

1 Objectives

Revealing of heavy metals anomalies in soil cover of Belarusian cities, determination of sources of soil pollution and valuation of completeness of the account of heavy metals emissions.

2 Methods

One of the most important indicators of the urban environmental condition is the level of soil contamination by heavy metals (HM). For an ecological estimation of a condition of environment data on the absolute contents of heavy metals in soils and their ratio with ultimate acceptable levels are usually used. However these data is not enough for diagnostics of the factors and sources of environment pollution. Revealing of geochemical anomalies of HM – haloes with concentration, which is reliably exceeding background values in view of their background variations are necessary.

In this article results of work on revealing anomalies of heavy metals in soils of Gomel (which is the second on a population and industrial output Belarusian city) are stated. As information base the data of pedogeochemical sampling of the city territory (approximately 700 samples of the surface horizon of soil) have served. This work was a part of Territorial Integrated Scheme of Nature Protection elaboration. Main sources of pollutants of the city environment are road transport and such industrial activities, as machine-building, chemical industry etc.

As a criterion of anomality for allocation of anomalies minimum abnormal concentration has served, which has been calculated on the formula:

$$Ca = x(b) + 3s/m,$$

Institute for Problems of Use of Natural Resources & Ecology, 10 Staroborysovski tract, 220114 Minsk, Belarus

J. Breuste, H. Feldmann, O. Uhlmann (Eds.)
Urban Ecology

where x (b) – average background contents, s – average background standard deviation, m – number of correlated samples. Thus belonging of samples to anomaly is determined not only of an element concentration in it, but also in adjacent points of sampling. Abnormal concentration for one (Ca = x + 3s), two (Ca = x + 2s) and nine (Ca = x + s) correlated samples were defined. The points with abnormal concentration were united in anomaly on an external contour of minimum abnormal concentration in view of number of correlated points.

3 Results

On a number of points with abnormal concentration to main accumulating elements belong Pb, Cr, Cu and Zn. Local anomalies also other heavy metals (Hg, Cd) and microelements form (also such specific as Y, La, Sr).

The abnormal concentration of mercury were uncovered in 8% of samples; areal anomalies of this element were not revealed.

The abnormal concentration of lead were uncovered in 33% of general amount of samples. The main part of it can be incorporated in five areal anomalies.

The anomalies of zinc were revealed in 28% of samples. Structure of soils pollution by this element is rather simple. Large anomaly in eastern part of city includes the zone of individual building up, and also territories of enterprises Gomselmash, Self-propelled Combines Plant and some others.

Structure of soils pollution by copper is rather mosaic. Abnormal concentrations were revealed in 39% of samples. However there are no large intensive anomalies; a few small nucleuses of pollution have a common rather extensive peripheral part. Such character of anomalies testifies to presence of several rather small sources of copper emission.

The abnormal concentration of chromium was uncovered in 34% of samples. The main part apportioned to large anomaly, which cover industrial zone in western part of the city, where the contents of chromium in soils amounted >205 mg/kg.

The analysis of genesis of anomalies has shown, that on the occupied area anomalies of aerial origin prevail. They are formed in direct affinity from enterprises-emitters as a consequence of deposition of large aerosol particles. Near a powerful source they are symmetric and comparatively great, at a weak – consist of separate abnormal points. On these anomalies haloes of pollution of other types are frequently imposed. The most characteristic anomalies of such type are anomaly of lead in central part of Gomel city, caused by vehicle emission, and chromium anomaly in western part, connected with emission of Centrolit. Characteristic anomalies are formed on territory of land-fills, treatment facilities, and also in places of garbage burning on vacant lands.

They are frequently rather intensive, however do not occupy significant areas. A variety of kinds of wastes has caused wide spectrum of chemical elements contained in them. In comparison with regional background value, solid wastes, storaged on the city land-fill, are enriched with copper in 65 times, with lead in 31 times, with chromium in 15 times, with nickel in 11 times and so on.

Hydrogenic anomalies are formed on geochemical barriers on a way of pollutants migrations with water flows (in particular, in places of a crude storm drain disposal). They are limited on the area and often not revealed during regular sampling.

Anomalies were marked which can not be interpreted as the result of technogenic influence: close to them enterprises are not present, attributes of pollution with wastes are away. The part of such anomalies can be interpreted as the consequence of geochemical heterogenity of the city soils.

4 Conclusion

The official account of emissions can interpret only pollution of soils with lead (more than 90% of its emisson is stipulated by vehicles). On the other elements the account is rather partial or lack of. On the basis of soil contamination a number of sources of pollution, not discounted by statistics have been revealed. The revealed structure of soil contamination with heavy metals, which reflects long-term level of air pollution, can be used for retrospective monitoring of urban environment and for optimisation of urgent monitoring system.

Use of Lichens for the Indication of Air Pollution Dynamics in Cities

Sergey V. Kakareka and Ludmila A. Kravchuk

1 Objectives

The objectives are:

- to reveal the spatial structures of epiphytic lichens in the territories of Belarusian cities and investigate their dynamics,
- to evaluate the relationship of these structures with the dynamics of air pollution and emissions.

2 Methods

Epiphytic lichens are widely used as the indicators of atmospheric air pollution in ecological practice. The experience of use of lichens as indicators on the basis of floristic parameters has revealed significant complexities in the interpretation of the spatial structures of epiphytic lichens in large cities. One of the main reasons is the insufficient degree of investigation of lichens' reaction to certain pollutants and their complexes in real city conditions. Scales of tolerance of species, which are widely used for ecological interpretation of information contained in composition of lichens sinusia, were developed without an account of the concentration of particular pollutants and their dynamics because of the high cost of such investigations. Monitoring of epiphytic lichens in cities can be an effective and cheap method of increasing the quality of pollution indication.

We carried out monitoring of epiphytic lichens on a two-level system. On the first level mapping of lichens (repeated at certain intervals) was used with a rather dense network of constant sampling plots without rigid binding. The second level consisted of a few rigidly fixed sampling plots described in detail, and connected with atmospheric air pollution control stations.

Institute for Problems of Use of Natural Resources & Ecology, 10 Staroborysovski tract, 220114 Minsk, Belarus

J. Breuste, H. Feldmann, O. Uhlmann (Eds.)
Urban Ecology

In 1993–1996 such networks were set-up in Minsk, Gomel, Mogylev and Svetlogorsk. Constant sampling point on the first level were selected with an interval about 1 km. The description of epiphytic lichens, with differentiation of species, was carried out on trunks of 10 trees of suitable perimeter (no less than 60 cm) with the use of a frame (10 × 25 cm) from 2 sides at a height of 130 cm. For each site lichen cover was calculated, the number of species, indexes of poleotolerancy and air pollution and appropriate maps were drawn.

Plots of the second level of observation were incorporated close to the Committee on Hydrometeorology air pollution monitoring stations, where systematic supervision is supposed to be carried out on the dynamics of changes in lichens sinusia connected with the supervision of air pollution.

3 Results

Data obtained during investigation of the first level network in Minsk in 1993 were compared to results of mapping of epiphytic lichens in 1984 (Shucanov et al. 1986). The investigation in 1984 revealed (at 50 sampling points) 35 species of epiphytic lichens. During the repeated mapping in 1993 (on 177 sampling points) 29 species have been revealed. In 1993 some species sensitive to pollution were not found -*Cetraria pinastri (Scop.) Rohl., C.chlorophila (Wielld.) Vain., Parmelia saxatilis (L.) Ach., Evernia furfuracea (L.) Mann.*, and tolerant species were also not found- for instance *Lecanora carpinea (L.) Vain.* Species not discovered during previous investigations were found in 1993 - *Candelariella vitelina Mull., Caloplaca pyracea (Ach.) Nyl.*

The results of the 1984 investigation showed that a main part of the city was characterised by strong air pollution: zones 1 (index of poleotolerancy IP > = 9.5) and 2 (IP- 8.92–9.49). Only the outskirts of the city were characterised as zone 3 (IP 8.34–8.91). Large parks (Pobedy, Gorky, Cheluskitzev) held zones 4 (IP 7.76–8.33) and 5 status (IP 7.18–7.75). The purest sites -zone 6 (IP - 6.6–7.17) occupied an extremely insignificant area.

The most essential difference of results of the investigation in 1993 from the previous one is the reduction of IP from the average 1–1.5. Accordingly there is an absent zone with levels of IP > 9.5 and almost none with levels of 8.92–9.49. The basic territory of Minsk now occupies zones 4 and 5. Besides the loss of separate zones, other areas have the configuration of zone borders.

The analysis has shown that the period between mappings is characterised by the tendency of volumes of emissions to reduce. The most essential changes are in sulphur dioxide emission (with 72,300 tonnes in 1981 up to 35,900 tonnes in 1984 and 17,900 tonnes in 1992). It is known that, for lichens, sulphur dioxide is one of the most toxic components. Therefore the interpretation of the reduction of values of an index of poleotolerancy (i.e. relative increase of abundance of a less tolerant species) as a consequence of a reduction in air pollution is quite well-grounded. This is in spite of the fact that the supervision in stations does not reveal precise trends of a level of sulphur dioxide air pollution; one of the reasons for this is that in 1991 the technique for determination of sulphur dioxide in the air changed which complicates the analysis.

4 Conclusion

Repeated investigations of epiphytic lichens in cities can be used to reveal the tendencies of rather rapid changes of a level of air pollution. It is especially valuable where there is a poor network of air monitoring stations which are unable to reveal a distinct trend in air pollution and where the technique for determining pollutants has changed. Mapping of epiphytic lichens in cities at certain time intervals can be an important element of urban environmental monitoring.

Reference

Shucanov A.S., Rykovski G.F., Antonov G.N., Maslovski O.M. (1986) Indicatsija zagrjasnennosti atmosfernogo vozdyha v g. Minske s pomoschju lishajnikov i mohoobraznyh. Vestnic BGU N2: 36–41

Vegetation as a Sink for PAH in Urban Regions

Achim Kuhn, Hans-Joachim Ballach, and Rüdiger Wittig

1 Introduction

Polycyclic aromatic hydrocarbons (PAH) play a significant role in the assessment of air quality in cities. The main source of air pollution caused by PAH in urban areas is motor vehicle traffic. The aim of this study is to investigate the accumulation of PAH in plant leaves in order to estimate the role of vegetation as a sink for PAH in urban regions.

2 Materials and Methods

3-year old poplar cuttings (*Populus nigra* cv. Loenen) were grown close to a main road in Frankfurt/Main, Germany which carries 60,000 vehicles per day. Leaf samples were collected in May, July, and October 1996. (For leaf sample extraction refer to Kuhn et al. 1998). The air concentrations of particle-bound and gaseous PAH were measured over one growing season (Niehaus et al. 1990).

3 Results and Discussion

The PAH composition in ambient air (PAH-fingerprint) is shown in Figure 1. Although the PAH concentrations in ambient air fluctuate by a factor of 3, the relative composition remains almost constant. The PAH-fingerprint is typical for a location with a high density of motor vehicle traffic (Khalili et al. 1995).

In the leaves of the test plants, the PAH-fingerprint of the investigated PAH differs significantly from that in ambient air (Fig. 2):

Department of Geobotany and Plant Ecology, Johann Wolfgang Goethe-University Frankfurt/Main, Siesmayerstraße 70, D-60054 Frankfurt/Main, Germany

J. Breuste, H. Feldmann, O. Uhlmann (Eds.)
Urban Ecology

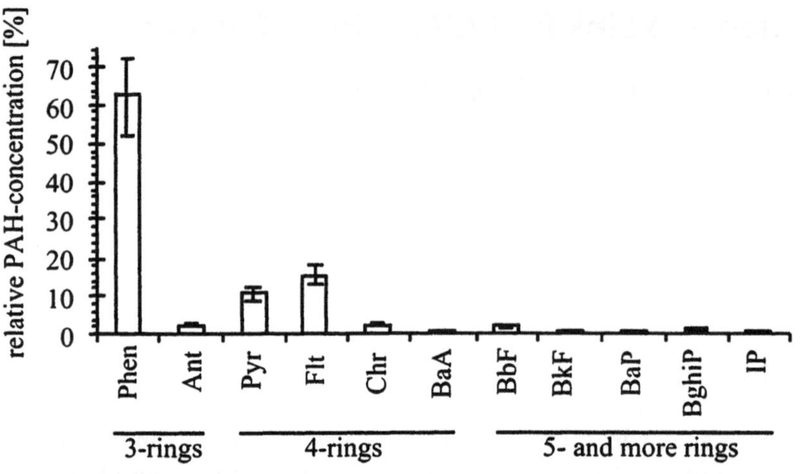

Fig. 1. Fingerprint of PAH in ambient air. 100% = Sum of all gaseous and particle-bound PAH. Values averaged from 24 single measurements from 24th April to 8th October 1996. Phen = phenanthrene; Ant = anthracene; Pyr = pyrene; Flt = fluoranthene; Chr = chrysene; BaA = benzo[a]anthracene; BbF = benzo[b]fluoranthene; BkF = benzo[k]fluoranthene; BaP = benzo[a]pyrene; BghiP = benzo[ghi]perylene; IP = Indeno[1,2,3-cd]pyrene

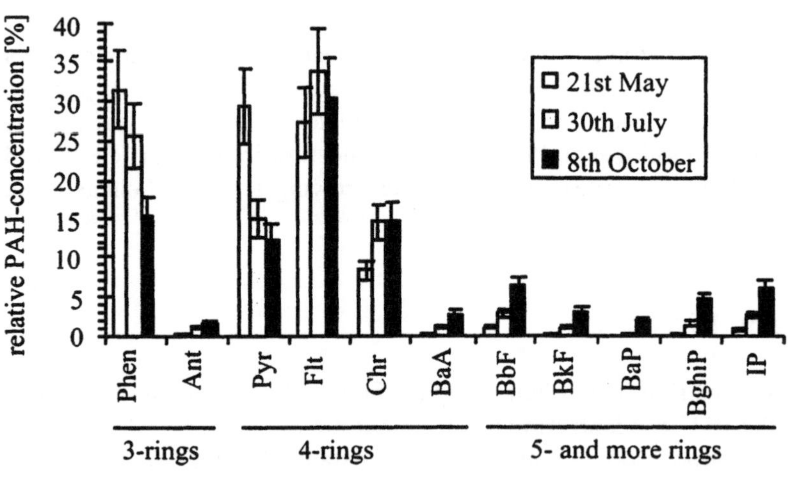

Fig. 2. Fingerprint of PAH in poplar leaves. 100% = Sum of all PAH. Values are averaged from 2 single measurements per sampling date

1. The PAH-fingerprint in the leaves changes during the growing season.
2. The relative content of the 3-ring PAH in the leaves is considerably lower than in ambient air and decreases in the course of one growing season (from 32.0% in May to 16.9% in October).

3. The relative content of 4-ring PAH in the leaves is considerably higher than in ambient air and only slightly decreases in the course of one growing season (from 65.6% in May to 60.6% in October).
4. The content of higher molecular PAH with 5- and more rings in the leaves rose from 2.4% in May to 22.7% in October.

The vapour pressure is the main factor controlling the accumulation of PAH in plant leaves. In comparison to the surrounding atmosphere, the content of PAH with a lower vapour pressure was greater in the leaves, whereas the content of PAH with a relatively high vapour pressure was lower.

Due to the lipophilic properties and large surface areas of plant leaves, and the fact that the most important PAH from an ecotoxicological point of view has a low vapour pressure, vegetation functions as an important sink for PAH and plays a significant role in the improvement of air quality in cities.

References

Khalili NR, Scheff PA, Holsen TM (1995) PAH Source Fingerprints for Coke Ovens, Diesel and Gasoline Engines, Highway Tunnels, and Wood Combustion Emissions. Atmos Environ 29: 533–542

Kuhn A, Ballach H-J, Wittig R (1998) Seasonal Variation of the Distribution of PAH in Poplar Leaves. Fresenius Environ Bull 7: 164–169

Niehaus R, Scheulen B, Dürrbeck HW (1990) Determination of Airborne PAH using a Filter/Adsorber Combination. Sci Total Environ 99: 163–172

Estimation of Potential Nitrogen Pollution from Urban Soils

Andreas E. H. Lehmann[1] and Karl Stahr[2]

1 Introduction

The examination and understanding of urban soils is a topic of primary imp ortance in aspects of sustainable city development because soils in urban areas vary the environmental quality in cities. The behaviour of nitrogen in urban soils was investigated (for the above reason) and an evaluation key was developed. The potential pollution due to nitrogen transfer from urban soils in humid temperate regions could then be estimated (Stahr et al. 1997). Published evaluation keys (AG Bodenkunde 1994 and AK Stadtböden 1997) as well as personal experience and personal studies were used to estimate the losses of nitrogen from urban soils. In situ incubations (3 sites, 9 horizons during 1994–1996) and laboratory incubations (Zöttl 1958) were carried out to assess the influence of temperature, soil structure and parent material on the potential risk.

Each curve of potential nitrogen mineralisation shown Table 1, p. 175 characterises a class of the main factor 'parent material' which can be easily differentiated. The number of points reached by the abbreviated evaluation key leads to the five class evaluation. Two special classes (A, I) qualify for a risk classing independent of the number of points allocated. In the case of mixed parent material the key should be used according to the portions of the different materials. The described estimation key is a proposal which should be improved by discussion and further examinations.

2 Evaluation of Factors

1. Parent Material
   The potential nitrogen mineralisation of typical soils from different parent materials is shown in Table 1, p. 175.

[1] Friedrich-Schiller-University of Jena, Department of Geosciences, Burgweg 11, D-07749 Jena, Germany
[2] University of Hohenheim, Department of Soil Science and Ecology, D-70593 Stuttgart, Germany

J. Breuste, H. Feldmann, O. Uhlmann (Eds.)
Urban Ecology

Table 1. Potential nitrogen mineralisation of typical soils from different parent materials

<table>
<tr><th>Criteria</th><th>Potential Nitrogen Mineralisation of a Typical Soil</th></tr>
<tr><td>1) Soils with >30 cm depth consisting (mainly) of
a) sewage sludge or harbour sludge
b) fresher household-waste with more than 80% earth
2) N-disposal-area
4 points</td><td>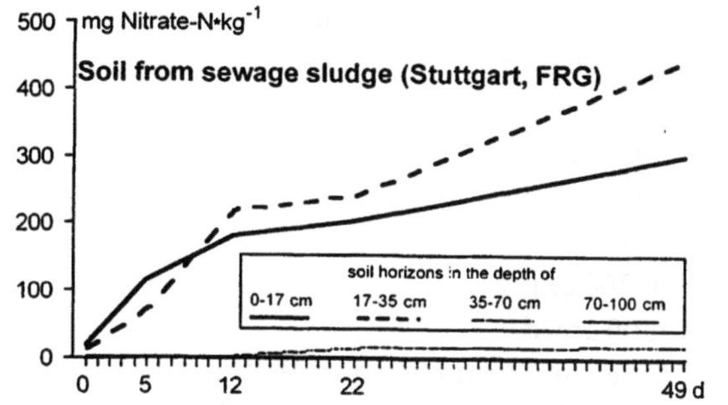
</td></tr>
<tr><td>Soils with toposils with contents higher than 5% organic carbon and having a clay content below 50%
3 points</td><td>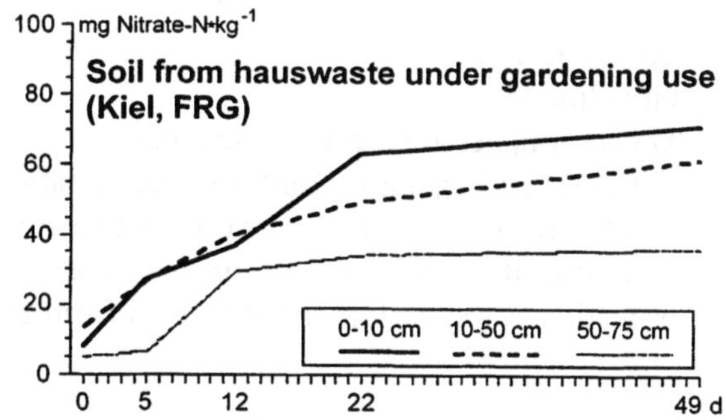
</td></tr>
<tr><td>Soils with contents of organic carbon below 1% in each horizon or permanently dry or frozen soils or with nitrification inhibiting contamination up to a depth of at least 30 cm
class I</td><td>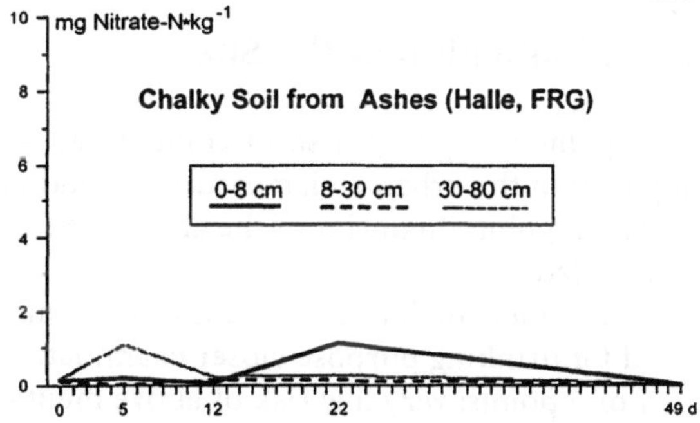
</td></tr>
<tr><td>Soils showing other features
2 points</td><td>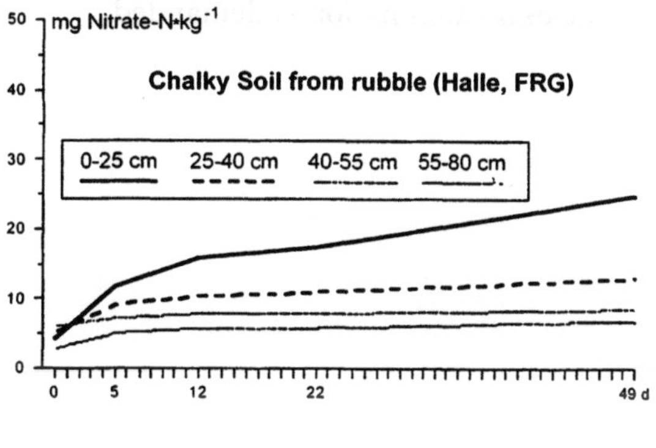
</td></tr>
</table>

2. Leaching	
Soil moisture balance	
if yearly soil moisture balance < field capacity	1 point
if yearly soil moisture balance < 100 mm	1 point
100–200 mm	2 points
200–300 mm	3 points
>300 mm	3 points
Conductivity	
<10 cm/d	1 point
10–300 cm/d	3 points
>300 cm/d	5 points
Stone content more than 10%	3 points
Lateral flow from adjacent soils	A
3. Local Climate	
city climate	1 point
4. Vegetation, Fertilising and Land use	
(Almost) no vegetation and no beaten path	A
Productive vegetation and young vegetation is moved from the site or beaten path or any other vegetation	−1 point
N-fertilised vegetation is not moved from the site	1 point

3 Final Evaluation of the Site

>12 points: very high risk of contamination of the groundwater by nitrate input from the urban soil, measures should be taken

10–12 points: high risk of the above mentioned type, closer examination is demanded

7–9 points: moderate risk of the above mentioned type, if groundwater is used for drinking purpose closer examination is required

I or 6 points: very low risk of above mentioned type

A: closer examination is demanded.

References

Stahr K, Lehmann A, Holland H (1997) Anionen. In: Blume H-P, Schleuß U (eds): Bewertung anthropogener Stadtböden. Schriftenr. Inst f Pflanzenernährung u Bodenkunde Univ Kiel 38, pp 66–100

AG Bodenkunde (1994) Bodenkundliche Kartieranleitung. E. Schweizerbart'sche, Stuttgart

AK Stadtböden (1997) Empfehlungen des Arbeitskreises Stadtböden für die bodenkundliche Kartierung. Sekretariat für Bodenbewertung, Kiel.

Zöttl H (1958) Die Bestimmung der Stickstoffmineralisation im Waldhumus durch den Brutversuch. Z Pflanzenernähr Bodenk 81, pp 35–48

Environmental Quality Monitoring in Urban Territories with Reference to Riga

Nataliya Matoshko, P. Berg and D. Veretennikov

1 Introduction

Riga is the capital city of Latvia and houses a third of the total population of the country; it is the main industrial centre and transport junction of the Republic. The shoreline of the Gulf of Riga forms the city's border in the north. The Daugava river flows throughout the city, from south to north. There are 2 lakes, Jugla and Kishezers, within the confines of the city.

The pollutant emissions from stationary sources measure approximately 16 thousand tons per year (15% of the total for the Republic), including 51% SO_2, 11% NO_x, 24% CO and 9% particulate pollutants.

The annual water consumption is 221 mil.m^3, with 57% being supplied by the Daugava river and the Jugla. Over recent years, a total of 182 mil.m^3 of waste water has been discharged per year.

2 Environmental Quality Monitoring

Environmental monitoring has been carried out by the national observational networks since the 1960–70's which provide observations in ambient air and surface water (by hydrochemical and hydrobiological indices).

The main principals of observations are: comprehensiveness and regularity, implications for simultaneous pollution, meteorological and hydrological measurements, and the functioning of common methods.

Air pollution observations due to human activity are carried out at 5 posts. Measurements cover the main pollutants – nitrogen dioxide, sulphur dioxide, particulate pollutants (TSP), carbon monoxide – and specific substances – formaldehyde, ammonia, phenols, hydrogen chloride, hydrogen fluoride, metals, etc.

Surface water control is being carried out at 4 measuring stations in Daugava and 2 measuring stations on each of the lakes. Samplings in the water

Latvian Hydrometeorological Agency, 165 Maskavas str., LV-1019, Riga, Latvia

J. Breuste, H. Feldmann, O. Uhlmann (Eds.)
Urban Ecology

bodies are made 6–10 times/year during the hydrological phases; oxygen, pH, salt composition, organic matter, nutrients, oil products, pesticides, heavy metals, hydrobiological (bacterio- and phytoplankton, benthos) and hydrological parameters are measured.

Characterisation of the status of the environment is made on the basis of the comparison analysis with the environmental quality norms; analysis of abrupt changes in the status of the environment; analysis of daily, monthly, seasonal and annual variations; establishment of long-term trends; complex air pollution potential, air pollution and water pollution indices.

2.1 Air Pollution Indices

For the status of ambient air the complex indices are used:

- I_4, Air Pollution Index, characterises contributions of the main pollutants (carbon monoxide, sulphur dioxide, nitrogen dioxide, TSP).
- APP, Atmospheric Pollution Potential, determines the frequency of occurrence of the 4 parameters that greatly affect the dissipation process: slow winds (0–1 m/sec), mist, surface temperature inversion (temperature increase with height) and stagnant air.
- Tendency in pollution level characterises the contributions of different pollutants to changes in urban pollution.

From the results obtained during the recent 5 years it may be concluded that at relatively stable total emissions and APP, changes in air pollution due to major pollutants were insignificant (Veretennikov et al. 1996).

The I_4 s analysis shows a fairly stable pollution level in Riga in 1990–1995 (for the main pollutants) with an increase in 1993–1994. Analysis of tendencies in pollution levels has revealed an upward tendency in nitrogen dioxide and ammonia and downward tendency in carbon monoxide (Lyulko 1997).

2.2 Water Pollution Index

The water pollution index (WPI) is derived from an equation that involves oxygen, biological oxygen demand, nitrite, oil products, metals.

According to WPI, the Riga's waters fall under Class II–III (conventionally clean – moderately polluted). Due to the curtailing of industrial production, some downward tendency has been observed in the hydrochemical pollution levels in Riga; the saprophyte number has decreased, but nevertheless remains high (Lyulko 1997).

3 Conclusion

- The observational system established in Riga allows complete information on environment quality to be obtained and defines tendencies in chemical and biological pollution necessary to make decisions on protection and rational utilisation of environmental resources.
- Analysis of air pollution indices shows a fairly stable pollution level in Riga in 1990–1995 (for the main pollutants) and increasing tendency of nitrogen dioxide and ammonia.
- According to the hydrochemical indices, the Riga city waters fall under Class II–III; judging from the saprobility data – under Class II–IV. Due to a reduction in industrial production some downward tendency has been observed in the pollution levels in Riga.

References

Lyulko I (1997) Environmental Pollution in Latvia. Annual Report 1996, Riga

Veretennikov D, Dubrovskaya R, Matoshko N (1996) Weather conditions and urban pollution (with reference to Riga). In: Gunlög Wennerberg (ed) European Conference on Applied Climatology. Norrköping, Sweden pp 180–181

The Role of Social Participants in the Communal Process of Climate Protection

Peter Moser

Local climate-protection has established itself in the last years as a matter of urgency in many European cities (e.g. Climate Alliance (Klima-Bündnis/Alianza del Clima), Local Agenda 21, Sustainable city). Although extensive studies of measures related to the technical and economic conception of climate-protection are already available, sociological and interdisciplinary questions concerning social inhibition and the role of participants in the process of climate-protection have hardly been explored at all. Within the framework of a thesis at the University of Osnabrück, a climate-protection model has been developed based on elements and concepts derived from environmental politics and from political field analysis which deals with a large number of related aspects and developments (see Fig. 1). The communal climate-protection process is made clear by means of analyses of basic structural elements. The analysis concentrates on the examination of the social and communal conditions for CO_2-reduction in the field of energy - especially through rationalisation of the use of energy and by means of regenerative forms of energy. For this purpose it was necessary to illuminate particularly the forms of action, the interests and influence of the participants in the process of communal climate-protection. In addition, the process of communication and the formation, function and importance of institutions for climate-protection and networks are examined. The role of climate-protection concepts, public relations and educational work as well as of participation are intensively analysed. The importance of establishing priorities and the difficulties involved in the aims of reduction of CO_2 are also considered. Following on from this (based on the case of Osnabrück) the structural framework, the beginning and the course of the processes of climate-protection, and the role of the participants - especially at the "Round Table CO_2 Reduction" - are reviewed. In conclusion, and as an essential recognition resulting from the research process, the hypothesis is proposed that the ecological crisis (especially the problem of climate change) brings about alterations in local political patterns.

University of Osnabrück, Dept. of Social Science, Am Turmhügel 14, D-49088 Osnabrück, Germany

J. Breuste, H. Feldmann, O. Uhlmann (Eds.)
Urban Ecology

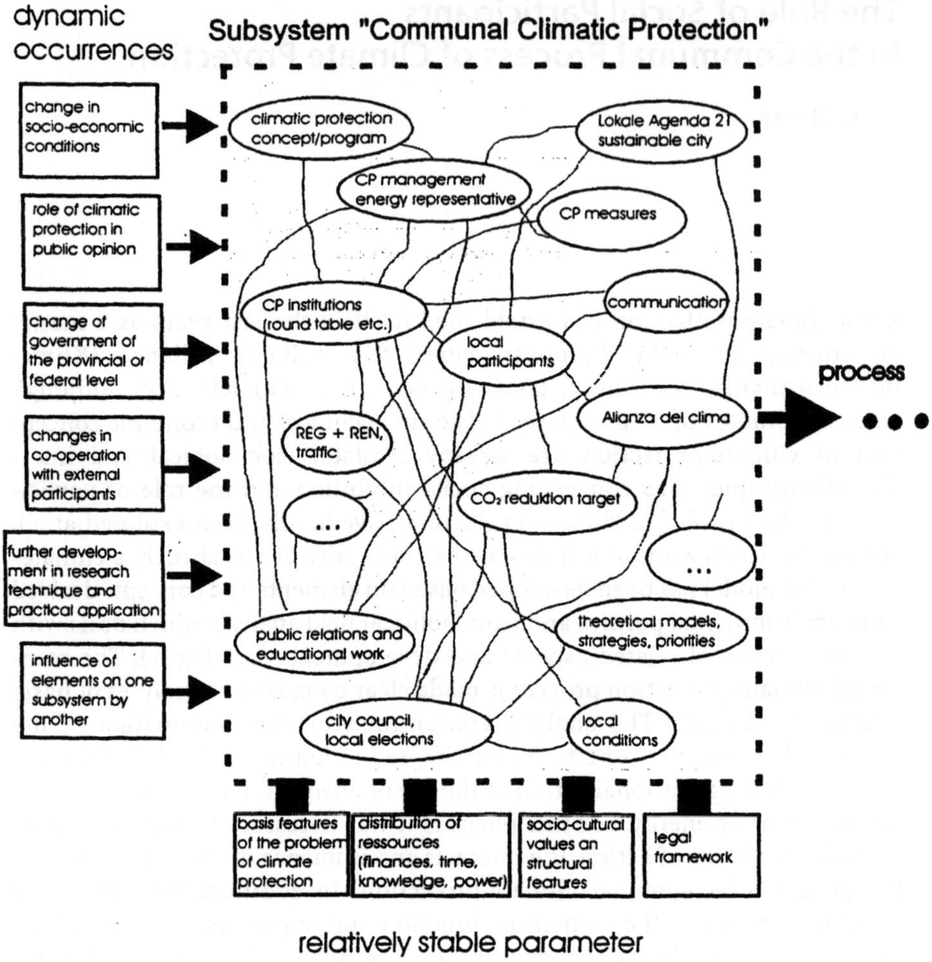

Fig. 1. A communal climate-protection model

Permeable Pavements – an Appropriate Method to Reduce Stormwater Flow in Urban Sewer Systems?

Birgit Münchow[1] and Michael Schramm[2]

1 Introduction

Impervious surfaces are of great environmental concern because of their disturbing influence on soil properties. The soil destruction has various consequences for the ecosphere. Impervious surfaces reduce the soil water content and the infiltration and storage capacity of the soil, as well as evapotranspiration. Therefore careful handling of all soil aspects is an important part of environmental protection. On the other hand, a sealed area causes accelerated runoff with higher peak rates thus the sealing of surfaces is a decisive factor in the design and construction of urban sewerage systems. Due to the lack of reliable data on the infiltration and runoff rates of permeable paved areas, an investigation of the permeability of different pavements was carried out in 1994, supported by the Department for Environment and Geology of Saxony.

2 Permeability of the Pavements

A variable intensity portable rain simulator was used on six urban car-parks and footpaths to determine the infiltration of stormwater events. The car park areas were covered by concrete block pavements with grass joints, concrete grid pavements with grass, and block pavement with drainage openings. The pedestrian areas were paved with bulk material (compacted sand clay surfaces with different content of fine grains) and a sand surface with a synthetic resin binder. The study sites, located in the city of Leipzig, differ in traffic volume and construction. Two ranges of rainfall intensity were tested for each pavement; the first experiment was conducted for an hour with 30–35 mm/h (dry run) and the second for half an hour with 70 mm/h (wet run). Figs. 1–4 illustrate the change in the infiltration rate for the investigated pavements.

[1] UFZ Centre for Environmental Research Leipzig-Halle, Dept. Urban Landscapes, Permoserstraße 15, D-04318 Leipzig, Germany
[2] IWAB – Consultants for Water, Soil, and Waste Management, Marie-Alexandra-Straße 27, D-76135 Karlsruhe, Germany

J. Breuste, H. Feldmann, O. Uhlmann (Eds.)
Urban Ecology

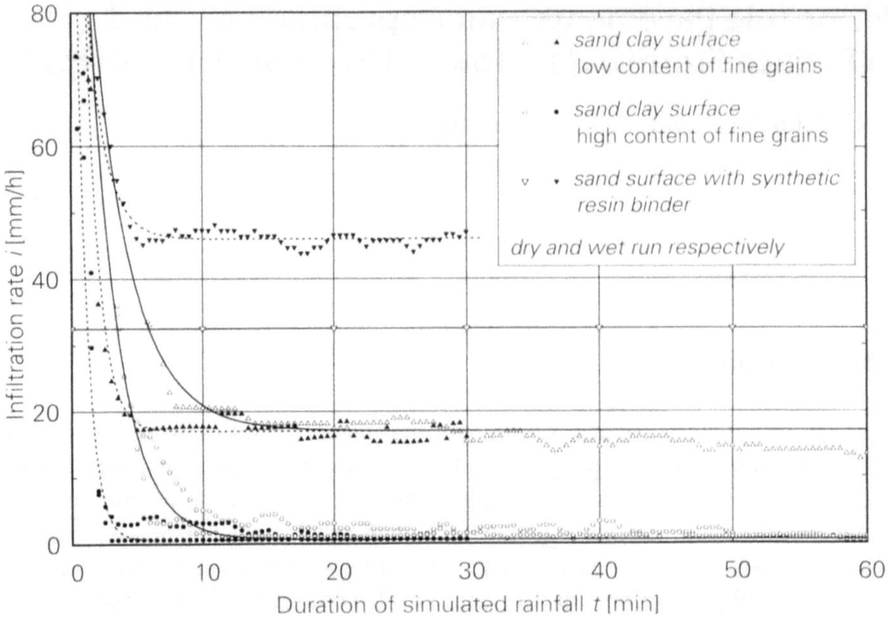

Fig. 1. Change in infiltration rate i(t) by compacted sand clay surfaces and sand surface with synthetic resin binder

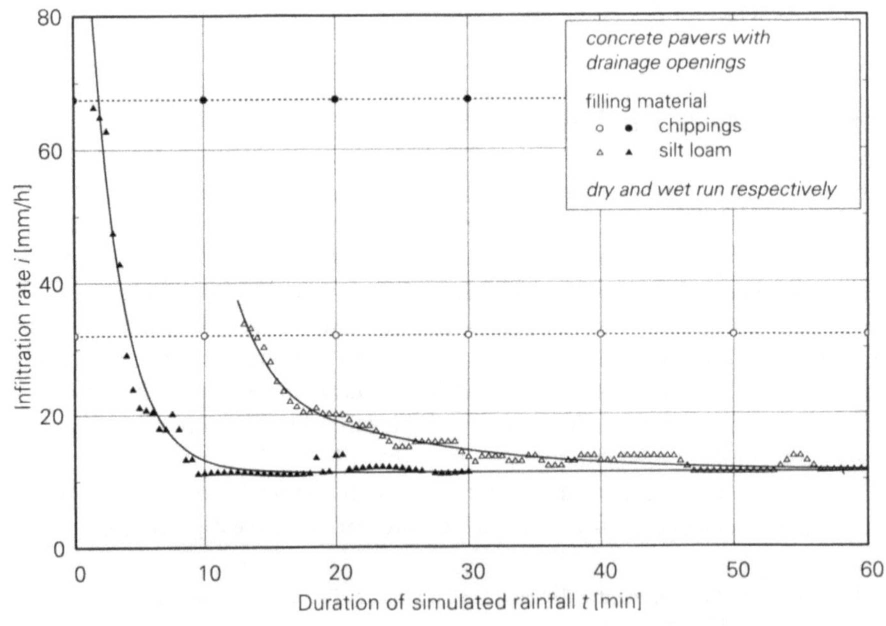

Fig. 2. Change in infiltration rate i(t) by concrete pavements with drainage openings, filling material chippings and silt loam

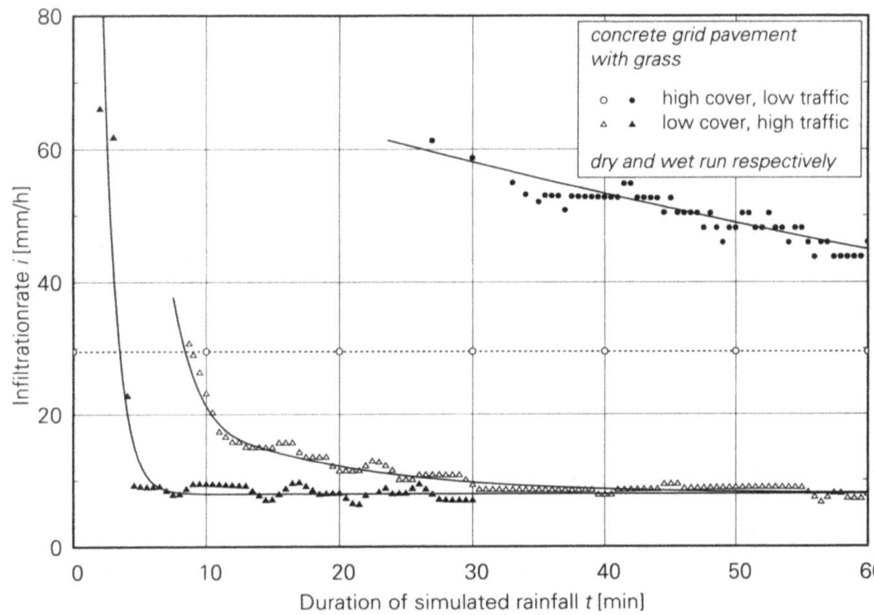

Fig. 3. Change in infiltration rate i(t) by concrete grid pavements with grass, high and low cover, low and high traffic

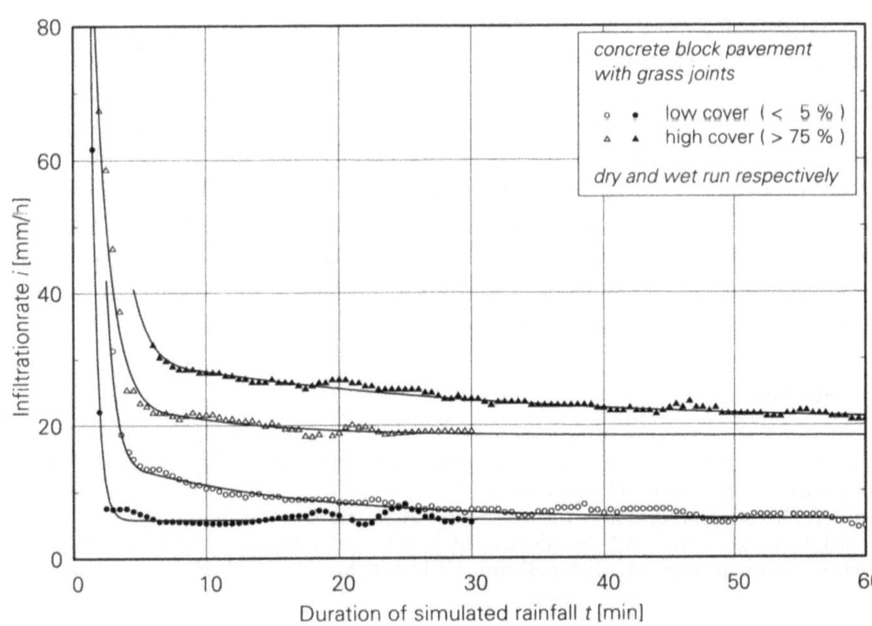

Fig. 4. Change in infiltration rate i(t) by concrete block pavements with grass joints, high and low cover

Compacted sand clay surfaces with a high content of fine grains show less permeability. A higher infiltration rate was measured on the surface with a lower content of these particles. On the sand surface with a synthetic resin binder most of the precipitation applied was infiltrated (Fig. 1).

For the concrete pavement with drainage openings filled with chippings no runoff was observed. On the same pavement with drainage openings filled with silt loam the infiltration was reduced (Fig. 2). A high infiltration rate was recorded on the concrete grid pavement with high grass cover and low traffic. On the car park with the highest frequency of cars the infiltration capacity decreased (Fig. 3). On the concrete block pavement with grass joints the density of vegetation also has a strong effect on permeability (Fig. 4).

3 Conclusions

Runoff increased on block pavements with grass joints and concrete grid pavements with grass with a decrease in vegetation density. The vegetation density depends on the traffic volume. The tests on block pavements with drainage openings indicate that the infiltration capacity depends on the type of filling material (Muth 1990). The lowest infiltration rate was measured on a footpath with a compacted sand clay surface with a high content of fine grains. This particle graded material shows the tendency to compact with time (Borgwardt 1995). The sand surface with synthetic resin binder enhances infiltration due to the porosity of the upper layer. On old pavements (cobblestones, flagstones, mosaic-stones) only small differences were observed in the infiltration behaviour as a result of the long term input of clay, silt and organic material. The permeability decreases markedly with age (Wessolek et al. 1997). This reduction in permeability can also be expected with the newly laid types of percolation-promoting pavements. The results of this study must be taken into account when selecting permeable pavements to reduce stormwater runoff and for the development of methods to assess the impacts of surface sealing on soils as well as the hydrology cycle.

References

Borgwardt S (1995) Untersuchungen der Durchlässigkeit von Bodenbelägen in wassergebundener Bauweise. Wasser und Boden 47: 11–16

Muth W (1990) Sickerfähige Pflasterbeläge. Neue Landschaft 35: 613–625

Wessolek G, Facklam M (1997) Standorteigenschaften und Wasserhaushalt von versiegelten Flächen. Zeitschrift für Pflanzenernährung und Bodenkunde 160: 41–46

Sana'a Urban Ecological Planning Guidelines

Stefan Norra

1
Aspects of Sana'a Ecological Assessment

Between 1967 and 1994 the population of Sana'a, the capital of Yemen, increased from 56,000 (Bertaud 1990) to 972,000 inhabitants (Statistical Department of Yemen 1995). The yearly growth rate was 10.8% between 1986 and 1994. At this rate, the population will have reached 3 million inhabitants by the year 2005. Sana'a's developed area expanded from 20.9 km^2 in 1979 to 94.5 km^2 in 1994. About 110 km^2 were planned for new urban quarters in 1995.

Sana'a is located in a basin at an altitude of around 2200 m. Mountain ridges block further development of Sana'a towards the west and the east. The predominant wind direction is north-east (Civil Aviation and Meteorological Agency 1993). The annual rainfall is about 250 mm/year (High Water Council/UNDP 1992). Wadi Sailah, an intermittent river, runs through Sana'a from south to north.

The urban growth in combination with the Sana'a basin's ecosystem has caused several ecological and environmental problems:

1.1
Air Pollution

Air pollution is caused by traffic, burning waste, industrial emissions, an overloaded sewerage system, lack of green areas, uncontrolled development of wind lanes, whirlwinds, large non-developed areas with no vegetation cover (dust source) in combination with the basin location. Extended haze and dust layers are able to cover the whole Sana'a area. There are no available data on the occurrence of Los Angeles type smog at Sana'a, but on the basis of recorded temperatures above 25°C, a relative humidity mostly between 40% an 56% and wind speeds below 3.9 kts (Civil Aviation and Meteorological

Institute of Petrography and Geochemistry, University of Karlsruhe, Kaiserstraße 12, D-76128 Karlsruhe, Germany

J. Breuste, H. Feldmann, O. Uhlmann (Eds.)
Urban Ecology

Agency 1993) and a relatively high UV-radiation due to altitude, smog might easily occur.

1.2
Water Pollution

The lack of sufficient sewage treatment facilities, scattered private cesspits and industrial sewage outlets endanger the water resources. Many private wells can be found within Sana'a. Foppen (1995) showed that the aquifers under the central and central-north parts of Sana'a contain high concentrations of chlorine and nitrates, probably caused by infiltration of domestic sewage via cesspits. Recharge by the return of waste water in 1990 was estimated to be 9.7 Mm^3, 34.1 Mm^3 in 2000, and 85 Mm^3 in 2010 (High Water Council/UNDP 1992). In 1990 studies showed that recharge was around one quarter of the total yearly recharge. In contrast, the ground water abstraction is estimated to be around 170 Mm^3 (High Water Council/UNDP 1992).

1.3
Destruction of Biotopes and Green Areas

Uncontrolled urban growth, grazing, woodcutting, lack of co-ordinated irrigation and water distribution, decay of terraces, careless handling of fuel and waste, etc., in combination with the small rainfall and a widely deforested area are causing the destruction of local biotopes and green areas. Wadis and urban wasteland are frequently crossed by off-road traffic and used for waste disposal. Sewage occurs in some wadi beds and causes the corresponding biocenosisses. Many public green areas are in poor condition, especially caused by the limited public budget, the restricted available amount of irrigation water, and the lack of environmental awareness.

Due to these problems, two missions were carried out by the author Yemeni – within the German project: "Planning Assistance for Urban Development" from 1/10/1993 to 31/12/1993 (Norra 1994) and from 28/5/1995 to 13/12/1995 (Norra 1996) to lay down guidelines for urban ecological planning in Sana'a.

2
Urban Ecological Planning Guidelines

An urban boundary should restrict by law further development beyond this line. This is necessary for the preservation of Sana'a's important surrounding areas and would provide wind lanes for refreshment of urban air, areas for groundwater recharge, potential afforestation sites and protection of existing green areas (tree and shrub covered areas, agriculture).

No development should take place towards the north-east (main wind direction) and the south-west (large tree covered areas).

A proposed Wadi Sailah protection plan should be further developed. This area is most probably able to function as an urban air refreshing wind lane, if laid out as an urban green area. Existing carpets of waste and sewage lakes have to be cleaned-up. Unauthorised buildings have to be removed (workshops, industries, squatters, slums). Existing agricultural areas have to be preserved and, if possible, extended. Traffic has to be limited. No thoroughfares should be constructed, only access roads as far as necessary. Wadi crossing roads should be constructed on low bridges to prevent traffic from driving in the wadi bed.

All wadis in and around Sana'a should be protected as wind lanes (mountain-valley air currents, main wind direction) and as areas for groundwater recharge.

Inner urban agricultural areas, gardens, and surrounding terraces, and tree and shrub covered areas should be protected. There are still many agriculturally used areas in the town. The surrounding mountains are often terraced. Many of these terraces have been abandoned and become increasingly run down due to the lack of agricultural management. Instead of developing these areas they should be planted with trees and if possible agriculture should be re-established.

Abandoned and non-developed areas (slopes and plain areas) in and around Sana'a are very often not covered by vegetation. This is caused by wood-cutting and grazing combined with the dry climate. Public parks are often in a poor condition whereas private gardens are often well kept. To deal with the problem of the lack of water, it is recommended that these areas are planted with native trees and shrubs that are adapted to the specific environment and wood-cutting and grazing management are established.

References

Bertaud AC in Kopp H, Wirth E (1990) Beiträge zur Stadtgeographie von Sana'a. Beihefte zum Tübinger Atlas des Vorderen Orients, Reihe B (Geisteswissenschaften) 95

Civil Aviation and Meteorological Agency, Yemen (1993) Digital data set

Foppen W (1995) Groundwater Quality of the Aquifers beneath Sana'a. In: Ministry of Electricity and Water, The Republic of Yemen (ed) Extended Workshop on the future of Sana'a Water Supply - Workshop Documents, Yemen

High Water Council/UNDP (1992) Water Resources Management, Options in Sana'a Basin. In: Final Report Vol. IX, Sana'a (Yemen)

Norra S (1994) Sana'a Urban Ecology Study. In: Ministry of Housing and Urban Planning/GTZ project report. Sana'a (Yemen)

Norra S (1996) Sana'a Urban Ecology 1995 Follow-Up Study. In: Ministry of Construction, Housing and Urban Planning/GTZ project report. Sana'a (Yemen)

Statistical Department of Yemen (1995) Census 1994

Geo-Ecological Investigations of Urban Soils in Modern Estates in Halle and Leipzig (Germany) – Geo-Ecological Inventory, Modelling and Planning in Urban Ecosystems

Martin Sauerwein

1 Aims

The first step in this process is the registration and valuation of the ecological location-characteristics of the examined surfaces and the analyses of the geo-ecological conditions. The manner and intensity of the different surface components, caused or altered by man and pollution, can then be characterised and statistical modelling of soil housekeeping inter-relationships with defined conditions (soil substrate, previous exploitation, time of building over etc.) can be carried out. Deductions made from geophysical and geochemical indications of the soil (pollutant) housekeeping levels led to the agreement of soil quality standards and quality aims.

2 Methods

In Halle and Leipzig 32 characteristic locations (150 soil samples) were selected using a GIS (Geographic Information System). Analyses were carried out for the standard soil parameters (pH-value, humus, calcium etc.), as well as typical urban heavy metals (Cd, Cu, Pb, Zn) and PAH. The exposure of the clean soil to the air for half a year and one year enabled the atmospheric input to be registered. The inputs and outputs of urban soils can be modelled to summarise the ecosystem (Sauerwein et al. 1995) -using the non-parametric statistic (Sauerwein 1997). Different extraction procedures were applied to determine the total heavy metal contents, the bio-available quantity and the cation exchange capacity.

Martin-Luther-University Halle-Wittenberg, Department of Geography, Domstraße 5, D-06108 Halle (Saale), Germany

J. Breuste, H. Feldmann, O. Uhlmann (Eds.)
Urban Ecology

3
First Results

The condition of the ground in the soil profiles can only be said to be natural in four exceptional cases. The soil structure of the profiles is very non-homogeneous, both horizontally and vertically. Due to the input of calcium through building activity the pH-values are generally greater than 6.5; the concentration of humus in the upper soil-layer is greater than 2%, due to human activity (e.g. gardening). The heavy metal analysis data show that almost the total contents, as well as the bio-available quantity, are within the permissible limits. The highest heavy metal contents in the upper layer are located in the 'oldest' soils. It is shown that the age of the urban locations, not the traffic, are reflected in the heavy metal pollution patterns of the upper soil-layer. The buffer stock of the soils is accounted for by the high absorption capacity.

References

Sauerwein M, Diaby K, Zierdt M, Frühauf M (1995) Geoökologische Untersuchungen zur Schwermetallbelastung städtischer Kleingärten in Halle (Saale). Hercynia N.F. 29: 291–314, Halle

Sauerwein M (1997) Zum geoökologischen Inventar und zur Schwermetallbelastung urbaner Freiflächen am Beispiel von Oberböden in Großneubaugebieten in Halle (Saale). Geoökodynamik XVIII: 133–164, Bensheim

Daylight in Housing Constructions

Hannelore Schweitzer and Steffen Kausch

1
Introduction

Aspects of the construction of buildings - directional positioning and distances from surroundings - irreversibly influence two factors of underestimated importance in their effect on quality of life - daylight and conditions of insolation.

Regulations of the states of the Federal Republic of Germany (FRG) concerning the possible (astronomical) *minimum insolation* for dwellings are not available. The *daylight* for dwellings is regulated in most of the Federal states by a minimum proportion of window openings to floor area (1:10 to 1:8) of lived in rooms.

More comprehensive minimum levels are included in DIN 5034 (German Industrial Standard). It was examined whether these recommendations were taken into account during the phase of building design.

Planning data of three new residential areas of a city in Saxony, Germany permitted an assessment to be made: in one case a detailed statement (A: 316 dwellings), in two cases a rough assessment (B: 370 dwellings and C: 118 dwellings).

2
Possible (Astronomical) Minimum Insolation and Daylight Conditions

It was examined if the duration of insolation achieves 4 hours in at least one living room of a dwelling at equinox according to DIN 5034. This was carried out on the basis of development plans as well as building and advertising documents by means of a graphic method. The results for residential area (A) are summarised in Table 1.

Technical University Dresden, Institute of Medical Microbiology and Hygiene, Löscherstraße 18, D-01309 Dresden, Germany

J. Breuste, H. Feldmann, O. Uhlmann (Eds.)
Urban Ecology

Table 1. Observance of possible insolation in residential area A

Dwellings		DIN-criterion accomplished		DIN-criterion not accomplished		main cause of insufficient insolation
size	n	n	%	n	%	
one-room	121	89	73,55	32[a]	26,45	NE-orientation
two-room	56	30	53,57	26	46,43	NE-orientation
three-room	106	101	95,28	5	4,72	obstruction
four-room	33	33	100	-	-	
sum	316	253	80,06	63	19,94	

[a] exclusive attics

For the residential area (B), the recommended minimum insolation of 4 hours is not achieved in nearly 50% of the dwellings. In the smaller residential area (C) this recommendation is met in all but three dwellings.

The causes of insufficient possible insolation are: one-side-positioning toward northern exposure of the (mostly smaller) dwellings, buildings too close together (mainly toward east resp. west) and in residential area (A) the use of very slim and low windows.

The recommendations of DIN 5034 on the quotient of daylight, as well as intensity of daylight, was examined.

The investigations were limited in this case to the sitting- and bedroom of a typical basement dwelling of residential area (A) equipped with especially small windows which are suspect from the hygienic point of view. The prescribed minimum proportion of window openings to floor area is adhered to.

The results for the sitting-room showed that the legally binding arrangements of minimum proportion of window opening to floor area does not guarantee sufficient daylight conditions. The quotients of daylight met the minimum value of 1% recommended in DIN 5034 at neither of the two reference points (Fig. 1). Furthermore the recommended intensity of daylight for the most unfavourable work-place ($\geq 300\,lx$) is barely achieved at reference point D_2 in the course of a year (Fig. 2).

3
Conclusion

At least the sitting-room of every dwelling should be equipped with sufficient large and insulated windows and achieve recommendations on minimum amounts of insolation and daylight according to DIN 5034.

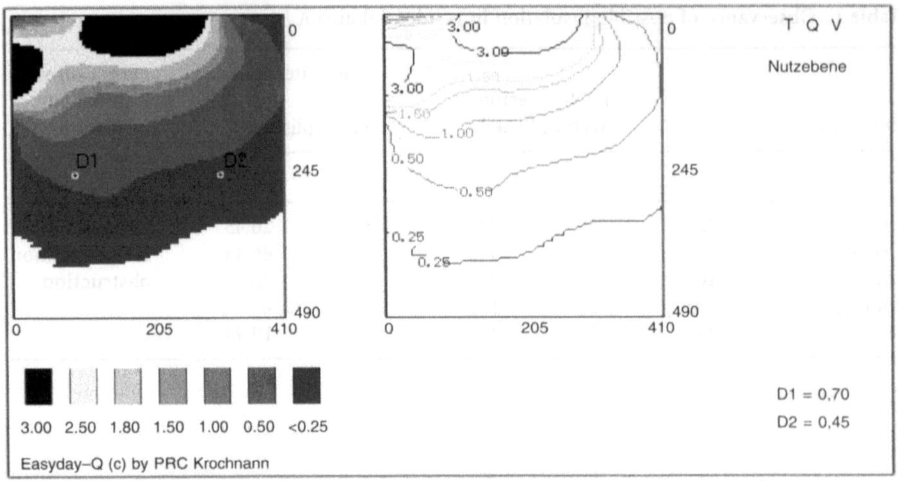

Fig. 1. Distribution of daylight-quotient in the sitting-room

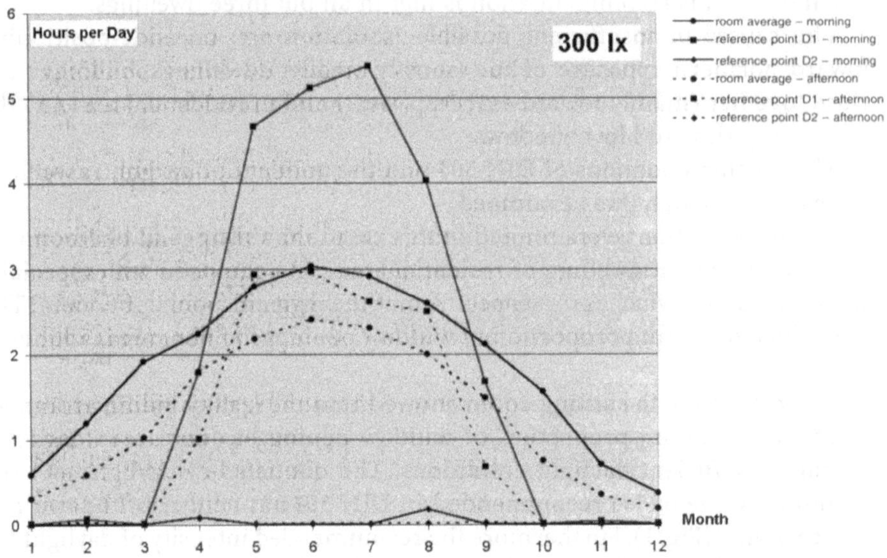

Fig. 2. Hours per day with intensity of daylight ≥300 lx in the sitting-room

It is in the interest of healthy housing conditions to establish legally binding minimum standards of insolation and daylight in living rooms because obligatory recommendations of standardisation are taken into account insufficiently during planning practice.

Statements on prospective insolation and daylight of dwellings should be discussed at the planning stage.

Transport, Transformation and Retention of Harmful Anthropogenic Substances in Urban Areas

Doris Stüben[1], Manfred Meurer[2], Karin Huck[1] and Stefan Norra[1]

1 Introduction

About 74% of the European population lives in urban areas (86% in Germany). Urban and traffic areas cover 11.3% of Germany. The rapidly increasing urbanisation causes more and more damage to urban ecosystems. The UNO carried out the 'Man and Biosphere' project MAB-11 and organised two "HABITAT" conferences in 1976 and 1996 which accorded great importance to urban areas for human lives. In Germany research in the field of urban ecology is well known, especially in Berlin (Sukopp 1990). Research has shown that the effects of damage to various compartments of the urban ecosystem are extremely complex. Consequences for urban habitats and the human population can only be guessed at. Preliminary studies at the University of Karlsruhe and at several institutes of the university have been undertaken as a prelude to an application for a Collaborative Research Centre (SFB 1661) to be founded by the DFG (German Research Association) under the leadership of Prof. Dr. D. Stuben. The main objectives of SFB 1661 are the investigation of the dynamics of anthropogenic substances caused by traffic and household emissions in, and between, different ecosystem compartments, the development of inter-disciplinary research methods and the development of predictive models for the behaviour of harmful substances as a result of different modes of urban development.

2 Case Studies of Traffic Induced Emissions

Detailed studies have been undertaken by Norra (1997) on the distribution of trace metals in urban soils adjacent to roads in Karlsruhe. The analysis of dif-

[1] Institute of Petrography and Geochemistry, University of Karlsruhe, Kaiserstr. 12, D-76128 Karlsruhe, Germany
[2] Institute of Geography and Geoecology, University of Karlsruhe, Kaiserstr. 12, D-76128 Karlsruhe, Germany

J. Breuste, H. Feldmann, O. Uhlmann (Eds.)
Urban Ecology

ferent soils and their layers has shown that individual sites can be differentiated by the occurrence of certain trace metals. However, contamination of soils adjacent to roads is induced not only by traffic. Soils can be transported to the sites during construction and can be contaminated by other sources of trace metals. Comparison of different trace metal concentrations give information about such situations. Lead and zinc concentrations in soil samples taken at different road sites in Karlsruhe and at a reference site show a linear regression [$r = 0{,}84$; $n = 90$] which is typical for traffic emission (Norra 1997); one site does not fit to this regression. Another intensive source of contamination besides traffic can be assumed. In addition to trace metals, the Platinum group elements have been analysed in street sediments, soils and vegetation along roadsides in urban areas (Cubelic et al. 1997). A typical Rhodium to Platinum ratio of 1:6 indicates emissions caused by catalysts. This ratio is also observed in roadside soils and street sediments. Palladium is also a typical catalyst-emitted metal which can be determined in these soils and street sediments. Street sediment samples of different locations beside roads show concentrations of between 45 to 200 μg/kg Pt, 7 to 31 μg/kg Rh and 3 to 12 μg/kg Pd. Special leaching techniques were tested to identify the mobility of trace metals within the street sediments and urban atmospheric particulate matter (Heiser et al., in prep.). It is shown that Zinc is highly mobile in urban particulate matter, whereas it is less mobile in street sediments sampled in Karlsruhe. The highest amounts of lead were extracted from the so-called organic fraction of street sediments and urban particulate matter.

Ground management in cities requires ecological concepts which are based on ecological stock taking. To study the traffic situation in the cities, it is necessary to investigate a narrow pattern of samples. The most important indicator for traffic-induced air pollution is the distribution of NO_x. It is important to map the distribution of NO^x versus time and to compare these data from the city to the surrounding area for differentiation. Studies in Luzern (Müller and Meurer, 1994) show that most of the sites sampled lie above the Swiss limit of 30 $\mu g/m^3$ NO_x. Traffic-induced emissions were observed in soil samples in Gaggenau and Luzern. The topsoil was characterised by trace metals concentrations (Cd, Cu, Pb and Zn) showing a direct relationship with the distance to the street (Meurer and Müller 1992). Apart from traffic, industry and households are the main sources of air pollution which cause problems for the health of trees within the cities. The mapping of tree vitality in urban areas is an indicator of the rate of pollution. Environmental data such are air pollution, soil contamination, tree vitality etc. can be collected and mapped using Geographic Information Systems; such data provide a sound basis for environmental management of cities.

References

Cubelic M, Peceroni R, Schäfer J, Eckhardt J-D, Berner Z, Stüben D (1997) Verteilung verkehrsbedingter Edelmetallimmissionen in Böden. Umweltwissenschaften und Schadstoff-Forschung, 9 (4)

Heiser U, Norra S, von Wagner M, Stüben D (in prep.) Chemische sequentielle Extraktion von Schwermetallen in Stäuben. Umweltwissenschaften und Schadstoff-Forschung

Meurer M, Müller H-N (1992) Erfassung der Umweltbelastungen in einem Stadtökosystem. Geograph. Rundschau, 10, 562–567

Müller H-N, Meurer M (1994) Verkehrsbedingte Luftbelastung in Luzern im Vergleich zum Umland. In: Luzerner Stadtökologische Studien, Bd. 8, Luzern, pp 119–137

Norra S (1997) Anorganische Schadstoffbelastungen in Stäuben, Straßensedimenten, Böden und Pflanzen entlang innerstädtischer Straßen am Beispiel von sechs Standorten in Karlsruhe. In: Karlsruher Ber. z. Geogr. und Geoökol., H 10, Inst. f. Geogr. und Geoökol., Univ. Karlsruhe (TH), Karlsruhe

Sukopp, H. (ed.) (1990) Stadtökologie Das Beispiel Berlin. Berlin

The Influence of Shanghai Urban Development on Urban Climate and Human Comfort

Zheng Jingchun and Wang Yuanfei

1 Some Features of Shanghai City

Shanghai is the largest industrial and commercial city in China with a high population density and level of urbanization. The population density ranges from 14,000 to 70,300 inhabitants/km^2 in urban areas to less than 1000 inhabitants/km^2 in rural areas, according to the *Statistical yearbook of Shanghai 1994*. In the Shanghai urban area the urbanization rate (defined as the percentage of built-up area in the urban area) is more than 80%. Only 1.14 m^2 of public green space is available for each inhabitant, thus the anthropogenic modification of land surface is very striking.

Due to the rapid economic development in Shanghai, the annual industrial coal consumption rose from 9.19 million tons in 1970 to 30.27 million tons in 1993 adding much man-made heat to the atmosphere.

2 Urban Heat Island in Shanghai

The most obvious urban climatic effects, such as urban heat island, dry island, rain island etc. can be identified from the meteorological data in Shanghai (Chou Shu Djen et al. 1989).

Shanghai city's urban heat island is very obvious. Based on the meteorological data of recently years, the mean monthly temperature difference between urban and rural areas (ΔT_{u-r}) is at least 0.5°C, the maximum difference is 1.5°C, occurring in autumn (October). Detailed statistics show that the frequency of the urban heat island is 27.8 days per month during the period 1991–95, i.e. urban heat island existed almost every day; the highest intensity heat island ($\Delta T_{u-r} \geq 3.0$°C) occurred mainly in autumn and in winter, occurring 3.1 days per month in these two seasons.

Department of Geography, ECNU, 200062, Shanghai, P. R. China

J. Breuste, H. Feldmann, O. Uhlmann (Eds.)
Urban Ecology

3
The Degree of Human Comfort in Shanghai City

The urban climatic conditions can also affect human comfort. Some widely used biometeorological indices such as temperature humidity index THI (the US Weather Service's human comfort index, Mather 1974), new effective temperature T_{sef} (Ellis 1972) and the wind chill equivalent temperature WCET (Schlatter 1981) are utilised here to assess the influence of climatic conditions on urban inhabitants in Shanghai.

Based on the criteria given by Mao Zhengdan (1992) we find that spring (April, T_{sef} = 15.96) and autumn (October, T_{sef} = 18.58) are two comfort seasons, but winter (January, T_{sef} = 9.85) and summer (July, T_{sef} = 26.13) are the two uncomfortable seasons. From November to April, the WCET are 9.16, 1.4, −2.0, −1.0, 2.3 and 10.1 respectively. Compared to the comfort degree given by WCET, January and February are chilly, March and December are cold, April and November are cool. This indicates that the winter in Shanghai lasts a long time and many more inhabitants feel uncomfortable in this season.

Wang (1997) found that when the daily maximum temperature is above 33.5°C and THI is greater than 29.5, extreme heat is indicated. The comfort degree in the two hottest months (July and August) is studied here by using THI. We find that during most of the day the degree of comfort is low. The frequency of comfort is only 1.6% in the middle of the night; although the air temperature is very low at this time a higher relative humidity keeps the THI high. Further analysis reveals that the urban heat island will always give rise to a great increase in THI thus the urban heat island in summer will make inhabitants experience much more discomfort than usual.

4
Conclusion

From the preceding discussion it can be concluded that the urban influences of Shanghai on climate are very pronounced. The urban climatic effect will be enhanced along with the development of the economy.

The degrees of human comfort in winter and summer are based on the biometeorological index as T_{sef}, WCET and THI. Spring and autumn are the two comfortable seasons, but winter and summer are uncomfortable. As discussed in the text, the high temperature and high relative humidity in summer and lower temperature with high humidity and wind velocity in winter are the main causes of discomfort in these two seasons.

References

Chou SD, Zhang C, Zheng JC (1989) Urban climate and regional climate. ECNU Press

Ellis FP, Smith FE, Walters TD (1972) Measurement of environmental warmth in SI units. Brit J Industr Med 29, 361–377

Mao Zhengdan (1992) New effective temperature and the climate comfort belt in China. Tropical Geography 12 (3), 249–255.

Mather, JR (1974) Climatology: Fundamentals and Applications, McGraw-Hill, inc. New York

Schlatter TW (1981) Weather queries. Weatherwise, 34 (6), 266–267

Wang Yuanfei (1997) The temperature humidity effects and human comfort in hot season in Shanghai city. Journal of ECNU (in press)

Improvement of Urban Environmental Conditions The Integration of Ecological, Economic, Social and Cultural Aspects

2

The Social Dimension of Urban Ecology

Hans-Norbert Mayer

1 Introduction

Urban ecology has to deal with two tasks. The first one is how to cope with the consequences of the industrial past. The old industrial society, which in former days created an inconceivable wealth for an incredible number of people, now withdraws from the cities leaving contaminated soil, unused industrial plants and polluted rivers. The emerging problems require technological solutions for which the necessary knowledge is not always available, and they create expenses for which the money has not yet been provided. Nevertheless, the ecological heritage of the industrial society is the easier task to cope with because it basically requires only technological knowledge and money. On the contrary the second task – the organisation of sustainable urban development – is more complex. In contrast to the repair of damage already done, the prevention of ecological damage is also a social task because it touches the way of life of urban citizens. The subject I would like to refer to is this second task of urban ecology. I would like to show why the achievement of sustainable urban development should not be considered to be just a matter of technological efficiency, economic instruments and planning competence, rather it should be considered to be a social and political challenge.

2 Definition of Aims

In what way do we achieve the aims of an ecological policy? The most common reason for the need for action is the indication of impending catastrophes. The warning signs of an ecological disaster reveal negative limits to further cultural development, but these do not give positive indications of which cultural standards mankind should attain to survive in a humane way. Natural sciences are only able to give negative answers: namely how it should

Carl von Ossietzky University Oldenburg, Institute for Sociology, Town Research, P. O. Box 2503, D-26111 Oldenburg, Germany

J. Breuste, H. Feldmann, O. Uhlmann (Eds.)
Urban Ecology

not continue. The way in which life should continue remains a question of political and social values and of culturally developed aims. That is one reason why the image of a new sustainable city cannot be derived from scientific calculations only.

The other reason is that it cannot be said that the natural environmental conditions of a city are pure nature, as it is almost always cultivated. The idea of having a balance of nature from which practical norms could be derived is fiction. Every state of balance in the urban environment is defined and desired by human beings and it requires constant human effort to keep or establish such ecological situations. Thus each decision concerning an environment worth living in includes, simultaneously, a decision about a certain style of living and working.

To summarise: natural sciences cannot relieve politics of the decision of what the sustainable city of the future is supposed to look like. The aims for the ecological conversion can only be developed out of a democratic debate within society, out of a discussion about a new image of the right way to live.

3
Global Relationships of Ecological Problems

The result of the social debate will not only have an effect on the ecological situation in Europe, but also far-reaching consequences for the ecological damage or alleviation in other parts of the world. It is common knowledge that today 20% of the world's population use 80% of the natural resources. It is also not a new fact that this is the reason why the main changes initially have to take place in the western industrial states as they are the main culprits of the world's ecological crisis. In ecological matters the north is enormously indebted to the south.

The question of urban ecology in the cities of the so called 'first world' is a completely different one to that of the cities of the so called 'third world'. For the European cities, on which I will concentrate, it is true that the environmental problems they have to solve are not only site specific, but also global. Local actions must be taken to ameliorate the global situation. To what extent, for example, the energy consumption should be decreased in order to stop the greenhouse effect, is of course disputed and many studies have been carried out on this subject. If we take the responsibility of the north seriously the problem requires a solution on a huge scale which can not just be achieved with better technology, but will also entail losses to the affluent society.

Technical growth of efficiency is important and indisputable, but it will not be sufficient to guarantee the necessary reduction of the ecological damage. It will be even less able to balance out the negative consequences of the further growth of prosperity in the western countries. Technical solutions cannot compensate for the ecological damage that would result from an increase of the standard of living for all people. Thus another factor will also have to be changed: sustainable development will not be successful if the way of life is not put up for discussion – this concerns specifically the way of life of the citizens of the western world.

At first glance this appears to deal with the affluent society, but it is not only about excessive luxury. The way of life that has developed in the course of urbanisation is also characterised by emancipatory achievements that are worth maintaining.

4 Characteristics of the Urban Way of Life

Today's way of life is primarily an urban way of life and the structure of European cities reveals its materialised form. It is the way of life of an urban consumer household that could not exist without the private market and the public infrastructure. Historically the consumer household is a very young phenomenon. It only started to gain acceptance with a large proportion of the population in the middle of the 19th century when urban industrialisation began. The division of residential and working places, the extension and density of the cities, as well as the functional segregation of the quarters resulted in the production of fewer services on the level of the single household. The urban households were dependent on an increasingly complex machinery of supply and waste disposal even in their daily domestic tasks. This way of life only became normal for the bulk of the population after the second world war due to economic growth. Today the urban consumer household is typical of all social classes in western Europe and it is also widespread not only in cities but also in rural settlements.

The reasons why it was so successful are not only the logic of the organisation of production, the pressures of urbanisation and the technological development. It is also because it seems that through this way of life people can become emancipated: the hope of liberation from natural and social constraints can be fulfilled by escaping to the cities. The city is a place where life is neither dictated by natural physical conditions nor by the rhythm of nature. Integration into the machinery of the city provides an opportunity to relieve the household workload and enables people to spend more time pursuing leisure activities; this is a positive step away from the working world. The division of publicity and privacy in the city was a precondition to an autonomous life and individual development.

Today we all know the ecological drawback of the way of life of the urban consumer household: the enormous increase of energy consumption and the consumption of resources, urban sprawl, industrialisation of agriculture, dominance of individual transport, mass consumption and an increase in the amount of garbage. The difficult task of urban ecology results from the fact that the elements of the way of life that threaten the basis of survival today are most closely connected with old promises of emancipation. Certainly, the emancipatory ideas of urbanity have not been realised in many ways, they have often been simplified or have even been turned around, but this does not mean that they have lost their subjective value for the urban citizens. An ecological policy that offends against such moral concepts and hopes will run into opposition. That is why, in the discussion about a sustainable way of life for urban citizens, there is more at stake than a reduction of needless prosperity and wastefulness.

5 Social Consequences of an Ecological Policy

With the following three examples I will show where possible conflicts could arise between the requirements of urban ecology and the achievements of urbanity: in the limitation of land use, in new ecological technologies concerning the household and the whole city, and in the environmentally correct behaviour in everyday situations.

One large problem of urban ecology is the unbroken, increasing use of settlement areas for house building; this causes destruction of the landscape and an increase of traffic. The continuous demand for more living space and building land is due to, besides growing prosperity, one significant reason – the reduction of the size of households. According to the statistical average in West European countries, the size of households i.e. the number of people living in one apartment, is decreasing. The smaller the household, the more living space is used per person. That is the reason for the paradoxical phenomenon that although there is a stagnation or even a decrease in population, the demand for living space is still increasing. Behind the tendency of forming smaller households there is a change in forms of living. Life is becoming more individualised, expressed for example in the wish of young people to leave the parental home as early as possible or in the increase in single households. This process has especially been led by women through their legal and economic emancipation. Virginia Woolf entitled her book about women's issues "A room of one's own" for a good reason.

This phenomenon concerns social trends that are difficult to control politically. They are based on an important achievement of urbanisation: the apartment as a place for individual life, strictly separated from the public. An ecological urban policy that wants to stop the increasing use of land thus has to fight against more than just an egoistic use of the landscape and hedonistic consumerism. This policy can become contradictory to basic values, such as the wish for an individual development and a private sphere protected from any form of social control.

A similar dilemma occurs with the introduction of ecologically orientated technology concerning the household or the whole city that contributes to a more efficient use of energy and water and to a reduction in domestic waste. The technological and constructive refitting of houses and the creation of new infrastructure facilities are important elements of the urban ecological conversion, but they will not make the cities easier to handle. On the contrary, lifestyles must be adjusted to fit in with the new installations and new forms of organisation. They require complementary practice of behaviour so that the saving potentials can really be made use of. On the other hand, incorrect behaviour can disturb their ability to function or reduce their effects. That is why the households will have to acquire additional knowledge, to know how to handle the ecological innovations correctly, they will have to take up more responsibility and they will have to be ready to invest more work and effort in their daily routine. In this sense ecological changes run the risk of offending against a further emancipatory promise of urbanisation: The hope for relief from work and obligations.

The modern city has always also been a machine that freed the individual from work and obligations necessary for the survival and the coping with the daily routine; that gave him the freedom and time for other self-chosen activities. This is the progressive logic of Fourier's Phalenstère, of the kitchen of Frankfurt, of Le Corbusier's living machinery and of the functionalism of the town planning. The estate building and the housing construction have, to a large degree, followed these principles of rationalisation until today. The urban consumer household has become, within this system, a station of continuous-flow into which consumption goods flow in and dirt and waste flow out. In that respect, its ecological difficulty is equally as visible as its relieving effect.

If now the ecological conversion of the cities demands more work, more knowledge and more responsibility from its citizens, then tasks that in the course of urbanisation had been taken out of the private household will partly return in a new appearance. This could especially affect the women. Many of the new demands in the household could mainly fall back on them, considering the still prevailing sexual division of labour. In any case further obligations restrict the attractiveness of the shaping of urban life. The readiness to take them up can not be expected without a second thought. Also in this conflict opposition can occur that impedes the possibility of majority support of an ecological policy.

Finally there is another contradiction to mention that is connected with a necessary change of the urban way of life. The exact transformation of the well-known motto: "think globally - act locally" by urban citizens assumes a kind of a new protestantic work ethic - now under ecological circumstances. A sustainable way of life means to carry out the described additional work and obligations in reality and regularly. In addition to the correct use of the new installations a permanent environmentally aware behaviour is demanded for a further reduction of the use of resources. For that, morality and self-discipline are necessary because it is about behaviour in private, often intimate spheres and it is about many little activities of daily routine. Things like prudence, conscientiousness, consideration, economising are demanded – this can hardly be controlled by punishments and bans.

Theoretically this can be described as a further push of internalised controls in the historical process. Norbert Elias analysed the process of civilisation as a process of an internalisation of external compulsions. The history of urbanisation is, as I mentioned, a history of the relief from obligations, but at the same time it is also the history of the setting up of new thresholds of shame and embarrassments which, in comparison to former rural ways of life, demand more abilities of self-control of physical and psychological reactions. It belongs to the ambivalence of urbanity that the liberties of the urban life have been bought with a certain kind of discipline on another level. The ecological question will further intensify this ambivalence. New environmental thresholds of shame and an universal conscience must be created. The only alternative to it would be a frightening system of observation and sanctions, if one wanted to guarantee from outside that each one also in his private life acts in the way global thinking requires it. This is neither practical nor desirable so urban citizens will have to change their way of life voluntarily. The aim of a sustainable city will push up the screw in the course of civilisation and this will not be possible to accomplish without disapproval and resistance.

6
Conclusion

These were three examples to illustrate my central thesis that the demands on the citizens can get into conflict with historically developed emancipatory promises of urbanity. There are really "good" reasons, namely legitimate reasons, for disapproval or resistance against the demands of urban ecology, because what the people want to protect of their way of life does not only amount to material interests like property, consumption or working place. Problems like the lacking of readiness for contribution, little environmental awareness, the rejection of ecological innovations will not be solved only by moral appeals, more information, better background conditions or new regulations, as long as the demanded changes of behaviour are in conflict with deeply rooted hopes and values. That is why a policy of urban ecology has to ask for these "good" reasons and has to deal with them in a systematic way.

I will not dispute that one has to deal with other very big obstacles as well. Of course interventions in the production area must take place, new infrastructures have to be created, planning premises have to be translated into action and market instruments have to be set up. This is difficult enough, but these measures will not be sufficient and many of them will only be enforceable in the end if an ecological policy is supported by a majority in the long run. That is why the first step toward a sustainable development means to find out which conflict lines and conflict points will be raised by the changes of the way of life. Then it is important to create a positive picture of urban life that brings into accord the ecological necessities with the preservation of emancipatory achievements.

On the political level, a democratic debate must take place about the aims of an ecological conversion. This will be a discussion about the image of which life actually is the right life and about the global responsibility of the citizens in this part of the world. On the concrete level of translating the single ecological measures into action, learning processes have to make it possible for people to find their own way toward changes of behaviour according to their individual lives and residential situation. Processes of learning require rights of participation, scope for decisions, opportunities for experiences and last but not least, the chance to make mistakes.

A change of behaviour will only then become reality if it seems not only necessary but also desirable for a majority. We need a new picture of urban culture that reconciles the advantages of urban life with sustainable development. If this is not successful, then the ecological conversion of the city will fail or it will require authoritarian, undemocratic solutions – both are not very pleasant alternatives.

Workshop: Policy Development Approaches 2.1

Local Agenda 21 Processes and Integrated Urban Development Planning in Germany

Sandra Lustig and Ulrike Weiland

1 Introduction

Some German towns and cities have begun Local Agenda 21 (LA 21) processes and are grappling with the difficulties of integrating the various aspects of sustainable development. The current debate on sustainable urban development is, in part, focusing on issues that were dealt with by *Stadtentwicklungsplanung* (integrated urban development planning, *STEP*) in Germany in the 1970's, but often fails to use the experiences gained from work carried out previously.

2 *STEP* in Germany in the 1970's

STEP is a voluntary type of planning which was never legally codified, and thus provided much opportunity for adaptation to local needs and conditions. Characteristics common to the different approaches to *STEP* include: responsibility being accorded to the local government; its application to the entire city; a comprehensive approach to all local issues and policies; common goals of guiding urban development and strengthening local government. Little has been published on the development, practice, and evaluation of *STEP* (see Lustig 1988).

In some cities, *Stadtentwicklungspläne* were prepared; these are comprehensive plans including detailed urban development goals and the measures for achieving them. In other cities sectoral plans for the whole city sectors or comprehensive plans for specific areas were drawn up without a holistic plan for the entire city. Elsewhere, *STEP* was considered to be a tool for initiating and co-ordinating policies without preparing a formal plan. *STEP* focused on areas where the local government, not the private sector, was to carry out the plan. The main topics were spatial planning, infrastructure, transportation,

Technical University Berlin, Institute for Management in Environmental Planning, Franklinstrasse 28/29, FR 2-7, D-10587 Berlin, Germany

J. Breuste, H. Feldmann, O. Uhlmann (Eds.)
Urban Ecology

housing, and business development. At the time when *STEP* evolved environmental protection was just emerging as a local government issue, and the concept of sustainability had not yet been developed; various organisational set-ups evolved for *STEP*.

In the context of political reform in the late 1960's/early 1970's, *STEP* seemed a good tool for guiding urban development. A decade later, the political climate was more conservative. The 1970's approach to planning was disparagingly called "planning euphoria", and the general feeling was that *STEP* had failed. The main points of criticism were that it was too cumbersome in its all-encompassing approach (see Ganser 1991) and that public participation slowed down decision-making. Another point, seldom voiced openly, is that a rational planning process may contradict politics. Where a detailed plan exists, there is less room for politicians to make decisions and achieve successes they can call their own. It was also pointed out that where *STEP* had had adequate resources and political support, it had been reasonably successful (see Lange 1983, Küpper 1983).

By the 1980's, some German cities that had conducted *STEP* abandoned the approach, either through a formal decision, or by cutting down its activities or staff. Most cities that still use *STEP* have reduced it either to sector planning or to planning for smaller districts within the city; little has changed in the 1990's.

3 Local Agenda 21 Processes in Germany

Local Agendas 21 (LAs 21) are long-term local programmes of action which work towards the goal of sustainable development. Their goal is to implement Agenda 21 on the local level in co-operative, consensus-oriented processes encompassing all affected actors (see Enquete-Kommission 1996; Difu 1997; Weiland 1997). The main actors involved are local government, local politicians, non-governmental organisations, and individual citizens. LA 21s are specifically intended to include marginalised groups of society but, in practice, it has been difficult to include them. While women's issues are seldom dealt with, more women take part in LA 21 than in traditional planning processes. Local businesses and their organisations rarely participate, and when they do it is only for a short time. Public awareness of LA 21 is low, with only a small percentage of the general population being conscious of LA 21 efforts, and a smaller fraction actively participating. Once an LA 21 is agreed upon, it is carried out by both public actors (such as local government) and private ones (such as businesses and individuals).

There is a widespread understanding of sustainability as an environmental issue in Germany, thus LA 21 processes are often located within the local Environment Department. As a result, LAs 21 tend to focus on environmental protection, especially global warming and energy related issues, as well as transportation, spatial planning, housing and building. Less frequently, they take on social and economic issues or the issue of achieving justice between the North and the South. The output of LA 21 processes is supposed to be a programme of action which sets priorities and uses a system of goals and objectives for the local level. Only a few LA 21 processes in Germany have produced such documents to date. It is still too early to assess the success of such processes, especially concerning their implementation.

4
Similarities and Differences

In order to determine what we can learn from the experiences with *STEP* for LA 21, it is useful to establish their similarities and differences.

Major similarities:

- Both tools were developed out of the feeling that a new approach to urban problems was needed, and were the local response to larger societal debates. Both were used at the local level without a legal requirement, and without a body of theory to draw upon. This explains the large variety in many aspects of *STEP* and LAs 21 in practice.
- Since neither tool is required legally in the German planning system, they operate independently of the traditional system of urban planning. This helps to explain their weakness within the decision-making system and their lack of adequate funding, which in turn lessen their potential for making an impact.
- Both *STEP* and LA 21 use a comprehensive approach to urban development, so both must deal with the trade-offs between comprehensiveness and manageability of issues.
- Although both instruments emphasise public participation and a long-term time fram on a conceptual level, in practice even public awareness of them is often low, and they are often concerned with short-term issues and decisions.

Major differences:

- LAs 21 share the goal of sustainable development, while *STEP* had no general goal in terms of content. While *STEP*'s procedural goal was to make planning more rational and to strengthen local government, LAs 21 are oriented towards achieving consensus. LAs 21 operate within the context of the Rio process, while *STEP* lacks an international context. In the 1970's, environmental protection was just coming onto the agenda, and the concept of sustainability had not yet been developed - both of which are issues of central importance to LA 21.

- *STEP* was located within local government, worked for local government, and included the public only within limited public participation activities. LA 21 is located both within and outside of local government, and is carried out by a broader spectrum of actors. To summarise, the public plays a more dominant role in LA 21 than in *STEP*, and local government's role has been altered.
- In principle, all local government departments were supposed to be involved in *STEP*, but in practice, it was difficult to achieve their full co-operation. Where LA 21 is located within the local Environment Department, a sector agency, its comprehensive approach is even more difficult to realise than that of *STEP*.

5 Learning from *STEP* for LA 21

While the similarities between *STEP* and LA 21 are considerable, we think that LA 21 is not just an example of *STEP* (in contrast to BMBau 1996). LA 21 goes beyond *STEP* concerning its goal of sustainability over a period of several generations, its international context, its focus on building consensus, and the inclusion of many actors outside local government. Although *STEP* cannot provide guidance in these matters there are other lessons to be learned.

LA 21 attempts to deal with many of the same issues of content and procedure as *STEP*. In addition, LA 21 has to grapple with the issue of sustainability which was not yet on the political agenda when *STEP* was in practise. It is generally agreed that achieving sustainable development is an extremely difficult and complex task; LA 21 is attempting to do what *STEP* sometimes failed to achieve, as well as guide urban development towards sustainability at the same time.

Even more than *STEP*, LA 21 is located outside of the traditional systems of urban planning and decision-making. While *STEP* was positioned within local government, LA 21 processes are even partially outside it. Neither *STEP* nor LA 21 are legally codified, which weakens their positions versus other forms of planning. Sector planning agencies, e.g. those responsible for transportation planning, can draw on greater knowledge, financial and institutional resources to prepare more detailed plans than comprehensive approaches like *STEP* or LA 21 are able to. These agencies can use that knowledge and those resources strategically to promote their own interests and even impede a comprehensive approach.

Considerable theoretical work on methods of *STEP* was conducted while it was practised. LA 21, too, requires theoretical support. Research should be supported at national and international levels, drawing on questions that arise at the local level and aiming to improve LA 21. Once LA 21 processes have progressed further they should be evaluated, and there should be a comparative analysis of LA 21 processes in different countries so that individual LAs 21 can benefit from the experiences of others.

The most important lesson to be learned from the experiences with *STEP* is to be realistic as to what LA 21 can and cannot achieve. Attempting to change the course of urban development towards long-term sustainability and to achieve consensus on a wide range of issues by LA 21 alone may result in frustration and disappointment. When assessing what may or may not be realistic for LA 21, it is essential to be conscious of the local power structure, including key players, and the local planning system, as well as the potentials and limitations of local government. If this knowledge is used constructively to choose the right issues and strategies, an LA 21 process will be able to contribute to urban sustainability.

References

BMBau (Bundesministerium für Raumordnung, Bauwesen und Städtebau) (ed) (1996) Lokale Agenda 21. A: Stand und Perspektiven der Umsetzung von Kapitel 28 in Deutschland; B: Übersicht über internationale Programme und Strategien. Schriftenreihe Forschung Nr. 499. Bonn

Difu (Deutsches Institut für Urbanistik) (1997) Erfahrungsaustausch von LA 21-Initiativen. Pressemitteilung. Berlin

Enquete-Kommission (1996) Anhörung der Enquete-Kommission 'Schutz des Menschen und der Umwelt - Ziele und Rahmenbedingungen einer nachhaltig zukunftsverträglichen Entwicklung' am 18.11.1996 zum Thema 'Kommunen und nachhaltige Entwicklung'. Fragen- und Sachverständigenkatalog und Stellungnahmen der Sachverständigen. Kommissionsdrucksachen 13/3 vom 24.9.1996, 13/3a vom 6.11.1996 und 13/3b vom 12.11.96. Bonn

Ganser, K (1991) Instrumente von gestern für die Stadt von morgen? In: Ganser, K, Hesse JJ, Zöpel C (eds): Die Zukunft der Städte. Nomos, Baden-Baden pp 54–65

Küpper UI (1983) Leistungspotentiale in der Stadtentwicklungsplanung. In: v.Lölhöffel D, Schimanke D (eds) Kommunalplanung vor neuen Herausforderungen. Bausteine für die Verwaltungspraxis der 80er Jahre. Birkhäuser, Basel

Lange H-G (1983) Funktion der Stadtentwicklungsplanung in den 80er Jahren. In: v.Lölhöffel D and D Schimanke (eds) Kommunalplanung vor neuen Herausforderungen. Bausteine für die Verwaltungspraxis der 80er Jahre. Birkhäuser, Basel

Lustig S (1988) Umweltschutz in der kommunalen Gesamtplanung. Master's thesis (unpublished)

Weiland U (1997) Nachhaltige Stadtentwicklung - Diskussionslinien und offene Fragen. In: Hübler K-H, Weiland U (eds) Bausteine für eine nachhaltige Raumentwicklung in Brandenburg - Berlin. Verlag für Wissenschaft und Forschung. Berlin

Community Energy Planning – Energy Action Plan of Brundtland Town Rajec

Ladislav Židek

Rajec became a "Brundtland" town in 1992. The Brundtland project was named after Gro Harlem Brundtland, the former Norwegian prime minister, who wrote an UN report in 1987 dealing with the idea of "permanently sustainable development." The idea of decreasing the cost of energy and its costs of production while improving the condition of the environment are important factors when working towards the attainment of a permanently sustainable way of life. By improving the conditions for life on earth, conditions are created for a fulfilling life for ourselves and for those who will come after us; this is the idea behind the Brundtland project.

The main goal of the project is to decrease energy consumption and the utilisation of renewable energy sources with the goal of improving environmental conditions. Implementation of this goal will contribute to the economic strengthening of the region and the creation of conditions that lead to a permanently sustainable way of life.

The task of this Energy Action Plan is to evaluate the current course of the project, to quantify the results achieved, and to suggest the direction in which the municipality of Rajec should proceed in order to implement the Brundtland town demonstration project.

Rajec lies in Northwest Slovakia. The city has 6100 inhabitants, out of 30,000 inhabitants in the entire region. The area is mostly forest, providing an abundance of energy material (wood) which is the mainstay of the project.

1 Reducing Energy Consumption

In order to fulfil this goal, the following projects have been implemented in the past four years:

- Changing of the heating delivery system in the housing estate SEVER. This included:
- the liquidation of two coal boilers;

Rajec City Council, Námestie SNP 2/2, 01501 Rajec, Slovak Republic

J. Breuste, H. Feldmann, O. Uhlmann (Eds.)
Urban Ecology

- a new two-channel system for heat delivery;
- installation of equipment for decentralised preparation of hot water in each building;
- meters and regulators for heat and hot water.

These measures resulted in an average energy savings of 30%.

- A Phare demonstration project was implemented in 1993 for insulation, provision of meters, and regulation of 24 flats; another 24 flats were improved in 1994.

Under this insulation programme, 25 individual households have been renovated between 1994 and 1997.

These measure resulted in an energy savings of more than 40%.

- Improvements at the House of Gratitude (a home for the elderly and handicapped) included:
- regulation of heating;
- installation of compact light bulbs;
- insulation of the ceilings and windows;
- installation of high efficiency boilers for heating and hot water;
- installation of thermostatic valves.

These measures should reduce energy consumption by 20%.

2 The Concept of Biomass Utilisation

In comparing the possibilities for using biomass in Slovakia, specifically Rajec, with Austria, Denmark, Germany and other countries, we see that it is possible that biomass, especially wood, could provide 10–15% of the energy used in the region for heating and hot water.

The most usable and accessible material is waste in the form of sawdust and wood pieces from saw mills, which can then be made into wood chips for combustion. For this process, it is important to learn about the storage, chipping, and drying of wood. Waste from forests is used only minimally (5%).

2.1 Salix Programme

In 1996, a local environmental protection group obtained a grant from the Environmental Partnership for Central Europe and from municipal funds for the planting of one hectare of the energy rich willow *Salix viminalis*. In total 24,000 seedlings of three varieties of *S. viminalis* were planted in Kriva na Orava in northern Slovakia, seedlings were also obtained from this area.

We are currently looking at the possibility of expanding the area to between 10 and 20 hectares, and using the willow to help with waste water treatment.

2.2 Biomass Combustion

In Rajec, we are looking for other locations where installation of wood waste boilers would be possible. Total output of the boilers in the foreseeable future will be about 10 MW and, using this calculation, wood will provide approximately 10% of the town's energy in the next five years.

In this project, the following activities have been implemented:

- wood waste boiler at the local Church School;
- conversion of a heating estate boiler room from coal to wood. The boiler room provides heat and hot water to 187 flats and a kindergarten, and we are looking at possibilities for adding more consumers to the heat network.

The basic output of the coal boilers was 3.05 MW in comparison to the new boilers for wood waste which have an output of 2.4 MW and an efficiency of 88%.

2.3 Conditions for Continuing the Biomass Projects

Projects for using biomass have, until now, been supported by the state and other funds, but these are insufficient. It is also necessary to overcome apathy and to obtain support from private citizens. It is simply not possible to create a project on an industrial basis and to show that it is financially viable with current commercial credit with a return time of four years.

For the implementation of these projects, it is necessary to create good conditions and a competent management team, who must carry out a complete analysis of the situation, from the source of wood waste to the collection of results.

3 Results Achieved

An analysis of energy consumption and emissions in 1992 was the starting point for the prognosis for future energy consumption and the burden on the environment. The implementation of the aforementioned projects, as well as others, has resulted in the following:

- total energy consumption in Rajec after four years was reduced by 6.4%. These savings fit in with the prognosis formulated in 1992.
- The structure of the consumption of individual sources of energy from 1992 to 1996 changed with the increased use of natural gas (from 39 to 46%) and wood (from 0.3 to 4%). Use of coke and coal was reduced from 44 to 31%. Electrical energy use grew only slightly.
- In comparison with the use of renewables in the 15 EU countries, Rajec comes close to the average figure of 5%. In Rajec, wood waste is used four times more than on average in Slavakia.
- A comparison of the emissions from 1992 and 1996 shows that these projects have made an impact, as emissions have been reduced by 49% in the last four years.

4 Prognosis for Development

The Energy Action Plan for Rajec was prepared with grants from the Slovak Ministry of the Environment, a Danish ministerial fund for eastern European countries, represented by the Danish Technological Institute. This Energy Action Plan is part of the project "the Reduction of Pollution from the Production of Energy in Slovakia."

In the plan, further trends in the region's development are outlined, and these trends are divided into three scenarios. For the long-term scenario of 20 years, energy use is predicted to decrease by 35% and CO_2 levels by 42%. The share of renewable energy sources for the energy used in the town should be increased to 10–12% of the total sources after five years.

Decreasing energy consumption and the use of renewable energy sources also has economic advantages for the region. In 1992, only 0.2 million Slovak crowns (Sk) remained in the region, but this figure increased to 7.9 million Sk in 1996 from the saving of energy and 1.9 million Sk from the use of local biomass.

According to the prepared prognosis, by 2015, the total figure could increase to 81.4 million Sk.

Implementation of these goals will go far towards attainment of the goal of the Brundtland project, and it will also help to involve the widest variety of inhabitants. At this point, the project cannot continue without the understanding and support of the majority of the town.

For this reason, the Foundation for Energy Savings and the Protection of the Environment in Municipalities has set up a programme "Integrated Programme of Environmental Education for Inhabitants." This involves different schemes for educating the next generation and their parents about energy saving. This year, 1500 children from the Rajec region were involved in the programme.

In 1997, the Municipality of Rajec received a grant from the European Commission for the project "Energy Efficiency in the Municipality of Rajec." The programme is part of the Ecos-Ouverture and Phare programmes. Our partners in this project are the Norre-Rangstrup and Ballerup municipalities in Denmark, and the Rheinsberg and Neuruppin municipalities in Germany.

The main goal of this project is to exchange political and administrative experiences and knowledge regarding the administration of municipalities. Special emphasis was placed on the concept of "green administration;" that is, to attain a state in which energy and environmental goals are part of everyday routine and general municipal activities, from parks to kindergartens, from energy supply to education. Increasing energy efficiency and the distribution of information regarding energy and the environment to citizens and schools also play a part in this project.

The results of this project should be as follows:

- transfer of experiences between EU and Eastern European countries;
- implementation of a pilot project for street lighting;
- education of children and adults about reducing energy use;
- reduction of negative environmental impacts;
- long-term expectations include a reduction in energy consumption and emissions by 25–50%.

The logic behind the Brundtland concept is to maintain and improve the condition of the environment. It is necessary to save the gift of our earth for the generations which will come after us. This model illustrates the idea of permanently sustainable development, as outlined by Mrs. Brundtland in 1987, which we aspire to in all our ecological, economic and political activities.

Targets for Sustainable City Development – The Example of Halle (Saale)

Martina Werheit

1 Introduction

Although the declaration of the Agenda 21 of Rio de Janeiro demands that the communities and citizens should have been consulted in order to achieve a Local Agenda 21 by 1996, targets of sustainability in an urban context are a relatively new subject in Germany. The Local Agenda 21 process requires targets and indicators to have clearly defined concept and measures, which point out the extent of sustainability (BMU 1992).

A concept of quality targets for the city of Halle will be worked out within the scope of a research[1]. A central aspect is to examine if and how sustainability in a local framework can be measured.

The aim of the project is to develop spatially modified targets for a sustainable city development within the pattern of environmental quality target concepts, which are already maintained in urban environmental planning in Germany (UVP-Förderverein 1995). The foundations are general aims derived from global and regional ecological carrying capacities, which have been modified for the city of Halle. Indicators related to characteristic structural areas will be identified (in accordance with the above background information), which can be transferred to comparable urban structural areas (Fig. 1). In this context ecological as well as economic and social components will be considered, according to the demands of the Agenda 21 (BMU 1992).

2 The City as a Part of the Global Ecosystem

The exploitation of nature on the one hand and the environmental stresses on the other hand do have limits in a finite global environment. This leads to

Martin Luther University of Halle – Wittenberg, Institute of Geography, Heinrich-und-Thomas-Mann-Straße 26, D-06099 Halle (Saale), Germany

[1] The research project "Identification of models and targets for selected urban structural areas of the city Halle/Saale concerning sustainable development" is being carried out at the Geographic Institute of the Martin Luther University of Halle.

J. Breuste, H. Feldmann, O. Uhlmann (Eds.)
Urban Ecology

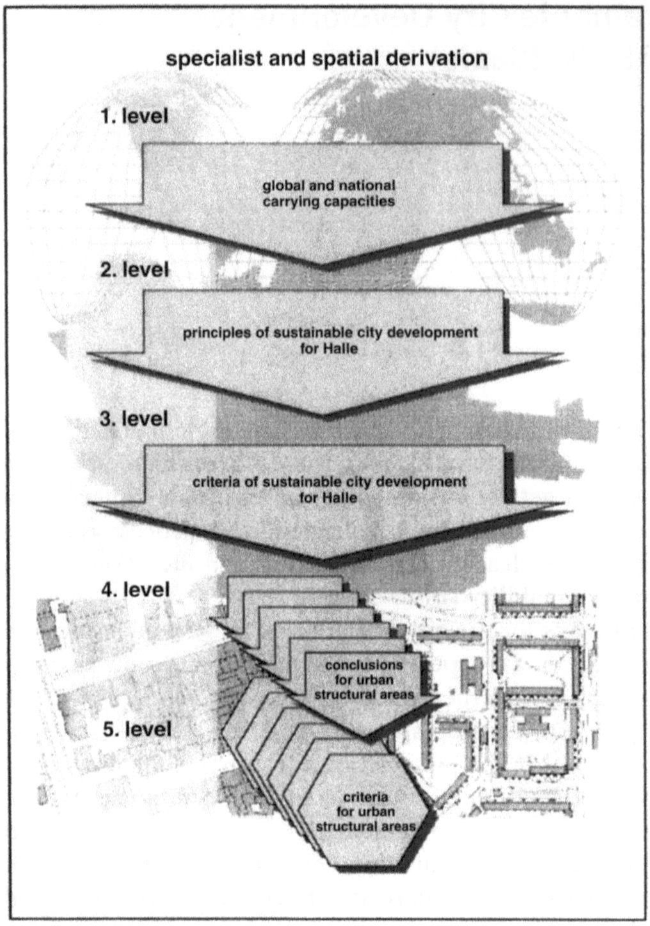

Fig. 1. Deriving of targets of sustainable development for urban structural areas (Werheit/Friedrich/Kühling 1998)

two principles of sustainable development, which are also valid for an urban sustainability process. The first principle refers to the use of resources: on the one hand the use of resources should be restricted by their ability to recover; on the other hand the consumption of finite resources should be reduced by replacing them with renewable resources or by substituting them with technical improvements. The second principle demands that the preservation of the natural stock requires a limitation of pollution inputs within its carrying capacity. (SRU 1994) Therefore the main consequences for a city are a reduction in the use of resources on the input side, and a decrease of emissions on the output side.

Within the cities there are four important requirements to apply to the two principles. Firstly, the way in which renewable and non renewable resources are consumed must be questioned. Secondly, a reconstruction of the processes of interchanges is required, especially considering the urban fringes, but also global interchanges. Thirdly, the reorganisation of urban structures of settlements and space is required for efficient consumption of energy, soil and material and an ecologically viable mobility. Fourthly, the integration of social and economic demands is an entirely new dimension of traditional community management in Germany.

3 Urban Sustainability Targets

In addition to qualitative aims, the definition of quantifiable targets are a central part of a Local Agenda 21 process as a basis for political decisions and actions, as well as for controlling the process. The four leading requirements already mentioned can be maintained as general goals for an ecological urban development. In contrast to these, self engaged measurable targets for a sustainable city development are rare. However, the derivation of existing targets in an urban context seems to be problematical, because the relationship between criteria and targets is often unclear.

The current qualitatively determined carrying capacities have been scientifically derived. The Wuppertal Institute has shown the limits of a sustainable development for Germany in a case study (BUND/Misereor 1996). Basing on global limits of environmental carrying capacities, the scientists have determined the "environmental space" of the German population in 2010 and in 2050[2]. Building on this kind of national indicator systems, conclusions for sustainability targets in the urban framework of Halle can be made. Proceeding from national targets it is necessary to ask what kind of indicators within the urban system can reflect the capacity limits respectively the urban environmental budget. Additionally, the development of suitable indicators has to be taken into account further criteria, e.g. a relation to measuring marks, a spatial and a time relation (see for more: Werheit 1996).

For the city of Halle eight central indicators have been chosen. The selection is based on indicators which point out the most important global and regional environmental problems caused by the specific city system of Halle. Apart from common environmental indicators, socio-economic indicators are chosen to specify the causes of environmental impacts (Fig. 2).

These indicators make the framework for the formulation of quality targets for urban sustainable development in Halle. Initial results show that to some

[2] The "Friends of the Earth Netherlands" have coined the expression of the "environmental space" and have determined it for the Netherlands.

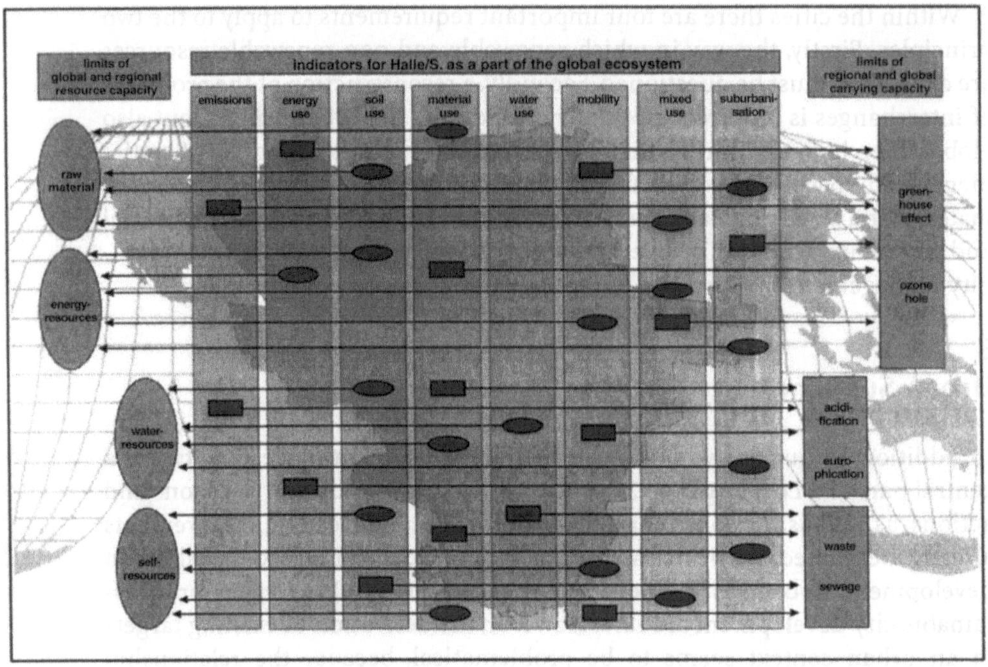

Fig. 2. Deriving of urban indicators from aims of sustainable development (Werheit/Friedrich/Kühling 1988)

extent aims on the local level cannot be derived from general national aims. For instance, the Wuppertal Institute demands a stabilisation of the spatial use until 2010, but in comparison to the average German city, Halle has a very high population density, so the increase in traffic and building areas per year can possibly be higher than the average level in other cities.

4 Outlook

The introduced target system will soon be presented to the public of Halle. The development of the concept is taking place in co-ordination with the community administration, so that the results can contribute to the Local Agenda 21 process in Halle. Hopefully it will start a political discussion towards a sustainable development, because unfortunately the Local Agenda 21 process in Halle is only just starting. It is desirable that this target concept responds to many actors and that, thereby, many inhabitants become conscious of their responsibility for the future generations.

References

BMU (Bundesministerium für Umwelt, Naturschutz und Reaktorsicherheit) (ed) (1992) Bericht der Bundesregierung über die Konferenz der Vereinten Nationen für Umwelt und Entwicklung im Juni 1992 in Rio de Janeiro, Agenda 21. Reihe Dokumente, Bonn

BUND/MISEREOR (eds) (1996) Zukunftsfähiges Deutschland. Studie des Wuppertal Instituts, Basel Bosten Berlin

Werheit M, Friedrich K, Kühling W (1998) Konzeptionelle Überlegüngen zur nachhaltigen Stadtentwickling. Der struktutelle Ansatz eines Qualitätszielkonzepts für die stadt Halle/Saale. In: Frühauf M, Hardenbicker U (eds) Geowissenschaftliche Umweltforsching im mitteldeutschen Raum, pp 139–146

UVP-Förderverein, Arbeitsgemeinschaft Umweltqualitätsziele (ed) (1995) Aufstellung kommunaler Umweltqualitätsziele. Anforderungen und Empfehlungen zu Inhalten und Verfahrensweisen

SRU (Der Rat von Sachverständigen für Umweltfragen) (ed) (1994) Umweltgutachten 1994. Für eine dauerhaft-umweltgerechte Entwicklung. Stuttgart

Werheit M (1996) Indikatoren für eine nachhaltige Entwicklung. In: Unabhängiges Institut für Umweltfragen (UfU) (ed): Informationsbrief 29, pp 10–14

A Normative Model for Urban Ecology Practice: Establishing Performance Propositions for Ecological Planning and Design

Mary-Ellen Tyler[1] and William T. Perks[2]

1 The Basis for a Normative Approach

The practice of urban ecology is concerned with how urban environments *should* function as built cultural-technological systems. In this context, applied urban ecology is essentially a normative activity. A normative approach is consistent with Simon's (1969) demarcation of the "sciences of the artificial". Unlike the natural or classical sciences, the "sciences of the artificial", which include the disciplines of planning, management, design, architecture and engineering, are focused more on how things should be rather than the analysis of how they are. This normative or prescriptive approach to the artifacts and artifactual systems of the cultural-technological world is characteristic of the design professions that deal with the built human environment. This normative versus analytical approach, is illustrated by planning and design. Both are processes that are practiced by synthesizing, reformatting and systematically adapting extant knowledge for the purposes of resolving societal or contextual problems. Normative propositions represent abstract approximations of reality that can only be evaluated in terms of their epistemological usefulness (Morales 1974). Normative performance propositions for ecological design are both valid and useful epistemological tools for urban ecology practice. However, *what* these normative ecological performance propositions for the built environment should be, remains an outstanding question.

A normative approach views urban ecology as an emerging design process involving form-function synthesis of ecological, sociocultural and technological systems in order to address human activity-centered environmental problems. In this design synthesis, ecological systems and human activity systems (including technology) become mutually interactive and interdependent 'wholes' in which the desired functioning of one system must be conducive to

[1] University of Calgary, Dean, Faculty of Environmental Design 2500 University Drive N.W., Calgary, Alberta T2N 1N4, Canada
[2] University of Calgary, Faculty of Environmental Design, Calgary, Alberta T2N 1N4, Canada

J. Breuste, H. Feldmann, O. Uhlmann (Eds.)
Urban Ecology

the desired functioning of the others. This urban ecology synthesis represents a new concept of urbanism guided by ecological systems principles and a new concept of ecological system function which incorporates technology. The notion of a 'built environment' now must also encompass the notion of a 'built ecology.'

However, to date, the notion of an urban ecology 'by design' remains a radical notion to both natural and social science researchers and the notion of a 'built' ecology is virtually heretical to those dedicated to the conventional wisdom of nature conservation. Similarly, ecology has yet to emerge as the dominant paradigm for design. The industrial paradigm and Le Corbusier's 'city-as-machine' metaphor have created the mechanistic prescriptions for urban space and its social function that dominate modern urban form regardless of geographical and sociocultural context. The purpose of modern science has been technical calculation not the establishment of normative design goals. As a result, the proposition of designing an urban ecology is currently as foreign to modern architecture as it is to modern science.

2 A Brief History of Canadian Urban Ecology

Over the last twenty five years, the Canadian environmental discourse has focused on the legal (regulatory) and public policy mechanisms for addressing regional scale 'point source' air, water, and soil pollution problems associated with industrial manufacturing activities and habitat and species loss related to regional scale natural resource exploitation. It has only been in the last ten years that an urban focus for environmental concerns has emerged around 'people-centered' environmental issues such as health, environmental quality and consumer behavior issues such as energy and water and household level recycling. Now for the first time, the human built environment is the primary focus of attention. For the first time, responsibility for environmental issues falls within the professional realm of urban designers.

However, the potential for urban ecological design is largely unrealized in Canadian urban planning and design practice. In a popular culture where socioeconomic status and success equates with excessive consumption of resources, Hahn and Simonis's (1991) pronouncement that contemporary urban form is an "ecological dead end" has to date fallen on deaf ears.

3 The Current State of Urban Environment and Sustainability Indicators

Within the last seven years the Canadian institutional discourse has focused on developing 'state of environment' indicators. The assumption being that the environmental impact of human activities (in the form of a stress – response relationship) and the 'quality' of human built environment can be monitored by quantifiable indicators. However, urban environment sustainability indicators remain disconnected from urban design. The opportunity costs of supporting large institutional data collection exercises at the expense of doing *in situ* urban ecological design and restructuring demonstration projects need to be seriously assessed.

Urban ecology is about the structural and functional synergies between human artifact systems and ecological systems. Ecological design practice gives built technological form to these human-ecological system synergies. In the end, the practice of urban ecology through ecological design will be farther ahead by knowing what works on the ground and the range of performance that can be expected in specific geographical and sociocultural contexts. Therefore, from a normative ecological design perspective, the next step in the evolution of urban environmental and sustainability indicators needs to be a methodological framework that focuses on the interconnections among individual and sectoral (ecological, socioeconomic, technological) indicators in order to generate increasingly integrated functional relationships that have implications for urban design and urban form response.

4 Four Factors and Four Performance Propositions for Ecological Design Practice

Based on the authors' work in the development of an ecological infrastructure concept for Winnipeg, Manitoba, Canada (Tyler et al. 1995) and a suburban re-design and re-structuring plan for the Edgemont district in Calgary, Alberta, Canada (Perks et al. 1995 and 1996), two major components of a normative approach to urban ecology practice have been identified. The first is the 4 Factors model, and the second is the development of four normative performance propositions for urban design. All complex system constructs have four characteristic factors: function, structure, behavior, and inter-relationships. Ecological systems and human sociocultural and technological systems manifest these four factors by analogy to this general system construct but in characteristic ways. Although they are manifested in different ways, all ecological systems (regardless of type) have energy flows and nutrient cycling (function), all ecosystems have vertical and horizontal stratification (structure), all exhibit community change over time (behavior), all have trophic webs and habitats (inter-relationships). Similarly, physical and spatial urban systems and human sociocultural and technological systems also exhibit specific and characteristic function, structure, behavior, and inter-relationships.

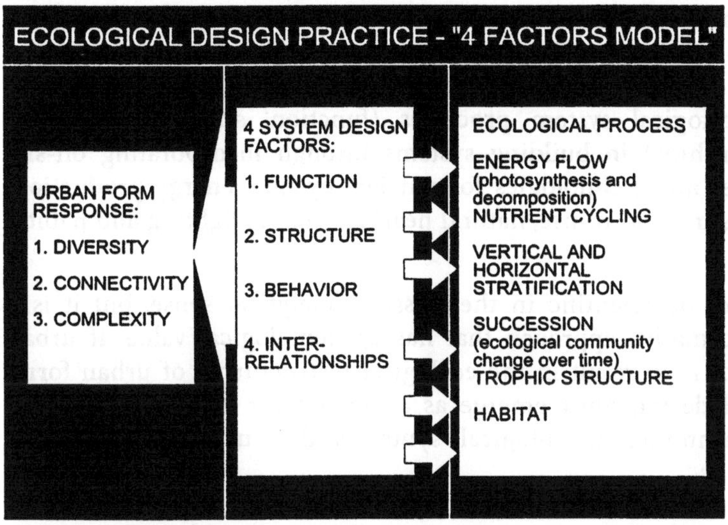

Fig. 1. Ecological Design Practice: "4 Factors Model"

All complex system constructs have four characteristic factors: function, structure, behavior, and inter-relationships. The 4 Factors framework illustrated in Fig. 1 provides a heuristic design and planning tool for metaphorically identifying, cross-matching, and interconnecting the function, structure, behavior, and inter-relationships characteristic of ecological systems with the function, structure, behavior, and inter-relationships characteristic of human-technological systems. For example, knowing how energy flow and nutrient cycling processes function in ecological systems, how can we identify a functional corollary in urban systems and then, how can we identify what possible alternative technological, spatial, and physical processes and forms are necessary to support this? This heuristic inquiry can also work in the reverse direction. While we can identify potential 'artificial' built or technological surrogates for ecological processes, we can also identify potential biological and ecological surrogates for 'artificial' technologies.

Urban form response that focuses on the design of diversity, connectivity and complexity can be directed by four normative propositions:

1. Maximize mix of housing types and layouts that give site specific attention to the configuration of ancillary buildings as elements that define and shape access to and interconnections between outdoor and rooftop spaces and common facilities that supplement the residential unit.
2. Maximize spatial and structural intensification of urban form elements using pedestrian circulation, time-distance, multiple use of space and functional interdependencies among housing, workplace and commercial services as key organizing elements.

3. Maximize land use mix and variations in site development standards that are based in community context and address goals for social activities, ecological function and their mutual interdependence.
4. Simulate ecological system processes (function, structure, behavior, inter-relationships) in building systems through incorporating on-site opportunities and technologies for nutrient cycling, energy production, water harvesting and re-use, natural heating, cooling lighting and habitat diversity.

Design may not be scientific in the classical analytical sense, but it is a form-giving normative approach that has epistemological value. If urban ecology practice is to result in the ecological restructuring of urban form, then ecological design must emerge as a method for giving form to the new human-technological-ecological synthesis that urban sustainability represents.

References

Hahn E, Simonis U (1991) Ecological Urban Restructuring. Ekistics 58: 348/349

Morales R (1974) A Philosophical Approach to Mathematical Approaches to Ecology. Ecosystem Analysis and Prediction. Proceedings of the SIAM-SIMS Conference. Alta, Utah

Perks WT, Kirby R, Wilton-Clark A (1995) Edgemont II: A Design Study in Sustainable Community Form. Centre for Environmental Design Research and Outreach (CEDRO), Faculty of Environmental Design, University of Calgary, Calgary, Alberta Canada

Perks WT, Wilton-Clark A (1996) Testing Consumer Receptivity to Sustainable Community Design. Canada Mortgage and Housing Corporation, Ottawa

Simon H (1969) Sciences of the Artificial, 2nd edn. The Massachusetts Institute of Technology, Cambridge

Tyler ME, Dunn S, Moore J, Robertson J (1995) The Ecological Infrastructure Concept and its Implications for Urban Form and Function. Dept. of City Planning, University of Manitoba, Winnipeg, Manitoba (unpublished)

Integrated Flow Management in European Urban Regions: Managing Technical Networks as Instruments for Environmental Change

Timothy Moss

1 Introduction

The aim of this chapter is to illustrate how the management of utility services for water, sewage, electricity and solid waste in different urban regions in Europe is shaping their potential to save natural resources and protect the environment. Utility services govern the flow of a substantial proportion of materials and energy used in cities. They draw on external water or energy resources, distribute these in the required quality and form to consumers and dispose of waste products in and beyond the urban region. It is true that the urban form creates openings for resource efficiency – for instance via combined heat and power generation – but the technical networks which helped build our cities have themselves contributed to the high level of resource use and related pollution.

The challenge to limit the environmental impact of utility services has in the past been treated by urban planners and network managers primarily as a technical problem which can be solved via the application of the appropriate technology. However, recent difficulties in meeting the demand for natural resources, in financing network modernisation and in planning infrastructure plants are suggesting that the root problem has just as much to do with the way utility services are being managed. It is clear that the structures, processes and logic of conventional network management are often not conducive to resource conservation and the prevention of pollution. In order to tap the enormous potential of utility services to improve environmental quality there is a need for innovation – not just technical and behavioural innovation, but *institutional* innovation. This search for new modes of infrastructure management is particularly poignant today, at a time when the liberalisation, deregulation and even globalisation of energy, water and waste services across Europe is creating new openings and undermining old strategies for environmental protection.

Institute for Regional Development and Structural Planning (IRS), Flakenstraße 28–31, D-15537 Erkner, Germany

J. Breuste, H. Feldmann, O. Uhlmann (Eds.)
Urban Ecology

This chapter explores the forces shaping the environmental management of technical networks in three sections: conventional forms of network management, new challenges facing the 'flow managers', and recent responses by the actors. For each section the aim is to pinpoint the forces working for or against the application of environmental technologies and practices. It is argued that the rapidly changing context of utility services is forcing urban managers to question the supply-driven logic of conventional infrastructure management and to seek a closer balance between supply and demand, involving greater interaction between service provider and user.

2 Conventional Network Management

Conventionally, management of urban infrastructure has been dominated by a public service ethos to provide adequate and secure services which are available and affordable to all. The overriding aim has been to build up supply capacity and extend the networks to meet rising demand, thus encouraging the development of centralised network systems. Under this supply-driven logic there has been very little interest in shaping the intensity, time or place where demand is routed. Most utilities, after all, have in the past enjoyed a territorial monopoly over local services. Consequently, there has been little social or spatial differentiation in the treatment of consumers, and interaction between service providers and service users has been predominantly one-way and top-down. Environmental improvements have followed regulatory pressures and have been interpreted in largely technical terms.

Within this general picture there are significant variations regarding regions and sectors. These depend primarily on the degree of local government responsibility, the regulatory framework, the market structure of each utility service and the dominant technique or techniques used. Whereas in the UK (prior to privatisation) infrastructure services were provided by distinct national and regional public bodies, in Germany and Denmark they remain a municipal responsibility operated by a large number of often minor actors; this applies particularly to the water/sewage sector. The electricity sector in the two countries is characterised by a hierarchy of operators from the supra-regional to the municipal level, whilst solid waste services - being independent of pipelines - have a more flexible structure of spatial organisation.

3 New Challenges For The 'Flow Managers'

The conventional features of infrastructure management, however, are coming under pressure from a variety of emerging challenges. These new demands or 'signals' include primarily: the liberalisation of utility service markets, the tightening of several environmental standards, new forms of environmental incentives, the high cost of network modernisation, competition between a growing number of viable technologies, uncertainty over future consumption patterns and over-capacity in some networks. Infrastructure management is thus currently undergoing considerable transformation on several distinct, but inter-related planes. The pace and nature of this transformation process varies considerably according to regional and sectoral specifications – the different degree of privatisation of utility services in the UK compared with Germany or Denmark is a clear example. The important point is that the emerging challenges are affecting all 'flow managers', whether public or private, local or supra-national, and thus tensions are created between these and conventional forms of infrastructure management.

If we 'read' these signals in terms of their impact on urban environmental policy we can detect both new opportunities and new obstacles to past practices. For instance, the current modernisation of water, sewage, electricity or waste infrastructure offers the chance not only to improve environmental performance but also to limit resource consumption through Demand Side Management, particularly where demand threatens to exceed existing capacity. Conversely, stagnating or unpredictable demand – as detected for the water and electricity sectors in Berlin and Copenhagen – questions the conventional supply-oriented logic which is based on steadily rising consumption. Here, the immediate impact may be in the opposite direction, maximising under-utilised existing infrastructure.

To take another example, new regulatory pressures and financial incentives to conserve energy and water, minimise waste and limit pollution lead to substantial environmental improvements. Yet it is difficult for service operators to meet the immediate costs of these measures when they are under strong pressure to lower service charges in an increasingly competitive market, particularly when consumption and sales are stagnating. Indeed, the commercial challenge of liberalisation is undermining the regulatory basis of several important environmental instruments of the past. In Germany, the principal tool for promoting the use of renewable energy for generating electricity – the „*Stromeinspeisungsgesetz*" – relies on the system of territorial monopolies to specify which power company is responsible for paying for electricity fed into its network from such sources.

4 Recent Responses by the Actors

The combination of these new challenges are changing the context of infrastructure provision, forcing network managers to re-think how resources and networks are managed. Our research illustrates how the different actor groups are responding to these signals through the way that services are being provided and used.

Under the new competitive climate utility companies are cutting operational costs and improving management efficiency. In the case of the larger companies they are also diversifying and widening their commercial interests into other utility sectors as well as into other countries. Again, the scale of this response depends on the particular commercial context; for example, the actions of privatised companies in the UK are driven by the overriding interest of their shareholders.

A new logic of differentiation in the treatment of customers is slowly developing as utilities seek to capture 'lucrative' customers, resulting in more reciprocal and dense relationships between producers and select consumers. Moreover, service providers are increasingly seeing commercial opportunities through encouraging key customers in areas of network stress to reduce consumption levels and use resources more efficiently. Measures of Demand Side Management are being adopted to avoid economically and environmentally costly infrastructure investment and to steer the level and timing of demand. Nevertheless, environmental improvements are determined more than before by commercial dictates; in areas of over-capacity the incentive of the operators to conserve water or energy is small.

In response to these trends, the bodies responsible for regulating the environmental and economic performance of the utilities are seeking new constraints and incentives capable of stimulating resource conservation and pollution prevention under the altered framework. These may take the form of taxes on resource use (e.g. on water abstraction, fossil fuels, landfill sites), performance standards (e.g. on the quality and price of services) or measures to strengthen the 'polluter-pays' principle (e.g. promoting the use of modern monitoring techniques).

The users, or consumers, are also developing an interest in utility services which in the past they took for granted. Some are registering their opposition to planned sites for new plants, or to the rising cost of service charges caused by network modernisation and customer differentiation. Many are seeking ways of reducing their own consumption levels and recycling waste, whether for environmental reasons or to offset the increase in unit service charges. Some are themselves becoming infrastructure 'providers', operating their own rainwater percolation systems or composting units.

5 Conclusion

This chapter has argued against the common tendency to reduce infrastructure management to an exercise in technological modernisation. A narrow focus on technical innovation is ill-suited to meet the multiple challenges confronting today's 'flow managers'. There is an urgent need to understand the complex processes and the rapidly changing structures shaping the kind of utility services provided and the ways they are used. An integrated approach to network management capable of anticipating the social, economic and spatial implications of recent developments would appear to be essential if the enormous potential of technical networks to improve environmental quality is to be exploited fully.

Our research suggests that urban managers need to recognise that utility companies are urban and regional players with a growing influence over local spatial development and local governance. The overall scope for urban policy intervention in the process of network management may be diminishing, but there are openings for public sector influence. In particular, closer co-ordination of network and urban planning enables areas of network stress to be relieved through Demand Side Management and areas of over-capacity to be the target of development, in the interest of both utility companies and city planners. This scenario depends, however, on new institutional forms of co-operation and consultation which, as yet, are only in their infancy in Europe.

Note: This chapter is based on a research project - "Technical networks as instruments of sustainable flow management: a comparative analysis of infrastructure policy and planning in European urban regions" - funded by the European Commission under the 'Environment and Climate' RTD-programme and conducted jointly by the IRS (project co-ordinator), the University of Newcastle (GB), the Technical University of Denmark (DK) and the planning consultants ELIUS (GR).

Thoughts Concerning a Theory of Sustainable Urban Development

Ursula Richter

1 Theoretical Basis

The needs of future generations must be taken into consideration when planning and using resources to satisfy present day needs. Theoretical discussion of such principles lead to numerous interpretations of *sustainable development* which are determined by two essential standpoints: the narrow interpretation of an ecologically sustainable development embracing purely ecological aims, and the more comprehensive interpretation, embracing the conflicting economic, social and ecological objectives (assuming a growth trend). There is currently no theoretical model which can serve as a reference framework for differing ideas; discussion subjects, mostly connected with development and business, are based on reviews of real-life conditions (Vornholz 1995). The main theoretical points are also revealed in the social sciences, i.e. initially from the side of a town critique as exposed by Jacobs, Mitscherlich, Barth, Häußermann/Siebel, Sennett, Hoffmann-Axthelm. The question is whether sustainability can be conceived of as an evolutionary concept.

One of the results of the German BMBF Research Association on Land Use Competition is: "What sustainability actually means becomes evident in myopically revolving plans based on empirical and scientific knowledge. However these changes correlate just as well with valuation as with sustainability. Sustained management of the economy is therefore a social process of learning and self-organisation" (Weise 1996a).

The sustainability criterion for ecologically compatible development precludes the reduction of the natural resource stock for which there are three elementary rules: first, the rate of depletion of resources does not exceed the rate of natural regeneration; second, the level of emissions does not threaten the assimilating capacity of the ecological system; and third, the constancy of the natural capital stock has to be guaranteed by the substitution of exhaustible resources.

IMU-Institute GmbH, Head of the BMBF Research Association on Land Use Competition, Pohlandstraße 19, D-01309 Dresden, Germany

J. Breuste, H. Feldmann, O. Uhlmann (Eds.)
Urban Ecology

2
Sustainability in the Urban Context

Important links for transferring sustainability concepts to the urban context are reversibility, regeneration, substitution and reduction. *Reversibility* seems an intractable problem when considering the ecology of urban areas; the trends of social and cultural considerations and decisions also need to be reversed. This requires tolerance, adaptability and flexibility, the ability to appreciate realities and the acceptance of reversibility as a valid principle. *Regeneration* means the ability to be restored or reactivated. *Substitution* means the possibility for substances and energies to be replaced through avoidance of use, applied technology and affiliated technologies incorporating a different urban engineering basis and organisation of the urban economy as well as *reduction* of effects, including those on the physical environment (Arlt, Siedentop 1996).

Sustainable urban development standards should not be seen from a purely scientific angle, rather, as a process tending to be open, integral and in all probability very dynamic; this differs from the current philosophy of basing actions on the implementation of plans. New resources must be used – i.e. people.

An altered, integral appraisal of nature and of humans is required, in particular regarding the prevalent philosophy of engineering which generates permanent pollution, has huge consequences on settlement structure and causes wastage of resources. Advanced engineering has a direct bearing on sustainability (Richter 1996) concerning the threat and destruction and also the potential solutions; current standards and quality should be retained whilst taking into account sustainability criteria. Other possible solutions include the alteration of the industrial manufacturing basis and the resulting, more "sustainable" prospects. The (25 year) "old" solution used by the social critics included localised urbanisation, zoning mixture, ecologically qualified etc..

3
Applicability to the Context of Town and Economy

A sustainable urban development concept requires ethics and rules to ensure the long-term conservation of the city without forgetting that the city is the place of cultural innovation, of trade and development. The integration of various actors in the urban milieu is thus of great importance.

The contemporary settlement pattern opposes what we would understand today as sustainable urban development. The use of land in European towns is neither resource-orientated nor renewable despite the alternative ways which are available and have already been used, as shown by examples in the

15 years of conversion towards ecological urban development. It encourages segregation and destroys social diversity as well as the diversity of functions due to new mixing and diversity. Functional de-concentration and segregation as *unwanted* ecological process are the spatial results of the interaction of the industrial division of labour and segmentation of functions.

The foundation of the consumer society, through the division of labour and specialisation, has always combined progressive segmentation of functions in space with continuing consumption of land; the potentials that begin to become evident due to the transformation thus assume a key position for sustainable urban development (Brake 1995). The resulting practice of municipal planning and economic actors have meant that there is a lack of consideration of the relevance of enterprises in a town; it is thought that business enterprises do not really belong to the town, but can be located anywhere, preferably where they cause the least disturbance (Feldtkeller 1994).

Unlike the philosophy of functional division of labour in space (based on the guiding principle of the Charter of Athens) which, until recently, determined the administration of planning, mixed use in town planning is now an important model for a generally acceptable sustainable development (BfLR 1996). Structural economic change is accompanied by a new awareness of municipalities' problems and the future development of the town as the location of production and innovation. Current research programmes on mixed-use areas (in particular BMBauExwost etc.) question to what extent and under what conditions this model can be improved on as a self-supporting process.

Through the results of the BMBF Research Association on Land Use Competitions a number of empirical findings were found with respect to the correlation of fitness for sustainable development patterns and economic innovative features based on theoretically derived criteria. The theoretical aspects include: *characteristics of the location* as small-component structures, diversity (density and functional mix), nature of the area (idiosyncratic), identity; *characteristics of the ecological carrying capacity* as ecological functionality (low in emissions, resource efficiency, cycle-orientated, renewing, human resource orientated); *characteristics of a sustainable economy* by differentiated regional added value in the context of an ecological business management following the on-site-principle, just as ecological management can support regionalisation at other places (Biehler et al. 1996). A sustainable economy would then be characterised by its integration of spatial-ecological and business-ecological fitness and innovation characteristics in an advanced socio-cultural (not just technically derived) process of innovation (Heller, Richter 1996; RAW 1995).

Sustainable economic activity would accordingly have to contain a differentiated regional added value which secures, develops or produces the future options (e.g. qualification, open land/nature in the town, resource efficiency).

This leads to the creation of reproduction potentials (Sassen 1991) of urban society which at the same time open up innovative, creative environments and increase citizenship and responsibility. Sustained economic behaviour in the town implies the prospect of *another* economy (Busch-Lüty). For the most part the economy-relevant character within the field of another economy has not fully been understood as *innovative surroundings* have been defined socially only in a technological context and not in the relevant decision-making processes in favour of new social self supporting mechanisms (Biehler et al. 1996). The empirical findings indicate a number of possibilities for re-integrating more specific economic activities into parts of the town and especially into dense, functionally-mixed quarters: in particular those with more advanced environmental engineering and hence a reduced degree of disturbance. Higher demands for stimulating surroundings (Jessen 1995) and even the greater demand for flexible workers (part-time workers) make functionally-mixed areas of town attractive. Functionally-mixed micro-locations are useful for certain types of business (or factories) and may lead to novel kinds of intra-urban structures. Land policies and moderation functions of a local innovation policy are of special importance for the transfer of technology (Richter 1998). An active treatment of ecological and sustainable economy should become the main constituent of municipal innovation policy for the future. The promotion of ecological operating methods and the development of preconditions for a sustainable economic concept for the most part requires a correct attitude to cost and benefit.

4
Derivation of Principles of Practice in BMBF Research Association on Land Use Competition

We have established six principles of practice for negotiating the requirements for ecological sustainability, economic efficiency and socio-cultural workability of urban development (Weise 1996 b; Biehler 1996): *internalising external effects, reducing the use of materials and components* (i.e. resources), *maintaining capacity of self-regulation, developing substitutes, allocating functions according to efficiency criteria* and *producing incentives for developing self-organisation* within a democratic process.

The main point is that urban structures should be rendered *ecologically sustainable,* both gradually and continually. The elements for this have been derived for example from the theory and practise of ecological urban renewal (Hahn 1992) and are well known but not necessarily applied. The final ingredient is „ecologisation“ and (if possible) „regionalisation“ of value-added chains (Spehl 1994).

Urban structures must also be made *economically efficient,* sustainable and workable. Sustainable urban development standards (for evaluation from an economic angle) form the basis for the internalisation of external effects; standards for suitable reproductive functions needed for a sustainable urban development should also be established. New markets for renewable merchandise should be developed through the alteration of basic conditions – for example the basic conditions which causes the false mobility price and its resulting environmental problems. Urban structures must be *usable socially* and *workable culturally* when considering the diversity of trading relations. The urban legacy for future generations may evolve from social discourse on „sustainability as value" for which the town requires the right to decide and act at borough and urban quarter levels (Nohl 1996); this is, after all, the basis for the social acceptance of the economic in-value-setting of ecological needs.

5 Conclusions

The economic evaluation principle of internalisation of external effects forms the basic principle for formulating land allocation models from global to regional. It can thus be assumed that a standard for reproductive functions could be formed, but incentives are currently not ideal. Kapp stated that future markets are lacking, as are market preconditions for taking advantage of the economic structural change with respect to the ecological urban renewal. Our research project investigates to what extent cost-benefit studies should be introduced into municipal practice to set a starting point for a more efficient and resource-conserving sustainable development for the urban society. It is important to identify the value of ecological productivity and to evaluate costs and benefits of the ecological functions. If possible, productivity of ecological functions with property rights should be produced. High transaction costs, public goods, external effects etc. may make this unfeasible, in which case policy instruments such as fiscal charges, taxes and rules should be developed. In both cases an approach is needed in order to specify and evaluate ecological productivity to give nature a fair chance to compete.

References

Arlt G, Siedentop S (1996) Grundzüge eines Sustainability-Konzepts für die Siedlungs- und Flächenentwicklung in Großstadtregionen. In: Brake K, Richter U (eds) Sustainable Urban Development: Ausgangsüberlegungen zur Theorie einer nachhaltigen Stadtentwicklung. Dokumentation der 1. Projektwerkstatt, Referat für Stadtplanung und Bauordnung, München 29. und 30. Juni 1995, BIS, Oldenburg pp 13–31

Biehler H (1996) Wie läßt sich eine nachhaltige Stadtentwicklung operationalisieren? In: Brake K, Richter U (eds) Sustainable Urban Development: Ausgangsüberlegungen zur Theorie einer nachhaltigen Stadtentwicklung, Dokumentation der 1. Projektwerkstatt, Referat für Stadtplanung und Bauordnung, München 29. und 30. Juni 1995, BIS, Oldenburg pp 93–96

Biehler H, Ramschütz E, Richter U (1996) Nachhaltige Wirtschaftsweise in der Stadt - Schlußfolgerungen für die Entwicklung von Indikatoren aus der Sicht der Mikrostandortentwicklung. In: Brake K, Richter U (eds) Methoden zur Analyse und Bewertung von Flächennutzungs- und Standortmustern. Nachhaltige Entwicklung von Großstadtregionen. Dokumentation der 2. und 3. Projektwerkstatt, Dezernat Stadtentwicklung und Bau Dresden, 14. und 15. März 1996, Dezernat Stadtentwicklung und Raumplanung Leipzig, 17. und 18. Juni 1996, BIS, Oldenburg pp 189–203

Brake K (1995) Nutzungsmischung: ein Baustein der "kompakten Stadt" - lebenswert und nachhaltig. Referat zur International Making Cities Livable-Konferenz (IMCL), Freiburg

BFLR - Bundesforschungsanstalt für Landeskunde und Raumordnung (1996) Städtebaulicher Bericht. Nachhaltige Stadtentwicklung. Herausforderung an einen ressourcenschonenden und umweltverträglichen Städtebau. Bonn

Feldtkeller A (1994) Die zweckentfremdete Stadt. Wider die Zerstörung des öffentlichen Raums. Campus, Frankfurt New York

Hahn E (1992) Ökologischer Stadtumbau. Konzeptionelle Grundlegung. Peter Lang, Frankfurt a. M.

Heller G, Richter U (1996) Ziele der Städte auf dem Weg in die nachhaltige Entwicklung - das Beispiel München. In: Brake K, Richter U (eds) Sustainable Urban Development: Ausgangsüberlegungen zur Theorie einer nachhaltigen Stadtentwicklung, Dokumentation der 1. Projektwerkstatt, Referat für Stadtplanung und Bauordnung, München 29. und 30. Juni 1995, BIS, Oldenburg pp 83–93

Jessen J (1995) Nutzungsmischung im Städtebau. Trends und Gegentrends. In: Bundesforschungsanstalt für Landeskunde und Raumordnung (ed) Informationen zur Raumentwicklung 6/7, Nutzungsmischung im Städtebau, Bonn pp 391–403

Nohl W (1996) Soziokulturelle Brauchbarkeit der Stadt und nachhaltige Entwicklung In: Brake K, Richter U (eds) Sustainable Urban Development: Ausgangsüberlegungen zur Theorie einer nachhaltigen Stadtentwicklung, Dokumentation der 1. Projektwerkstatt, Referat für Stadtplanung und Bauordnung, München 29. und 30. Juni 1995, BIS, Oldenburg pp 37–45

RAW - Referat Arbeit und Wirtschaft (1995) Wirtschaftspolitisches Konzept der Landeshauptstadt München (Stadtratsbeschluß; Grundlagen für die Perspektive München), Munich

Richter U (1998) Ein Lehrstück auf dem Weg in eine nachhaltige urbane Entwicklung. Produkte für den Umwelt- und Ressourcenschutz - eine Konversionsinitiative der MBB GmbH und der Stadt Augsburg. IMU Studien 22, Munich

Sassen S (1991) The Global City. Princeton/NJ, New York London Tokio

Spehl H (1994) Nachhaltige Regionalentwicklung. ARL-Forschungs- und Sitzungsberichte: Dauerhafte umweltgerechte Raumentwicklung, Hanover, Bd 212: 69–98

Vornholz G (1995) Die ökologischen Ziele im Sustainable Development-Konzept. Eine ökologisch tragfähige Entwicklung. In: Nutzinger HG (ed) Nachhaltige Wirtschaftsweise und Energieversorgung. Konzepte, Bedingungen, Zusatzpunkte. Metropolis, Marburg pp 82–115

Weise P (1996a) Selbstähnlichkeit, Nachhaltigkeit und Typisierung von Flächennutzungen. In: Brake K, Richter U (eds) Methoden zur Analyse und Bewertung von Flächennutzungs- und Standortmustern. Nachhaltige Entwicklung von Großstadtregionen. Dokumentation der 2. und 3. Projektwerkstatt, Dezernat Stadtentwicklung und Bau Dresden, 14. und 15. März 1996, Dezernat Stadtentwicklung und Raumplanung Leipzig, 17. und 18. Juni 1996, BIS, Oldenburgpp 61–67

Weise P (1996b) Nachhaltige Entwicklung aus ökonomischer Sicht In: Brake K, Richter U (eds) Sustainable Urban Development: Ausgangsüberlegungen zur Theorie einer nachhaltigen Stadtentwicklung, Dokumentation der 1. Projektwerkstatt, Referat für Stadtplanung und Bauordnung, München 29. und 30. Juni 1995, BIS, Oldenburg pp 45–55

From *Sustain*ability to *Create*ability? – Analysing *Local Agenda 21*-Initiatives

Roland Söker

1 Agenda – Approach

Local Agenda 21 is a term which probably less than 1% of the inhabitants of a city like Leipzig will know about, which is part of the problem.

Roughly translated, Agenda 21 means what has to be done for the 21st century. The term results from the Agenda 21, one of the final documents of the UN-Conference on the Environment and Development five years ago in Rio.

In section 1 and 2 the Agenda 21 gives information about *what* has to be done; in section 1 social and economic dimensions (combating poverty, changing consumption patterns etc.) and in section 2 – conservation and management of resources. The central conceptual approach of the Agenda 21 is Sustainable Development, going far beyond urban ecology.

The conceptual approach is multi-dimensional: the classical triangle-*ecological, economic and social dimensions*, and cross-sectional- *the international dimension*, in terms of fair North-South-relations and the *inter-generational dimension*, giving coming generations a liveable earth and enough elbow-room for their future.

The central perception is, that development has to take place in the north!

Beside what has to be done the Agenda 21 formulates an approach of *how* that has to be done; in section 3 the methods of strengthening the role of major groups are related and in section 4 the means of implementation.

Throughout the whole Agenda 21 the role of the civil society is recognised. Sustainable Development cannot be administrated by decree. It is a bottom-up process.

Chapter 28 points out the role of local authorities: Each local authority should enter into a dialogue with its citizens, local organisations and private enterprises and adopt a local Agenda 21 (LA21). Through consultation and consensus-building, local authorities would learn from citizens and from

anshiftung ggmbh, Daiserstraße 15 Rgb., D-81371 München, Germany

J. Breuste, H. Feldmann, O. Uhlmann (Eds.)
Urban Ecology

local, civic, community, business and industrial organisations and acquire the information needed for formulating the best strategies.

The key-words are:

- dialogue/consultative process,
- learning from citizens and local civic organisations,
- consensus-building.

These key-words name the major challenges to a LA21. It is difficult, because: dialogue, learning and process contradict with traditional local administration practice, and consensus building is inconsistent with traditional political thinking of parties and organisations.

Sustainability is not just living after adoption of new regulations and accepting limited resources. Sustainability can only be a development model if it sets creativity free and enables people use their skills in this area. Thus Sustainability can be translated into 'Create-ability'.

2
Agenda – Transfer

Before switching from approach to practice, I would like to give you some information about AGENDA-TRANSFER in a short profile:

AGENDA-TRANSFER - *Profile*

Project:

Clearing-house for applied futures (CAF)

Financed by:

Ministry of Urban Development, Culture & Sports and Chancellery North Rhine Westfalia for 3 years (July 1996–June 1999)

Leading idea:

Learning from experience and best practice

Tools:

- brochures & education-materials,
- Info-Service *Stadtgespräche (= talk of the town)*,
- workshops & seminars,
- initial lectures,
- network of co-operating partners in NRW.

3
Agenda – Practice

3.1
Focus on NRW

NRW is a Bundesland in the west of Germany. It has 18 million inhabitants and includes not only the heavily industrialised Ruhr-area, but also rural areas such as Münsterland and Sauerland. I would like to focus on Germany and especially NRW, because that is the region of Germany AGENDA-TRANSFER is mainly working for.

It is difficult to quantify a local Agenda. Some cities call a document a local agenda, but I doubt that it has anything to do with the Agenda 21-approach I mentioned. Some cities give good examples of a local agenda-process, but they do not name it as such. One indicator is a resolution of the city council, because it states the official political declaration we want a local agenda. Of course, that does not ensure action afterwards.

Anyway, a look can be taken at how many city-councils have made a Local Agenda 21-resolution so far. About 60 are already known about in the whole of Germany, but there might be more. I can give you the results of a survey we made recently for NRW. When we started one year ago, there were 4 councils that had a resolution: Wuppertal, Hagen, Düsseldorf and Bocholt. Today, one year later, there are 17 resolutions of city councils and 10 resolutions of committees on environment or planning. In addition there are about 20 other municipalities that have resolutions in preparation. About 100 representatives of municipalities came to our conference in May 1997. Bearing in mind that there are 396 municipalities in total in NRW, and that Germany is committed to undertake the Agenda 21 1996, the number recorded is not very high. One reason is that there was no support by the German Government and only very slow support by the associations in the municipalities.

3.2
(First) Results of Local Agenda-Initiatives in Process:

The local Agenda 21 process in Germany is at an early stage so there are only a few results. The consultation process has just, or not yet, started in most of the cities. Nevertheless I can point out some example of the first results of the Agenda-initiatives in process.

Innovative Campaigns
Raising of awareness is one of the most important and most difficult challenges of the Agenda-process. In some cities there are special Agenda magazines. The most popular campaign has been the summer campaign ('97) in

Munich, inviting the people to take part through a widespread variety of actions. It is organised by the NGO Green City, but financed by the administration and through sponsoring. In Germering, a small city close to Munich, there is another example of the concept of a city-wide Agenda-related artwork on different public places in the town.

Innovative Participation-Models
There are different concepts of working-structures like public forums, round tables with citizens and experts, advisory boards representing the civil society, and steering committees or Agenda-teams.

Reform of Urban Administration
In Münster an Agenda-bureau with four new concerns was established. Many administrations have cross-sectional comities.

Application of Innovative Methods
Connected to the 'new' working-structures are 'new' methods, such as future-workshops or other visionary methods, external moderation or mediation or events for special target-groups, like women and youth.

Innovative and Inherited Projects
A huge number of innovative projects for implementation are 'the living part' of the Agenda-process. It can also be a reinvention of old ideas, developing them to a new stage and making them more attractive.

3.3 Problems of Local Agenda-Initiatives in Process

Some of the main problems are

- participation of the economic actors,
- participation of people who have not been organised,
- doubts concerning implementation,
- contrary development of other sectional planning,
- lacking linkage to the current political topics.

4 Agenda – Outlook

In conclusion, Local Agenda 21 is a local innovation process that aims not only at the interdisciplinary approach of the idea of sustainability, but also at the action-oriented approach of the idea I named 'create-ability'. 'Creatability' is an important step from an instructive society to a learning society. The interactive exchange between discussions of visions, ideals and models for the future and implementation of projects and social innovations (and vice versa) is the specific attraction of a local Agenda-process. As far as I can see, LA21 is the only broader process which is working in this direction.

Since all the LA21-Initiatives in Germany are at a very early stage of development, an evaluation of their success is impossible and counterproductive. Much still has to be done in order to overcome the LA21-problems named above and to obtain sustainable and creative local conditions.

Building a Totally Functioning and Ecologically Vivid City – A Chinese Way Towards a Sustainable City

Rusong Wang and Dan Hu

1 Holistic Thinking: Understanding the Urban Interaction

For thousands of years Chinese have been striving for a harmonious relationship among Tian (heaven or universe), Di (earth or resource) and Ren (people or society), and advocating that Man and Nature become one. The result is a systematic set of principles for managing the relationship between man and its environment, including Dao-Li (physical relationship with the universe, geography, climate, etc.), Shi-Li (planning and management of human activities, such as agriculture, warfare, politics, medicine etc.), and Qing-Li (ecological ethics, psychological feelings, attitudes, motives and values towards the environment). The Yin and Yang theory (negative and positive forces play upon each other and formulate all ecological relationships), Wuxing theory (five fundamental elements and movements within any ecosystem promoted and restrained by each other), Zhong Yong (things should not go to their extremes but keep equal distance from them or take a moderate way) and Feng-Shui theory (Wind-Water theory expressing the geographical and ecological relationship between human settlements and their natural environment) are some of these principles (Wang and Ye Oi 1991). Ten principles of eco-cybernetics were summarised from the ancient Chinese human ecological philosophy (Wang and Hu 1994) which may fall into three categories: competition for efficient resource and available eco-niche; symbiosis between man and nature, among different groups of human beings, and between any human ecological unit and its upper level ecosystem; and self-reliance to sustain its structural, functional and process stability through self-organisation and recycling. Any weakening of these mechanisms will cause a decline in the cities' sustainability.

The city is a kind of artificial ecosystem which is dominated by human behaviour, sustained by natural life support systems, and vitalised by ecological processes. It was named as Social-Economic-Natural Complex Ecosystem

Research Centre for Eco-Environmental Sciences, Chinese Academy of Sciences, 19 Zhongguancun Road, Beijing 100080, China

J. Breuste, H. Feldmann, O. Uhlmann (Eds.)
Urban Ecology

in China (Ma and Wang 1984). Its structure is expressed as a complex between human beings and their working and living settlements (including geographical, biological and artificial environs), their regional environment (including sources for material and energy, sinks for products and wastes, pools for buffering and maintaining) and their social networks (including culture, organisation, technology and so on). Its function includes production, consumption and sustaining. The traditional development view pays attention to only two types of functions, production and consumption, neglecting the sustaining function. This third and most important function ensures sustainable resource supply, assimilation, steering and buffering of man, resources and the environment, and the vitality of self-organisation and self-regulation in human society. According to Lao Dan, a famous ancient Chinese philosopher, this sustaining function is "such a thing which seems to come forth from nowhere, and yet it penetrates everywhere". "It is formless, shapeless, vague, indefinite, imperceptible and indescribable, always changing, and reverting to the state of nothingness" (Yang 1968). This nothingness is a kind of accumulated energy described by H. T. Odum (Odum 1983), a kind of steering force or QI in Chinese word, a kind of higher hierarchy information or orderliness, or a kind of "holy spirit" across time and space.

To measure this nothingness, we check its internal mechanism of competition, symbiosis and self-reliance; its quantity, shape, form and order; its structural wealth, functional health and people's faith in it; its efficiency, equity and vitality; the balance among the four driving forces of energy, money, power and spirit; and the human interference of technology, institution and culture, both internal and external, upper and lower scaling, in the long and short term, centralised and decentralised.

2 Ecopolis Development: Planning the Sustainable City

In recent years, a campaign of ecopolis development was spontaneously initiated in some Chinese cities and counties, which aims to achieve sustainable development of human settlements through ingenious exploitation, utilisation and management of local resources, setting up an ecologically sound institution and influencing people's behaviour. The essential idea of ecopolis development is to plan, design, manage and construct the ecosystem's function of production, living and sustaining according to ecological cybernetics. The ecopolis construction includes four stages of self-learning, self-evaluation, self-regulation and implementation from technological, institutional and behavioural aspects (Wang 1991).

The ultimate goals of the ecopolis development are wealth, health and faith. Wealth measures the structural state of the monetary assets, natural assets (mineral, water, forestry, soil, air and biodiversity), human resource and social resource (institution, arts etc.). Health measures the functional relation: human health, ecosystem health, and the risks and opportunities of human beings and their life support systems; faith measures the behavioural mode: attitudes, traditions, customs, perceptions, ideas, concepts, values, believes, ethics and life styles. Systematically responsible planning, total functioning engineering and ecologically vivid management are the three determinants for sustainable urban development.

2.1 Systematically Responsible Planning

System here means not only now but also in the future; not only this region or sector, but also other regions and sectors; not only economic, but also ecological development; not only its structure, but also the function and process; not only wealth, but also health and faith; not only effects, but also feedback; not only efficiency, but also equity and vitality; not only production and consumption, but also a service function.

Responsibility here means obligation not only to oneself, one's family or employer, but also the ecosystem as a whole in which one lives, consumes and relies on; not only today, this century or this generation, but also tomorrow, in the next century and for future generations.

Here the key for systematically responsible planning is to plan and design the service function of the urban ecosystem. There are two kinds of service function, one is for inner city services which include the function of resource supply, products and wastes acceptance, transportation and communication, spatial containment, safety and cultural sustainment; another is for regional services which include the purification of pollutants, climate buffering, soil cultivation, biodiversity conservation, water reserving and landscape harmonisation.

2.2 Totally Functioning Engineering

There are several kinds of technology which involve the relationship between man and his environment: environmental technology, cleaner technology, nature conservation technology and eco-technology. Environmental technology aims at reduction and treatment of pollutants released by human activities; cleaner technology aims at establishing a cleaner production process so that fewer pollutants are produced, while eco-technology is to arrange systematically the whole process (steered by humans) from resource exploitation, production, consumption to recycling in an ecological way in order to sustain an optimum function of the total ecosystem (Mitch 1994).

Totally Functioning Engineering(TFE) encourages ecologically sound, economically productive, behaviourally acceptable and systematically responsible technology characterised by:

systematically designed technology through integration of various single technologies, of high technology and traditional low technology, of primary, secondary and tertiary industries, of hardware (technical innovation), software (institutional reform) and mindware (capacity building);

ecologically efficient technology through encouraging higher levels of processing and comprehensive use of local resources, zero emission of hazard wastes into the ecosystem, having long term and large scale positive environmental impacts;

profitable technology having higher economic efficiency, lower input/output ratio, higher market potentials and higher diversity and flexibility of renewal of products.

3 Ecologically Vivid Management

The vitality of an ecosystem depends on its self-regulating ability. A natural ecosystem optimises its function through competition, symbiosis, and natural selection while an artificial ecosystem regulates its function through a government plan and chain-linked administration. The lack of self-regulation in an artificial ecosystem often leads to erroneous decision making. Town administrators, used to cause-effect reasoning and mono-objective management methods, do not or cannot understand and manage the effects of time-lags, regional impacts and information feedback of a dynamic, diversified, and large scale ecosystem. These circumstances illustrate the need to develop a heuristic instrument of self-learning for decision makers to help them understand their system, learn its dynamics and cybernetics, and find an appropriate way of leading the human ecosystem towards ecopolis (Vester and Hesler 1980). The vitality includes the vigour of the natural ecosystem: wind and water (i.e. flowing air and water, fertile soil and the healthy hydrological cycle) and the vigour of flora and fauna (i.e. the rich green coverage and the biological diversity); the vigour of the production system: the renewable resource using ratio, eco-niche competition capability, technological innovation creativity and production diversity and flexibility; and the vigour of human beings: the eco-maturity of policy makers, the eco-awareness of industries and citizens, the sensitivity of information feedback, and the flexibility of institutions.

Human beings have been dreaming of an ideal society for thousands of years. The sustainability study in China does not seek to build an ideal model for demonstration but rather to look for a feasible way and healthy process to achieve sustainable development. A sustainable city probably will not become a utopia, but it could certainly be changed from a physical complexity to an alive ecosystem. If all cities or any human ecological units fully understood the dynamics and cybernetics of their own systems and acted spontaneously with nature, the whole world could realise a sustainable development.

References

Ma S, Wang R-S (1984) Social-Economic-Natural Complex Ecosystem. Acta Ecological Sinica 4(1): 1–9, 1984

Mitch WJ (1994) Ecological Engineering: A friend. Ecological Engineering 3 (20): 112–115

Odum HT (1983) Systems Ecology-An Introduction. John Wiley & Sons

Vester F, Hesler A (1980) Sensitivitatsmodell, Regionale Planungsgemeinwchaft Untermain, Frankfurt

Wang R, Hu D (1994) From economic prosperity to ecological sustainability – a theoretical and practical concern of sustainable development in China. Journal of Environmental Sciences, Vol 6, No 4: 389–401

Wang R (ed) (1991) Human Systems Ecology. China Science and Technology Press, Beijing

Wang R, Ye Qi (1991) Human ecology in China: Its past, present and prospect. In: Suzuki S, Borden RJ, Hens L (eds) Human Ecology Coming of age: An International Overview. Free University Brussels Press, Brussels

Yang Jialuo (1968) Truth and Nature, Hong: International Publishing Company

An Integrated Urban Framework for Cities, their Inhabitants and their Environment in the Information Society

Wolf Dieter Grossmann and Andreas Rösch

1
Information Societies, Environmental Issues and Urban Restructuring

1.1
Information Society and Urban Transformation

During the last 100 years cities have increasingly mirrored the necessities and attitude of the industrial age. Now „Europe is shifting towards an information-based economy, where networks and network infrastructure play as significant a role as did the rail networks in transforming the European economies in the last century.“ ISPO 1994 The growing information potential, in particular the number of participants in information networks, is dramatically changing the economy and life styles which, in turn, is affecting structures, and the role and importance of cities (see e.g. Knight 1995). The comparative advantages of rural areas compared to cities change, as elaborated by Wuermeling (1995). Mitchell (1996) questions whether cities have a future as „teleworking“ can partially replace commuting as it allows one to work, trade or act from any distance. Urban conglomerations are often no longer attractive to citizens although they still provide more and better jobs than rural areas; urban-rural interactions adapt to the change in lifestyles.

Given the high unemployment in Western Europe, a shift to an Information Society is the highest political goal of the Commission of the European Union (Bangemann 1997), as information-based companies create numerous new jobs. This can be seen through a near doubling of jobs in the U.S. from 1960 to 1990 or in Oceania (OECD 1994). Information-based companies have dramatically different urban, environmental, infrastructure and qualification demands compared with established industries. The outdoor environment of cities, i.e. forests, green and agriculture areas, also changes due to the

UFZ Centre for Environmental Research Leipzig-Halle, Group Regional Future Modelling Permoserstraße 15, D-04318 Leipzig

J. Breuste, H. Feldmann, O. Uhlmann (Eds.)
Urban Ecology

information potential. Cities of the 21st century will correspond to a post-industrial society (Naveh 1995), in the same way that urban areas of this century corresponded to an industrial society.

1.2
Co-Dependence between Sustainability and Information Based Economies

The transition to information societies is one of two major developments that change the role and structure of urban conglomerations. The other is the task of sustainability, in particular the oxymoron of sustainable cities. Sustainability requires better products and services with far lower consumption of materials, energy and land area. Such schemes need sophisticated equipment and elaborate management – both components of an information society. As an information-based economy could benefit the environment and in turn prosper, due to the increase in environmental concerns, improved competitiveness, new offers and the provision of more jobs, a node begins to unravel: the economic pressure for new jobs drives environmental activities.

1.3
Opportunity for an Integrated Framework of Living, Economy and Environment

Old industries tend to consume nature and to enforce a separation between housing and work. People and politicians alike are ready to accept severe and lasting unfavourable changes to cities if this is the price for new jobs. Information based economies can manage with very few resources, small areas and little transport, therefore they facilitate a new role for nature around and within cities. Information based companies are incredibly demanding with respect to urban attractiveness (Multhaup and Grossmann 1997) whereas old industries endanger their location if they become environmentally demanding. We have applied this principle in our case studies.

In order for urban revitalisation to take off, people must be enabled to understand how to use the new possibilities of the information potential for their jobs and businesses (Grossmann et al. 1997). Once they understand this, people become ready to look at its other implications. In particular, they understand how much more environmentally demanding they are forced to become if they want to keep their key staff.

In the next step we elaborate how their work will change, their housing, their requirements on the outdoor environment, the qualification and characteristics of staff needed for the new business and so on. This creates a lobby for urban and environmental revitalisation through information based economies. One result of our case studies was the request by managers „for an outstanding ecological revitalisation of the city and its environment to make the city competitive in the Information Society", (Grossmann et al. 1997) which will eventually require new urban and landscape design. One of the tools underlying this work is a list of the 28 „new fusions".

2
Fusion between Human Activities

Whereas during the industrial era urban and regional planers referred to the paradigm of the separation of 'human activities and functions' ("Charta of Athens" 1933), nowadays the synergy of fusion between these becomes more important, e.g. in areas such as tele-work, tele-learning, tele-medicin etc. Fusion is important for two reasons: 1) it influences urban types, structures, functioning and dynamics; 2) it is a mainstay of the "new economy". Although there is some literature on information and its relation to environmental issues, see for example Radermacher (1996), we found no literature on „new fusions".

Eight main human activities are used to show the application of fusion: working, caring for everyday needs including shopping, learning and training, holiday, recreation, spare time, housing and mobility. The fusion between working and housing signals the reversal of the long-term trend of increasing separation of working and housing. Fusion occurs between the most unlikely combinations of main activities. One example concerns holidays and learning. The present rapid change and obsolescence of knowledge and skills requires lifelong learning (see Rösch and Grossmann 1997). Part of this learning process will be carried out in pleasant environments; holidays may be extended to allow for learning. 28 types of fusions are possible between eight human activities. Moreover, each of the eight human activities has to fuse with environmental issues (Marvin 1997). We have elaborated all types of fusion for consultancy and urban affairs.

Fusion can have detrimental effects. For example: people have a time budget for travel but if tele-work decreases the necessity to commute, people have free time in their travel budget which they might use for extended drives to see friends, the so-called „counter-intuitive behaviour of complex systems"; (Forrester 1969) or "rebound effects"; (Radermacher 1996). Sophisticated incentives and disincentives must be developed to control negative environmental impacts by fusion. The following overview outlines a few types of fusion.

Housing and recreation: recreation at home becomes possible in pleasant, remote locations which affects the urban – rural relationship. Housing areas can be so well integrated into natural landscapes that the sites fool remote sensing images into indicating forest areas, although these are advanced low density settlements. In a seminar on remote sensing the participants developed a low density mixture of settlements in National Parks for the purpose of reducing the harm caused by arsonists.

Environment and housing: integration of housing into the environment occurs as larger plots of land are available at remote locations. This allows a new city architecture of extensive, however environmentally compatible, urban sprawl, a haunting option.

Holidays and working: for small entrepreneurs holidays become possible for the first time in decades because they can remain linked through cellular phones connected to notebook computers connected to their business. Business and holidays can be combined. Many types of business can thus move out of cities into attractive, remote areas.

Housing and learning: computer networks allow the home to serve as a laboratory of a university anywhere in the world, as a lecture room, a universal library, a glyptotheque, a pinakotheque. Part of the lifelong learning process will happen at home but the attractiveness of cities with respect to learning and education may prevail due to the concentration of physical sites of higher education institutes and access to research institutions (see Cheshire 1995).

Learning and recreation: learning on advanced levels becomes possible even in the most remote areas due to computer networks and new satellite links (e.g. Iridium, operational end of 1998- see Wired 6/1997). This might result in the migration of many types of learning to recreational environments outside of cities or in very remote areas, „learning on holiday". One example is the spare time complex E2000 (Source: Internet) established by Nolan Bushnell. One result of the „Alternative Landscape Plan for Visselhoevede" (see Grossmann et al. 1997) is the offer of „Holiday and Internet on the farm" which allows farmers to earn additional income.

Mobility and economy: due to netbased activities, mobility is no longer a main prerequisite of economic activities. This has manifold implications for urban traffic planning (more decentralised, individualised traffic will complicate traditional planning, emphasis must be put on fostering efficiency, i.e. lowering resource use). Spatial de-coupling of activities becomes possible, in the extreme case through working on a boat, including video conferences or remote computer-based planning.

Consequences and implications: heavy industry has often caused high environmental pollution. In all types of fusion the extremely high environmental standard requested by people in the information society is a common factor. Advanced business sites mirror these high demands, e.g. Sophia Antipolis in France, near Nice or the Warwick Science Park in UK near Coventry. In both parks, one third of the area has economic uses and two thirds have to remain semi-natural landscapes. Within the economy area only one third may be covered by buildings and infrastructure. Thus, at most about 10% of the soil is sealed.

3 Conclusion

The information society seems to imply a redefinition of many aspects of life and the economy. This requires a huge number of inventions and innovations which cannot be managed by a small elite from above but need a high number of creative people, a "creativity from below". In human history the atmosphere of a city has always influenced the ideas in that city, the spirit of inventiveness and boldness for new endeavours. These are based on its science, culture, social life and its predominant inhabitants. If the creativity of millions of people becomes so central, cities should support and mirror this phenomenon. They could trigger people to be creative and will, in turn, be altered by their creative environments.

In the present era the attractiveness of a city will stem from the human desire for joy, pleasure and adventure. In spite of the new environmental concern, shaping of the urban environment will only, to a small degree, be for the sake of nature. Thus, as new equipment, new products and new management schemes decrease the environmental pressure, the main factor in shaping urban conglomerations becomes even more exposed and more dominant: that of attitude, responsibility and spirituality. Information is not knowledge, and knowledge, although vastly superior to information, is not wisdom. It is wisdom that would allow the building of humane cities, allow extensive revitalisation of urban green areas and curb mobility. In none of our case studies or workshops did any participant ask for „cities for people". People do not dare to feel or think this need, to say it, or develop it. However, from our study into future information societies it became painfully obvious that no technical „fix" will heal human longings. This will not change in the information society. The secrets and mysteries of cities remain intertwined with human nature (see Jacobs 1993).

Since new types of fusion between human dimensions and new creativity from below are going to change urban life and planning, it is important to know more about these new possibilities and the challenges of information societies on urban systems. The new possibilities underline that human desire and human attitude are the ultimate factors which decide on the homeliness, sustainability and viability of cities.

References

Bangemann M (1997) Lecture. EU-Conference: Mainstreaming the Information Society, February 1997, Brussels

Cheshire P (1995) A new phase of urban development in Western Europe? The evidence for the 1980s. Urban Studies, Vol 32, No 7: 1045–1063

Forrester JW (1969) Urban Dynamics. MIT Press, Boston

Grossmann WD, Meiß K-M, Multhaup T, Fränzle S, Rösch A (1997) Alternativer Landschaftsplan für eine kleine attraktive Stadt. UFZ-Bericht 13/1997, Leipzig.

ISPO 1994 Green Paper: "Liberalisation of telecommunications infrastructure and cable television networks of the European Union". (www.ispo.cec.be/...)

Jacobs J (1993) The Death and Life of Great American Cities. 3rd ed (1st ed 1963), New York

Knight RV (1995) Knowledge-based Development: Policy and Planning Implications for Cities. Urban Studies, Vol 32, No 2: 225–260

Marvin S (1997) Environmental Flows, Telecommmunications and the Dematerialisation of Cities? Futures, Vol 29, No 1: 47–65

Multhaup T, Grossmann WD (1997) Vier Fallbeispiele informationsbasierter Wirtschaft - Folgerungen für Umweltqualität. Submitted paper, Leipzig

Mitchell WJ (1996) City of Bits. MIT Press, Boston

Naveh Z (1995) Transdisciplinary landscape ecology education and the future of post-industrial landscapes. IALE World Meeting 1995. Bolletino deu Museo di Storia Naturale della Lunigiana, Aulla, Italy, Vol 9 Suppl pp 13–26

OECD (1994) OECD Jobs Study. OECD (ed), Paris

Radermacher FJ (1996) Telework: Its role in achieving a sustainable global economy. Ulm (Internet: http:// www.faw.uni-ulm.de/deutsch/...)

Rösch A, Grossmann WD (1997) Structural Change, Creative Milieu and Regional Planning. Accepted by Environment and Planning

Wuermeling J (1995) Informationsgesellschaft: Neubelebung des ländlichen Raums? Der Landkreis. Zeitschrift für kommunale Selbstverwaltung

Workshop: The Experimental Dimension: The Citizens Perception 2.2

Citizens – The Building Material of Sustainable Cities

Piet Beekman

1 Introduction

In 1985 the Faculty of Architecture and Urbanism of the Eindhoven University of Technology was requested by the Gedaref town council – Sudan to assist in planning the development of the town. Gedaref was overpopulated by a high influx of refuges and displaced persons caused by the drought and civil war. Most of the services and facilities became overloaded, especially water and health. There was also an urgent need for infrastructure and building locations.

It was soon discovered that the management capacity of the town council was very weak and the size of the management staff was far below the required level.

An important question was therefore how to find some main issues of such common interest and importance that they could guarantee the participation of as many citizens as possible.

We have learnt that co-operation is a general condition for sustainable town-development and we concluded that town planners should attune their proposals to the (possible) patterns of participation. Moreover, the town council and the town planners should learn to see their plans and proposals first of all as instruments to structure and build up the network of the urban society. In such a weak municipality co-ordination should have a higher priority than initiation. The municipality can only initiate activities in which people are interested and which they have requested. A plan for development should be seen as an instrument for co-ordinating the interests of the citizens.

Building up civil society has a lot to do with the co-ordination of activities of citizens. Living is a "consumptive activity" and thus unsustainable. A "productive activity" is always more active and sustainable as it means a continuous challenge for people in a social as well as an economic sense. Active inhabitants are the fuel of such an urban factory. Building up the civil

Eindhoven University of Technology, O. Box 513, 5600 MB Eindhoven

J. Breuste, H. Feldmann, O. Uhlmann (Eds.)
Urban Ecology

society will transform cities into sustainable, productive urban factories. "Urban ecology" is a productive process in a kind of competition; sometimes it is "sweet and friendly looking", but mostly it is a struggle for survival of every creature in the urban environment.

2 Planning for a Poor City

The project for the urban development for Gedaref town goes far beyond the simple creation of a suitable plan. From the start, one has to co-ordinate the simplest powers, knowledge and means of contributing to a realisation of the plan. Principal guidelines in this project are the idea that town-planning is capable of motivating people to participate in the development of their town, and that the relevant development plans must be based on this notion.

It renders a "development related" character to the actual job. The skill is thus becoming uncoupled of its tight technical rationality and it can be evaluated for its suitability in improving living-conditions.

Gradually it has become apparent that there is more latent potential available in Gedaref than was expected, in spite of the fact that more than 60% of the population is illiterate and living under the UN-survival level. This is most likely due to the fact that environmental conditions play a less dominant role here than in western Europe. People are firstly asking for a more confident future and they are motivated to contribute to that on an every day basis; having income is the highest priority. The relief of urban poverty can be effected only through a general increase of the productivity. This demands not only the introduction of facilities from the municipality, but also the possibilities for expansion of employment in order to enable the people to obtain the required income. A problem is that the authorities are falling short of their public duties due to their limited capacities. The size of the municipal organisation has not kept pace with the developments, while systems and methods still date back to the time of the British Protectorate and the equipment available is rather poor.

The upgrading of the administration is so crucial for project implementation and maintenance, that in all the project activities priority is given to the managerial context. This policy does not only concentrate on the functioning of the municipal government, but focuses on the development of models of transition of public and private administration by including patterns of participation and self-government. In Gedaref we have experienced that an improved municipal administration may be composed of a governing-structure with various levels which are directed from the public sector to private bodies. Planning and administering a poor city therefore will mainly become a co-ordinating activity.

3
Participation

The aid projects will be developed along two lines:

1. a planning line for the urban and technical aspects, which is supported by the faculty for building and construction of the Technical University Eindhoven;
2. a managerial line, which provides for the management support on the municipal level, supported by the department for management support of the Municipality of Eindhoven.

Amongst these two lines of "planning" and "managerial" support, a line has been projected for "participation", which mediates between the two. Participation is an indispensable condition for the development of Gedaref.

The projects which we are implementing provide not only an objective, but also a vehicle to contribute to:

1. the urban development, meaning that each project contributes and adds value to the building of a civil society;
2. an amelioration of the functioning of the urban management, i.e. when shaping a project, the creation of a management structure is aimed at;
3. the emancipation of the population by promoting the institutionalisation of co-operation and being aware of the capacities needed to realise the objective.

Participation in the projects by the parties concerned is a very important aspect, and is a condition for sustainability. Moreover the improvement of the process of the urban development – considered from either an urban-development viewpoint or from a governmental viewpoint – will be of little or no effect, if the population does not participate.

Of course the objective of the participation is focused on the initiatives coming from the population. We cherish the principle that it is not our task, but those of the inhabitants, to formulate, apply, realise and preserve the development.

Our role is just of a temporary nature.

4
The Learning Process

The learning process of the Gedaref project has developed from being confronted by an extremely poverty-stricken condition. Our experience has shown that under such circumstances, it is best to expose the most essential values.

The state of welfare in which we are living apparently does not only stand in the way of our social contacts, but it is also a smoke-screen against the principles of a sustainable future.

The richer part of the world is also becoming confronted with the problems, which are comparable with those of the developing nations.

Unemployment, an ever expanding "black market", the informal sector and corruption are growing alarmingly.

The Third World seems to be unexpectedly nearby, when we see the thousands of homeless people in our cities who rely on garbage for nutrition. The arrival of legal- or illegal immigrants are adding to this vision. Our welfare-state is beginning to showing cracks. We are also told that our governments are scaling down their social support programmes for the aged and the destitute, thus these people must attempt to take care of themselves.

In comparison with the Third World we are opulent, but this is where the problem lies. The difference is that people in the Third World have "grown up" with scarce resources and poverty, while it seems to have taken us by surprise and we do not know how to cope with being on both sides of the hedge, in order to tackle the problems.

The experience in Gedaref is that poverty can solve its problems itself, provided that people remain involved in the discussion of their future. In such a way the urban policy could contribute constructively to a solution of the poverty problem. In addition, the experience in Gedaref is teaching us that it would be beneficial for the entire city if everyone is "kept on board" and is able to consolidate their own position because poverty is the greatest threat to a sustainable future.

Is There Such a Thing as the Aesthetics of Sustainable Cities and What Could It be Like?

Werner Nohl

1 Introduction

If we look around the world, we will have severe difficulties in finding a sustainable city. If we go into details, we might find urban elements, like private and public open spaces, which already function in a sustainable way. I do not want to give a comprehensive definition of urban sustainability, but, with regard to urban open spaces, I assume that sustainability will be higher, the more people can use such spaces in a free and unconstrained manner, and the more nature can grow wildly and spontaneously. It is my belief, that the knowledge of the aesthetic functioning of these urban elements could help to develop an understanding of the aesthetics of future sustainable cities, and gain inspiration for necessary and proper structural changes.

2 The Aesthetic Qualities of Allotment Gardens

As a suitable example of private open spaces I have chosen allotment gardens. Originally introduced as gardens of the poor in the 19th century, they now often serve as leisure gardens. Each lot, furnished with a cabin, comprises approximately 300 m². They are mostly arranged in groups of 40 to 200 lots. As a rule, the members are organised in an association which takes care of common tasks. In their social structure the gardeners often represent a cross-section of the total city population. In Germany the number of garden lots is estimated to be more than 2 million (Fig. 1).

As I expounded in detail elsewhere (Nohl 1987), three basic aesthetic needs - for home, for freedom and for careful handling of resources- play an important role. To begin with the last, the *careful handling* of resources and nature is a fundamental aspect of allotment garden aesthetics. It often leads to a special attitude, which might be labelled *reutilisation creativity*. This means

Workshop for Landscape and Free Space Development (Werkstatt für Landschafts- und Freiraumentwicklung), Stockäckerring 17, D-85551 Kirchheim (Munich), Germany

J. Breuste, H. Feldmann, O. Uhlmann (Eds.)
Urban Ecology

Fig. 1. Allotment gardens

people like to alter the design of their lots with things which are unusable elsewhere, or which were constructed for another purpose. Old bathtubs might serve as water reservoirs, and panes of an old bus function as covers for hotbeds. Cabins are constructed out of demolition materials, and bottles are used to edge flower beds. There is almost nothing which allotment gardeners cannot use, thus diminishing the consumption of resources in an aesthetically conspicuous way.

The second important aesthetical aspect of allotment gardens is due to the need for *home*. The aesthetics of allotment gardens is not understandable without the effects of this marked yearning for home. It is expressed in at least two different forms: familiarity and beautification. *Familiarity* comprises the well known, the usual, and the traditional. It may be most spectacular in the use of elements, which incorporate the "good old days". The handcart, the wagon wheel, the milk can, antique styled vases, buckets and baskets – these are typical objects which signal the need for home. *Beautification*, the other form, indicates the inexhaustible wish of the gardeners, to embellish their lots. Thus, flowers might be found all over the garden, walls are greened with plants, or the gardens are decorated with miniaturised landscapes, such as alpine ridges, moors, heath landscapes etc.

Freedom or *independence* is the third need with peculiar aesthetic effects in allotment gardens. It is less strong yet intensive enough to be detected quite easily in special stylistic elements of the allotment gardens. The unusual, the unfamiliar, the strange, but also the fantastic, the dreamlike, and the chaotic are typical expressions for this need. The freedom seeker, first of all, finds happiness in foreign and unknown situations. In his garden we therefore find signs of "non-identity" everywhere. In this way, the aesthetic need for freedom is also a need for the *alien*.

This aesthetic need for freedom and the alien appears in at least two forms. In many gardens there is an accentuated preference for the bizarre-scurrilous, and on the other hand a liking for the fairy-tale world. A *bizarre-scurrilous* design is often obtained by use of familiar things in an unfamiliar situation and vice versa, for example a wooden wheel on top of a gate, a bleached cow skull at the cabin entrance, or a tree root remodeled into an electric lamp. The *fairy-tale world*, is visualised in a multiplicity of topics too. Thus the enchanting world of dwarfs and gnomes might be found in many gardens as well as romantic castles, magnificent palaces, or lonely farmsteads.

To put it in an aesthetic-philosophical way, one can say that all three needs owe their existence to the search for a better world, symbolised in the nature of gardens. The need for home is oriented more to the past. It tries to seize hold of a lost home, using metaphors of beauty and familiarity. In opposition, the need for freedom is more future-oriented. Using dream and fairy tale scenes, it tries to make a mark, which points to an independent and self-organised life. The need for careful handling of nature, finally, refers aesthetically to the fact, that all better life has to be realized in a world whose resources are limited, and therefore have to be saved for other people and future generations.

Here we see the connections of such aesthetics with *sustainability*:

- Self-organisation means that people are allowed and prepared to participate actively and according to their own plans and needs in many daily processes. This will enlarge the pool of intelligent solutions considerably, and encompass the chance to save and recover resources and materials in a creative way. At the same time it forms a solid foundation for a new aesthetical approach in sustainable cities, which will be characterized first of all by a fertile co-operation between experts and city dwellers: participatory aesthetics.
- If the example of allotment gardens does not fail, the aesthetic needs for home as well as for the alien will be of particular significance in sustainable cities. The aesthetic form of sustainable cities will probably be determined by the apparently paradoxical wishes, to be at home, and at the same time, to be in places where one can gather new experiences.
- Furthermore, sustainable cities will reflect in their aesthetic structures that the protection of nature and resources belongs to their cardinal goals. The allotment gardens suggest an aesthetic approach, which expresses the careful and economical handling of nature and resources, especially the creative re-use of old structures, elements, and materials.

3
The Aesthetic Qualities of Derelict Sites in Cities

I now want to address the aesthetics of derelict sites in present cities, accessible to the public. They exist, because their original urban use (industrial, commercial, traffic, or even residential) was given up some time ago due to economic reasons. The combination of the wild growing vegetation and the relics of former use (ruins of buildings, relics of factories, rotten tracks, building rubble, waste material etc.) often impart a peculiar appearance to these sites. If such abandoned land is within reach of residential areas and not contaminated, it is mostly used by people as additional public open space. Particularly, kids, teenagers, dog-keepers, joggers etc. belong to the frequent users of derelict sites (Fig. 2).

If people use these sites, they normally leave behind visible traces which will be erased by nature, if activities are not executed regularly. Nature wipes them away, and the process of use can start afresh. This successful partnership seems to show that abandoned sites aesthetically work like a symbol for a world, in which man and nature are reconciled with one another (Nohl 1988). Thus, in the eye of the beholder, the experience of derelict sites could open a prospective horizon of a better life in the future. I suppose that much of the aesthetic joy elicited by derelict areas, can be explained by this predominantly symbolic meaning of such places.

More than allotment gardens derelict sites are places in which people try to fulfill their needs for *freedom* and the *alien*. Here it is not necessary to build up strange or even scurrilous structures for this purpose. The combination of spontaneous nature and human relics and ruins creates mysterious and excit-

Fig. 2. Derelict site

ing visual scenes. On the other hand, users of derelict sites often leave behind fancy and strange traces. Kids build bold tree-houses, dog-keepers beat secret paths and so on. Therefore the aesthetic experience delivered by derelict sites is predominantly governed by disharmony, disarray and chaos, instead of harmony, order and unity. On the other hand the aesthetic need for *home* can be met here, too. Users often arrange familiar situations by carrying with them tables, chairs, sun-shades and other things for outdoor use. Sometimes a single object like a blanket, a mat, or a towel is used to indicate a home situation, a secure territory. Of course, a vegetable garden in a derelict area is especially apt to give the feeling of a "home place". Although these things are obviously introduced for practical reasons, they work at the same time as aesthetic signals. They tell that some traces of home are needed to feel physically and mentally good and secure, even or particularly in the urban wilderness.

As we have seen, the derelict sites - and that is aesthetically decisive - support the self-organisation of people. Since such places are officially considered as less valuable, people may act here in a very autonomous way, and they are able to leave their marks without having feelings of guilt. In derelict sites spontaneous development of nature and permanent erosion of hard structures generate a quality of space, which might be called *provisional* (Nohl 1990). The provisional allows and encourages creative handling of space and nature, and if it is to remain a provisional arrangement, nature must be allowed to live its own life as well. In a provisional arrangement normally neither does man overpower nature, nor nature take over man. In fact, abandoned sites gain much of their aesthetic charm by this quality of the provisional.

As is the case with the allotment gardens, the aesthetics of derelict sites might be used to approach the aesthetics of future *sustainable cities*, in which self-organisation of people and the protection of nature and resources are of decisive value. Thus the aesthetic discussion of derelict sites brings up the suggestion again, that in sustainable cities people will be much more eager to participate actively in the aesthetic production and organisation of their daily life. There will be strong aesthetic needs, like those for home and for the alien, which must be satisfied to a high degree in cities themselves, if they are to be sustainable. We also might conclude from the analysis of derelict sites, that sustainable cities will have to reflect symbolically, that the protection of nature and resources belongs to their cardinal goals.

4
Concluding Remarks

I do not want to say that derelict sites and allotment gardens are typical aesthetic expressions of sustainable cities. Both types of open space belong to the contemporary city, which, as a whole, is not sustainable, but the examples show that under conditions of self-organisation people will find themselves in a situation, where they can learn to make use of nature in a way which is beneficial for them, as well for nature. This conciliatory interrelation between man and nature is the aesthetic message, which the sustainable city might one day express.

References

Nohl W (1987) The Aesthetics of Home Separated Gardens in Germany: Traces of Participatory Aesthetics. The Journal of Architectural and Planning Research Vol 4 No 3: 212–227

Nohl W (1988) Open Space in Cities: in Search of a New Aesthetic. In: Nasar JL (ed) Environmental Aesthetics - Theory, Research, & Applications: 74–83. Cambridge University Press, Cambridge New York

Nohl W (1990) Gedankenskizze einer Naturästhetik der Stadt (Sketching a Nature Aesthetics of the City). Landschaft + Stadt 22(2): 57–67

Objective and Subjective Indicators to Evaluate Quality of Life (QOL) in Two Districts in the Stockholm Region[1]

Vânia Ceccato and Folke Snickars

1 Introduction

Quality of life is a very complex concept which has often been associated with numerous others concepts such as level of living or people's satisfaction with their conditions. It is hard to have a consensus about what QOL is, specially because QOL has been studied based on so many different approaches. Nowadays when people are so mobile, it becomes even more complex to define QOL because we live in a place, work in another and may spend our leisure time in another place of the city. Given such multi-dimensionality of the subject QOL, it has been chosen to focus only on one dimension of QOL, the *living* quality of life. The objective is to present some preliminary results from an one-year project which has as the main objective to analyse the living quality using objective and subjective indicators in two residential areas in the Stockholm region.

One aspect that is usually forgotten in QOL studies is the spatial dimension. It is known that people daily move around in their residential area and/or outside of it, it is possible to say that different persons use their residential areas in different ways, as a function, for example, of their life styles. How can these differences in movement patterns lead to differences in how people perceive their living place? For example, how do people define the boundaries of their residential areas?

2 Method

Two residential areas in Stockholm region has been chosen, Djursholm (5600 inhabitants) and Hökarängen (7400 inhabitants). Djursholm is a typical upper

Royal Institute of Technology, Fiskartorpsvägen 15 A, S-10044 Stockholm, Sweden

[1] This project has been financed by funds provided by Lars Lundbergs Stipendiestiftelse and Riksbyggens Jubileumsfond *Den goda Staden*. The presentation of this work at the International Conference on Urban Ecology has partially been sponsored by Knut and Alice Wallenbergs Stiftelse.

J. Breuste, H. Feldmann, O. Uhlmann (Eds.)
Urban Ecology

class residential area while Hökarängen is a predominantly low income area where several services no longer exist.

A database containing objective data (statistical data) and maps was created using a desktop software. At the same time, a questionnaire was elaborated in order to get the subjective data, that is, how people evaluate their living place. The questionnaire contained twenty pages, including maps. An identical questionnaire was sent to the areas based on a random sample. It was 200 questionnaires for each area. The questionnaire was composed of seven parts based on the theoretical framework for the *living* live quality dimension:1. Background questions, 2. Housing, 3. Outdoor environment, 4. Accessibility, 5. Security, 6. Social links, and 7. General evaluation of the QOL. Contables and Logistic regression have been used to analyse the influence of the background variables (age, gender, citizenship, etc) on the evaluation.

3 Some Preliminary Results

3.1 Satisfaction with Services in Hökarängen and Djursholm

Do people living in a place with better services and infrastructure evaluate their living place more highly than those who do not? Both areas seem to be generally satisfied with their services or another kinds of infrastructure that they have in their living areas. However, people living in Djursholm are in general more satisfied with their services than people in Hökarängen. In this case, there is a correlation between what they really have there in terms of services supply and how they evaluate them, in other words, there is a correlation between the objective conditions and the subjective evaluation. Those services that are lacking or deficient are those with which people are more dissatisfied, for instance, the case of pharmacy in Hökarängen. However, there is still a significant percentage of interviewed population that are satisfied or quite satisfied with them even if these services do not exist anymore. How can this be explained? According to the literature, people who live in depriving living conditions may have lower expectations and with time become *accommodated* with their current living situation even if they are aware that they could have better conditions. It may explain in part the satisfaction of a certain group in Hökarängen. How can we define a person as accommodated? It is known that it can be related to age (for instance, older people generally tend to have lower expectations than youths) how many years the person has lived in the place and educational level. An answer can also be that *this high level of satisfaction* in this area reflects the views of those who remain living in the area. People who are dissatisfied with the QOL in their living place, may simply move to another place perceived by them to be better. It is one of the problems using subjective data because they fail to indicate what the evaluation of those who do not live there anymore is. Contrarily to those who are accommodated with their situation, there is the dissatisfied group. Those who are dissatisfied tend to have high expectations which are nurtured, for instance, by information about the reality as whole. In this process, there is a factor of the fundamental importance, that is the *comparison that they establish between what they have and what other's have* (Abaleron 1987; Oppong et al. 1988). *The others* can be on an individual scale, comparisons between a person and his neighbours or his relatives and, at the collective level, comparisons between two residential areas. This comparative perception nurtures the mechanisms of needs, wishes, satisfactions or frustrations. In Djursholm they have all the basic services in the centre. 90% of the interviewed population has access to car against to 47% in Hökarängen. It means that in Djursholm they can have easy access to other centres, having other choices. The higher dissatisfaction level in Hökarängen can be seen here not only as a reflection of lack of services but also as an unbalanced access to resources and opportunity to make choices in life in general. Only 30% want to remain living in Hökarängen if they have the possibility to choose another place to live, against 75% in Djursholm.

3.2 Type of Tenure and Satisfaction with Housing

The two areas differ greatly concerning housing tenure. While Djursholm is composed only of 25% of rental housing, in Hökarängen rental apartments constitute about 85% of the total. One question that arises here is to what extent housing tenure could influence satisfaction with housing? The results show that type of tenure seems to have some influence on how people evaluate housing. This correlation is clearer in Hökarängen than in Djursholm. People who own their houses in Hökarängen, where the most common type of tenure is rental housing, tend to be more satisfied with their houses than in Djursholm, where housing ownership is the most common type of tenure. What can explain this fact? It is known that housing ownership is the most desirable type of tenure in Sweden. Another answer for this as Oppong et al. (1988) writes is that people's subjective evaluation is very dependent on the relative conditions enjoyed by those who live in the immediate vicinity. It also means that if a person had enough resources to invest in a house, this person also had a choice where and how he would like to live. People who choose to live in Hökarängen, buying a house there, have already made a favourable pre-appraisal of the residential area.

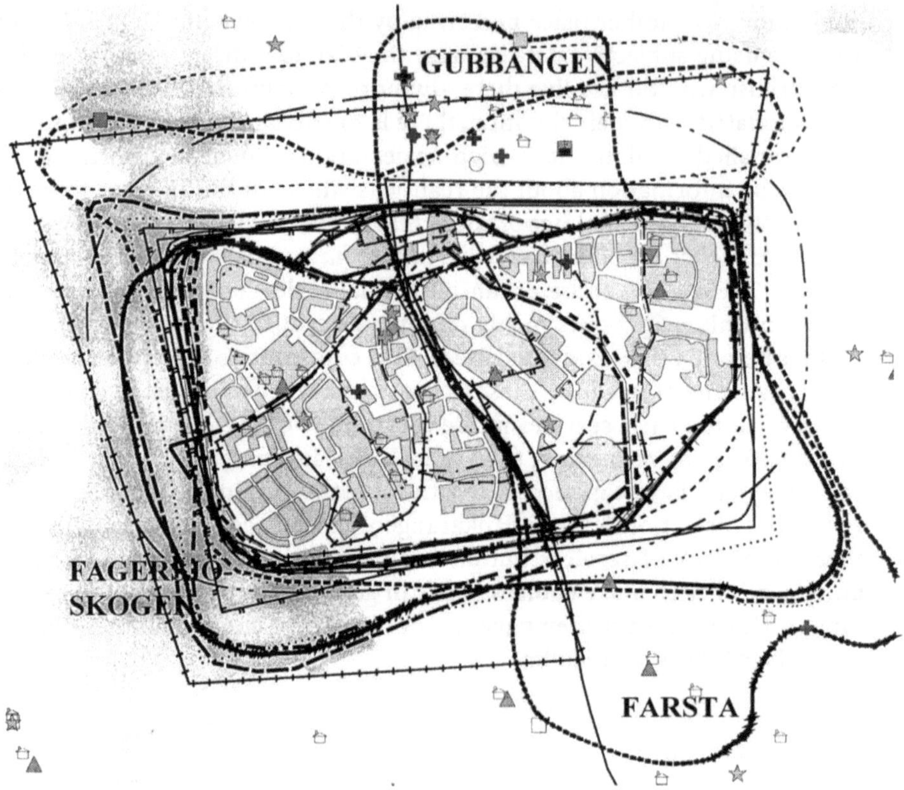

Fig. 1. The boundaries of Hokarängen according to the inhabitant's perception. The symbols represent different kinds of services

3.3
The Mental Map of the Living Place

The Fig. 1 shows how people perceive the boundaries of Hökarängen. Those who perceive the boundary smaller than the official one mostly divide Hökarängen in two parts using the underground line as the cutting line: Western and Eastern Hökarängen, considering only one of these parts as Hökarängen. Others that also perceive the boundaries of Hökarängen as smaller than the official ones, just make a circle surrounding their houses. They generally live in multi-family houses. Those who perceive the boundaries as larger than the official ones take the official boundary into consideration but also some neighbouring residential areas. These neighbouring areas offer many services that do not exist in Hökarängen, so this boundary can be clearly associated to the use of services (schools, pharmacy, library, etc) and/or green areas that are located nearest Hökarängen. Similar results were found for Djursholm. The mental maps reflect the daily used and known spaces.

4 Concluding Remarks

1. People living in a place with better services and infrastructure evaluate their living places more highly than those who do not, even if in both areas the population affirm to be, in general, satisfied with their living place.
2. Type of tenure seems to have some influence on how people evaluate housing. This correlation is clearer in Hökarängen, where the most common type of tenure is rental housing than in Djursholm.
3. The perceived boundaries of the studied areas are different from the official ones. The mental maps seem to be a clear reflection of the *used* and *known* spaces.

Acknowledgements. I am very grateful to the researcher Danuta Biterman, at Socialstyrelsen, Stockholm, for many discussions and editorial advice during the preparation of this study.

References

Abaleron CA (1987) Condicionantes objectivos y percepción subjectiva de calidad de vida en areas centrales y barrios o vecindarios. Revista Geografica, Sao Paulo, 5/6:103–142, 1986/1987

Oppong JR, Ironside RG, Kennedy LW (1988) Perceived quality of life in a centre-periphery framework. Social Indicators Research 20: 605–620

Interdependencies Between Ecological and Social Problems in Residential Areas

Sigrun Kabisch

1 Introduction

In the framework of urban ecological change, subjective factors such as the attitudes and the behaviour of the inhabitants are increasingly becoming factors of strategic importance. Therefore, urban planners must know and act upon the preferences and needs of the population concerning their housing conditions.

Sociological investigations in urban areas underline the importance of the natural environment when residents decide where they want to live. In this sense the discussion is part of strategies to achieve practical steps to greater sustainability in urban development (European Commission 1996).

In this paper, results of surveys on satisfaction with the housing conditions and on intentions to stay or to leave the quarter, with respect to natural environmental conditions are analysed and compared. The aim is to verify the hypothesis, that features of the natural environment such as green spaces, clean air and no noise pollution have to be considered on the same high level as the elements of the social and built environment. Only with such a close connection is it possible to stimulate the people's desire to live in these places.

2 Selected Aspects of Urban Development after 1990 in East Germany

Since 1990, one of the most important problems of large cities in east Germany, especially of Leipzig, has been migration from the city centre to the surroundings. After German reunification and the establishment of specific tax-allowances for housing construction, new suburban residential areas were built; more are still under construction. The results of a survey of the

UFZ Centre for Environmental Research Leipzig-Halle, ÖKUS – Dept. of Ecological Economics and Environmental Sociology, Permoserstaße 15, D-04318 Leipzig, Germany

J. Breuste, H. Feldmann, O. Uhlmann (Eds.)
Urban Ecology

municipality show that the main reason for the new building areas is the leaning towards ownership of an one-family house in a green area with low noise and air pollution. The negative consequences are growing land consumption and increasing traffic accompanied by air and noise pollution. Furthermore, many of the migrants feel a loss of the urban lifestyle. Now there is a trend for residents, who have lived as tenants in new apartment houses outside the city, to migrate back into the city. This is driven by the increasing number of new or well renovated apartments, beautifully situated, with moderate prices or rents, that are coming onto the housing market.

There is still an imbalance. Between 1990 and 1996 the City of Leipzig lost 54,000 inhabitants, 12% of the population.

The strong out-migration tendencies have serious consequences for the social networks in the inner parts of the city as primarily young families disappear from the neighbourhood. There is a rising number of empty apartments – renovated and not renovated – inside the city, which leads to the dilapidation of buildings and the decay of social networks leading to rising crime rates. In 1995 there were 34,835 empty apartments, most of them in a very poor state of repair (Stadt Leipzig 1996). A more detailed analysis indicates that exactly the same districts that have the highest number of empty apartments, have also had a very high rate of migrants in recent years. The two primary reasons are unclear ownership titles in the quarters with a high percentage of apartment houses dating back to before 1918, and houses being in a poor state of repair. In other quarters, where there are clear titles, there are deficits in renovation activities, inside and outside the house. Finally, there are a growing number of empty renovated or newly built apartments because of the high rents or prices for condominiums.

The question is, how is it possible to improve the attractiveness of residential areas to avoid continued massive out-migration, and which role do ecological features play as pushing or pulling factors?

3
Research on Attractiveness Differentiation of Urban Areas

This question is part of a research project dealing with the differentiation of attractiveness of residential areas with respect to the segregation tendencies in east German cities. As a first step a "Social-Atlas of the City of Leipzig" (Kabisch et al. 1997) was created. On the basis of statistical data from different sources, more than 50 maps with the same scale were produced. They show the social and housing structure, features of the social infrastructure and land use patterns of the city of Leipzig, after the reunification of Germany. Thus, there are maps embracing the 49 districts of the city showing population density, change of the number of inhabitants, unemployment rate or age structure of inhabitants. With respect to the housing structure there are maps on ownership, age and state of the residential buildings and number of empty apartments. A third set of maps comprises maps on medical care and commercial institutions. In a fourth section, the maps show structural types of land use patterns, selected re-development areas and green spaces and water areas inside the city. Finally a synthesised map contains so-called "types of social spaces"; it ignores administrative borders and identifies instead typical structural patterns with similar residential building and social structures.

The interpretation of the maps in connection with empirical studies and the utilisation of urban-sociological research has led to initial results that can be used for projections of further social differentiation of urban neighbourhoods. The idea was to combine features of the social structure with features of the built structure and with the natural structure (like green spaces or air quality). The maps show the differentiation of districts with high and with low attractiveness. They allow conclusions to be made as to where social problems might be expected. However, the results are on too general at the spatial level (49 districts) because of the statistical data base. To obtain more practical results and recommendations, it is necessary to deepen the investigations in selected areas and to combine the analysis of statistical data with the interpretation of survey results.

4 Description of Different Building Structures and their Residents

In a recent survey on residential attachment to different residential areas in one district, the influence of the environmental conditions was investigated. The selected district, Stötteritz, is located in the south-eastern part of the city. It has about 12,000 inhabitants and an area of 3.8 km^2. In the initial stage of the project, the physical structure of the housing situation – building structure, ownership, empty flats and buildings, green spaces, traffic situation, noise pollution – was described. The data were put into a Geographical Information System (GIS). The map of land use patterns of this district shows that there is a large amount of green space, but it is not evenly distributed.

Five different types of residential building structures were identified. They are each located in spatially well-defined areas. Every structure is connected with a special quality of green space and noise pollution due to traffic.

Type I consists of old blocks dating from the beginning of the century, little greenery near the residential buildings, much traffic, relatively high noise and air pollution, a large deficit of renovation activities, many empty apartments (30%), but also many commercial and service shops. For years the quarter has been loosing inhabitants. The social structure of the population is quite mixed, but especially young families with higher incomes tend to leave. The average income is lower than in the other types. *Type II* is characterised by impressive villas and large apartments with a lot of greenery in the immediate surroundings. This quarter has a very good image, it is renowned for its beauty. Deficits of renovation activities due to unclear titles, and some empty apartments are also found in this quarter, as is a lot of through-traffic. These factors have a negative effect on the quarter's image. Some inhabitants intend to leave the quarter. More than half have good professional positions with high incomes. *Type III* consists of detached single and two-family houses in settlements dating from the twenties and thirties. Every house has a garden, consequently there is a lot of greenery. There is only a small amount of traffic, no noise pollution and the private houses are in a good state of repair, thus there is high satisfaction with the housing conditions. There are no migration tendencies. The residents have lived there for decades, and the average age of the inhabitants is higher than in the types described before. The incomes are above average. *Type IV* comprises blocks of flats dating from the twenties and thirties with four floors, which are partly renovated. There is a lot of greenery, especially in the courtyards. There is through-traffic in some streets, affecting the housing quality. Some residents with higher incomes intend to leave the blocks because of the lack of renovation activities. The incomes correspond to the average income. *Type V* is characterised by pre-fabricated apartments, dating from the early sixties. The apartments are quite small (60 m^2), but well renovated and well equipped. There is much greenery and places for relaxation in the surroundings. There is no through-traffic. The residents are very satisfied with their housing conditions. Many tenants have lived in their apartment for thirty years, thus the average age is very high. The incomes lie above average.

5
Environmental Conditions and Residential Attachment

A survey was carried out on the basis of the five different types of residential buildings. The aim was to find out the importance of environmental factors for the attachment to the residential area. A sample of 301 respondents was taken, consisting of five groups according to the five types.

The survey results show a very high attachment to the district Stötteritz as a whole and to the quarter. Of the respondents 65% said, that they completely happy in Stötteritz and 78% in their quarter. This result shows that there is a differentiation in the evaluation between the whole district and the individual residential quarters. Furthermore, attachment is much higher in the neighbourhoods with a high environmental quality than in neighbourhoods with lower quality (Tables 1 and 2).

The migration tendencies from the quarters with little greenery are higher than in greener areas. Many more respondents in type I intend to leave the

Table 1. Satisfaction with environmental conditions (using scale from 1 = very satisfied to 5 = not satisfied, value 1 and 2 summarized, in %)

Building structures	n	Greenery	Noise	Cleanliness
I - old blocks	72	49	27	12
II - villas	38	91	41	68
III - detached houses	60	98	74	76
IV - older apartments	66	89	58	42
V - pre-fab apartments	65	96	81	62
total sample	301	84	56	50

Table 2. Feeling fine in the residential quarter, N = 301, in %

Building structures	Feeling fine	With restrictions	No
I - old blocks	44	50	6
II - villas	81	19	
III - detached houses	96	4	
IV - older apartments	75	25	
V - pre-fab apartments	95	5	

Cramer's V = 0,34907 p = 0,00000

quarter than in the other types. Unsatisfactory environmental conditions, lack of renovation activities and cleaning up of the area, and expectations of rising rents after renovation are the main reasons. Already, the concentration of low income groups is higher than in the other quarters. This result indicates social differentiation within the district. There are different images of neighbourhoods according to building structure, state of repair of the buildings, the visible green spaces and the visible social structure. In contrast to type I, type II obtains support in further development as it is a quarter where higher income groups will live. Nevertheless, the attractiveness of this residential area with a lot of greenery is partly affected by traffic, causing noise and air pollution. The residents complain about these negative effects, but the building structures with large and well equipped apartments, and the location near a big park are very important preconditions for further improvement of the housing quality on a high price level. The other three types have good chances of stabilising the number and the social structure of inhabitants.

The results of the survey show a close link between high attachment to the residential area and good ecological quality. The main reasons of attachment are many and varied green spaces, no noise pollution, well renewed residential buildings and cleanliness. There were no differences between private owners living in detached houses with gardens and tenants living in apartment houses.

6
Concluding Remarks

It is clear that several reasons cause attachment to a residential area. The primary ones being visible environmental conditions, such as green spaces and resting places, and perceptible ones such as clean air and no noise pollution.

Finally, out-migration processes from the quarters with a high building density, with buildings in bad state of repair and with a lack of green areas could provide a chance to improve the housing quality. One step in this direction is the demolition of old residential and commercial buildings in a bad state of repair and reuse of the empty areas as parks, playgrounds and green spaces. The aim is to bring the green landscape into the city!

This can lead to improvement of attractiveness of these quarters and to growing attachment to the residential quarters while stabilising the population on a lower level. I can underline the ideas of Mr. Sukopp in his keynote presentation on the necessity of green spaces near apartment buildings.

References

European Commission (1996) European sustainable cities. Report by the Expert Group on the Urban Environment, Brussels

Kabisch S., Kindler A., Rink D. (1997) Sozialatlas der Stadt Leipzig. UFZ-Umweltforschungszentrum Leipzig-Halle GmbH, Leipzig

Stadt Leipzig (1996) Gebäude- und Wohnungszählung 1995. Leipziger Ergebnisse im Überblick, Leipzig

The Meyer'sche Häuser (Meyer's Houses) in Leipzig – an Example of Careful, Socially Balanced Urban Revitalisation in East-Germany

Dipl.-Ing. Markus Lang

1 The Meyer'sche Häuser – A Pioneering Project of Social Housing in Germany

Having achieved a good name through its trade fairs, dating back more than 700 years, Leipzig underwent a period of rapid industrialisation in the second half of the 19th century which caused an increase of the population from 40,000 to about 625,000. This resulted in land speculation and a desperate shortage of housing among the poorer inhabitants. Consequently, several industrialists and intellectuals drew up schemes to improve housing and living conditions.

This happened in 1900, precisely when the *Foundation for the Construction of Low-Cost Housing* was set up by the owner of the publishing house *Bibliographisches Institut*, Hermann Julius Meyer (1826–1909).

He and his architect Max Pommer (1847–1915) developed and brought into being a social housing strategy aiming to be a counterpart to the speculative transactions of the housing investors in Leipzig – 30 years before social housing became a topic on the political agenda in Germany.

The social and economic principles behind *The Meyer'sche Häuser* were as follows: buying low-priced ground on the outskirts of the city; setting-up urban architectural ensembles with economical ground-plans and construction costs; calculation of rents based on income and prime costs; keeping administration costs low; establishing community facilities; the usage of inner courtyards as parks and allotments.

Although the foundation was not expropriated during the Third Reich and socialist era, it was unable to work independently. Since all housing projects were at the mercy of the public authorities, maintenance could not be performed and living conditions lagged behind growing needs. This was a fate shared by almost all late 19th century housing in Leipzig, for in the 1970's state housing policy had adopted different priorities, particularly the construction of pre-fabricated housing estates on the outskirts of the city.

Office for Urban Renewal (ASW), City of Leipzig, D-04092 Leipzig

J. Breuste, H. Feldmann, O. Uhlmann (Eds.)
Urban Ecology

Since German reunification, Leipzig has been faced with structural and economic redevelopment and social problems such as the loss of 85,000 jobs in industry; the need to repair 196,000 of the total of 262,000 flats; the architectural monostructure of large housing estates on the outskirts of the city; a technical infrastructure decades behind present-day standards; huge industrial estates, shopping centres and housing estates competing with each other in the suburban area.

2 The Lindenau Estate

The foundation continues to be named after its founder, and now manages 2700 dwellings on four estates known as the Meyer'sche Häuser. The first and oldest estate was set up in Lindenau, in the western part of Leipzig (see Fig. 1).

This estate was constructed between 1888 and 1898 and contains 52 buildings with 501 flats, a kindergarten and a separate laundry-house. The original costs ran to a total of 2.6 million Reichsmark.

Its architect, Pommer, designed four different types of buildings all having a basement, a ground-flour, two or three stories and a loft with garrets. Each apartment has a separate hall which contains one or two rooms and a separate kitchen. The toilets are located outside the flat, half-way down the stairs.

Fig. 1. Aerial Picture of the Lindenau Estate. Source: Stiftung Meyer'sche Häuser Leipzig

The urban structure is determined by modified block structures, clearly defined and huge, with architecturally designed entrances which open up to the interior of the block, providing better air circulation and the possibility of gardening. 15,000 m^2 of the total of 22,000 m^2 have been reserved for walk-ways, yards and 202 allotments, each of them offering 40 m^2 to be leased to the tenants (at that time: families of regular size of the working class).

The only chance to open a new chapter in the foundation's chronicle happened to come through the *Wende* (re-unification) and the changes in economic, social and political terms which enable housing and urban renewal to be funded through the state.

3
The Modernisation Project of the Lindenau Estate and the Tenants' Participation

When looking at the various problems (see part 1), the Meyer'sche Häuser are eminently suitable for careful, socially balanced urban revitalisation using existing housing stock. Acting in conjunction with the city administration and a team of architects, the foundation has drawn up rehabilitation programmes for all four housing estates. In addition to structural aspects, these plans consider social, ecological and architectural preservation issues. All these projects have been presented and discussed at a number of tenants meetings so that they can be tailored to the residents' needs. Since 1992 there has been a foundation-wide working citizen's association which supports all the actions and tries to assist the tenants.

The aims of this 10-year project in Lindenau are firstly to enable existing tenants to remain in their flats, and, secondly, to upgrade properties in order to attract more younger families with children.

The rehabilitation costs are very high due to the decades of neglect. However, excessive rent increases to cover these costs would contradict the foundation's principles. Therefore expensive alterations to the existing ground-plans are being avoided and old building components are being reused. Flats are also being enlarged by incorporating the corner turrets and through the conversion of lofts.

Housing is also being made affordable by allowing tenants to perform some of the work themselves. All such work is chosen from a list contained in the self-help catalogue and carried out on a contractual basis.

Since the installation of bathrooms, showers and indoor lavatories reduces the living space, special attention is devoted to maximising the layout of the green inner courtyards. All courtyards are still kept free of cars and have common facilities such as gardens, allotments, playgrounds, washing lines and benches. Other aspects of the outdoor programme include climbing plants on facades, the installation of rainwater cisterns and the building of nesting sites for rare species of birds.

The modernisation of all 500 flats in Lindenau began in 1995. Two of the six sections are already complete: 26 buildings with a total of 208 flats have already been modernised; the average size of a flat increased from 48.5 to 55.2 m^2; the total costs are 22 million DM (1915 DM/m^2); rents climbed from 5 DM/m^2 to 10–11 DM/m^2 after completion of the work.

4 Conclusion and Outlook

The first results can be analysed. Modernised apartments, new roofs, painted facades and improved common facilities have achieved their main goal - to conserve the whole estate and to offer a better quality of life for the tenants.

There is some criticism to be mentioned. The strategy of modernisation is based on the existing tenant structure and medium sized flats - according to arguments over the ease of renting the apartements of the foundation. It is not focused enough on the future or desired tenants (e.g. handicapped people or big families).

Secondly, self-help of tenants does not play the desired key-role. This is mainly related to the reduced incentives for the inhabitants (physical work cannot be replaced by one's own capital, the rent saved by working in a self-help project will be deducted from the housing subsidy or the unemployment benefit). A more important factor is that the spare time of the people has changed completely: those who have enough (e.g. senior citizens) are too old or have to tackle with more substantial problems (e.g. unemployment) and those who are employed and/or young do not have the time, and housing issues are secondary to them.

Furthermore the federal, state and local subsidies and loans have been cut dramatically which has caused a slower progression of the project.

Additionally, only a part of the planned ecological measures have been carried out - mainly due to the high costs and a lack of experience.

Last but not least, the changes in the different housing market sectors and the competition between them have a huge influence on the Lindenau estate.

Nevertheless, in general the revitalisation project of the Meyer'sche Häuser is a good example of contributing towards a more environmentally, economically and socially compatible development of neighbourhoods and cities.

Beijing's Socio-Spatial Structure in Transition

Chaolin Gu[1] and Christian Kesteloot[2]

1 Introduction

It is a popular belief that socio-spatial disparities have diminished compared to their levels in the past and compared to cities in market economies within socialist countries (French and Hamilton 1979). However, a new urban phenomenon - social polarisation and spatial segregation - has taken place in Beijing since the implement of China's economic reform and open-door policy in 1978, especially through the urban reforms since 1984. An enormous floating population began to migrate from rural to urban areas due to an erosion of the pillars of the migration controls, such as the employment system, the resident registration (*Hukou*) system, the grain rationing system, the urban housing system and so on (Chan 1994, Gaubatz 1995, Gong 1995). The original socialist equity in an urban classless society has taken a beating due to the large gap in the welfare system between the urban inhabitants and the rural population under the specially designated Chinese resident registration system. As a result the urban socio-spatial structure has been altered to fit in with the urban reform processes. The pace and the extent of changes have been rapid and substantial in the China's capital city - Beijing. Four main periods or patterns are briefly summarised in this chapter to describe a shift in the orbit of Beijing's socio-spatial structure.

2 Traditional Socio-Spatial Structure (1271–1949)

Beijing's traditional urban socio-spatial structure is different from that of western countries (Chang 1977). The socio-spatial structure was formed from the principle of the "ideal" planned layout in the book of Artificers (*Kao Gong Ji*) which was written about 2500 years ago (Hou 1986a, 1986b). This norm

[1] Chinese Academy of Sciences, Institute of Geography, Beijing 100101, China
[2] Catholic University Leuven, Institute of Social and Economic Geography, Leuven 3001, Belgium

J. Breuste, H. Feldmann, O. Uhlmann (Eds.)
Urban Ecology

defined urban spatial pattern as an uniform, regular and hierarchical square, surrounded by a city wall. Within the city there should be nine longitudinal and nine latitudinal thoroughfares to express the urban socio-spatial structure; the centre was the palace, with the administrative centre at the front, the main market and commercial centre at the back (*Qian Chao Huo Shi*); the memorial park of emperor ancestors was on the left, the worship site of the gods of soil and grain on the right (*Zhou Zhu You She*), and the residential area around the centre. The old Beijing city was built according to this rule and became a more famous metropolis, called *Dadu* city in the *Yuan* Dynasty (1271–1368 AC). The layout of *Dadu* followed a well-defined central point, the Central Tower, and the north-south central axis. The palace city was located in the centre. The residential quarters were divided into 50 wards around the palace city. Social or racial segregation had certain distinguished characteristics. The residential areas were inhabited by people of imperial lineage, who always lived closer to the palace city than others.

Beijing was affected later by the European industrialisation and technological revolutions. In general, Beijing kept the traditional Chinese urban socio-spatial structure for more than seven centuries until the foundation of the new People's Republic of China in 1949. The spatial structure of the pre-1949 Beijing city, which was a walled area of 63.2 square kilometres, can be divided into two parts: the old inner city and the old outer city. The old inner city was the Imperial City, which contained the Palace, large lakes with their adjacent parks, temples, libraries, examination halls, public building and some rich residential areas (*Hutongs and Siheyuans*). The Palace city marked the location of the centre of the Imperial City. The major gates, courts, royal residences and the market were arranged along a north-south planning axis. The road network of the Imperial City was constructed in a chessboard style system. The old outer city, which was located in the southern part of the old Beijing city, and adjoined the old inner city, was the residential city for the common Beijing people. The social structure of the city showed a distinct social and racial segregation: (1) the social segregation. The Palace City had been named the Forbidden City and was encircled by a high purple wall. Only four gates formed linkages with the whole Imperial city. (2) the racial segregation. The old outer city was built in the *Ching* Dynasty for the purposes of racial segregation. All *Han* Chinese were evacuated from the old inner city and moved into the old outer city after the *Manchu (Ching)* overthrew the rule of the *Ming Dynasty*. In late *Ching*, the old outer city reverted to "ghettos" that mainly consisted of some poor residential areas and crowded shopping centres.

3 Transitive (Socialist) Socio-Spatial Structure (1949–1978)

During the 1949–78 period, some reconstruction was carried out in the old inner city of Beijing (China's capital in the new socialist era) in order to show the huge achievements of the vigorous socialist system (Sit 1995). At the same time the most rapid urban expansion occurred in the urban fringe owing to the fast urban growth of the state capital (Kesteloot 1992).

In the old inner city, firstly, *Tian'an Men* Square was enlarged and reconstructed; ten main buildings, such as People's hall, Chinese History Museum, Nationality Hotel and Beijing Railway Station etc., were established along the *Changan* Street and formed a new east-west urban development axis which contrasted with the old north-south Imperial Axis. Secondly, Beijing also started another urban renewal process during that time. In the first 20 years, urban renewal put its main stress on the open space. Its chessboard-like system of streets, *Hutongs* and the garden and water system did not change until 1974, but some *Siheyuans* (quadrangles) were demolished or changed their function. During the "Cultural Revolution" period, the units were permitted to build housing within held urban lands to solve a housing shortage problem, as a result, many simple one-story houses were built in the courtyard of *Siheyuan*. Fortunately, because of the old road system, most imperial or noble residences with low densities, and gardens of the *Ming* and *Ching* Dynasties were preserved so that the urban renewal did not change the function and framework of the inner city. In the urban fringe, a fast spatial sprawl has occurred since 1953. The urban built-up area was increased from 109 sq. km in 1949 to 340 sq. km in 1978. Between 1953–57 some newly established manufacturing industries concentrated mainly in the outer rim of the old city and formed some new built-up areas adjoining the old city; in 1958 a master plan of Beijing was made in the light of the socialist ideology to eliminate three great disparities in human society. New planning development areas were split off from the old city and formed 12 "scattered clusters" in its inner suburbs. The function of the "cluster", in fact, was the urban-rural unity, in which its main economic sectors included both industry and agriculture, and its residents had both workers and peasants; of course, some open spaces were kept among these "clusters." Until the end of 1970s Beijing's spatial expansion always followed this kind of "scattered clusters" model. As a result, Beijing's socio-spatial structure was enlarged from the old inner city to the inner suburbs through the construction of these new industry-residential areas, public transport networks and commuting to work.

Since new jobs are always accompanied by industrial development, the new "scattered clusters" spatial structure also resulted in the appearance of the new social structure. This new socio-spatial structure was expressed by the old inner city and some new industry-residential areas in the inner suburbs.

As a result, "old Beijing people" still like to remain in the old inner city; "new Beijing people" tend to settle down in the newly built-up areas around the old inner city. A large amount of public housing was built by state-owned enterprises and government branches in inner suburbs and employees obtained their housing from their working units. Some other people, who worked in main urban industrial areas of 12 "scattered clusters," may choose to live in the old inner city, in the new inner suburbs or in new outer suburb settlements adjoining industrial areas.

The urban land use structure varied widely because changes in the spatial and social structure took place, main residential and industrial areas extended side by side along the periphery of the old city, educational and research areas concentrated on a few plots. The whole city sprawled outward along the radial main roads to the urban fringes and the inner suburbs.

4
Reforming Socio-Spatial Structure (1978–)

Owing to the transformation of the social and economic system in China during the recent decade, a reform of the urban socio-spatial structure has occurred in Beijing. In spite of the short reform period compared with China's long history, shifts of the urban socio-spatial structure are so fast and deep that they affect either the reorganisation of city layouts or the whole urban society. In the old inner city, even if the framework of the city remains the same as before, the urban land-use structure is being reorganised and some new urban functions are added. In the suburbs, the urban spatial expansion is not only linked to some new forces and mechanisms, such as the real estate market, the modification of public housing and home ownership programme, as well as the development of urban services and high-tech industries, but is also mixed up with the suburbanisation process created by the rich and powerful upstarts from the urban central area and the poor and weak new immigrants from rural areas. Beijing's urban fringe is becoming an active region where the fastest changes in the socio-spatial structure occur as well as the majority of the urban problems and the biggest disparity between the rich and the poor (Gu 1995).

In the old inner city, massive urban renewal and reconstruction has taken place since 1978. There are two main reasons for this: (1) Beijing, historically, lacked a Central Business District (CBD) so that most of the financial and business buildings were located and dispersed in areas outside *the 2nd Ring Road*. In order for Beijing to be able to carry out its international metropolitan goal and develop its service industry, there is a strong demand for office, commercial and service lands; (2) the old inner city consists mainly of old houses of wood and bricks, one-storey courtyard houses of the *Ming* and *Ching* Dynasties, in which a water tap, toilet, kitchen and heating are usually shared with several families or not provided, Sit (1995) described it as a "slum area".

In Beijing urban renewal of the inner city was pushed and accelerated by two forces in 1984: (1) a huge demand and need to launch new institutions (especially international political, economic and cultural institutions, governmental or foreign invested and joint-venture enterprises' liaison offices) and (2) a reopened real estate market. The implementation of the urban land use compensation policy also speeded up the process of the urban renewal. In 1990, Beijing Municipal Government launched its urban renewal plan for the inner city. Eight renewal projects have been carried out. Some important shopping centres and squares that embody the urban functions and traditional style and features, such as *Wangfujin, Dongdan, Xidan, Xisi, Xingjiekou, Beixingqiao, Caishikou, Zhushikou, Ciqikou etc.*, were redesigned and reconstructed. A new shopping centre has appeared quickly in the *Wangfujin - Dongdan* region.

In the urban fringe, the outward pattern of urban sprawl becomes more and more evident. The whole city is expanded outwards along the radial main roads to the urban fringes and the inner suburbs. The built-up area increased from 340 sq. km in 1980 to 426 sq. km in 1994. The urban non-agricultural population increased from 4.78 million in 1980 to 6.104 million in 1994. However, statistical data showed that the total population in the old city only decreased by 50,000 people between 1978–94. Obviously, a massive suburbanisation, i.e., migration from the inner city to the suburbs, has not yet occurred.

5 Conclusion

The biggest change in Beijing during the last decade was in its social structure. Beijing's new poor rural immigrants and high income upstarts are rebuilding an urban stratification and socio-spatial division by the way of social segregation. For the social structure, the new rural immigrants tend to concentrate in the urban fringe since some surplus and cheap rooms exist for rent in this area. The *Zhejiang* village around *Muxuyuan* has become a large new slum area, while some villa areas along the *Jingtang Road* from the Asian Game Village to *Xiao Tangshan* in Beijing's north-eastern suburb are forming a new richer residential zone.

With regards to the urban spatial structure, manufacture and service industries have decentralised to some urban areas of cheaper labour/land (for example, urban fringes) while spatial centralisation of larger service units and restructuring of urban functions has led to renewal of the old inner city.

This paper was written while GU Chaolin was a visiting professor at the Institute of Social and Economic Geography, Catholic University of Leuven.

We would like to thank the ISEG staffs at Catholic University of Leuven for their assistance in GIS and researching much of the literature.

References

Chan KW (1994) Cities with Invisible Walls: Reinterpreting Urbanization in Post-1949 China. Oxford University Press, Hong Kong

Chang S-D (1977) The Morphology of Walled Capitals. In: Skinner GW (ed) The City in Late Imperial China. Stanford University Press, California

French RA, Hamilton FEI (1979) Is there a Socialist City. In: French RA, Hamilton FEI (eds) The Socialist City. Wiley, Chichester pp 1–22

Gaubatz P (1995) Changing Beijing. Geographical Review Vol 85 No 1: 79–96.

Gong H (1995) Spatial Patterns of Foreign Investment in China's Cities: 1980–1989. Urban Geography Vol 16 No 3: 198–209

Gu C (1995) China's Urban Housing System in Transition, Journal of Chinese Geography, Vol 6, No 2: 16–38

Hou R (1986a) Evolution of the City Plan of Beijing, Third World Planning Review, Liverpool University Press Vol 8: 5–17

Hou R (1986b) The Transformation of the Old city of Beijing, Harris CD (ed) World Patterns of Modern Urban Change, The University of Chicago, Chicago pp 217–239

Kesteloot C (1992) Some spatial implications of socialist planned economies, Acta Geograplica Lovaniewsia, Vol 33: 277–281

Sit VFS (1995) BEIJING: the Nature and Planning of a Chinese Capital City. John Wiley & Sons, Chichester

Workshop: Awareness Raising, Information and Educational Methods 2.3

Promoting Sustainable Urban Environments by Continuing Education for Local Authorities

Maik Adomßent

1 Local Authorities in Germany: Maintainers of Sustainable Urban Environments?

1.1 Agenda 21 – a Global Action Plan for the Century of Cities?

In the Agenda 21, the United Nations' action plan for the next century, local authorities are seen as important maintainers to fulfil the objectives for a sustainable development in the future. Chapter 28 of the Agenda describes them as '... the level of governance closest to the people, they play a vital role in educating, mobilising and responding to the public to promote sustainable development' (Quarrie 1992).

In the long run, evidence will have to be produced about the importance of a sustainable suburbanisation as a means of responsible use of natural resources and cross-connections on the national, regional and local level (so-called 'glocalisation'). Many people already tend to call the forthcoming hundred years the 'century of the cities'. This is true for Germany, where two thirds of the citizens live in urban areas (StBA 1996).

1.2 Different "Roles" of German Communities in Environmental Protection

Until the seventies, activities related to the environment were only seen as low priority tasks, subordinate to fresh water supply or other traditional urban duties like sewage and waste management. Meanwhile, management tools like rural and urban planning or the establishment of information activities such as eco-counselling demonstrate the communities' contribution to environmental protection. In constructing, operating and maintaining economic, social and environmental welfare, German cities and villages

University of Lüneburg, Institute for Environmental Communication, Rotenbleicher Weg 42, D-21335 Lüneburg, Germany

J. Breuste, H. Feldmann, O. Uhlmann (Eds.)
Urban Ecology

have become administrative and executive keystones, playing different "roles" in communal environmental protection. For example, they implement and establish EC-directives, national and Federal environmental policies and regulations. Furthermore, they have legislative rights on their own, e.g. for the empowerment of construction plans or tree protection measures. Environmentally sound procurement or energy-saving buildings are possibilities that can be implemented with the commitment of authorities and inhabitants.

Although the growing importance of communities for a sustainable future is constantly underlined, counterproductive backlashes always have to be taken into account. Primary discrepancies stem from the ongoing reduction in financial support for environmental aims because of the alarming decline in households in the communities.

It is not possible to estimate whether or not our cities are able to cope with present and future demands for sustainable urban environments without taking a look at the qualifications and abilities of the personnel of local authorities.

2 Vocational Qualification: Situation and Educational Demands

2.1 Signs and Traces of Environmental Management Qualifications

In Germany, the qualification structure on the local level tends to be rather heterogeneous. The environmental department staff in most cases are a kind of melting pot for a colourful mixture of different professions. Although there is an urgent need for personnel with a dual qualification to be able to cope with both administrative legislation and environment related problems, this "generalist type" is still hard to find. These circumstances become even more alarming in rural communities and small towns, which form the majority of local authorities in Germany. The most important reason for this problem is the ignorance of most communities, who did not formulate any basic qualification guidelines for their environmental personnel. Considering this lack of (additional) environmental competencies, the question arises whether at least the current supply of continuing education, being more flexible and less formal than before, is able to cope with the deficiencies of the past.

2.2 Missing Links Between Reality and Future Needs

Due to the systemic character of environmental problems and their impact on different environmental media, jobs in the field of environmental protection generally ask for an *entire* (i.e. trans-disciplinary) type of qualification including:

- the ability to think and work in 'trans-boundary' dimensions together with other related disciplines;
- the capability to replace reactive with anticipating attitudes (future orientation rather than looking at the past);
- the development of communicative skills for effective and sustainable work both inside and outside local authorities.

A look at all the different offers in environmental continuing education provokes the anxious question whether or not these requirements can be fulfilled. Although there is a huge variety of environmental oriented continuing education (for details see references in Adomßent 1997), offers that have been tailored to suit the needs of people engaged in ecological and environmental work for local authorities are difficult to find. More alarming than this quantitative "bottle-neck" is the widespread lack of basic quality standards in continuing education. These serious deficiencies can be summarised in two different ways:

1. Concerning contents:
   - perpetuation of specialisation and selective perception of problems in particular disciplines, whereas 'trans-boundary'/multi-disciplinary approaches are neglected;
   - ongoing domination of single-minded teaching with a stress on technical subjects and solutions, most often the offers show a preference for „end-of-pipe"-principles,
2. Concerning organisational structures and didactic criteria:
   - though offering considerable relevant experience, people without higher degrees (university or specialist college) are rarely admitted;
   - only few courses can be carried out alongside normal work;
   - in most cases, the important relationship between contents and intervening didactic needs („learning ecology") is neglected.

This unsatisfactory situation led to the creation of a new distance learning approach of the Institute for Environmental Communication at the University of Lüneburg which will be described in the following section.

3 Bridging the Gap with a New Educational Approach

3.1 Aims and Conceptual Framework

The idea of "Environmental protection management for local authorities", a project supported by the German Federal Foundation for the Environment (Deutsche Bundesstiftung Umwelt), is to develop a distance learning offer that might be able to bridge the educational gap of qualification facilities for communities.

In the first term, participants receive an outline of environmental protection in local authorities. In the second term there is a focus on special problems that have been recognised to be of primary importance. Students have to decide which of the following fields they want to specialise in:

1. Landscape and urban ecology, recreation
2. Energy and traffic management and water supply
3. Waste and waste water management and soil protection.

In a final project during the third term, all the knowledge has to be applied to develop a mobility concept for designing a brochure on the Agenda 21 or for installing an information system. Depending on the complexity and required expenditure, team work is possible.

Looking at all 30 study-letters from a didactic point of view, three types of knowledge can be distinguished:

- *Basic knowledge* introduces environmental protection in communities and frames the understanding.
- *Problem related know-how* deals with specific environmental problems of communities.
- *Instrumental expertise* is considered to be most important within the study concept. The participants shall be enabled to cope with most commonly occurring problems with the help of instruments which integrate the environmental dimension into the organisation of communities.

Additionally, the students come together in weekend meetings two or three times a term. These seminars deal primarily with examples and applicable methods, whereas the self-teaching texts tend to cover the theoretical learning part of the course.

To minimise one of the disadvantages of distance education – the shortcoming of face-to-face communication between students and teachers- modern telematic techniques are used to ease organisation and to bring benefits for the participants' job-efficiency.

3.2 From Trans-Disciplinary Education to Urban Sustainability

A closer look at the curriculum reveals the focus of the project's design on the Agenda 21, reflecting the discussion about sustainable development and the efforts to transfer environmental management instruments from the contexts of enterprises to communities' conditions. For a start towards a sustainable urban future, this ideal of desirable qualifications should be valid at least for employees in higher positions. They should have an in-depth knowledge of the use of instruments and methods which take the environmental dimension into account and lead to integrated decisions and actions.

One central affirmative proposition that can be derived from the Agenda 21 is that people are as much part of the problem as they can be a part of the solution. Consequently, problems on the local level can only be solved co-operatively or, as Piet Beekmann (in this book) points out: 'in the end it is citizens forming the "building material of sustainable cities"'.

References

Adomßent, M (1997) Kommunaler Umweltschutz: Weiterbildung für die Zukunft. Umweltmagazin 26(1/2): 28–32

Quarrie, J (Ed) (1992) Earth Summit 92: The United Nations Conference on Environment and Development, Rio de Janeiro 1992, London Ch.28 pp 200

StBA (Ed) (1996) Statistisches Jahrbuch 1996 für die Bundesrepublik Deutschland. Bonn

Ecology as Urban Culture

Detlev Ipsen

The title of this chapter contains a thesis which leads away from the description of city life that has been common up to now, namely the separation of urbanity and urban culture from nature and ecology. Generally, sociologists talk as though the city were a social area that excludes nature (Häußermann, Siebel 1989). Cities are seen as being emancipated from nature. Empirical evidence initially supports this. Is it not the city that turns night into day? Is it not the city that represses and regulates nature? Yes it does, but the night remains night (measured in temperature, humidity and light levels) and the water in the pipes is still water. The development of a specific relationship to nature belongs to the city just as much as enlightenment and the dynamism of the monetary economy. We are currently in a phase in which our relationship to nature in the city is being reconsidered as the 19th century model (which is still with us today) is now endangering the chances of future development of the cities (Hoffmann-Axthelm 1993, Lipietz 1991). I will try to formulate briefly the theoretical thoughts and some empirical findings in seven propositions.

1. The city can be understood as consisting of a number of different and conflicting projects which become one through urban culture.

When we speak of urban culture, we do not only mean the cultural functioning of the city or that the culture of a city stands in opposition to its economic or social life. We believe that urban culture is a system which penetrates not only the most varied institutional areas, the economy and lifestyles (*Lebenswelt*), but also nature in the city. It is easy to see a city, especially a modern one, as being composed of a number of projects. Two projects of a modern city stand out: the city as a socio-political project which identifies closely with enlightenment and the French Revolution (freedom, individuality, solidarity and openness are features of this project) and the city as an economic project, closely identifying with capitalism and its leaning towards innovation and universality (Simmel 1958, Giddens 1996). These projects conflict both with each other and the associated relationship with nature and

University of Kassel, AEP (Working Group for Empirical Planning Research), Mönchebergstr. 17, D-34109 Kassel, Germany

J. Breuste, H. Feldmann, O. Uhlmann (Eds.)
Urban Ecology

require integration to achieve a positive field of tension. The system of urban culture infiltrates the *Lebenswelt* and nature in equal measures, as well as allowing an antagonistic integration of the city to enable it to become a meaningful unit. Thus, when we speak of ecology as urban culture, we mean the ways in which nature in the city is connected to the city's economy, politics, lifestyles and physical form.

2. The modern city developed a relationship to nature in the 19th century which was based on power over nature, distance from nature and compensatory mechanisms to cater for people's longing for nature.

The dualistic approach of seeing ecology and urbanity separately is itself an expression of a specific relationship to nature in which the nature's inherent value has been forgotten (Mayer-Abich 1986). A number of ideas, concepts and visions developed in the cities of the 18th century which had the effect of nature becoming ever more bureaucratically regulated and kept under control. Building on the absolutist approach of maximum power over nature in park architecture, the new relationship to nature can be seen especially clearly in the planning and building of the hygienic city of the 19th century. The principles first became clear in the management of water in the cities. It was no longer well but systems of pipes which served households and industry with water. Sewerage systems were built which made the collection of excrement from houses and use of it as a fertiliser in the market gardens on the edge of town obsolete. Water disappeared from the minds and responsibilities of the inhabitants. Technical and bureaucratic systems came between the inhabitants and nature in the city. Once nature was lost sight of it was effectively expelled: country landscapes and nature were locked into opposition with the cities. The process of civilisation, in mastering the external nature, also obtains control over the inner nature. This power over nature and distance from it corresponds with the natural human desire for nature. Through walking and mountain climbing and through nudism and nature protection the "out of town nature" is being won back. Garden cities and people's parks, allotments and open air baths are expressions of longing for nature from within cities (Ipsen 1992).

3. The maintenance of the social balance between work and money through a higher use of nature is sustaining today's ecological crisis.

The tensioned relationship between the Utopian vision of a good and luxurious life in the city and the logic of capital growth in the 19th and early 20th centuries is born out by the social realities of impoverishment and class battles in cities of this era. The fascination with "Fordism" that transcends all levels of political thought is all too apparent. Work and money have a common interest in the ability for people to achieve a luxurious and comfortable lifestyle. The idea of a successful relationship with nature is being privatised by a grand coalition of those interested in profit and consumption. This idea can be seen in detached houses with gardens, in the greenery surrounding tower blocks, and in holiday trips. The rationalisation of agriculture and the concentration of supplies by large food multinationals has quickly brought down the costs the necessities of life. The import of cheap energy and the disregard for the self-cleaning potential of the soil, air and water are what have made ecology, and thus nature, a subject in the city project once again. So far neither the existing social groups, nor the resulting political parties have been able to achieve a social balance through town planning proposals, as the lines of conflict are too complex and touch every point in our lives.

4. The break between ecological values and ecological behaviour lies in a dilemma caused by the structural framework.

It is obvious that there are differences between values and behaviour. The Christian church's confessional is a way around this anthropogenic predicament. However, when it comes to environmental awareness we are mainly talking about the gap between the person with his/her own observational powers, experiences and motivations and the person as part of the world. This can be seen, for example, in a survey we carried out in Dresden and Frankfurt a.M. Asked about the problems present in the city only 4.1% explicitly mentioned environmental problems. Even if one includes the group that cited traffic as a problem (11.5%) we are still left with a relatively small group (15.6%); there are good reasons for seeing traffic as a problem with nothing to do with ecology (safety, use of road space for other purposes, noise). In comparison, when asked directly about environmental problems, each person could name countless problem areas. The majority (57.5%) do not have any personal experiences with environmental problems and almost everyone hears about environmental problems from the mass media. Everyone agrees for the need to behave in an ecologically sound way. The media commentary makes it possible to create a divide between morality and individual experiences, interests and environmental consciousness. The things that one wants tend to have unwanted consequences. The conflict between the equally strong wishes and counter wishes results in a rigid separation of the levels of consciousness (Ipsen 1994).

5. Moral appeals for ecologically sound behaviour can lead to a permanent loss of credibility.

It is probably not always possible to make the move from proposed standards to practical behaviour; the opinion-making elite tries to bridge the gap with constant moral appeals. However, in this case, morals come to grief when it comes to reality; equally strong norms and standards stand in contradiction with one another. Values of civilisation and modern understanding of "a good and decent life" are in stark contrast with the idea of a long-term, sustainable development for oneself and one's children. Our qualitative studies on water use have clearly shown this (Heidenreich, Glasauer 1996). The predominant standards of hygiene are contradictory to a sparing and above all protecting, less chemically intensive use of water. As long as a white collar, clean shirt and an odourless body are essential for the participation in business and public life our use of water can only marginally change. Attempts at differentiating the use of water (i.e. the use of different water for cleaning, hygiene and for food) fail thanks to the technically and bureaucratically standardised supply of water. The utilities only supply one sort of water (high quality) for all types of use. In such a situation of structural disjunction, appeals to people's general morals (saving, planning, going without) run the risk of creating a permanent loss of credibility. This helps neither people's well-being nor nature's ability to regenerate itself.

6. If we can bring the term landscape back to the cities we have the chance to make ecology a fixed part of urban culture.

When one speaks of landscape in connection with the city today, one generally means countrifying the city (Hoffmann-Axthelm 1993) whereas in the 1920's architects meant the development of garden cities (Lucarelli 1995). Today the architect thinks of distance and dissolution, contrasting with the picture of a dense and compact city (Sievers 1997). I, however, prefer to think of the direct use of the term landscape for the city. Landscape is the combination of natural relationships, the moulding of these relationships through work and the development of an aesthetic of space. Why should we call village communities and the work of farmers a landscape (since they are beside, and formed by nature) but not a city community, technical work and the topography of such an area? One could then ask what a city was if it could not be described as a landscape. What else could city waters, the use of topography to create the skyline, urban flora and fauna, city soils, air and water possibly be? The association of the term landscape with the countryside lies in the understanding of, and behaviour towards nature in the 19th century that saw the modern city as emancipation from countryside living and nature. Landscape became an expression in the minds of migrants who had left their fields and villages behind and who looked back with a shudder but also a simultaneous longing for the countryside. When we manage to view the landscape of a city as a project and when a city reclaims a harmonious relationship with nature in the same way that it is open to new things, eager for fashion and economic innovation, individuality and solidarity, then ecology will become urban culture, a part of what makes a city a unit (Rüttgers 1997, Sievers 1997, Böhme 1989).

7. Ecological aesthetics as urban culture can help us to see the city as a landscape as long as they respond instinctively to the contradictions between

the individual projects in a city and the fractures in an urban relationship with nature.

Nothing shows the power of aesthetics more clearly than the aesthetics of products, without which capitalism would never have gained any power over the people (Welsch 1993). Aesthetics touches emotion and cognition to the same extent, it results in recognition and motivation. Ecology will not become a driving force in the development of society and our cities in particular without aesthetics. However, we believe that ecological aesthetics should not split the world into the beautiful and the necessary if it is to lead us to seeing cities as landscapes. Ecological aesthetics must attack the contradictions between urban landscapes and the other projects of social, political and economic life. It must turn to the visible things, the city stream and the hill, as well as to the invisible, the supply and removal networks, to the chemical processes that we can neither smell nor taste. Such aesthetics will have the power of bringing things together as a result of their instinctive actions, even when they sometimes only show that here there is no bridge.

References

Böhme G (1989) Für eine ökologische Naturästhetik, Frankfurt a.M.

Dahl J (1984) Der Unbegreifliche Garten und seine Verwüstung – über Ökologie und über Ökologie hinaus. Stuttgart

Giddens A (1996) Konsequenzen der Moderne. Frankfurt a.M.

Häußermann H, Siebel W (1989) Ökologie statt Urbanität in Universitas 6, pp 514–525

Heidenreich E, Glasauer H (1996) Wasserkultur-Texte 5/7, Kassel

Hoffmann-Axthelm D (1993) Die dritte Stadt. Frankfurt a.M.

Ipsen D (1992) Stadt und Land – Metamorphosen einer Beziehung. In: Häußermann H, Ipsen D, Läpple D, Rodenstein M, Siebel W (eds): Stadt und Raum, Pfaffenweiler

Ipsen D (1994) Umweltwahrnehmung und Umgang mit Wasser in Agglomerationsräumen

Jpsen D (1994) WasserKultur Texte 5, Kassel

Lipietz A (1991) Zur Zukunft der städtischen Ökologie. In: Wentz M (ed): Stadträume, Frankfurt a.M.

Lucarelli M (1995) Lewis Mumfort and the Ecological Region, London

Mayer-Abich KM (1986) Wege zum Frieden mit der Natur. Praktische Naturphilosophie für die Umweltpolitik, München

Rüttgers J (1997) Kunststück Zukunft. Die Zeit 12 14, p 62

Sievers T (1997) Zwischenstadt, Braunschweig

Simmel G (1958) Philosophie des Geldes, Gesammelte Werke Vol 1, Berlin

Welsch W (1993) Die Aktualität des Ästhetischen, München

Scenario Techniques as a Tool for Supporting Endeavours to Analyse Complex Problems in Urban Ecology

Karl-Heinz Simon[1] and Anna Musinszki[2]

1 Introduction

The project, to which the scenario activities contributed, is called „WasserKultur" (Ipsen et al. 1998). As part of research activities on urban ecology the project brings together issues from civil engineering, geography, sociology, town planning, and the arts. Hence, research teams are located within different disciplinary „cultures", such as natural or social sciences; this creates certain difficulties with respect to communication among project teams and to a common topic of project elaboration.

In this chapter the following points are shortly discussed: (1) some properties of the scenario method, (2) some details of the process of scenario definition and evaluation, and (3) a few selected components of the scenario "a multitude of waters in the city" and the four dimensions in which elements can be grouped together.

2 Scenario Techniques

Scenario techniques have a long tradition as a method for supporting activities of town and landscape planning as well as in engineering and economics. Despite such a tradition, no repertoire of standardised tools exist, although several projects have dealt with computer-based tools. It is widely accepted that scenarios are very helpful tools in dealing with complex problem situations as well as in supporting the search for solutions where groups of people are working together (Forschungsverbund Lebensraum Stadt 1994).

From a methodological point of view three types of scenarios can be distinguished: trend, alternative and contrast scenarios. In our project the third

[1] University of Kassel, Centre for Environmental Systems Research, Kurt-Wolters-Straße 3, D-34109 Kassel, Germany
[2] University of Dortmund, Faculty for Regional Planning, August-Schmidt-Straße 10, D-44221 Dortmund, Germany

J. Breuste, H. Feldmann, O. Uhlmann (Eds.)
Urban Ecology

type of scenario has been developed, a description of the future state of the water supply system in urban areas. The discussion processes focused on what ought to be reality in the future, not on what will emerge from trends which are already present. Having made the decision to describe the future state of the system, it was also not necessary to describe in detail developmental steps which would be suitable to reach this future state.

One of the main tasks of the sub-project „scenarios" was a mediation of group processes to elaborate a common version of the scenario definition. In order to achieve such a definition, open discussions were combined with preparatory and finishing work in a small project group. Rudimentary techniques of knowledge acquisition, representation and structuring, as known from systems analysis, can be applied here.

Special attention has to be paid to flexibility and openness to guarantee that different „types of knowledge" can be integrated equally, i.e. it did not seem to be adequate to carry out a straight formalisation or quantification of knowledge.

When discussed in more general terms, the approach can be seen as part of a much broader modelling approach, with knowledge acquisition and knowledge structuring being the centre of interest.

3 Elements of a Scenario „Multitude of Waters"

As a starting point of scenario definition (Musinszki, Simon 1996) several problem dimensions have been analysed comprising the main difficulties with recent water usage. These dimensions are:

1. Aspects of quantity and quality: no water supply problem exists in general (at least in most of the cities of „the North"), the type of resources used and the problem of water quality, however, cannot be denied.
2. Aspects of transparency: on the city's territory water is not (or hardly ever) visible as a natural resource, nor are its technical and ecological consequences.
3. Expertise: water is – with respect to resources and distribution – treated by experts only.
4. Aspects of change: "the water question" is embedded in a bundle of dynamic processes continuously altering city structures and demand patterns.

With the above problems in mind concepts and ideas can be collected which might function as essential components of a future state of the system in which the problems are no longer present.

A few examples of elements of the scenario "multitude of waters" are arranged in three groups as follows. Needless to say, these few elements represent only a small portion of the concepts and ideas of which the scenario definition is composed.

First group of aspects (resource issues): several different sources of water should be used in the future: local wells, rain water cavities, water from rivers. To allow utilisation of different resources protection of natural systems and groundwater ought to be in practice nearly everywhere in the city. Another more technical aspect is that cascades of water applications and closed cycles have been established (see Fig. 1 for some of the components of such a concept).

Second group of aspects (technical arrangements): in this group technical aspects such as a broad application of conservation techniques, options like water-free toilets, and organisational basic requirements to support public utilisation of devices, e.g. in washing centres, and the installation of a second line-system, in some places, for grey-water distribution.

Third group of aspects (social and political issues): the idea is that resource issues and technical arrangements alone are not sufficient to change the system fundamentally; social and political constituents also have to change. Ideas for this are: establishment of water co-operation committees or new political actors such as a water consultation group and arbitration boards. Elements of civil society will be of greater importance than today and all the actors concerned are expected to work together, for example in city-environs-conferences.

The three groups show that aspects from natural sciences, engineering and social sciences have to be integrated to achieve a system state where at least the capability for dealing adequately with the problems mentioned has been

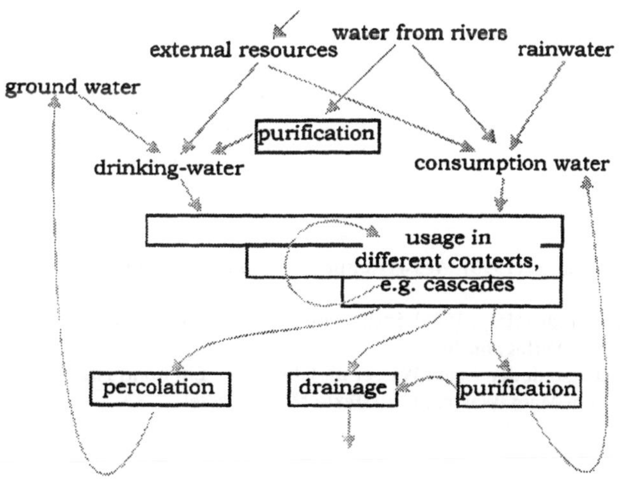

Fig. 1. Essential components of the resources-related part of the scenario

produced. The essential elements of such a system in the future are: decentralisation of sinks and sources and local networks and line-systems, visualisation of ecological and technical arrangements, participation and co-operation, differentiation of water usage.

The following four dimensions can provide assessment criteria to obtain a basis for evaluating the fitness and adequacy of concepts with respect to their capacity to solve the mentioned problems with water treatment (see Musinszki/Simon 1996 for more details):

1. Availability of the resource water: ubiquitous < > limited.
2. Regulation characteristic: administrative < > one of civil society.
3. Perspective: particular < > integrative.
4. Aspects of usage: unified < > differentiated.

4
Résumé

Some of the outlines of a future water system in urban areas were drawn up in the scenario. The main idea is to have realised more diversity in the future and to achieve this by establishing decentralised system components where convenient. Technical, ecological and socio-political aspects have to be integrated and brought together to form a new coherent system of dealing with water in the future. A system change with such far-reaching consequences will not be an easy task but many of the components have already been developed and concepts and prototypes are to hand, even with respect to social and political facets of the system. The main task is to integrate many of these components to build a new system which has the power to survive even in a changing environment. Not all concepts and ideas are equally qualified. In addition to ecological constraints and the claim that there must not be an extended energy input or application of chemicals (from a sustainability point of view) there must be a simultaneous guarantee of the continuation of the functioning of the city with all its implications and preconditions.

References

Ipsen D, Chichorowski G, Schramm E (1998) WasserKultur – Urbanität, Technik, Ökologie. Analytica Verlag, Berlin (in prep)

Forschungsverbund Lebensraum Stadt (Hrsg. 1994) Szenarien und Handlungswege. Mögliche Zukünfte der Stadt. Ernst&Sohn Verlag, Berlin

Musinszki A, Simon KH (1996) Die Stadt der vielen Wasser im Jahr 2023 – Ein Kontrastszenario mit Beispielen aus Frankfurt am Main. WasserKultur Texte Nr. 18, Kassel

Social Limitations of Sustainable Water Consumption

Herbert Glasauer

When we talk about the necessity of ecologically sound behaviour we often complain about the gap between environmental consciousness and ecological behaviour. Rather than trying to improve pedagogic concepts, we should be thinking about the following central questions: is the gap between environmental consciousness and ecological behaviour an individual problem of a lack of self-discipline or is it a structural problem? Is it possibly a mental block resulting from a conflict between wishes and abilities? I would like to present my thoughts in the form of seven theses, taking our problems with water as an example.

Thesis 1: The subject "water" plays a nearly irrelevant role within people's everyday lives.

At a first glance this thesis seems to be diametrically opposed to our experiences. If the subject of saving water is mentioned either in discussions or in empirical studies, almost everyone agrees that we should be sparing in our use of this resource. However, despite this broad agreement, problems associated with water hardly play a role in people's everyday lives. When asked about environmental problems in their city that were relevant to them, the subject of water came fourth in Frankfurt and Dresden behind air pollution, traffic and rubbish, and was mentioned by only 8% of those surveyed. Water hardly gets a mention in our qualitative interviews on descriptions of everyday household life. Whether it be personal hygiene, drinking coffee with friends or colleagues, cooking, doing the washing or keeping the house clean, it is only in the rarest of cases that people are aware of the water use associated with the activity. In the organisation and implementation of the daily routine dealing with certain tasks and demands are foremost in people's minds, not the use of water.

Thesis 2: Households see only a few possibilities to reduce their internal consumption of water.

In our qualitative research some water-saving activities were mentioned comparatively frequently: lowering the amount of flushed water used in toilets,

The University of Kassel (GhK), Gottschalkstraße 28, D-34109 Kassel, Germany

J. Breuste, H. Feldmann, O. Uhlmann (Eds.)
Urban Ecology

purchasing water-saving appliances and the well-loved example of turning-off of the tap while brushing your teeth. Although considered dramatically more difficult, first steps can also be seen in reducing water use in the areas of personal hygiene and the washing of clothes. As soon as a person's various and individual everyday routines are seriously questioned, one can expect hurdles and opposition. Much larger savings in household consumption are due to relatively recent social developments seen in the break up of traditional family structures, the increasing percentage of women with careers, the growing importance of fast food and eating outside the house, the growth in modern, experience-orientated swimming baths etc. Compared to the use of water outside the house, the internal use of water is being increasingly reduced as a result of these changes.

Thesis 3: The development of plans for water-saving consumption have failed largely for reasons of confusing or conflicting central cultural norms at the level of the individual.

Why does one group of people have to take a shower more often than another? Why do they change their outer clothing more frequently than others? Why do some others only feel comfortable in their home if it is thoroughly cleaned every day? The answers we get show ever more clearly that individual standards of cleanliness have very little to do with personal environmental consciousness, but everything to do with other, equally positively thought of norms such as being tidy, clean and well-groomed. The cultural models of cleanliness and smartness, repeated daily and demanding conformity, are the guarantees of integration in society and of social identity. Behavioural conflicts on a personal level are then the result of trying to conform to the opposing norms of cleanliness and tidiness on the one hand and of water saving on the other.

Thesis 4: A sustainable consumption of water could easily be realised if this goal is encouraged through incentives instead of punishment.

The most usual form of billing for the use of water in a household provides no incentive whatsoever for individuals to be sparing in their use of the resource. Especially in city-typical blocks of rented flats, sparing use by one occupant can be made irrelevant by the excessive use by someone else, while, simultaneously, the excessive use by that person is hardly paid for by that person themselves, as the extra costs are split up among all the occupants.

When additionally looking at the price increases imposed by the amenity bodies, which are justified as a result of lower water use by households while the fixed costs for the utility remain the same, one can see that this causes irritation at the very least.

Thesis 5: The development of ecological consciousness requires information campaigns by the mass media, but these campaigns also cause irritation because they are partly wrong, incomplete, inconsistent and can also be diametrically opposed to people's perceptions.

The image of water scarcity in people's minds originates from media representations. Parched earth and dried-up springs combined with references to southern Europe and the "Third World" are not suitable to describe our drinking-water problems, which I would like to call water scarcity among abundance. Equally spread through the media is the comprehensive knowledge of the amounts and types of pollution of surface waters and the potential dangers to flora, fauna and human health.

As a direct and personal experience of environmental damage is generally not possible, we hear from the news that a sudden reddening of the waters in the Rhine is a potential danger to our health and not an aesthetic natural spectacle. We read in the newspaper that we should not bathe in our rivers and lakes any more, or as recently, that in some cases it is now "safe" to do so once again.

Media reports on environmental problems are the essential prerequisites for building up the potential for environmental consciousness, but the conflicting expert opinions are creating growing confusion rather than a variety of viewpoints among the public at large.

It is also possible that personal experiences in a household can be in direct disagreement with generally publicised opinions in addition to the conflicts within the media itself. For the inhabitants of Frankfurt, water is almost everywhere in the environment around them, conflicting with the media images of water scarcity. Almost everyone in Frankfurt knows of their river called 'Main', the 'Nidda' and other smaller streams and lakes. Even within the built-up areas they are brought into contact with water in the form of fountains, ponds, water channels etc. The recent floods which reached as far as the historic centre of the city will still be in many people's minds.

Thesis 6: An ecological policy which aims at a sustainable consumption of water has to take note of the diverse vital social meanings of the resource within the everyday life of the people. If not, such a policy will be confronted with disapproval and resistance.

A political viewpoint which only understands water as being an essential natural resource is one which will turn out to be extremely short-sighted and problematic in the long-term. It may indeed be ecologically questionable to use valuable and scarce water of the highest quality simply to wash away human faeces, but it is just as problematic to impose water-saving regulations that completely ignore the variety of culturally orientated meanings that this resource has for people in their everyday lives. Apart from being a vital resource, water used "carelessly" has the power to bring people pleasure, feelings of comfort and relaxation, as well as a general lust for life. Campaigns which only concentrate on the miserly use of the resource will come up against either open or hidden opposition among those targeted, as they know of the many positive experiences that they have had in their daily lives through the "careless" use of water.

Thesis 7: A new water-policy has to show pleasure-orientated perspectives as alternatives for the water-saving consumption within the household instead of enforcing and perfecting their pedagogical stingy conditioning.

Significant savings of scarce drinking water are not going to be achieved by saving water, rather through the differentiated use of water of differing qualities for different tasks. This will reduce the demand for water of the highest quality. Appeals for the saving of water must be combined with the construction of places that allow the public to "carelessly" use water. Here it would be possible to usefully combine the pleasurable experience of "wasting" water with education on the use of water that is of lower quality. This is why it is important to take on board the positive experiences that people have on holiday or in their free time, as in the very obvious developments towards a new pleasure from water as seen in the many modern swimming baths with wave makers, water chutes etc. The still sceptical individual attitudes towards water of a lower quality could be broken down and changed by taking action in this area.

The current image of hygiene and cleanliness that has worked itself into people's actions and feelings during the history of civilisation did not originate in the home, it originated in public organisations such as schools, the army or in hospitals and public baths. This process was not just a case of individuals painfully adapting to externally-dictated standards. There were certainly pleasurable elements and individual rewards through a climb in social standing. Such positive, pleasurable elements are largely missing in the current water-saving campaigns.

Urban Brooks: Task and Pleasure for Engaged Citizens

Ludwig Tent

1 Brooks in the Urban Landscape

Water plays an important role in urban ecology. Brooks were formerly used mainly as a transport medium for wastewater and a fast drainage system for rain; recent changes indicate a more holistic approach (Schumacher, Thiesmeier 1991, White et al. 1993, DVWK 1996). Running and stagnant waters and their surroundings nowadays are not only seen as pathways for fauna and flora to enter cities, but also as recreation and adventure areas for the citizens (Ipsen et al. 1998, Probst 1993, Reichmann 1997). Rain water schemes are increasingly being planned as visible, partly dynamic structures (Geiger, Dreiseitl 1995).

To serve these multiple purposes qualitative improvements have to take place, not only with reference to water quality, but also to the morphology/structure of waters and their adjacent areas. The aspects of potable water, the routes of wastewater and the handling of rainfall are not only tasks to be dealt with by administrators, planners and engineers, but are interesting themes for the public as well. Co-operation in water related themes in schools, in adult education courses and active participation, e.g. in 'Bachpatenschaften' (brook sponsorship schemes), can lead to citizens of different ages feeling more familiar with their place of residence.

2 Characteristics of the Running Waters' Habitats in the Borough of Wandsbek

The total length of running waters in the Borough of Wandsbek (14.8 ha, 390,000 inhabitants) is about 300 km. Up until the last few decades they have been altered for housing and other construction purposes so that most of them are straightened and deepened canals filled with mobile sand and mud.

District Office (Bezirksamt) Wandsbek/GU 40, Robert-Schuman-Brücke 8, D-22041 Hamburg, Germany

J. Breuste, H. Feldmann, O. Uhlmann (Eds.)
Urban Ecology

Bank vegetation has often been dominated by the grassland of parks. Hydraulic engineers in the past talked about having no running waters, rather open rain water pipes in the urban surroundings.

For the limnologist these stretches are headwaters in a landscape formed by the glacial ages, characterised by a stable stream base with gravel and boulders, being thus potential salmonid biotopes with cold summer water (Schumacher 1991). The bank vegetation normally consists of alder wood plants. Unfortunately the knowledge about the potential of species richness and natural fish production (Nielsen 1994) has not only been lost for such waters, but also for those of rural areas (Tent 1998). The Wandsbek brooks, like most urban running waters nowadays, are inhabited mainly by roach and perch with a few other mainly cyprinid fish species (cf. Schulz, Meyer 1995 for the City of Braunschweig).

This discrepancy reveals a huge amount of work needed to change today's status to a more sustainable environment.

3 The Idea of Brook Sponsorship

The poor status of brook habitats and the willingness of engaged persons to lend a helping hand led to the idea of brook sponsorship in the eighties (Hansen et al. 1991). The 'Bachpatenschaften' offer exciting activities for urban citizens to watch and improve their 'streams on the doorstep'. A very informative presentation from an activist's point of view is given by Owens (1993) for a US-wide engaged group 'Trout Unlimited', showing all aspects of success and drawbacks of their work.

Tasks of brook sponsors vary with personal interest and prior knowledge and the condition of the stretch of water. Observing the brooks and their biological communities, surveying and exploring damage, suggesting remedial action, removing refuse and artificial bank protection systems, planting trees, introducing gravel beds and current deflectors etc. are examples of the wide variety of actions carried out. Brook sponsors also give exhibitions with presentations of invertebrate life to other citizens.

4 Ten Years of Experience

Information to help these groups of laymen with different measures is provided by the local administration in co-operation with nature conservation organisations (Baubehörde Hamburg 1992, Hansen et al. 1991, Tent 1993). The seven Hamburg boroughs give environmental advice on the spectrum of aquatic topics as well as on other themes. The vision is to bring the water cycle

Table 1. Changes in brook sponsors from 1990 to 1997

	1990	1997
schools	7	23
nature conservation groups	8	4
junior nature conservation groups	7	6
interested individuals	7	24
neighbourhood communities	4	8
other (allotment assoc., local history assoc.)	2	9
	35	74

back to the people's minds, thus promoting the feeling of responsibility in the social system. Technical aspects, such as saving potable water, and local activities to enhance the trickling of rain water are coupled with nature conservation and limnological aspects, e.g. the elevation of low flow levels in urban brooks. Knowledge about the actual and potential inhabitants of the waters is dispersed via evening lectures and brochures. The documented species change after brook improvement work from ubiquitous invertebrates to indicator organisms of lively running waters, such as mayflies and caddis flies (stone flies are still scarce), is a real reward for the participating citizens.

During about ten years of brook sponsorship in Hamburg, significant changes were noted in the character of the participating groups (Table 1, Fig. 1). In the first few years nature conservation groups 'occupied' long stretches to show their interest in stream restoration. Depending on the shift of stake holders and activists it became clear that they could not actively care for this huge distance. Nowadays schools, families and citizen groups dominate, with stretches being up to 500 m long.

Financial assistance, when needed, has been provided to a limited extent by the local authority. The flexibility of the sponsors, however, to find persons and firms willing to give donations, and the chance of the sponsors' supervisor to receive money from trusts interested in environmental protection has improved activities in recent years.

5
Today's Situation and Future Aspects

In 1997 in the Borough of Wandsbek more than 70 brook sponsorship schemes existed. More than 700 individuals from pupils to the elderly feel responsible for their 'brook on the doorstep' and are active in improving urban waters and their surroundings. Information contacts have been established by several groups on a regional and an international scale using offers

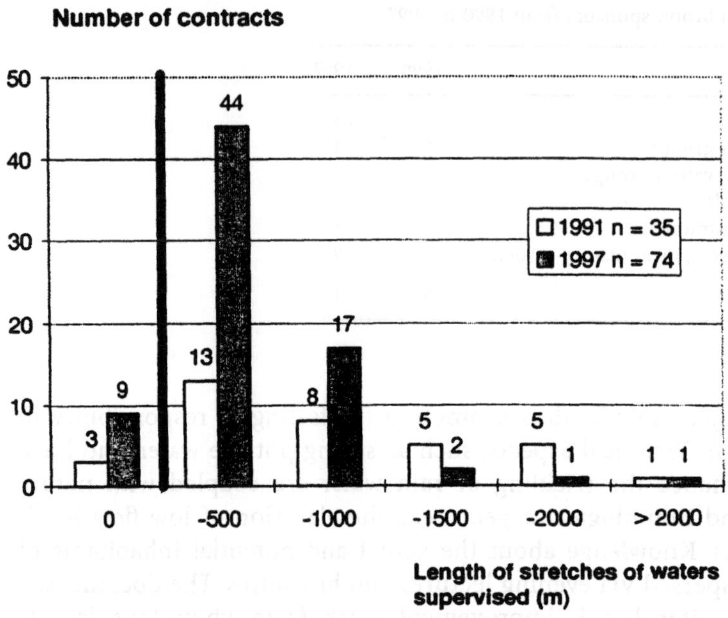

Fig. 1. Changes in brook sponsorships 1997 vs. 1991
0 = stagnant waters

like GREEN (Global Rivers Environmental Education Network) and GLOBE. (Global Learning and Observation to Benefit the Environment). All of this can be considered as an important aspect in Agenda 21 activities (which have not been evenly developed in Germany up to now (Ax 1996, Schaeffer 1995) and support for the improvement of urban society.

The different situation in the Hamburg boroughs reveals, however, that continuing mental and practical support is necessary to keep these activities alive. The Wandsbek area is characterised by a skilled supervisor in the administration helping the brook sponsors to answer both scientific and practical questions, as well as in struggling through the administration's jungle. Unfortunately the supervisors have been financed by Job Creation schemes, thus the administration seems to be unable to ensure that at least a part-time post will be allocated to the supervision of brook sponsorship schemes for the future. With the supervisor posts being terminated in the other boroughs, the sponsorship declined almost to zero. The current situation in Wandsbek will change in the year 2000 and it is to be awaited whether some alternative will be developed to keep brook sponsorship alive.

Acknowledgements. Special thanks go to Cathrin Couwenbergs, the enthusiastic supervisor of brook sponsorship in the Borough of Wandsbek, to Helmut Neuling for calculations and the production of figure 1 and, last but not least, to Friedrich-W. Ritzmann, head of the Wandsbek Water Authority, for his engagement in inter-disciplinary work on urban brooks.

References

Ax C (1996) Agenda 21: Hamburgs mühsamer Weg ins 21. Jahrhundert – Empfehlungen an den "Worthaltenden Bürger-Meister". In: GERMANWATCH (ed.) Lokal handeln – global denken. Konkret Literatur Verlag: 42–54

Baubehörde Hamburg (1992) Bachpatenschaften – Grundwissen, Tätigkeiten, Beispiele. Freie und Hansestadt Hamburg

DVWK (1996) Urbane Fließgewässer – eine städtebauliche Chance. DVWK-Materialien 2/96, Wirtschafts- und Verlagsgesellschaft Gas und Wasser mbH, Bonn

Geiger W, Dreiseitl H (1995) Neue Wege für das Regenwasser: Handbuch zum Rückhalt und zur Versickerung von Regenwasser in Baugebieten. Oldenbourg München Wien

Hansen H, Langbehn J, Mayerhöfer H, Tent L (1991) Bachpatenschaften in Hamburg – beispielhaft dargestellt an den Bezirken Altona und Wandsbek. In: Schumacher H und B Thiesmeier (eds) Urbane Gewässer: 464–467

Ipsen D, Cichorowski G, Schramm E (1998) Wasserkultur. Beiträge zu einer nachhaltigen Stadtentwicklung. Stadtökologie Bd 2. Analytica-Verlag, Berlin

Nielsen J (1994) Vandløbsfiskenes Verden – med biologen på arbejde. Gads Forlag, København

Owens OD (1993) Living Waters – How to save your local stream. Rutgers University Press, New Brunswick New Jersey

Probst W (1993) Naturerlebnisräume in der Stadt – mehr Freiheit für die Natur, mehr Freiheit für kreatives Spielen. Geobot. Kolloq. 9: 59–67

Reichmann S (1997) Aufgetaucht. Oder: Wo "verläuft" der Kaitzbach im Bewußtsein von Dresdner Akteuren? WasserKultur, Urbanität – Technik – Ökologie, Nr. 8: 16–20

Schaeffer R (1995) Erfahrungen bei der Umsetzung einer neuen Wasserpolitik. Difu-Materialien 10/95: Wasserwirtschaftspolitik in den Kommunen – Ansätze für eine nachhaltige Entwicklung: 201–226

Schulz R, Meyer L (1995) Zur Fischfauna strukturarmer innerstädtischer Gewässer am Beispiel des Bürgerparks in Braunschweig. Braunschw. naturkdl. Schr. 4: 755–767

Schumacher H (1991) Limnologische Vorgaben und Bewertungskriterien zur ökologischen Verbesserung urbaner Fließgewässer. In: H Schumacher und B Thiesmeier (Hrsg.): 16–36

Schumacher H, Thiesmeier B (1991) Urbane Gewässer. Westarp Wissenschaften, Essen

Tent L (1993) Ecological improvements for lowland streams in Hamburg Wandsbek with special emphasis on "Bachpatenschaften" (brook sponsorship schemes). In: White KN et al. (eds)

Tent L (1998) Reconstruction versus ecological maintenance – improving lowland rivers in Hamburg and Lower Saxony. In: Hansen HO, BL Madsen (eds) River Restoration '96 – Proceedings – International Conference arranged by the European Centre for River Restoration. National Environmental Research Institute, Silkeborg, Denmark

White KN, Bellinger EG, Saul AJ, Symes M, Hendry K (1993) Urban Waterside Regeneration, Problems and Prospects. Ellis Horwood, New York London Toronto Sydney Tokyo Singapore

Rufisque – Surviving City in Senegal

Ritva Kivikkokangas-Sandgren

African cities can be divided into six categories according to their history and cultural background (O'Connor 1983): truly indigenous, Islamic, colonial, European, dual and hybrid. The truly indigenous cities might be cult centres, or centres of political power, administration, trade and crafts. Islamic cities have a grand mosque and a main market as a terminal for trans-Saharan trade. Rufisque is a typical colonial city from the 19th century. These European cities were built by colonialists for their own residence. The dual city is an example of two cultural traditions without any integration. The hybrid city combines both indigenous and alien elements with some integration of cultural elements.

In Africa the urban population is currently growing at a rate of over 5% per year. More than 70% of African urban areas are completely excluded from public water supply, sewage systems and household refuse collection. The growing cities often lack formal city planning and the organised infrastructure of a Western city. Community participation by active self-help groups is the main survival strategy in most squatter areas in Africa. This is also the case in Rufisque, (Wiese 1995.)

1 The History and Spatial Structure of Rufisque

Rufisque is a West African colonial city of 200,000 people on the Cap Vert Peninsula, 30 kilometres east of Dakar, the capital city. The city was named Rio Fresco (or freshwater river) by Portuguese sailors in the 15th century. In the 1990's the city could hardly be described as a freshwater city, rather a depressing place which faced serious problems of sanitation and dumping of refuse. Most areas have no plumbing because of lack of concern by the City Planning Office and the City Council Office.

The situation in Diokoul, an ancient suburb consisting of eight low-income districts, shows a positive example of joint collaboration by local people and

University of Helsinki, Department of Geography, PL4 (Yliopistonkatu 3), 00014 University of Helsinki, Finland

J. Breuste, H. Feldmann, O. Uhlmann (Eds.)
Urban Ecology

of development co-operation to solve the sanitation problems. This project was carried out in co-operation with the Scheme of Community Involvement and Appropriate Technology from 1990 to 1994, (Gaye, Diallo 1995 and Sharp 1994.) Rufisque combines five areas of different socio-political and environmental elements: the colonial centre (European square formula), squatter areas (spontaneous, fast growing settlements), traditional fishermen's village (Diokoul), public areas comprising markets, and the coastal area (with erosion problems). The colonial centre is the most important area of business, administration and services. The city is unique in having horse-drawn carts and colourful street markets of rural and urban small business. The city has 30 suburbs (Fig. 1). Squatter settlements with informal infrastructure and no formal legislation were built up by newcomers. The coastline has reduced by 300 meters since 1940 because of the rise in the sea level. The groundwater table is very close to the surface and the overflowing of dirty water from the street channels is a continual problem. Socio-spacial problems in the city comprise: clay soil which easily becomes fluid during the rainy season,

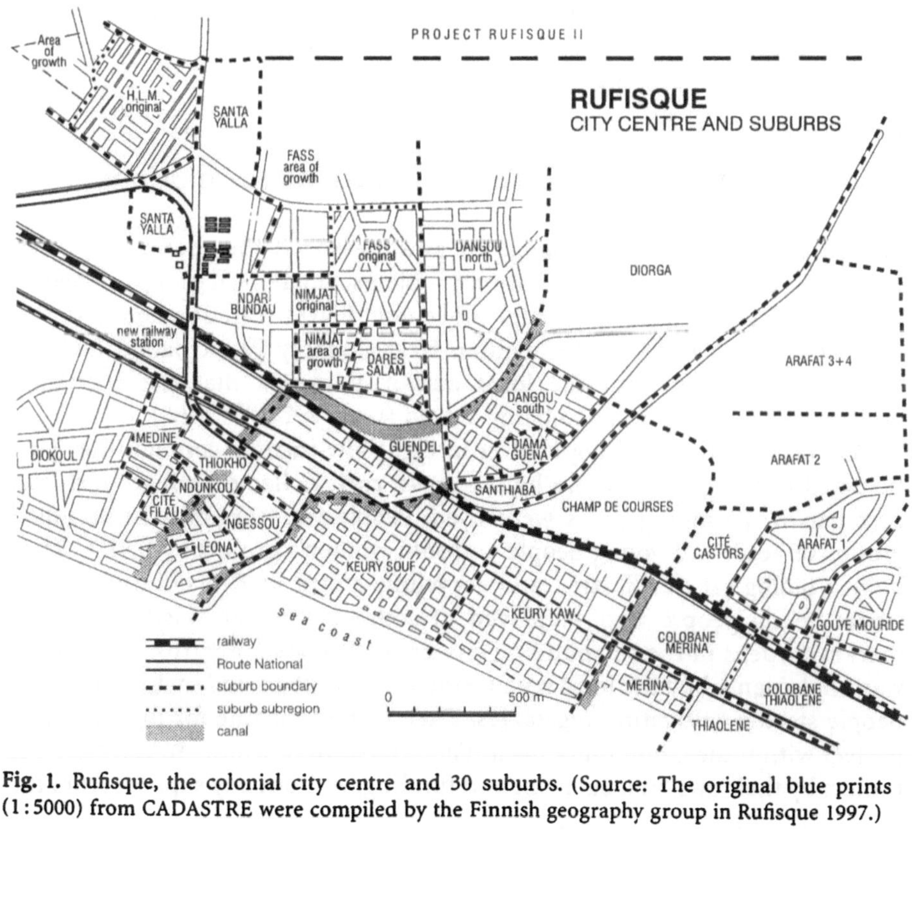

Fig. 1. Rufisque, the colonial city centre and 30 suburbs. (Source: The original blue prints (1:5000) from CADASTRE were compiled by the Finnish geography group in Rufisque 1997.)

vulnerability of the groundwater table to pollution, erosion, pollution of the coastal zone, and the ignorance of the City Council Office of environmental problems and legislation.

2 The Change in Shared Environmental Attitudes

People like to live in colonial houses. The degraded sanitation system in old, dilapidated houses is a growing problem (Urban Renewal Group Report 1996). The first concern of the people is their personal appearance, and they are commonly ignorant about the public environment. When moving from a village to a city people lose their sense of shared environment and social control. In the city different tribes and cultures live next to each other, each having a different idea about the environment. A sharp break in traditions takes place. After moving to a fixed urban environment, people neglect the public environment. Dirty streets filled with stagnant water symbolise the general feeling of powerlessness. There is a strong hierarchy among people. The cultures of men and women are strictly separated. People are not used to demanding public services and good governance from their often corrupt politicians. The sanitation problems of suburbs are totally the responsibility of the citizens themselves (Urban Renewal Group Report 1996).

3 The Solving of Local Problems Through Micro-solutions

In answer to the urban challenge of powerlessness, the community itself started to solve sewage and sanitation problems. Small local groups of people were organised by the local people's movement ENDA (Environmental Development Agency) and by the Canadian Host Country Participation Fund in 1990. Collaboration with the local authority means that the sanitation problem is now on the way to being solved (Gaye & Diallo 1995).

One third of the total population in eight low-income communities participated in the project where horse-drawn carts collected refuge, and low-cost narrow plumbing pipes were used. People were organised into small groups of adult and young people and into women's groups by ENDA. They built a biological purification and recycling centre in Diokoul and Castor, where young people were employed to compost organic waste for use in market gardens in the surrounding city areas. The idea of agro-urbanisation was implemented by the local management committees and by ENDA. Young people started cultivating vegetables. They earned income for funding their project within the community itself. The groups were willing to continue the project by their own management and funding, without any support from the

City Council. About 30% of households involved in the scheme were headed by women. They also played an important role in activating the male-headed households by encouraging men to get involved in solving problems.

The community empowerment by appropriate technology and by the democratic participation was created by the collaboration of the people's organisation and the activities of the self-help groups. The protection of the environment was perceived as an integral part of everyday life. The awareness that active attitudes could solve local waste problems was instilled into the small groups. These activities do not solve the problem of good governance and management on the political level, but they do show the capability of local people to take care of their city environment.

References

Gaye M, Diallo F (1995) Urban Management. Best Practices Programme. Habitat II secretariat, Dubai

O'Connor A (1983) The African City. Hutchinson, London

Urban Renewal Group Report (1996) Rufisque field project, Senegal. Helsinki University of Technology, Finland

Sharp R (1994) Senegal. A state of change. Oxfam, United Kingdom.

Wiese B (1995) Senegal. Perthes Länderprofile. Justus Perthes Verlag Gotha.

Workshop: The Integration of Environmental Dimensions in Land Use Planning and Development: Practical Examples 2.4

CEMIS Indonesia: Community-Based Assessment and Monitoring of Living Conditions with the Help of a House Card

Charles Surjadi[1] and Andre Dzikus[2]

1 Introduction

The implementation of Chapter 7 on Human Settlements of Agenda 21 provides great opportunities for Governments to address the complex environmental and health problems of cities and expand their productive bases. A Community-based Environmental Management Information System (CEMIS) has been developed by UNCHS (Habitat) to promote effective management approaches which will assist Governments to address the complex problems of cities and the sustainable development of human settlements.

CEMIS, which is currently being applied in low-income communities of Accra (Ghana) and Jakarta (Indonesia), has been developed as an open and flexible information and management system. Its main objective is to empower communities to monitor and assess their own living environment without external support and, based on information gathered, develop and manage requisite interventions. CEMIS has been developed to be as simple as possible and only make use of technologies appropriate for its host community. The CEMIS process involves the following steps and modules.

The first module is *community preparation and mobilisation* in which community leaders are identified. The expected output is an increased community awareness of the inter-relationship between housing and environmental health.

The second module focuses on *environmental risk assessment and monitoring*. The communities, through self-assessment, are able to identify, prioritise and monitor human settlement conditions and related environmental health risks.

The third is on the *assessment of technological options* in which communities are enabled to select appropriate technologies. Besides technical information, communities also require information on installation, operation

[1] Atma Jaya Catholic University, Center for Health Research, 14440 Jakarta, Indonesia
[2] United Nations Centre for Human Settlements (Habitat), P.O. Box 30030, Nairobi, Kenya

J. Breuste, H. Feldmann, O. Uhlmann (Eds.)
Urban Ecology

and maintenance costs, environmental and health impacts as well as options on how to manage the technology.

The fourth is *assessment of effective demand* in which the resource base of the community is determined by the community itself. It assesses the community's willingness and commitment to contribute towards specific inventions.

The fifth module is intended to enable communities to *prioritise human settlement interventions* based on the use of a criteria catalogue. The criteria catalogue includes information on the ranking of environmental problems and related strategic human settlement interventions. A further prioritisation of human settlement interventions conducted was based on available technological information and options and available resources.

The sixth module is *community action planning and implementation* in which guidelines for planning, implementing and managing intervention are provided.

The seventh module is on *monitoring and evaluation.* This module provides tools for monitoring and evaluating human settlements intervention in communities.

The last module strives to *strengthen the institutional framework* developed by the project and the capacities developed to sustain CEMIS and the improved living conditions (Dzikus 1994).

2 Methodology

As a part of Module 2, CEMIS in Indonesia has developed a house card as one methodology for community-based environmental assessment and monitoring. It is an ideal tool for community self-surveys or self-reporting. So far little experience has been gathered in community self-surveys in low-income settlements; the few good examples that exist and function emanate from the area of community health. Indonesia is the country with the longest experience in conducting community self- surveys (for further information on the topic see UNCHS (1995) and Dzikus and Surjadi (1995)). In Indonesia the development of the house card was based on nine principles:

- Community-based: the information system is managed and maintained by the community itself after initial training of community volunteers and leaders.
- Simple: it is easy for ordinary people to implement the community self-survey and to interpret data collected.
- Self-assessment tool: the household itself can easily assess their living conditions.

- Significant: the information is important, useful and interesting.
- Specific: the house card is measures what it aims to measure.
- Sensitive: it should indicate when there is a relevant change in the conditions.
- Objective: when used by different people in the same situation, the house card should give the same results.
- Low cost: it should be affordable by low-income communities.
- Feasible: it should be possible for low-income communities to complete a house card without major external support.

During the trial of CEMIS in Indonesia, a house card was developed which covered nine aspects of the living environment of low-income communities. The following table shows the criteria according to which inhabitants can classify their living conditions into good, fair or bad/poor.

The house card is printed on thick folio card (Table 1). The front page contains questions on the nine aspects of the living environment of the household, as well as information about household demography, on the husband and wife, number of adults and children under five living in the household and the address of the household.

Table 1. Information on living environment on house card

Information	Good/Green	Fair/Yellow	Bad/Poor/Red
1. Toilet	toilet with septic tank	toilet without septic tank	no toilet
2. Ventilation of kitchen	there is ventilation and smoke from kitchen can easily escape the room	there is ventilation and smoke from kitchen does not easily escape	no ventilation
3. Ventilation of the bed room	there is sunshine and air can enter the bedroom	only limited sunlight but air can enter the bedroom	no ventilation
4. Humidity of the house	not damp and no mould on walls	moist but no mould on walls	moist and there is mould on walls
5. Drainage around house	no mosquito larva and good run off	mosquito larva and some run off	stagnant
6. Solid waste disposal	have closed garbage bin	have open garbage bin	no garbage bin
7. Source of drinking water	Connected piped water	water vendor	well water
8. In-house crowding	$>10m^2$/persons	$6–10m^2$/persons	$<6m^2$/persons
9. The presence of stagnant water surrounding house	no stagnant water	there is stagnant water but no mosquito larva	there is stagnant water with mosquito larva

The field-trial of CEMIS in Indonesia is being conducted in two low-income communities of Jakarta, Kali Anyar and Kali Baru. Altogether 21 community volunteers were trained on how to fill in the house card and tabulate the data collected. The majority of volunteers were women aged 15 to 59 with a large proportion not having finished senior high school.

After the training the volunteers practised the process of gathering, tabulating and analysing data in sample areas in the two communities. After the training the volunteers were requested to fill in an evaluation questionnaire on the user friendliness of the house card. The majority found the house card easy to understand and use.

Once all the house cards were filled in by households the data was collected and compiled at the house of the local leader. The same house card can be used to compile the data. Under each living environment category the total number of households living under good, fair or poor conditions can be summarised. This enables the community to aggregate data and to obtain easily an overview of its living conditions.

3 Results

Within one month 417 households had completed the house card, 130 in Kali Anyar and 417 in Kali Baru. It was found that ventilation of the kitchen and bedroom was a problem in Kali Anyar. In Kali Baru the lack of a toilet was a major problem. In both communities in-house crowding affected one third of the population.

4 Discussion

All too often information on living conditions in communities is collected by researchers, "outsiders", using so called participatory rapid appraisal methods. These methods are comparatively costly as they rely on expensive external researchers and only involve, if at all, the community to collect the data. The collected data is often analysed and interpreted outside of the community and very often not even fed back to the community.

The findings of the field-trial of CEMIS in Indonesia show that it is possible to build capacity in a community to assess and monitor its living environment itself by collecting data with house cards through community self-surveys. The collected information is specific to the needs of the people, sensitive to changes over time and is objective, as it is collected by the community itself. This makes it highly significant for community-based decision making. This process has not just empowered the community to assess and monitor their living environment but also to decide on essential interventions. As an immediate priority the community of Kali Anyar has decided to address the issue of ventilation in kitchens and bedrooms, whereas Kali Baru is exploring the possibilities for constructing toilets. Both communities are trying to reduce in-house overcrowding.

The house card was perceived by the community as easy to understand, simple to fill in and a tool which makes it easy to attain information on living conditions. If this community-based information system were extended to a large number of communities and aggregated at a higher administrative level this would provide governments with a cheap and reliable way of collecting data for city wide geographical information systems (GIS).

References

Dzikus A (1994) Summary Overview of Framework for Community-based Environmental Management Information System (CEMIS); In: UNCHS, Mempera & Kskp (1994): Laporun Seminar CEMIS; Publikasi No 11, Jarkata

Dzikus A, Surjadi C (1995) Willingness to Pay: Assessing Community Water Demand. Habitat Debate, November 1995, Vol 1, No 3: 13–14

UNCHS (1995) CEMIS Module No. 4: Guidelines for Assessing Effective Demand of Communities for Environmental Infrastructure; UNCHS (Habitat), Nairobi, HS/373/95E, ISBN 92-1-131287-6

The Environmental Impact of the Spatial Restructuring of Retailing in City Regions of Eastern Germany

Axel Bergmann and Claudia Leonhardt

1
Introduction

After unification and the subsequent introduction of a market economy, the location pattern of retailing experienced a dynamic transformation in eastern Germany. It was characterized by an expansive development of large-scale and non-integrated retail stores (shopping centers, hypermarkets etc.), a "thinning out" in scattered locations, and a diminishing importance of urban centres (cf. Kulke 1997). This paper deals with the environmental consequences of this spatial restructuring process, focussing upon the problems of traffic growth and land-consumption.

2
Impact of the Spatial Restructuring on Traffic

The restructuring of the "retail landscape" is generally conceived to be traffic generating. The thinning-out process of the supply network increases the minimum distance to the next store, while non-integrated locations require very long trips and can often only be reached by car. However, is the cause-effect relationship really that simple and if so, to what quantitative extent do location aspects account for traffic growth?

Figure 1 displays one of the most important results of the empirical studies adressing these questions. It shows the relationship between the average length of automobile shopping trips and urban population density. Note that trip lengths are markedly shorter in towns with higher population densities. This can be plausibly explained by taking into account the spatial pattern of retailing as an intermediary variable. It is known that the market potential of densely populated neighbourhoods fosters the establishment of food retail stores, and it has been proven that a dense network of grocery/food stores can reduce the frequency of automobile shopping trips and even more

Institute for Ecological Space Development Dresden (IÖR), Weberplatz 1, D-01217 Dresden, Germany, Tel. 0351/4679-0, Fax:-212

J. Breuste, H. Feldmann, O. Uhlmann (Eds.)
Urban Ecology

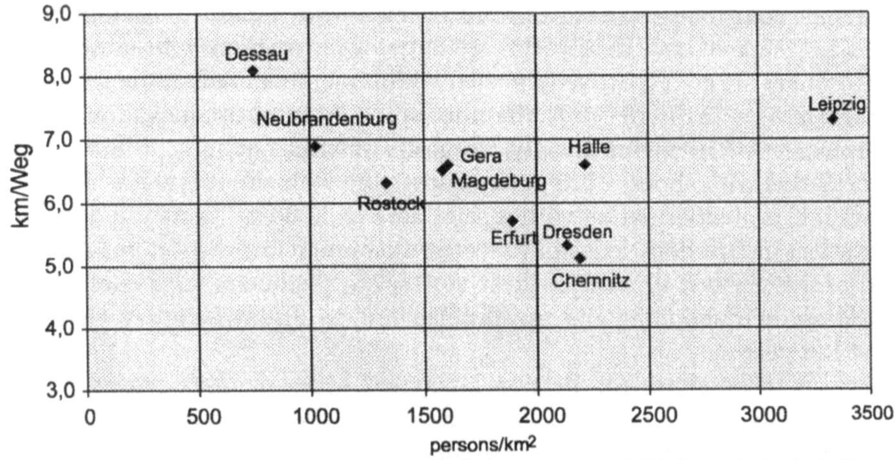

Source: SrV 1994, TU Dresden, Institut für Verkehrsplanung und Straßenverkehrstechnik (survey with Ø 400 households interviewed/town)

Fig. 1. The relationship between Population density and the average length of automobile shopping trips 1994

significantly the average trip lengths (Gertz et al. 1993). Consequently, compact settlement structures contribute indirectly to low levels of automobile shopping traffic.

As revealing as the "rule" are its exceptions. As can be seen, the average trip lengths in Leipzig and Halle are far longer than one would expect from their high population densities. The deviations can be interpreted as the consequence of an extremely unbalanced retail structure in this conurbation. Nowhere else have so many large and non-integrated shopping centres been established, among them the "Saale Park", with over 100,000 m^2 sales area it is the biggest of its kind in Germany. These large-scale establishments have huge catchment areas and potentially impair the development of smaller, integrated retail locations. The resulting extra traffic can be estimated by a simple model calculation: Leipzig has approximately 500,000 inhabitants, and according to the SrV-survey, the average automobile trip frequency for shopping purposes comes to 0.22 trips per person/day. Hence, there are about 100,000 vehicle trips per day. If the average trip length could be reduced by 2 km to the level of Dresden (cf. Fig. 1) then 200,000 vehicle-kilometres would be saved every day, or more than 50 million kilometres every year.

However, the relationship between urban form, retail structure and traffic is more complex and can not be fully understood by means of a simple correlation analysis at the scale of entire towns. Consequently, some case studies have been carried out to analyse the traffic impacts of individual retail establishments. Four food retail stores were selected with respect to location type and size: 'Magnet' and 'KaufMarkt' are large hyper-markets with more than 5000 m^2 sales space respectively, located within regional malls at the Dresden city fringe. The 'SachsenForum', the third study location, is a neighbourhood shopping centre in one of the city's largest residential suburbs. It includes a supermarket with about 740 m^2 and a superstore with about 3,400 m^2. Finally "Markant" is a supermarket with a sales area of 1,200 m^2 located in a shopping street in Dresden-Neustadt (new town), a municipal district built at the turn of the century. In front of these stores, 722 customers were questioned regarding their shopping and travel behaviour on Friday/Saturday, October 16/17 last year.

As was to be expected, the non-integrated hyper-markets proved to be comparatively more traffic-intensive due to the lower percentages of non-automobile customers and longer trip distances, calculated as linear distance between the starting point and the destination. In terms of average vehicle-kilometres per customer the non-integrated hyper-market 'Magnet' exceeds the neighbourhood centre 'SachsenForum' by a factor of ten, and the supermarket 'Markant' by a factor of about 30. Other empirical case studies have revealed comparable differences between integrated and large-scale, non-integrated food stores (e.g. Holz-Rau 1991). The results of such comparative studies are often countered by the argument that the customers of large stores buy comparatively larger amounts and less frequently, and that shopping centres offer an opportunity for 'one-stop-shopping' (Gerlach 1996, BMV 1995). Yet, even if traffic intensity is calculated as automobile traffic per DM of groceries purchased, the differences remain impressive (see Fig. 2).

But does this necessarily mean that the opening of a new large establishment generates new traffic? In order to be able to adress this question we compared the customers´ shopping and travel behaviour before and after the opening of the new stores. The results were to some extent surprising.

To begin with, the new locations effected a rather small change in the modal split. Most of the drivers interviewed stated that they had also travelled by car to their former store of main importance, and the bulk of pedestrians had done their shopping this way before. Moreover, switches of modes of transport cancelled each other out. With regard to trip distance, the locations' influence upon travel behaviour was more relevant. The food stores of the integrated neighbourhood centre 'SachsenForum' shortened the average trip length substantially, while the hyper-market 'Magnet' increased it. However, the increase by 'Magnet' was not as high as expected, and no increase at all could be observed for the other non-integrated hyper-market. The reason behind this surprising result is that many customers from Dresden had previously shopped in retail stores even further away from their home, namely

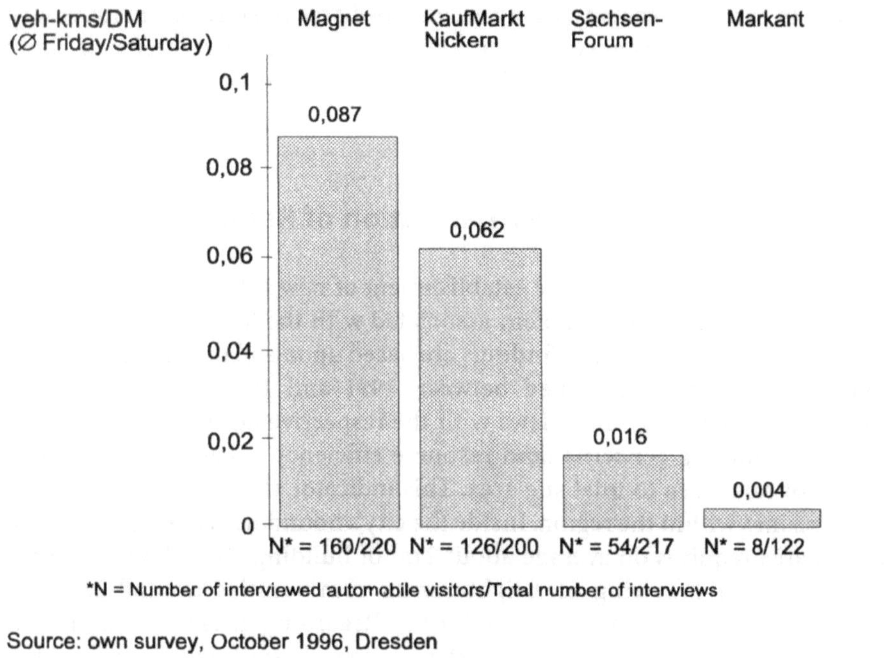

Fig. 2. Average Traffic Volume (veh-kms) per DM sales volume

greenfield sites constructed in the first phase of new economic development after unification.

What can be inferred from these results of the case studies? The traffic-generating effect of retail establishments may be overrated if it is only deduced from simple comparative studies between different types of stores, as is so often done. Whether a new project will generate extra traffic has to be assessed individually and with due attention to dynamic aspects of travel and shopping behaviour. Not the size but the location and relative market position of a store is crucial to the argument concerning generated traffic, notwithstanding the fact that large-scale, non-integrated retail establishments are associated with an immense traffic volume.

Taken together, the empirical findings admit some general conclusions for a traffic-minimising planning policy. The key objective should be to keep, or develop, an attractive and easily accessible food supply network and to prevent excessive out-of-town retail developments. An overall urban policy in favour of compact settlements can contribute considerably to this aim. Secondly, planning policies have to acknowledge that in an affluent, highly mobile world, consumers do not depend on nearby retail stores but choose their destination more and more with regard to aspects of attractiveness. Since prices are supposed to be the dominant criterion for the attractiveness of stores in eastern Germany, easily accessible discount stores seem necessary to circumvent long-distance shopping drives. Finally, where a city of short pedestrian trips is not feasible, planning should strive at least for a city of short drives.

3 Land Loss Due to the Suburbanisation of Retailing

Land loss in consequence of the establishment of new suburban retail centers is another environmental problem associated with the locational restructuring of retailing. The following findings are based upon a plan analysis of large-scale retail projects established between 1991 and 1996 in the region of Dresden, augmented by interviews with the respective developers.

The first finding concerns land resource efficiency calculated as the proportion of sales area to total site area. This indicator shows significant variations by zones within the region. Inside the city's boundaries one square metre of sales area requires on average about 2 m^2 of building land, the ratio drops for sites in the first ring of neighbouring communities around Dresden to 1:5 , and to 1:10 in the second ring. Regarding these figures, it can be generally concluded that the suburbanisation of retailing is linked with a disproportional growth in the demand for building land.

A second relevant point besides the quantitative aspects of "space consumption" is the deterioration in environmental quality on retail sites. From the total area of all establishments ("Sondergebietsfläche" or special use areas) over 70 % had previously been agricultural land or cultivated area, and only somewhat more than 20 % had been built-up areas. The dynamic expansion of retail capacity thus took place predominantly at the expense of open land which is becoming a more and more scarce resource in urban agglomerations. Moreover, a detailed analysis of the ground coverage revealed that 71% of the surface had been sealed by buildings, streets, parking facilities etc. (Arlt et al. 1997). Summing up, the suburbanisation of retail locations ranks highly as a cause for urban sprawl and the loss of open land. In general, the bigger, more car-oriented and less integrated a retail establishment is, the more land is required per unit of sales area.

4 Final Remarks

Regarding out-of-town developments, retail legislation in Germany is considered to be among the most strict in Europe. Nevertheless, new large-scale retail establishments which clearly violate essential planning objectives are established again and again. In order to better safeguard environmental concerns the Federal Ministry for Regional Planning, Building and Urban Development is planning to make environmental impact assessments (EIA) compulsory for large-scale retail stores. This could make the environmental consequences of retail establishments more visible and thus promote a more effective integration of environmental aspects into planning policy. However, like other regulatory instruments, EIA do not tackle the driving economic forces for the spatial restructuring process. Additional economic disincentives like a soil sealing tax or parking fees which internalise the external environmental costs of out-of-town locations seem necessary to check excessive suburbanization.

References

Arlt G, Lehmann I, Stutzriemer S (1997) Die ökologische Dimension des Stadt-Umland-Verkehrs. IÖR-Schrift 20. Dresden

BMV (Bundesminister für Verkehr) (ed) (1995) Forschung Stadtverkehr. Reihe Auswertungen, Teil 8 Verkehr und Umwelt. Bonn

Gerlach J (1996) Dezentrale Versorgungsstrukturen und umweltorientierter Verkehr – ein Widerspruch? Internationales Verkehrswesen 48 (No 9): 12–15

Gertz C, Holz-Rau C, Rau P (1993) Verkehrsvermeidung durch Raumstruktur – Personenverkehr. In: Enquete-Kommission des Deutschen Bundestags zum Schutz der Erdatmosphäre (ed) Studienprogramm, Vol. 4 Verkehr. Bonn

Holz-Rau C (1991) Verkehrsverhalten beim Einkauf. Wechselwirkungen zwischen Siedlungsstruktur und Verkehr. Internationales Verkehrswesen 43: 300–305

Kulke E (1997) Case Study. Effects of the economic transformation process on the structure and locations of retailing in East Germany. Journal of Retail and Consumer Services 4: 49–55

Sustainable Regional Development in the EXPO Correspondence Region of Anhalt-Bitterfeld-Wittenberg (Germany)

Fried M. Burkert, Volker Kleinschmidt

1 Introduction

The region of Anhalt-Bitterfeld-Wittenberg, the most central of the three administrative districts of Saxony-Anhalt, consists of:

- old industrial areas, often with contaminated soils (e.g. chemical industry),
- internationally important biosphere reserves (e.g. the Middle Elbe Reserve),
- cultural sites of international importance (e.g. the Dessau Wörlitz country park),

It also faces severe socio-economic problems caused by the fast transition from declining old industries to growing new industries as well as changes in other economic sectors.

Two years ago the region of Anhalt-Bitterfeld-Wittenberg was chosen as one of the EXPO 2000 correspondence regions in Germany. This background has been used as a basis for the preparation of a regional development concept. The aim of this concept is to define all strategies and targets of the regional development in this area. It is to serve as a basis for funding in excess of 300 million ECU provided by the European Union's structure funds and some 100 million ECU supplied by Federal or state funding programmes. The authorities involved in this field are:

- the district authorities and the regional state authorities (in this case the Ministry of the Environment of Saxony-Anhalt)
- the EXPO 2000 GmbH Saxony-Anhalt, and
- a regional forum consisting of district authorities, the regional state authorities, the communities, the chambers of industry and commerce as well as women's rights and environmental initiatives etc.

PRO TERRA TEAM GmbH, Mainstr. 2, D-06846 Dessau, Germany

J. Breuste, H. Feldmann, O. Uhlmann (Eds.)
Urban Ecology

2
Agenda 21 and Quality Objectives

The regional forum, which was founded in 1995, in co-operation with the scientific advisory committee (appointed by this forum and housed at the Bauhaus in Dessau) has decided that the regional development concept is to be supported by a regional AGENDA 21 covering environmental, social and economic quality objectives. The following environmental quality objectives have been defined in the draft version, taking into consideration the three factors of sustainability:

- no further increase in carbon dioxide emissions (a reduction of 50% has already been achieved since 1990),
- no additional sealing of soil surfaces without the de-sealing of other surfaces,
- termination of the suburbanisation process, leading to a higher consumption of natural resources and an increase in traffic,
- the direction of new industrial developments towards old industrial areas or areas with a low environmental sensitivity, and
- an expansion of regenerative energy use (water, wind etc.).

Not all developments will be controlled by data bases on physical land-use as described in chap. 3, as some developments are merely significant from the point of view of energy or raw materials (Werheit 1997). The Agenda 21 will reflect upon the main fields of the economy, ecology and the society in the region (Fig. 1). It will set environmental quality objectives to provide the basis for an environmentally sound economic development in the region of Anhalt-Bitterfeld-Wittenberg (Enquete-Kommission 1997).

3
Databases

A highly developed digital geographic environmental database is used to achieve environmental control of the economic development. This database includes for example CORINE landcover, information on nature reserves, IBA/FFH areas, regional plans and community land-use plans (Schauerte gen. Lüke 1995).

There are different types of utilisation and spatial distribution. Three urban, industrial centres have been defined, the first of them being Bernburg, the second the capital Dessau and the third the Bitterfeld/Wolfen region. Up until the late 1980's the "chemical park" of Bitterfeld/Wolfen was one of the largest industrial districts in Europe, and there were a number of large open-cast mining areas in the region of Bitterfeld/Wolfen.

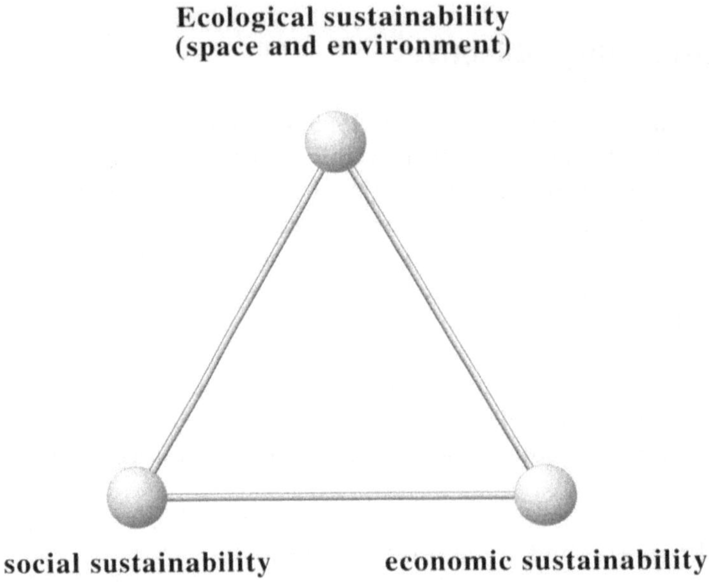

Sustainable development and the satisfaction of society´s needs include the social, ecological and economic sustainability, depicted here as a magic triangle (BUWAL 1996)

Fig. 1. Three dimensions of sustainability

The region has been divided into different nature reserves, the most important of these being the Middle Elbe biosphere reserve which covers the area from the town, Elster in the east, along the Elbe to Magdeburg in the northwest. There are significant drinking water reserves here; the unavoidable conflict between intensive agricultural land-use and water reserves is still a focal point.

To highlight problems which need to be solved on the route to a sustainable regional development, we have to take a look at the location of contaminated soil and suspect areas. The entire area contains more than 1500 locations which have been classified as such which reflects the historical distribution of the industrial areas (Regierungspräsidium Dessau 1996).

4
Pilot Projects for a Sustainable Regional Development

Between February and July 1997, the regional development aims for the Anhalt-Bitterfeld-Wittenberg region were defined by the local participants in 15 so called "participation meetings" including more than 1000 actors. During this participation process, the bodies involved were at all times open to suggestions by different political or social groups (PRO TERRA TEAM & ISW 1997).

The results will be compiled in a list consisting of 40 to 50 prioritised projects. By the end of 1997, an additional complementary sustainability duty checklist will also be developed; the conditions required to achieve sustainability should be listed for each of the projects. The following projects are being intensively discussed at the present time.

One project is a technology centre near the town of Köthen. The innovative nature of this project is guaranteed by forward-looking power supply technologies, an extremely low energy consumption and the possibility of system management experiments. Impulses for the whole region, especially their building projects, will be derived from this project as far as possible.

The second example refers to the marketing of agricultural products from the region (MRLU 1996). The following aims have been defined:

- an increase of direct marketing by means of new strategies,
- the establishment of a regional marketing organisation,
- the labelling of regional agricultural products for better identification, and
- the inclusion of the regional catering trade as an important consumer.

Another interesting project is a concept for the replacement of building material by recycled or down-cycled materials. To achieve this, a so-called regional "soil and building rubble exchange" is to be established. This exchange serves private companies as well as public institutions, e.g. the region's own existing waste management associations.

5
Problems and Final Observation

Naturally significant problems accompany the project. The various concepts of a sustainable regional development are as diverse as the number of different political and social groups involved. Conflicts between economic and ecological interests are to be expected. For example: during the last seven years several thousand hectares of prime farmland were used to improve the infrastructure and to satisfy the legal requirements of the development areas. This decrease in agricultural land naturally resulted in heavy losses for the farmers, which caused even more conflict as the region of Bernburg in the western part of the area is one of the most fertile areas in the whole of Europe.

The time schedule for a sustainable regional development, with respect to the chosen projects, is difficult to determine. The development has to be a continuous process over many years, strictly speaking over at least two generations. Without doubt most of the suggested projects will help to solve current problems or to initiate developments, but only for a restricted period of time. The ultimate aim must be to create and initiate projects which guarantee a lasting sustainable development.

The results of this regional development process will be presented during the EXPO 2000.

References

Bundesamt für Umwelt, Wald und Landschaft der Schweiz - BUWAL (Hrsg.) (1995) Elemente für ein Konzept der nachhaltigen Entwicklung, Diskussionsgrundlage für die Operationalisierung des interdepartementalen Ausschusses RIO. Bern

Enquete-Kommissions Zukunftsfähiges Sachsen-Anhalt (1997) Thesen für eine Landesagenda 21 für Sachsen-Anhalt

Ministerium für Raumordnung, Landwirtschaft und Umwelt des Landes Sachsen-Anhalt - MRLU (Hrsg.) (1996) Studie zur Umstrukturierung der Landwirtschaft im Land Sachsen-Anhalt, Magdeburg

PRO TERRA TEAM, Institut für Strukturpolitik und Wirtschaftsförderung (ISW) (1997) Entwurf Regionales Entwicklungskonzept Anhalt-Bitterfeld-Wittenberg, erstellt im Auftrag vom Ministerium für Raumordnung, Landwirtschaft und Umwelt, der Landkreise und kreisfreien Stadt Dessau, Dessau

Regierungspräsidium Dessau (1996) Regionales Entwicklungsprogramm für den Regierungsbezirk Dessau 1995. In: MinisterialBlatt Land Sachsen-Anhalt Nr. 22 vom 15.04.1996

Schauerte gen. Lüke N. (1995) Das automatisierte Raumordnungskataster des Landes Sachsen-Anhalt. In: Tagungsband 3. Deutsche ARC/INFO Anwenderkonferenz. 200–203

Werheit M. (1997) Regionales Qualitätszielkonzept für die Korrespondenzregion der EXPO 2000, Dessau-Bitterfeld-Wolfen-Wittenberg, erstellt im Auftrag der EXPO 2000 Sachsen-Anhalt GmbH, Halle (Saale)

Vienna is Different (!?) – a Critical View of the Current Development of City Limits and Landscape in the Vienna Urban Area[1]

Wolfgang Pfefferkorn[2]

1 Introduction

The opening up of eastern Europe and Austria's accession to the European Union in 1995 caused enhanced growth expectations in the Vienna Urban Area. This has led to the problem of securing and developing free space and the issue of city limits becoming particularly topical in Vienna.

In this chapter, after a short review of the development to date and a summary of the current stage of this discussion, some conclusions on the development of city limits and the securing of landscape and free space in urban regions will be drawn.

2 The Development to Date

2.1 Historical Review

„Starting out from a settlement on a southern subsidiary of the widely ramified Danube, Vienna grew more or less concentrically towards the south-west. Until the 19th century, the Danube formed a barrier to the north-east."(Vienna Municipal Council 1987). It was not until the Danube was regulated in 1870 that the construction of strong bridges and thus a closer link between the city and the villages north of the Danube was possible. Urban development, however, continued to concentrate in the west and south. After the second world war the districts across the Danube started to grow rapidly due to large urban housing projects.

The system of landscape and free space that formed during this development is marked by large landscapes forming a green belt (Vienna Woods,

[1] This article is an updated and shortened version of a lecture given by the author at the conference „Urban Ecology and Nature Conservation" in November 1995 in Paris

[2] Regional Consulting ZT GmbH, Schloßgasse 11, A-1050 Vienna, Austria

J. Breuste, H. Feldmann, O. Uhlmann (Eds.)
Urban Ecology

Bisamberg, agricultural landscape in the north-east and south), completed by the band-like element of the Danube (Prater, Danube Island, Lobau) and some radial structures.

2.2
The Urban Development Plan (STEP) 1984

In the seventies, preliminary work was started on the first Vienna urban development plan, which was resolved in 1984. It reflects the curbed growth expectations, the main emphasis being on revitalisation within the densely built-up area, and on moderate new construction within defined development axes. Agriculture was to play a major role as the carrier of free space in the areas between the planned axes.

As in most European metropolitan areas, urban growth in Vienna in the seventies and eighties took place mainly in the outer areas or suburban regions. This non-compliance with the recommendations of STEP 1984 led to the free space in the development axes becoming even more built up than was planned. This means that one of the major objectives of this urban development plan with regard to the limitation of urban growth in fixed development axes, and the securing and development of open spaces between these axes, could not be achieved. The causes of this development will be discussed in more detail later.

2.3
STEP 1994

"In the late eighties, soon after the first urban development plan was resolved, which did not provide for any considerable expansion outside the fixed settlement axes in the medium term, under-employment of the construction industry, the threat of shrinking economic growth and increasing international demands on the city with a tendency for incomes to decline all came together. The recommendations of STEP 1984 to limit built-up areas were a contradiction to satisfaction of the home-made construction needs, which had already served as an economic motor in earlier phases." (Moser 1995).

The opening of eastern Europe and Austrian accession to the EC were to bring about the higher political force that would lead to consensus and which was later glorified as the "second period of promoterism" (and seems, meanwhile, to have become a thing of the past). STEP 1994 is clearly marked by growth expectations; a population growth of the Vienna Urban Area from 2.26 million inhabitants in 1991 to ca. 2.53–2.68 million inhabitants in 2011 (+12.0–18.6%) is estimated (RC 1997).

As the development so far shows, Vienna did not even succeed in controlling the limits of the city, securing the existence of agriculture in the urban region and building up a stable system of free space during the phase of curbed growth in the eighties. In view of the current (meanwhile curbed) growth euphoria of the nineties, this seems even less possible.

2.4
The „Decentralised Concentration Concept"

The Concept for Decentralised Concentration in the Vienna Urban Area and Lower Austria is a result of the discussion about city expansion and urban sprawl. Based on a strong backbone of regional public transport, the main idea is to transfer the growth from the first belt around the city to regional centres in Lower Austria. The distance between these centres and Vienna City is about 15 to 60 km (Fig. 1).

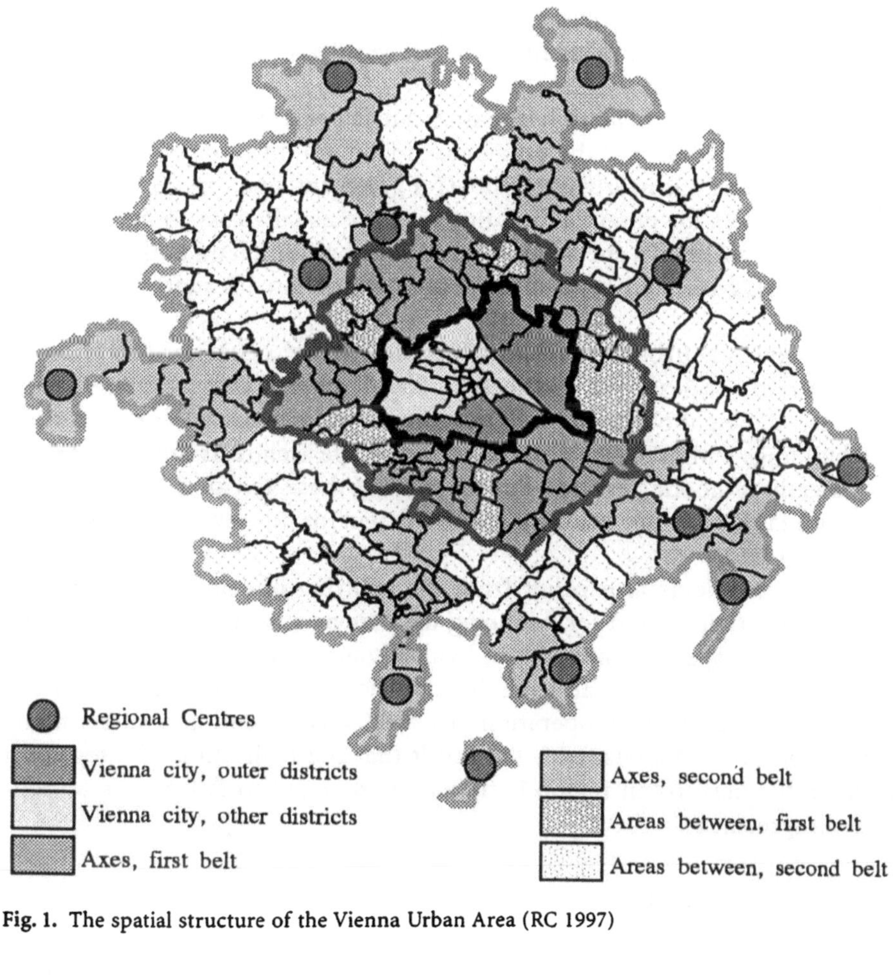

Fig. 1. The spatial structure of the Vienna Urban Area (RC 1997)

The analysis of the impacts of this transfer of population and working places from the first belt show that without a radical modification of present land market and land policy conditions in Vienna and Lower Austria, the transfer-capacities are too small to change the problems in the immediately adjoining urban area (ibid.).

3 Specific Aspects of the Current Development

3.1 Institutional Causes and their Effects

There are three main declarations concerning green space policy, the "Vienna Green Belt" of 1905, the "Green Land Declaration" of 1986 and the so called "1000 ha Programme" for the north-eastern part of the city of 1994. They all intend to give preference to nature and free space over other land use interests, but their implementation as legally binding instruments is unsatisfactory.

This lack of implementation is mainly due to present Viennese land policy. The main problem is that the provision of land – a difficult problem already for building land – is nearly impossible for green and open space. While interested developers for potential building land are usually easy to find (the city of Vienna established a special fund to acquire building land in the middle of the eighties), green and open space planning lacks an adequate sponsorship structure.

„Consequently urban expansion areas are often characterised by reduced, miniaturised scales: the projected large green axes become mere avenues, . . . neighbourhood parks become stand-in lawns between two high-rise buildings and many suggestions of landscape architects are transformed into merely symbolic gestures. But the total of these 'left-over lots' is mostly neither qualitatively nor quantitatively sufficient for the new expansion areas" (Proksch 1996).

Specific problems also result from the splitting of planning and implementation activities, and for financing the care and maintenance of existing or future free space among the individual municipal departments as well as between the municipal council and the districts. A further serious deficit is the lack of co-operation between Vienna and the surrounding regions of Lower Austria, with the result that the realisation of integrating plans in the peri-urban region is frequently doomed to failure due to political disagreement.

3.2
Conceptual Causes

In addition to the insufficient legal and financial security of free space planning interests, the poor quality of many landscapes and free spaces is also due to the deficiency of conceptual discussion on the aspects of quality of free space. Many examples in the past and, unfortunately, the present show that green areas frequently degenerate into pure set pieces, and often serve only as the backdrop for architectural staging. The social and ecological aspects of free space are frequently ignored.

4
Conclusions

Taking the shortcomings into account, the following improvements are considered to be vital for the Vienna Urban Area as well as for other cities:

4.1
Organisational Level

- Vienna requires a land policy that includes sponsorship of green and open space.
- Legally binding instruments for landscape and free space planning have to be integrated into the present planning system.
- There must be massive improvements of the personal and financial resources in the relevant departments of the Vienna City Council.
- The problems of division of labour between the authorities have to be solved.
- Free space planning concepts must become mandatory for urban projects.
- Framework conditions for a close co-operation between the city and the surrounding region of Lower Austria must be improved.

4.2
Conceptual Level

- The accessibility of existing green areas and free space must be improved.
- Peri-urban agriculture as a main carrier of free space has to be stabilised.
- Social aspects have to be integrated into free space planning concepts that emphasise multifunctional use much more than today.
- The importance of ecological and social basic research, in particular with regard to the planning and implementation process, has to be enhanced.

References

Doubek C, Holzinger E (1995) Grünraum für Wien. Anforderungen an eine soziale Grünraumplanung und -versorgung. Eine Studie des Österreichischen Instituts für Raumplanung im Auftrag der Kammer für Arbeiter und Angestellte für Wien. Vienna

Moser P (1995) Chancen der Begrenzung des Stadtwachstums. Zolltexte 1/1995. Vienna

Proksch T (1996) Landscape Planning in the Context of Urban Planning in Vienna. Vienna Municipal Council, MA 18: Wien, Grünes Netzwerk - Vienna, Green Network - The State of the Art. Vienna

Regional Consulting (RC) (1997) Ausbaustrategien für das ÖPNV - Netz in der Ostregion. Im Auftrag der PGO. Vienna

Planning Model for the Realisation of an Eco-city – The Case of Taejon Metropolitan City

Kwi-Gon Kim

1 Introduction

The most pressing global environmental, economic and social issues that we will face in the next century will be in the cities. The inclusion of environmental dimensions into urban land use planning and development can be seen in terms of the development of an eco-city which provides an integrated solution to the global social and ecological crisis.

This chapter reports on an UNDP ecopolis research project that has been undertaken by the research team, directed by the author. The result reported here needs to be seen in its context; the Taejon metropolitan city was selected because it is the first city to prepare a comprehensive eco-city plan in Korea.

2 Research on the Eco-city

It has been argued that a focus on urban ecosystems may suggest the start of a constructive approach to planning. There is a difference between traditional and emerging methods in the inclusion of a broader range of environmental variables and environmental effects and the greater sophistication with which they are modelled. This has resulted in a wealth of definitions of the eco-city (Register 1987, Urban Ecology Australia 1993, Platt et al. 1995).

In Korea, the eco-city received increased interest in the early 1990's, mostly initiated by the Ministry of the Environment; Kim executed a study on the eco-city in Taejon metropolitan city (Kim et al. 1996). In his report, Kim analysed an ecosystem situation in Taejon and suggested strategic issues and an eco-city plan for the city. Kim also defined the eco-city at the strategic and action level.

Seoul National University, College of Agriculture and Life Science, Department of Landscape Architecture, Suwon, Korea 441 744

J. Breuste, H. Feldmann, O. Uhlmann (Eds.)
Urban Ecology

One can argue that the integration of environmental considerations into urban planning and development is self-evident. However, the realisation of Environmentally Sound and Sustainable Development (ESSD) principles using eco-city concepts and methods is not that simple.

3
The Development of Eco-city Planning Guidance

The UNDP Eco-city project (directed by the author) has developed an eco-city planning guideline (Seoul National University et al. 1996) to help local governments to establish eco-city planning. Tasks to initiate eco-city planning and a full outline of its procedures are contained in the report on eco-city planning guidance which is available from the author on request.

4
The Application

Having spent some time reviewing the general theoretical background and meaning of the eco-city, as well as identifying the basic stages through which the eco-city planning proceeds (Kim 1996), the eco-city plans of Taejon metropolitan city as a case study will now be presented to test the validity of the guidance.

4.1
Study Area

The study area, Taejon, is located in the centre of Korea, within a region called Chungchung (the city of Taejon, 1995).

4.2
Eco-city Plan

The eco-city plan for Taejon was prepared in 1996 by the Seoul National University research team supported by UNDP, Ministry of Science and Technology of the Republic of Korea, and DAEWOO company. Generally known as the eco-city plan, it sought to anticipate the forces which create cities and facilitate their ecologically sound and stable development. It recognised the closed connection between biotope systems and the arrangement of land uses and the need to consider these two basic elements together.

5 goals were defined which have been the guiding principles for the eco-development of Taejon:

- Creation of the urban structure in harmony with the environment.
- Self-reliant, circulatory, stable and diverse functions of the city.
- Easy movement and access, not only for man, but also for plants and animals.
- Efficient and ecological use of resources.
- Environmental awareness and participation, with particular emphasis on the promotion of lifestyles for the co-existence of man and nature.

It is necessary to control the expansion of urbanised areas to maintain a balance between the conservation of pleasant surroundings and development. The best eco-techniques available were carefully taken into account during the preparation of the eco-city plan.

4.3
City Structure

The eco-city plan identified a number of key structuring principles which would meet its goals and define the character of the city, as well as provide the framework for the city's sustainable development. The plan was prepared to enable the idea of a compact city to be realised.

4.4
Green Network System and Ecosystem Strategy

A green circle and green stations are connected to create a green network through green pathways or ecological corridors. Individual areas vary widely in character and function within the continuous structure of the biotopes according to their location, accessibility, specific features and the nature of the surrounding development.

4.5
Eco-city Services and Facilities

The provision of eco-city services and facilities such as eco-bridges, district heating systems and allotments has been fundamental to the development of the Taejon eco-city. The main feature of the 'Sewage Treatment Plan' is to install small-scale sewage treatment plants along the streams. Treated water will be discharged into dry streams.

4.6 Residential Areas

A new eco town-in-town has been proposed as part of the Taejon eco-city Plan. It was intended that it should become a strong ecological focus point. The original planning philosophy encouraged a biotope system through the area, subject to the constraints from buildings, access and parking. The main features of the plan include:

- connection of 2 separate mountains by an ecological corridor which passes through the town;
- utilisation of a stream as another ecological corridor;
- creation of a circulatory water system throughout the town;
- making maximum use of solar energy;
- proposal for a wildlife park;
- facing of housing sites to the south to obtain maximum light.

5 Conclusion and Suggestions

In supporting a change to eco-city development, the right type of development at the right time should be promoted in the right place. The Eco-city Plan proposed in this study provides for a more ecologically sound and sustainable urban development than is possible under conventional urban planning and design methodologies. Further research is needed on environmental budgeting for evaluating extra expenditure which is to be argued for on environmental grounds. Another important question is how the proposed eco-city plan contributes to the sustainability of Taejon. Research is therefore also required on environmental performance, on performance indicators and criteria by using a sustainability evaluation matrix.

When a city is developed or re-developed, according to the eco-city planning methodology, this result should be compared and analysed with the results obtained through traditional planning methods in terms of environmental loads, as well as socio-economic or psychological aspects.

Acknowledgements. This research was funded by UNDP, DAEWOO Company and the Taejon Metropolitan Government. The author gratefully thanks all members of the research team, including his post-graduate students at the Environment and Ecological Planning Laboratory of Seoul National University for their work.

References

Kim KG, UNDP, MOST, Daewoo Company (1996) A study on the creation of eco-city in Taejon metropolitan city. Taejon Municipal Government

Kim KG (1996) Future image of Taejon metropolitan city in the year 2016 with particular emphasis on co-existence of human beings with nature. Paper presented at the Second United Nations Conference on the Human Settlements (Habitat II) in Turkey, Dialogue What cities will look like in the 21st century

Platt RH, Rowntree RA and Muik PC (eds) (1995) The Ecological City. The University of Massachusetts Press, Amherst, pp 7

Register R (1987) Eco-city Berkeley: Building cities for a healthy future. North Atlantic books, Berkeley, California, pp 35

Seoul National University, UNDP, MOST and DAEWOO Co. (1996) Co-city planning guidance. Seoul- in English and Korean

The city of Taejon (1995) The city of Taejon administration

Urban Ecology Australia Inc. (1993) The Halifax Eco-city Project: The world's first piece of eco-city. Adelaide Australia, pp 2

Ecological Aspects in Suburban Renewal and Experiences in Finland

Mervi Johanna Hiltunen

1
Urban Structure in Finnish Towns

By European standards Finland is a rather sparsely populated country, with fairly small cities located far from each other. The population is 5 million and the country itself one of the largest in Europe. The only metropolitan area is the capital region of Helsinki. Urbanisation in Finland began late, but since the 1960's has been extremely rapid. Today about 70% of all Finns live in urban environments. During the urbanisation wave, new housing was mainly built up in neighbourhood units, far from city centres. The migration movement of the 1960's and '70s was caused by changes in the economic structure: forestry and agriculture could no longer offer as many jobs as before, while industry and services in urban areas were expanding rapidly. The urbanisation process was closely linked with modernism and functionalism, building became industrialised providing preconditions for fast and efficient housing projects. Sufficiently cheap areas of land were available on the outskirts of towns. The building of neighbourhood units set off the dispersal of urban structure; Finnish suburbs typically cover a small land area. Around big cities they are normally isolated blocks of flats and in smaller towns mixed housing stock is built. Today about 1 million people, nearly 20% of the population, live in suburbs. (Ministry of the Environment 1996, Uuskallio 1997).

2
Suburban Residential Areas

Although the housing stock in Finnish suburbs is relatively young (20–30 years old) it now needs profound renovation. Most of the houses were originally built very quickly, with insufficient technological knowledge and their surroundings are unfinished and monotonous. Suburbs gained a bad

University of Joensuu, P. O. Box 111, FIN-80101 Joensuu, Finland

J. Breuste, H. Feldmann, O. Uhlmann (Eds.)
Urban Ecology

reputation from the start and they have been studied extensively, mainly by sociologists and planners. The degree of social segregation is low in Finnish towns and there are no urban residential areas which could be classified as slums. According to the latest studies the economic recession of the early 1990's may, however, have changed the social balance in cities (Ministry of the Environment 1996). Persistent mass unemployment and increasing social problems are today the greatest threats to positive futures of urban areas. As a result of centralisation of the trade branch many neighbourhood units are loosing nearby services. Public intervention and action is necessary in order to make suburbs better living environments and socially functional communities.

3 Suburban Renewal

Since the late 1970's there have been independent improvement projects in suburbs, but in 1995 Ministry of the Environment and National Housing Fund of Finland started to co-ordinate a comprehensive development programme for suburban residential areas. At the moment social and technical improvements are taking place in 49 of about 300 suburbs. Renewal projects are funded through state loans, and the Housing Fund concludes an agreement with town municipalities and local housing corporations. Local, long standing commitment has been set as a precondition for the loans. In most of the suburban renewal projects social and cultural aspects of sustainability have been given priority. Social sustainability refers to the strengthening of social identity and the individual's capability to conduct his or her own life. Renewal projects are seen as preventive action where new social and technical innovations can be experimented with.

Neighbourhood development requires a new planning approach, seeking to establish co-operation practices between property owners, inhabitants, various sectors of municipal administration, developers and planners. The improvement of new local planning methods and participation of residents have been encouraging in many suburbs. Renewal projects have opened "living-rooms" for residents to meet and share common interests. Very often the municipal authorities working in the project move their offices next to these meeting rooms to maintain closer contact to residents. Experiences have proved that commitment of local authorities and politicians is most important in order to change administrative culture and get residents involved. A three year project is much too short a time for a real change. It seems that local politicians seldom realise this, and are eager to cut down the municipal funding after the technical renovation.

4 Ecological Experiences and Aspects of Suburban Renewal

Aims and practical experiences of suburban renewal projects also include ecological goals and solutions. Urban renewal can be considered as an ecological act; to use existing urban infrastructure is always more ecological than to build completely new areas. It is rather easy to plan and carry out ecologically experimental housing projects in city regions where no building has taken place. There is no doubt about the necessity of such areas, but to develop the existing urban environment in an ecological direction is a more challenging task. Instead of designing alternative and utopian models for ecological urban structure, we need to accept the old community structure and try to improve it.

The latest planning doctrine stresses the necessity of increasing the density and mixed-use of zones in urban sustainable development. These ideals have also caused critique (Jessen 1996). Increasing density is not an ideal solution to current problems of urban structure. It should not be carried out in the areas which are intensively built already, such as city centres (Rådberg 1994). In Finland the population density in built-up areas is very low (530 people per sq km) in comparison to Swedish towns, where urban density is three times higher (Ministry of the Environment 1996). Therefore increasing the density is quite reasonable in many neighbourhood units. Although the infrastructure in Finnish cities is comprehensive in coverage and high in standard, it is inefficiently used due to its dispersed structure. Increasing the density of suburban areas can hence be seen as ecological. There is not much danger of higher urban density causing lower quality of life in suburbs (Breheny, Rookwood 1994). On the contrary, complementary building brings more residents into a suburb and increases the efficient use of infrastructure as well as profitability of local services. The surrounding countryside remains as an area for recreation and protection of urban biodiversity.

According to Lodenius (1995) Finnish suburbs have the following potential strengths, which are often forgotten when talking about problems and the future of suburbs.

Technical and economic resources: these rather new residential areas hold a significant portion of national wealth and capital.

Social resources: the residential local community is a powerful but undervalued resource in decision making, local administration and production of social services.

Ecological resources: relatively efficient suburbs, situated in the surrounding countryside and based on public transport, rational infrastructure and services, form a more ecological starting point than a dense city or dissipated scattered settlements.

Without exception the rhetorical phrase "sustainable development" is mentioned in the plan of each suburban renewal project. In most of the cases they are empty words, but there are, however, a few projects where sustainability and ecological aims have been taken more seriously. In Mikkeli, south-eastern Finland, every phase of the renovation of old block houses was ecologically assessed. Energy saving, water consumption, used materials and recycling are examples of the evaluated issues. In Vaasa, on the west cost, a suburban renewal project focuses on two matters; renewal of an old suburb and building of a new ecological village next to the suburb. In "Sun village", as the project is called, ecological innovations such as active and passive solar energy, a two water system and gravity based air conditioning are used. Both private and communal houses will be built in the village. The town of Joensuu, in eastern Finland, has become famous for its well-organised suburban renewal project. Last year the suburb received extra funding from the European Union's URBAN-structural funds. Ecological aspects are revealed in the detailed assessments of the condition of the houses before renovation, in hindrance-free building principles, projects dealing with increasing density, and in participation of residents.

The Association of Finnish Architects is compiling an information bank on ecological planning methods and practical experiments carried out in Finland. Nearly all of the experiments so far concern detached houses or eco-villages. Although there are some individual experiments on ecological renovations of residential blocks, the subject has been given minor importance. Again it is presumably easier and more interesting to build new than renovate old. What could be the reasons for this? In the suburban context the expenses of renovations have to be kept at the lowest possible level, in order to avoid too high an increase in housing costs. This concerns both rental and owner-occupied houses. To repair mistakes made in the 1960's and '70's is expensive. Building firms and constructors can lower their prices only by reducing building costs. This automatically leads to pressing timetables, poor quality and sloppy work. There seems to be no time or inclination for thorough planning and ecological reconstructive building! It is obvious that savings of this kind are very short-sighted. Other reasons can be found in the taxation system, in the merger of building companies and even in actions taken by government organisations. Finland has a long tradition in the production of social services, therefore it has been easier to take up social matters in suburban renewal than to pursue technical or ecological innovations. During the urbanisation wave traditional skills to build healthy, wooden or stone houses were lost. Heavy industrialisation of the building sector, new materials and techniques caused a break in the building tradition. Today we are paying for the mistakes and those who want to build ecological houses are trying to learn the old practices again.

Although there is an increasing interest in ecological building, comprehensive discussion and research on ecological renewal of suburbs has so far been regrettably low. In the renewal of suburbs a holistic approach is needed both at local and national level. Politics, decision making, lifestyles, town structure, as well as building and planning methods should all be taken into consideration simultaneously. This demands a change from sectoral administration towards a conversational and interactive planning. On a large scale it is a question of a change from modern functionalism towards ecological complexity. This change has begun. The steps taken are small and local, but their future value is wide and global.

References

Breheny M, Rookwood R (1994) Planning the Sustainable City Region. In: Blowers, A (ed): Planning for a Sustainable Environment. A Report by The Town and Country Planning Association. Earthscan, London

Jessen J (1996) Der Weg zur Stadt der kurzen Wege – versperrt oder nur lang? Archiv für Kommunalwissenschaften 35(1): 1–19

Lodenius S (1995) Avoin Lähiö – terve lähiö. Ekologisen täydennys- ja korjausrakentamisen mahdollisuuksia. [Open suburb – healthy suburb. On possibilities of ecological complementary and reconstructive building]. In: Lapintie K, Koskiaho B, Ikonen T, Tiilikainen V (eds) Ekopolis. Ekologisen kaupungin juuria etsimässä. (Ecopolis. Looking for ecological roots of the city). Gaudeamus, Tampere

Ministry of the Environment (1996) Sustainability as a Challenge. Finland's National Report to the Second United Nations Conference on Human Settlements (Habitat II). Finland

Rådberg J (1994) Termitstack eller ekobyar. Plan 48(4): 201–207

Uuskallio I (1997) National Overview on Distressed Urban Areas in Finland. Ministry of the Environment. Land Use Department. Series: The Finnish Environments 85. Edita, Helsinki

Adelaide and Whyalla: The Practice of Urban Ecology in Two Australian Eco-city Projects

Paul Francis Downton

2 case studies describe and analyse the experiences of the community non-profit education and advocacy association 'Urban Ecology Australia' and the ecological architecture and urban design company 'Ecopolis', in integrating ecological, economic, social and cultural issues through eco-city programmes in South Australia.

1 The Halifax EcoCity Project

In Adelaide, 1991, members of Urban Ecology Australia (UEA) gathered together to find an inner-city site for the development of 12 ecological dwellings. In 1993 the old Adelaide City Council (ACC) Depot was vacated and the 2.4 hectare site made available for redevelopment. Through Ecopolis, UEA sought to both design *and* develop what they came to term a 'piece of eco-city' (Downton 1995).

Intended for 800 residents, their proposal for a car-free, mixed-use development contrasts with conventional Australian urban form. Innovative earth construction, renewable energy, water capture and re-use, and on-site sewage recycling are integral to this development. The goal of the project was not just to create an 'environmentally-friendly' development but to address numerous issues related to development processes and the impact of the city on its hinterland. These include:

- exploring 'community-driven' development and social equity in development,
- challenging status quo power relationships and questioning assumptions about the purpose and cost structures of the development industry,
- exploring various forms of tenure and community management structures,
- linking urban and rural development and advocating holistic urban planning.

Centre for Urban Ecology, 84 Halifax Street, Adelaide, Tandanya Bioregion, SA 5000, Australia

J. Breuste, H. Feldmann, O. Uhlmann (Eds.)
Urban Ecology

Centre for Urban Ecology. Operated by local people since late 1996, ALCUE provides environmental information to the region and exhibits eco-city material. Community workshops ranged from ecological architecture, urban design and bioregion identification to a children's workshop for a 'Council of All Beings'. Members of the public underwent a 'crash course' in urban design and were encouraged to produce designs based on Kevin Lynch's 3 city types. They insisted on producing 'organic' designs in contrast with the ubiquitous Australian grid, calling into question the relevance of that colonial form for modern Australia.

Each workshop was supported by an exhibition and booklet. The first booklet 'Whyalla, Why Not?' introduced the eco-city idea with a short story set 25 years from the start of the EcoCity Whyalla project, making the point that the direction for the future will be set *now* (Downton 1996). A competition for school students to continue the story met with some success. The process of workshopping and setting up ALCUE was intended to engage the city's community in envisaging a future with economic renewal based on ecological development.

Aspects of the Whyalla project drawn from or inspired by the UEA/Ecopolis experience with the Halifax EcoCity Project include:

- community-based development and locally run Centre for Urban Ecology;
- a Project Team representing local interests and key players in the development;
- use of recycled materials and low-cost techniques for presentation of information;
- an 'urban organic' architectural style;
- design and development based on 'Ecopolis Development Principles' (see below).

2.1 Ecopolis Development Principles

1. *Restore degraded land*: urban development to restore the landscape.
2. *Fit the bioregion*: human settlements to fit natural cycles of the region.
3. *Balance development*: development balanced with 'carrying capacity'.
4. *Halt urban sprawl*: stop development from consuming the landscape.
5. *Optimise energy performance*: generate and use energy efficiently.
6. *Contribute to the economy*: create and promote economic activity.
7. *Provide health and security*: create healthy and safe environments.
8. *Encourage the community*: cities are for everyone.
9. *Promote social equity*: equal rights and access to services and facilities.
10. *Respect history*: respecting the past whilst looking to the future.
11. *Enrich the cultural landscape*: diversity creates vital cities.
12. *Heal the biosphere*: repair and support processes which maintain life.

Whyalla's project has elicited surprising support. April 1997 saw an interfaith blessing of the site with an Aboriginal ceremony, the first Australian 'Terbum' Ceremony, and a joint service by the Anglican, Roman Catholic and Uniting Church. Other faiths were present with spiritual commitment to the eco-city concept. The first buildings will include an Anglican ecological church and a Buddhist ecological temple. A disabled organisation is re-locating its headquarters to the core site and the Council is commissioning housing sub-divisions.

The UEA-Ecopolis strategy has been to promote an holistic vision to integrate social, economic and cultural initiatives with eco-city projects as catalysts for systemic environmental improvement in cities and their bioregions (Downton and Munn 1995, 1996). After 6 years, evidence of success includes:

- vision and commitment expressed in cultural activities, e.g. Interfaith Service,
- participation of one of Australia's largest non-bank financial institutions,
- Centres for Urban Ecology sustained by the community in two cities,
- Halifax site being decontaminated and Bourne Court commencing construction,
- foundations in place for key buildings on the Whyalla EcoCity Core Site 1997,
- infrastructure spending committed for Whyalla Core Site by the City Council,
- UEA presentations of the eco-city concept around Australia and overseas,
- eco-city projects in educational programmes for establishments from primary schools to universities,
- Ministerial Commendation for UEA volunteer work in 1996, and 'World's Best Ecocity Award' from Ecocity Builders Inc. in 1995 (Peeks & Register 1997),
- 1997, the Federal Government giving a grant to UEA, recognising it as the primary national community environment organisation on urban issues,
- the inclusion of the Halifax EcoCity Project on the UNCED database (UNCED 1996),
- CUE routinely receiving overseas visitors (including delegations from Korea, Malaysia, Singapore, Sweden & the Philippines) and hosting overseas trainees. These trainees typically work for 3 to 6 months and have included students from Denmark, Germany, Malaysia and the USA – contact UEA for information.

(I would like to thank Professor Santosh Ghosh of the Centre for the Built Environment, Calcutta, for presenting this paper on my behalf at the conference.)

References

Arkin L, Crenshaw R, Walter B (eds) (1992) Sustainable Cities – Concepts and Strategies for Eco-City Development, Eco-Home Media, Los Angeles
Downton PF (1995) The World's First Piece of EcoCity, CUE, Adelaide
Downton PF (1996) Whyalla, Why Not? & booklets #1–7, CUE, Adelaide
Downton PF, Munn D (eds) (1996) Urban Ecology Annotated Bibliography (Second Edition), Centre for Urban Ecology, Adelaide
Peeks B, Register, R (eds) (1997) Village Wisdom – Future Cities, Ecocity Builders, Oakland
Roelofs J (1996) Greening Cities, TOES/Bootstrap Press, New York
UNCED (1996) Habitat 2 Good Practice Database CD-ROM, UNCED

Leipzig-East Regional Project

Ekhart Hahn

1 Model Project for an Ecologically Oriented Urban-Rural Development

Responses to the ecological, social and economic crises of the modern city are developed and tested in this regional project. Special attention is given to the relationship between the city and surrounding countryside. The foundation of this model project consists of 12 individual projects that began in the politically dynamic years between 1989 and 1991. These efforts are generally characterised by a close co-operation between citizen initiatives and city administration, and instead of following the standard urban design practices have established sustainability as a primary objective. The themes of the various projects range from ecological agriculture and neighbourhood development, to the creation of an ecological-cultural centre as a forum for public activities. Overall the project is guided by three principles, organised in three project areas (Fig. 1).

1. A practical application and further development of current understanding of urban ecology in its dealings with buildings and neighbourhoods along with a networking of individual projects on a comprehensive and regional basis (project area I).
2. A renewal of urban-rural relationships that emphasise sustainability and quality of life: cycling and recycling of materials, protection of valuable landscapes and cultural sites, ecological agriculture marketed directly in the city, recreation (project area II)
3. The demonstration of economic development and employment strategies that are ecologically oriented and based on local and regional resources (project area III).

In summer 1993 the city of Leipzig obtained approval from the European Commission for running the project and received the largest sum awarded by LIFE Programme up until that time (2143 Mil. ECU funding with a total

Paul Lincke Ufer 30, D-10999 Berlin, Germany

J. Breuste, H. Feldmann, O. Uhlmann (Eds.)
Urban Ecology

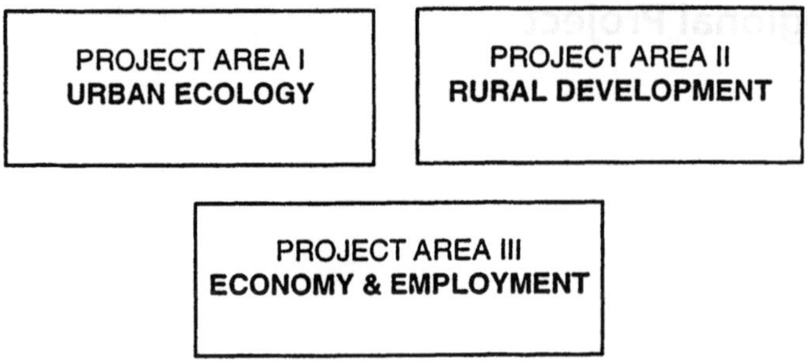

Fig. 1. Three project areas

project budget of about 9 mil. ECU). Additional support was received for subprojects from the Bundesbauministerium in Bonn (Federal Building Ministry, EXWOST Programme), and from the THERMIE Programme of the European Commission.

Urban Ecology (project area I) was conceived as the core of the Ostraum Project. In the following presentation some main ideas and results concerning this project area will be presented.

2
Green Radial Eilenburger Bahnhof

The former Eilenburger Bahnhof (railway station) in east-Leipzig offers a unique opportunity to demonstrate the Genius Loci of the area. The Bahnhof, including the former railyards, stretches over two kilometres, connecting the downtown with the city border to the east. Here, at the edge of the city, this urban section of land opens out onto the eastern countryside and the culturally and historically attractive landscapes (Fig. 2)

Four areas totalling 16 ha are to be transformed step by step into a multifunctional urban-rural Green Radial. A guiding principle is that as one moves from the city gate to the countryside the open spaces will change from having an urban to a rural character. Where as the eastern area would offer room for more open and free structures, at the western end, which is the urban pole, urban forms would be represented by the architecture and the water and green spaces. The neighbourhood park with the Ecostation and the Anger/Crottendorfer railway would be situated in between. The arrangement of these four different Bahnhof sections is to be structured according to the respective histories, including varying functional characteristics and social contexts. An emphasis is placed on the experiences one would find when crossing the entire Bahnhof, and in particular this means the expression of

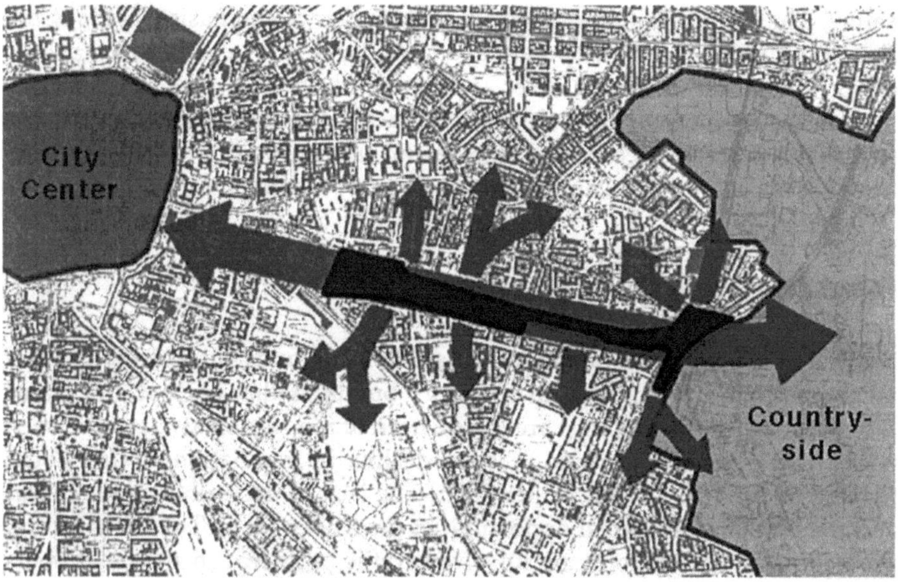

Fig. 2. Green corridor connecting city and countryside

the urban-rural connection. It is of course important that the Green Radial be inviting for bicyclists and pedestrians to make the trip from city to countryside or vice versa. Thus the complementary and contrasting features of city and countryside are to be presented in new ways that are visible and can be experienced. The sequencing of the four Bahnhof sections also offers a wide variety of design possibilities. For example, diverse forms and structures are envisioned for the green spaces which would consist of park-like areas, natural and biological zones, and artistically shaped experience zones.

Areas are planned to be suitable for flora and fauna as well as for people; path connections and protected areas for bicyclists and pedestrians, play grounds, adventure and „wild" areas for children and youth, zones for development of sensory abilities (Kükelhaus) as well as ecological and study gardens in connection with the planned Ecostation/Green Workshop.

2.1 Citizen Participation and Project Development

The Green Radial Eilenburger Bahnhof is not to be developed as a park designed exclusively by contracted firms, but rather as the result of the long-term, co-operative efforts of a great number of people working together with nature. The Green Radial and accompanying parks emphasise, as does the concept of Le Roy, the development of a multi-layered system, in which people can communicate with each other as well as with the elements of nature.

After the beginning phase of project initiation, the development is expected to reflect a process of continual change and transformation, with the objective of realising an increasingly diverse system over time.

3 Neighbourhood Development

North and south of the Eilenburger Bahnhof one finds turn of the century neighbourhoods, typically with five story buildings that enclose and frame entire blocks. These buildings are in dire need of renovation as they have been neglected for several decades. A clear decline in population is also documented in the city district of Reudnitz. The lack of accessible green spaces and the lack of recreational opportunities have been identified as significant problems in addition to many unsolved traffic problems in the neighbourhood. By improving the housing conditions, the residential population could be increased by about a third and in addition there are many empty lots which could be built on.

The Green Radial is seen as an important starting point for the demonstration of ecological renovation and renewal in the surrounding neighbourhoods. These model renewal projects are to be developed in close co-operation with local residents and property owners, city administrators and business representatives, and other local concerned groups. A primary objective is a support for citizen participation in connection with the creation of the neighbourhood "Ecostation".

Important neighbourhood development objectives are:

- Access by proximity: creation of appropriate conditions that will enable pedestrians and bicyclists to reach as many destinations as possible which are part of their daily activities and routines (work, shopping, school, recreation etc.).
- Reconstruction of streetscapes to suit changing mobility patterns and alternative uses.
- Mixed uses and appropriate density: integration of a diversity of urban and natural experiences and a small scale mix of uses; natural experiences are especially significant for children and elderly people.
- Attractive and healthy environments developed in the local neighbourhoods; new local qualities reduce the incentives for people to have to travel great distances for pleasure and relaxation.
- Demonstration of new technical and social concepts for reducing CO_2 emissions; considered here are model concepts for using local renewable

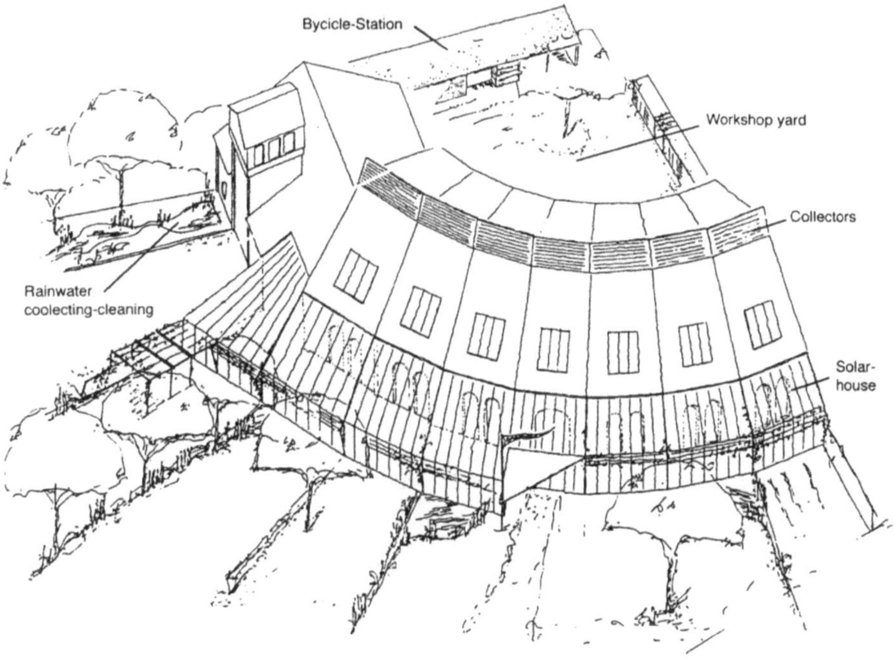

Fig. 3. Design proposal for ecostation

resources (energy, water, materials, etc.) and their widest possible application in the neighbourhoods.
- A phased reconstruction and renovation of the buildings with respect to the needs of the residents and their changing lifestyles; application of the most current building technologies with regards to ecology and energy efficiency; appropriate water and waste concepts and other measures for building and greening of courtyards.

4 Ecostation

The planned Ecostation/Green Workshop in the former Lokshuppen (locomotive shed) on the Eilenburger Street plays a key role in the concept of the Green Radial and the adjacent model neighbourhoods (Fig. 3).

The Ecostation is a place for neighbourhood democracy, information and communications as well as being a meeting place. Here one can find institutions supporting self-help activities and new forms of co-operation between citizens, city government, local schools and clubs, and representatives of the local economy and other concerned groups. The Ecostation is a place for exchanging information and for community decision making. It is a place for initiating and directing projects, such as the realisation of the Neighbourhood Park and the Green Radial and the development of demonstration projects in the adjacent neighbourhoods. In the Ecostation one finds discussions, festivals and performances, presentations, exhibitions and seminars. Also to be found in the Ecostation are workshops and other programmes which encourage citizens to be active, a natural foods restaurant and a reading cafe as well as counselling for ecological nutrition. It is also hoped that offices from environmental organisations, local colleges and consumer groups will find a place either in the Ecostation or in the neighbourhood.

Reference

Hahn E, LaFond M (1997) Local Agenda 21 and Ecological Restructuring. A European Model Project in Leipzig. Science Centre Berlin, WZB papers FS II: 97–408

Improvement of Living and Environmental Conditions of Informal Squatters in Hanoi

Nguyen Huu Dung

Vietnam is implementing the renovation process of its economy according to the market mechanism with the help of the State. The economy is growing with a high speed of GDP (about 9–10% per year, Ministry of Science, Technology and the Environment, 1997). The growth in construction and rebuilding of big cities is about 10–12% per year (HAU/MoSTE 1996, MoSTE 1997). The market economy is also exacerbating the situation of social classification. Poor people are encountering difficulties and lack of dwelling houses. Their living areas and the adjoined environmental conditions, are problems to which the urban planning and the environmental management of the country must pay attention. In big cities such as Hanoi and Ho Chi Minh City, slums and poor squatters' quarters are creating dangers for the urban ecological system.

1 The Present Situation of Some Informal Squatters in Hanoi

The total areas of Hanoi is about 913 km^2 comprising 40 km^2 inner city and 873 km^2 suburbs.

The population of Hanoi Capital (according to incomplete figures) is 2.4 million, including about 1 million city dwellers. The population density reaches up to 25,000 inhabitants/km^2, without taking into account hundreds of thousand of visitors and workers from other provinces and irregular residents. The remarkable number of users in Hanoi is challenging the service system to carry out a huge amount of work to ensure good communication, environmental hygiene and protection of the landscape in the city (SIDA 1993).

The construction and urban management of Hanoi City is a burning problem. Many public areas are being infringed on and used indiscriminately, leading to the degradation of the city. The population growth pressure causes social disorder and insecurity, lack of dwelling houses and the undermining

Hanoi Architectural University, Research Centre for Architectural Indoor Climatology and Environment (RCAICE), Ministry of Construction, Dal hoc klen truc Hanoi-Tranh Xuan, Hanoi, Vietnam

J. Breuste, H. Feldmann, O. Uhlmann (Eds.)
Urban Ecology

of the city's landscape, all of which in turn affect the planning and development of the city. Squatters' quarters are made up of spontaneous agglomerations of people who have settled on public land; these have no minimum infrastructure. The population component is complex and different situations can be found. The most salient squatters' quarters are those of the Thanh Nhan precinct- Hai Ba Trung district and Trung Liet precinct- Dong Da district (the site of the old landfill). These are illegal settlement quarters without direct management by the local authorities. Most of the population living in these squatters' quarters have immigrated from other localities according to the parachute method or through illegal buying. The population constituents here are chiefly unemployed manual workers, other workers, small dealers, and city service workers with unstable incomes. They belong mainly to the urban poor group. Only about 10–16% of the households residing here have the legal right of asylum with land use certificates, the remaining 84–90% of them is illegal. It is also here that the concentration of social problems such as prostitution, drug addiction, stealing etc. occur (IDRC 1996).

The urban infrastructure in these areas is extremely poor: rain and surface water flows naturally in open sewage drains which are criss-crossed by lanes and scattered with illegal houses built in a disorderly fashion. The low incomes of the population in these areas, and their very small houses, made of temporary and patchy materials are stinging points in the city.

2 Problems of The Shelter and Environmental Conditions in These Quarters – Harmful Impact on the Ecological System

2.1 Poor Infrastructure

- The poor transport system in the squatters' quarters are self-induced. The lanes are narrow and uneven, some of them are only about half a metre wide. They are low and waterlogged, even on sunny days; on rainy days they are flooded.
- The sewage system is unavailable here. The waste water flows into the lanes and some flows into the neighbouring lake in the Thanh Nien park; stagnant water spreads everywhere, even indoors.
- The clean water for cooking in these quarters is provided (and sometimes sold by pails) chiefly by some drilling wells. Piped water is unavailable and over 80% of the population have no safe water supply; they have to obtain clean water from outside the area.
- The electricity network here is very temporary and dangerous. There is no main network, only a small electric wire connected to a main electricity metre for the whole hamlet, other households then connect it to their own

houses. The money paid for electricity is divided equally amongst households and collected monthly by unruly persons. Few means of communication and information are available.
- Coal, kerosene or wood are the fuels mainly used for cooking (SIDA/SEI 1993).

2.2 Real State of Dwelling Houses

- The houses in squatters quarters are temporarily covered with thatch, bamboo boards, oiled paper or leaves, and are built with earthen walls.
- The houses are narrow and low. Most of them have no separate toilet and cooking areas or, if there are any, they houses are compartmentalised by bamboo boards.
- Some of them are always waterlogged.
- Some of them do not have enough covering structure and insulation materials so they are unable to protect against the winter cold and the summer heat, especially in rainy and stormy seasons.
- Most of them have a lack of windows for natural light and ventilation so they are always dark and humid.

2.3 Environmental Pollution

- The air is heavily polluted by smoke and dust from cooking fuel.
- The population living here has no idea about sanitation and hygiene. The solid waste is thrown around the area and even into the lake in the Youth's park, causing air (smell) and water pollution.
- There is no stretch of green vegetation to regulate the environment, except for some low wild bushes used as fences, which are inhabited by mosquitoes and pathological vectors.
- Due to the unavailability of sewage and drainage systems the area is always waterlogged when it rains (SIDA 1993).

3 Solutions

3.1 Planning and Reconstruction of Squatters' Quarters

According to the planning direction and the land use map of Hanoi city, the two squatters' quarters in Trung Liet and Thanh Nhan are located within the public land area which is planned as parkland for the Dong Da and Hai Ba

Trung districts. Thus it is imperative that the inhabitants living illegally here are moved to other areas. After recovering these land areas, it is necessary to conduct measures to design and plan the conversion of these areas into parks with trees, creating new recreation areas for the city as well as notably contributing to the improvement of the ecological climatic system of the region and other neighbouring areas.

Before these parks are built, it is necessary to clean the pond and lake system of the parks to recover their initial landscape quality. People owning land property in the dismantled areas must be reasonably compensated with houses built in new areas or with money.

3.2 Migration of the People in the Squatters' Areas

New quarters for squatters will be planned in an area of 10,000 m^2 allocated by the city authority (Vinh Tuy is an area in the south of the city – Hai Ba Trung District 1995) These new houses and apartments will be allocated or rented to appropriate target people from the squatters' quarters. For those who are not able to afford a house immediately, they would be able to pay after a certain time, those who are able to could buy them at once. The buildings are to be built with low priced facilities. Low storied houses (one or two storied) should have solid foundations so that later, when people have enough money, they will have the possibility to build further levels. These apartments should have areas for economic activities. This is important for poor people, as they need to be able to generate income to enable them to improve their living conditions.

When building multi-storied houses (3–5 stories) low priced houses should be built for rent or for sale by paying in long term instalments. Houses of this kind should have a garden behind them with a ground floor for storing work implements (income generation). The garden could also be used for service activities, especially for those in the category of KT2 households (those living in one area, but having household registration in another).

Poor people tend to be community minded, rather than individualistic, this should be taken into consideration when building these houses, e.g. each apartment should not have several small rooms, rather a multipurpose one for carrying out different activities and as a reception room.

In the planning of new dwelling areas, the building of minimum infrastructure should be included in the project to create good living conditions for the people. The State may partly support the budget and the rest would be contributed by the community. It is necessary to build low priced houses for rent, or for selling in instalments, or to urge every house building project to reserve certain areas for rent, with high storied houses and the minimum facilities.

When planning the location and design of the newly built houses or apartments for the urban poor it is necessary to respond to their job requirements and the ability to generate household income to overcome their poverty.

References

Hai Ba Trung District - Office of Housing and Land Management (1995) Project on rehabilitation 1995

HAU - MoSTE (1996) Master Plan of Energy Conservation - Survey on Building Sector

IDRC Project (1996) Improvement of Shelter and Environmental Conditions for the Urban Poor SIDA/SEI - Assessment of Environmental Problems of Hanoi - Final Report. Hanoi

Ministry of Science Technology and Environment (MoSTE) (1997) Country Report to the 19 Th. Special Session of the general Assembly on Implementation of Agenda 21- June 1997

Preservation of Environmental Conditions in Cities: A Case Study of an Ecological, Economic and Social Approach in the Lake Area on Pokhara Valley (Nepal)

Girija Prasad Gorkhaly

Nepal, sandwiched between India and Tibet, lies between 65 and 8848 metres above mean sea level and is blessed with a large biodiversity.

Pokhara, 200 km from Kathmandu, lies at the centre of the Western Development Region and is one of Nepal's favourite tourist destinations. It lies in the lap of the magnificent *Annapurna Himalayan* range with the peak of *Machcapuchre* (meaning fishtail) in the background. It has six lakes and an altitude of about 700 m above sea level.

Since it was declared the centre of the region in 1972, Pokhara has grown tremendously (at present the population growth rate is 7.8% per annum). The economic activity has also been increasing, mostly through tourism and the remittance of ex-army personnel. Although the economic opportunities have improved, the cost of improving the quality of life is becoming very high.

The continued growth (without adequate and appropriate environmental planning) has caused an increase in water borne diseases, the problem of solid waste removal, crowding and loss of natural scenery and amenities. In order to transform this change into a favourable one, informed and committed guidance and management from government and non-government institutions at the national, district and local level are necessary.

Planning of *Phewa* lake, the biggest of the six in the region, took place in 1995. *During the planning the biggest challenge was to tackle the conflict of interests- social versus economic.* The need for planning was triggered by the concern over the deteriorating ecological condition of the lake which started even adversely affecting the national economy, due to the negative impact on tourism.

The conditions in Nepal must be understood: very rapid population growth in the urban centres (around 5% p.a.), lack of a strong planning force and political commitment to provide long term benefits, and, above all the lack of strong and committed implementation.

It is obvious that the reason for deterioration in the lake and its surroundings is the increase in incompatible human activities in its watershed. In order to control this increase in incompatible human activities (whilst

Ministry of Housing and Physical Planning, Dept. of Housing and Urban Development, Barbar Mahal, Kathmandu, Nepal

J. Breuste, H. Feldmann, O. Uhlmann (Eds.)
Urban Ecology

keeping ecology and economy in mind), one proposal was to maintain large plot size, to control density, to keep the built-up area to a minimum and to attract activities which require a high environmental quality. *This means the encouragement of bigger hotels and other tourist services.* This is directly against social interests, at least at the first glance, for it causes displacement of many of the poor people.

Supporters of ecological and social interest proposed to restrict any construction except for huts to prevent displacement of 'the poor' and to preserve the lake, but it is argued that if restrictions were put on all constructions except huts, the land price will be held down artificially. Under the prevailing rapid population growth the area will be gradually taken over by 'the poor' who are not wealthy enough to care for the fragile environment of the nationally important lake. On the one hand this would prevent appreciation of the property of 'the poor' but on the other hand would force them to remain in a degraded, but potentially fruitful environment.

If large hotels were permitted under strict environmental control, the population density would be controlled and also the environment would be maintained by the investors themselves as their (hotels') survival is based on it. In the face of inefficient and difficult implementation of rules and regulations, it is more difficult to control the informal sectors than big formal sectors. This became a highly controversial issue, attracting not only municipal attention but even causing arguments between the district and national governments. The national government was interested in protecting the lake without thinking about who will benefit and at whose cost, whereas the local government was very much concerned with this and the implementation of the results of the planning.

We have seen many examples of restrictive planning and planning for 'the poor' (as they say); plans are often unrealistic and, in the long-run, of no benefit to the poor. If the planning is for 'the poor' then these people should be educated and be allowed to make their own decisions. One thing to be remembered is that usually they cannot invest in the future as that is too far ahead of their current concerns.

Ultimately, it was decided to allow construction of up to two floors with the following restrictions of built-up (surface sealing) area:

1. First 20 m from the lake frontier: no construction at all.
2. 20–100 m from the lake frontier: 20% of the plot.
3. 100–200 m from the lake frontier: 30% of the plot.
4. 200–300 m from the lake frontier: 50% of the plot.

Applications for hotels, with very sound environmental designs, have already been sent in and their construction has begun. With the successful functioning of these hotels, it is hoped that many such environmentally sound activities will be attracted to the area and thus will help to preserve the lake as well as benefit the local poor. Some of the population who decide to sell the land will benefit from the increase in land value, others who remain will be accrue benefits through the increase in quality and quantity of job opportunities.

It was observed that in the areas where small plots are located the density is high, the environment is poor and the ability to plan and enforce plans and regulations are meagre. In contrast, where the plots are big and formal hotels are located, the maintenance of the environment is much better. Of course, there were frequent charges that the big hotels are causing the degradation of the lake environment, but they are often disproved. It is difficult to believe the charges, as the hotels' survival is based on the conservation of the environment; it is sometimes small, informal and non-professional hotels that cause the environmental degradation.

The planning has been evaluated to be successful, although it is not free from comments from many planners who support social interest more than other aspects. Similar work has been undertaken in *Begnas* and *Rupa* lake areas. In contrast to the 20 m buffer zone in the *Fewa* lake, in *Begnas* and *Rupa* lake areas the buffer is 100 m as the areas are more rural and so the land required is still easily available. In addition to the land use management, there are some other proposals: these are to conserve the governmental open space and forest land, to construct checkdams to reduce the silt loads entering the lakes, to construct a barrage in the *Rupa* lake (as in the *Begnas)* so that the area and life of the lake is increased, as well as at the same time improving irrigation facilities around the lake.

References

Oli KP (1996a) Guidelines for Phewa Lake Conservation, NPC and IUCN, Katmandu

Oli KP (1996b) Environmental Study of Nepal's Begnas and Rupa Lakes, NPC and IUCN, Katmandu

Science Applications International Cooperation, USA and New Era, Nepal (1993) ADB Final Report on Environmental Protection Study of Phewa Lake, Pokhara, Nepal

Poster 2.5

From Garrison to Town

Ellen Böttcher

1 Project Summary

In 1993, the Soviet Army vacated its base in Grimma, south-east of Leipzig in Saxony, releasing an area of about 2.5 km^2 for use by the German public.

1.1 Strategy

Instead of providing agricultural land for the development, the local government supports the strategy of using the potential of a built-on site. A feasibility conversion study for one section of the area was commissioned that took not only the usual urban development aspects, but also environmental examinations into account. Ecological aspects were taken into consideration in the form of dealing with special waste, disposal of sewage and rainwater concepts; these apply to conversion sites due to the military use and the availability of large areas of developed space. It was necessary not only to examine the vegetation on areas where the barracks stood, but also the various stages of succession on a manoeuvre site.

1.2 Planning Instruments

The objective of designing a sustainable conversion process can only be achieved by careful analysis of the site and integration of the environmental impacts.

A green strategic plan, in accordance with the development plan, postulates landscape and open space planning objectives while forming the basis for subsidiary planning and management concepts.

This plan shows specific measures for the development and design of open space, the development of the urban fringe and the implementation of green strategies in town planning.

Frauenstr. 11, D-04668 Grimma, Germany

J. Breuste, H. Feldmann, O. Uhlmann (Eds.)
Urban Ecology

Fig. 1. The situation in the northern part of the area after demolition of the barracks. An important objective is to maintain all trees. June 1997

1.3
Implementation

From the planning process it became clear that it is possible to retain some structures in Grimma, for example trees, the main road-network and sewage pipes. Residential, commercial and recreational opportunities are to be provided, whereby the oldest barracks are to be preserved.

The concept for the military training area includes the conservation of biotope complexes and natural succession, thus supporting long-term landscape and biological diversity at a reasonable price.

For a long period the military area existed as an untouched terrain beyond the control of the German authorities. The area must thus change from being an "island" near the town, to a part of the community.

2
Time Schedule

- 1993: The Soviet Army vacated its base in Grimma
- 1993: The Federal Ministry for Building promotes structural concepts 'EXWOST' (Experimental house-building and town planning programme).

The town of Grimma took the initiative to use this programme for the development of the military area.

The main foci are:

- conversion;
- how to use former military land for civil purposes;
- design of a new part of the town (town planning scheme);
- implementation of the necessary facilities of the town such as service sector, housing and leisure amenities.
- 1994: The town of Grimma acquired an area of about 40 hectares for developing a new section of the town.
- 1995–1997: The period for working out a development plan and a green strategic plan to obtain the right to build.
- 1996: Elaboration of a plan for the development of this area (for traffic, water, energy . . .); demolition of the barracks and garages.
- 1997: The development of the first section starts (Fig. 1).

Inhabitants' Response to Ecological Issues Affecting their Daily Lifestyles

Susanne Guldager and Christine Nuppenau

The participation of inhabitants is a central part of the changing process towards sustainable development. The process is slow and challenges old habits. A survey carried out in the housing scheme Skotteparken in Ballerup, Denmark in 1997 shows that the inhabitants in general have understood the message of recycling resources and implementing other ecological actions. Unfortunately, it seems that the interest declines when getting down to activity and action.

1 In General

Values of society are constantly changing. Today environmental issues and sustainable development are a main focus as manifested in international documents such as the Brundtland-report of 1987 and the Agenda 21-programme passed in Rio in 1992. In accordance with the Agenda 21-programme a sustainable development requires that the citizens take an active part in the changing process. This process requires an adjustment of present standards and behaviour and only when the superior socially determined messages have been adopted by the citizens, participation and action can be expected.

The Danish Forest and Landscape Research Institute and the Danish Building Research Institute are at present working on a project with the aim of contributing to the practical, theoretical and methodical development of the urban ecological potential related to the green space of urban areas. Skotteparken, a housing scheme from 1992, is part of the study.

2 Questionnaire Survey

One of the issues in the project is the inhabitants' response to ecological initiatives implemented in Skotteparken. Their attitude to and commitment in

Ministry of Environment and Energy, Danish Forest and Landscape Research Institute, Hoersholm Kongevej 11, DK-2970 Hoersholm, Denmark

J. Breuste, H. Feldmann, O. Uhlmann (Eds.)
Urban Ecology

the ecological initiatives have been studied through a questionnaire survey in January 1997. 95 questionnaires were sent out, equal to the number of occupied flats in Skotteparken. The reply rate was 50%.

The survey reveals that the inhabitants of Skotteparken are, in general, in favour of the implementation of ecological issues. Water recycling and composting of organic waste are given particularly high priority which indicates that the citizens have a growing concern about ecological values and sustainability. This interest constitutes an essential potential for future sustainable development.

3
Skotteparken – a Housing Scheme in Denmark

Skotteparken is situated on the urban fringe of Greater Copenhagen. The scheme comprises of 100 flats with approximately 185 inhabitants. As well as a large green public area, most of the inhabitants have access to a small private garden. In the project design different ecological initiatives such as local water treatment, common kitchen gardens and sorting and composting of organic waste, were proposed and introduced by the project "Green Skottepark".

3.1
Sorting and Composting Waste

For a year the inhabitants in Skotteparken have been sorting their waste with the purpose of composting the organic waste in the area. Unfortunately, due to a technical error, the compost system has not been working for a longer period and, besides this, some inhabitants have found the chosen sorting system unsuitable for the small kitchens. Despite this, the survey shows a basic interest among inhabitants to sort their waste – although they contemporaneously admit not always doing so.

That expressed interest does not always lead to action is supported by calculations which show that in reality only 9–18% of Skotteparken's total amount of organic waste is collected. Experience from earlier similar surveys shows that this discrepancy is a commonly known phenomenon.

3.2
Reusing Rainwater

Collecting rainwater for the gardens has not been systematized in Skotteparken. Even so, some inhabitants collect rainwater and lots of others would like to. About one third of the inhabitants are, in addition, open to uses of the water other than watering, e.g. for washing cars or flushing toilets.

3.3 Green Spaces in Skotteparken

The green spaces in Skotteparken are used very traditionally, primarily for recreation purposes. Only a few households, the most active being those families with children, take interest in vegetable growing and make efforts to grow their own. A couple of essential and practical problems in this area are that the soil around the built-up area is highly compacted and the possibilities of watering are limited.

4 Inhabitants' Participation is of Vital Importance

The survey in Skotteparken has shown, that urban ecological methods must be adjusted to the local conditions if initiatives to develop the inhabitants' interest for urban ecology are to be a success. The initial phase can be especially difficult, as it requires active participation and tolerance of the inhabitants. Therefore, it is important that the inhabitants are part of the process from conception to realization.

As far as possible, urban ecological solutions should also be free of charge to the inhabitants; e.g. only a few inhabitants in Skotteparken find it fair if the environmental improvements lead to an increase in rent. Urban ecology should also not be time-consuming, even though some inhabitants declare themselves ready to use 1–10 hours/week on green projects in the local area.

Investigations of the Acceptance of Selected Green Areas in Leipzig City

Joachim Hoffmann and Klaus Kabisch

The problems of the use of, and acceptance by, the population of green areas in Leipzig have been investigated. Three areas were selected as examples for year round continuous non-participative observations of visitors to the park and for parallel on-going household surveys in the surrounding area (standardised in writing), these were:

1. A small green area (0,66 ha) in the new housing estate Grünau-North with a dominating lawn,
2. The Arthur-Bretschneider-Park in Eutritzsch. A "classical" bourgeois park (7,3 ha) created at the turn of the century with a representative old stock of trees and a pond (0,74 ha); it dates back to the years of rapid industrial expansion and is surrounded by detached family houses, old buildings and renovated old buildings,
3. The "leisure time park" Rabet, a green area (3,9 ha) with a large extent of surface sealing. It is situated in the middle of an old building area with a number of dilapidated buildings and with new buildings on the eastern border.

The integration of the residential area sphere, the size and the structuring of the green areas, the main emphasis of use (investigated using a catalogue of criteria) and the traversing of the green areas served as a basis for the comparative studies. At the end of the observation year the results showed the following significant tendencies of use:

- 60% of all the observed persons only passed the green areas (Fig. 1).
- The most important active form of use (with 25%) was the visiting of the playground by children.
- On average 59% of all the visitors were children and youth, 32% adults and 9% pensioners. No relationship evolved between the age pyramid of the population in the catchment area of the investigation space and the age of visitors to the green areas (Fig. 2).

University of Leipzig, Interdisziplinary Institute of Nature Conservation and Environmental Protection, Augustusplatz 10/11, D-04109 Leipzig, Germany

J. Breuste, H. Feldmann, O. Uhlmann (Eds.)
Urban Ecology

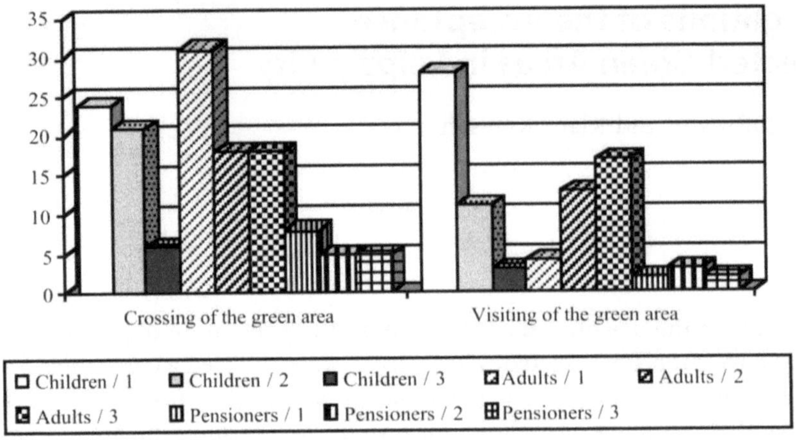

Fig. 1. **Observerd persons in the green areas (1994/95) (1-Grünau-North, 2-Arthur-Bretschneider-Park, 3-Rabet)**

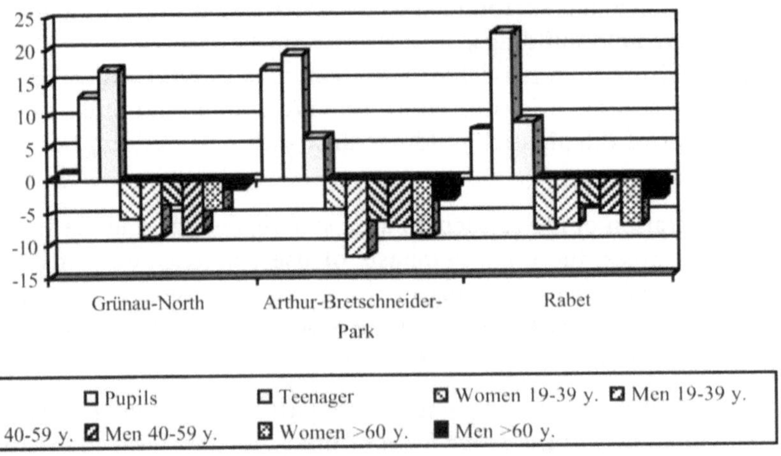

Fig. 2. **Differences between parts of the visitors of green areas and the age pyramid of the popultion in the catchment area of the investigation space (1994/95)**

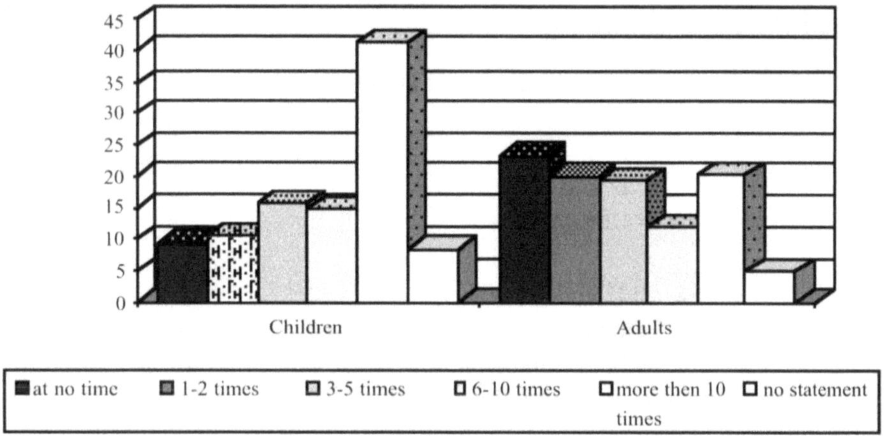

Fig. 3. Visits of the green areas per month (1993/94) (Results of the opinion polls)

- The facilities (kindergartens, day nurseries in schools) situated in the surrounding area, as well as the cultural facilities existing in the green areas (park stage, disco) had a strong influence on the number of visitors and their age structure.
- From the regional point of view the acceptance of the investigated areas declines clearly in the order of Arthur-Bretschneider-Park - "leisure time park" Rabet - small green area in Grünau-North (Fig. 3).

Social Atlas of Leipzig

Sigrun Kabisch[1], Annegret Kindler[2], and Dieter Rink[1]

The Social Atlas of Leipzig is the first atlas which considers the social and urban ecological structural changes in a large east German city after the unification of the two German States in 1990. It is an instrument for the sociological analysis and evaluation of urban development. The main objectives are as follows:

1. spatial registration, cartographic representation and assessment of these social consequences that accompany the abrupt transition from the centrally managed economy in the former GDR to the free market economy in an urban frame;
2. description of the social-spatial differentiation of residential areas by means of selected social, political and economic factors including the actual land use, as well as structural, infrastructural and environmental characteristics;
3. characterisation of the actual social situation in the city of Leipzig by registering the social conditions and their changes in time, their cartographic registration and interpretation.

The analysis and interpretation of data and maps led to a synthetic result – the definition of nine types of social spaces. These types indicate the distribution of similar social groups in comparable building structures with like natural environmental conditions. There is evidence for the hypothesis, that interdependencies between the built, social and natural environment exist also in east German cities.

In consequence, urban areas with initial gentrification or filtering down processes including a concentration of poverty can be identified. The results show that in former old working class quarters in the inner parts of the city social problems become serious, e.g. in those with the highest rate of unemployment. This situation is closely connected with a high rate of empty and dilapidated residential buildings and bad natural environmental conditions like poor quality of green spaces and high air pollution.

UFZ Centre for Environmental Research Leipzig-Halle
[1] ÖKUS – Dept. of Ecological Economics and Environmental Sociology
[2] ALOE – Dept. of Applied Landscape Ecology, Permoserstaße 15, D-04318 Leipzig, Germany

J. Breuste, H. Feldmann, O. Uhlmann (Eds.)
Urban Ecology

This result contradicts the opinion, that the 20 year old mono-structural and very densely built housing complex with about 80,000 inhabitants on the western outskirts of the city must be the area with most of the social problems.

The work on the social atlas will be continued. Sociological investigations in selected parts of the city of Leipzig are planned in order to obtain more detailed information about the inhabitants' perception of their residential environment.

The atlas consists of two parts, text and cartography. The text is divided into:

1. description of objectives, structure and methodology of the social atlas,
2. discussion of roots and symptoms of segregation tendencies in east German cities,
3. characterisation of nine types of social spaces inside the city of Leipzig,
4. analysis and evaluation of initial stages of social-spatial differentiation in Leipzig through the interpretation of several maps.

The cartographic part includes 49 maps:

23 maps describe the socio-demographic structure, 15 maps deal with the residential structure, 6 maps represent the social infrastructure, 3 maps the land use and 1 map the types of social spaces.

The atlas serves as a scientific basis to identify both common features and differences in the development of several urban areas, exemplified through the city of Leipzig. The methodology used can also be transferred to other cities.

The Social Atlas of Leipzig has been created for a broad audience – for urban and regional planners, investors, different authorities, scientists dealing with urban development and, last but not least, for the residents.

The Urbanisation of Slums in Belo Horizonte – Slums, Slums Profile

Cristina Menezes da Costa Croce, Maria Cristina Ramos de Carvalho, and Maria Rita Franco Gomes

1 Introduction

Data (source – IBGE):

- total area of Belo Horizonte .... 197 km²
- total area of slums .... 15 km²
- total urbanised slum area .... 5 km²
- total number of inhabitants of Belo Horizonte 2.160.000
- total number of slum inhabitants 350.000
- Age of Belo Horizonte(1997) .... 100 years
- oldest slum's age .... 100 years
- minimum monthly salary .... U$ 128.50
- climate of the Belo Horizonte .... tropical humidity
- city's topography .... variable to steep
- slum's topography .... variable

The slums are the result of the urban development and migration as well as many different kinds of social-economical and cultural conditions. Another point of the beginning of them in Belo Horizonte is the lack of districts for low class. These are a disorganised occupation without basic infrastructure in private areas and in areas of risks which were unoccupied. And, most of time the slums are situated very close to downtown, one reason that create problem to remove the people that live there.

2 Profile

Disorganised occupation with unstable buildings and unhealthy living conditions causing environmental degradation.

The buildings are usually unstable and the slums are situated in inappropriate areas, for example:

Avenida Contorno, 8.000 / 608 – Santo Agostinho, 30.110-120 – Belo Horizonte – Minas Gerais – Brasil

J. Breuste, H. Feldmann, O. Uhlmann (Eds.)
Urban Ecology

- topography with steep gradient,
- water reservoir areas,
- unused areas ignored by urban organisations. (Non-edificaendi areas)

There are alleys of all kinds and uses. The lack infrastructure results in open sewage, open water pipes, clandestine electrical, connection and untreated water.

3
Consequences

The disorganised occupation causes:

- Domestic waste and garbage blocking the drainpipes in surrounding areas,
- soil and underground water contamination,
- soil erosion / landslides,
- changes in climate and drinkable water flow,
- environmental pollution,
- fires,
- diseases and epidemics.

4
Results of the Urbanisation Plan

The City Hall created a law to urbanise the slums, trying to supply all needs of the inhabitants. And, with the urbanisation, the slums have services such as:

- urban streets (where is possible),
- legalization of property,
- gas service,
- drinkable water,
- light system,
- sewerage system,
- garbage collection,
- telephone,
- mail delivery,
- schools and kindergartens,
- first-aid centres,
- squares for recreation.

Urban Ecological Problems in the Landscape of an Old Industrial Region. Case Study: Resita

Doina Mihalca and Martin Olaru

1
Introduction

Environmental factors must be recognised and taken into account to organise and systemise space in a way which will facilitate long-term development. The decay of this old industrial region of Banat, which holds the centre of balance in Resiţa town, is profound and affects both the urban landscape and the quality of life.

A peculiarity of the urban structure of Resita is the combination of the settlement and industrial uses without a systematised plan.

2
Causes of Ecological Deficiencies in Resita

The following are among the causes of the ecological deficiencies:

- The relief: this acts as a restrictive element for the expansion of the built-up area and as a favourable factor for the pollution owing to the siting of the town in a depression. The capacity for automatic air purification is diminished and the polluted substances can stagnate for a long time.
- The climatic characteristics: frequent thermal inversions in winter, the predominantly calm wind conditions (64.4%) and the low speed of wind which blows towards the axis of the valley where the town lies, are favourable factors for air pollution.

The concentration of gaseous toxins in the air shows values which are within the admitted limits with the exception of some accidental emissions of SO_2, emitted by CSR in the year 1992 (Fig. 1). This was because pollutants were released directly into the air without filtering. The highest values of sediment forming powders were registered both at mean and at maximum measured values (Fig. 2).

West University of Timisoara, Faculty of Chemistry-Biology-Geography, Geography Department, Pestalozzi 16, 1900 Timisoara, Romania

J. Breuste, H. Feldmann, O. Uhlmann (Eds.)
Urban Ecology

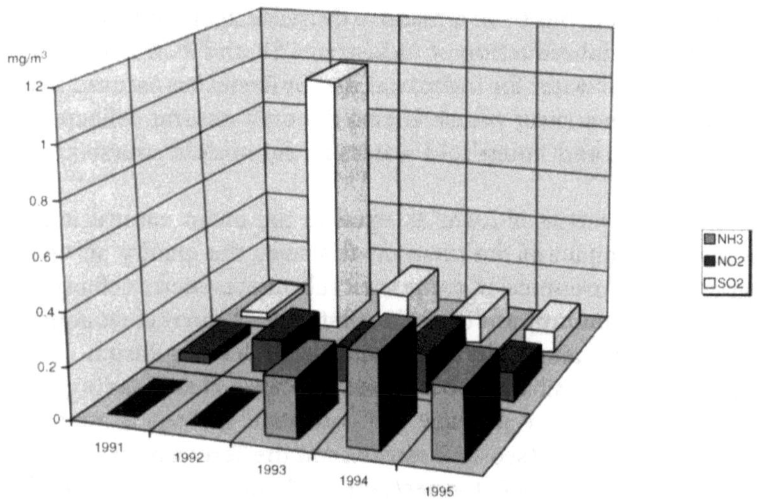

Fig. 1. Maximum concentration values

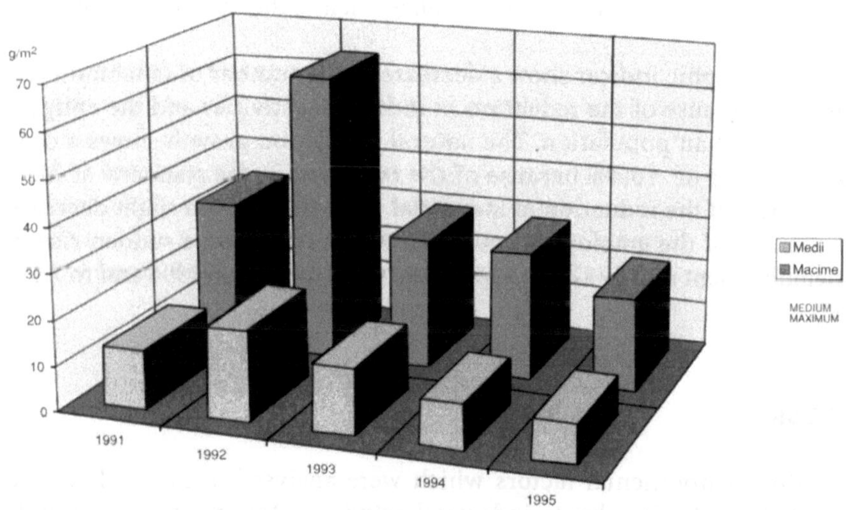

Fig. 2. Concentration values measured for a polluting substances (1991–1995) of sediment dust (1991–1995)

In the last few years there has been a massive decrease in air and water pollution because of the partial reduction of industrial CSR and ICMR activities.

The drinking water and water for industrial uses in Resita are assured from the upper basin of Barzava river, which is also the only natural collector of used industrial, meteoric and household waters in the middle course of the river valley.

The flow of the used waters of $2.4\,m^3/s$, equal to the mean natural multiannual flow, shows the impact of the town on this river, the quality of which is strongly degraded. The presence of some toxic substances such as chanures, phosphors, phenyls, ammonium, and organic substances mean that no aquatic life can be found for 15 km along the river course after leaving Resita.

The urban landscape of Resita is also degraded because of soil loading with a mixture of very old and newer garbage. The industrial garbage is made up of sterils, slags, ashes, slams etc. (see CSR sterils), and the Terova mining pond. The household garbage of 20,000 tonnes/year is deposited in the nearby residential zones in unsorted deposits and is a major source of pollution.

The large area of forests surrounding the town, although they are affected by air pollution (about 547 ha are under the threat of drying out), constitute a source of oxygen and provide a favourable micro-climate.

Most of the town's pollution is due to the main industrial economic agents: CSR, ICMR, PRESCOM (communal management), the thermal plants and automobiles.

The demographic indices show a decrease in the number of inhabitants by 15% in 1995 because of the reduction in industrial activities and the emigration of the German population. The natural population growth shows a continuous decrease of −0.4% because of the reduction in the standard of life.

As an effect of the reduction in industrial units after 1990, a slight decrease in the number of the employees can be seen, and after 1992 a sudden rise in the unemployment rate to 12.4%, which decreased to 5.3% in 1996 and to 7.5% in 1997.

3 Conclusions

Among the environmental factors which were analysed, most of them are favourable for pollution due to industrial activities. Only the forests provide a source of natural regulation in the hinterland of the town, supplying one of the basic values for future ecological development.

References

Data from the Environment Agency Caras-Severin County
Data from the Meteorological Service Resita
Ianos I (1986) Consideration sur l'organisation de l'espace urbain du municipe de Resita, Revue Roumaine de Géologie-Géophysique et Géographie, Tome 30

'Habitat Ecology' IM-192 & IM-32TP as a Small Town of the Future

Ivica Mladjenović

Habitation in family or collective houses with atriums is the theme (also the doctoral thesis) which the author of 'Habitat Ecology' has pursued for 30 years. The advantages of constructing apartments with their own winter gardens on the roof of the building or next to it have been researched on two different models of collective atriums housing (with similar plans).

I am not quite certain if today we can accept the following premise: the cities of the 21st century will probably retain the urban matrix of the same cities from the past centuries. The reason for a slight doubt rests on the fact that in all likelihood the present motor and pedestrian traffic will be essentially different from the traffic at the end of the second decade of the next century.

Today, as passengers, we cover the continental distances between two cities or two regions by buses, cars, trains or ships. We use planes to span great distances as fast as possible. Walking is a necessary evil.

When one looks out from the low flying plane to a densely populated region, a grid of numerous roads is first seen on which thousands of buses, cars or trains travel. By 2030 all those roads will be used by a ten times smaller number of buses, cars or trains. People will move in a standing or sitting position individually in special space suits, working through their own energy-powered propulsion. They will not have to use roads, but will move in a way similar to that of the American missile Tomahavk. Petrol will no longer be used. Tankers will be turned into ship-hotels.

Obviously, the network of roads will be superfluous. On large roads with two lines over 20 m wide vehicles will use only one present line for both directions. On the other line, and on the surface beside it, linear cities will be built. In these linear cities the 'Habitat Ecology' project will considerably change the housing design.

Cities of today are designed in such a way that housing is separated from buildings which serve other purposes. Industrial sites are usually situated on the outskirts of the city. Administration buildings, trading and cultural centres,, science, arts and sport centres are found in the city centres. Sometimes the groundfloors of residential buildings are used for shops, etc.

Architect, Bulevar Avnoja 113, ulaz 1, 11070 Beograd, Yugoslavia

J. Breuste, H. Feldmann, O. Uhlmann (Eds.)
Urban Ecology

'Habitat Ecology' IM-192 is a community, village or city in which all the functions of life are united in one place. 'Habitat Ecology' IM-192 includes a community, village or town with at least 100 and at the most 10,000 inhabitants (Fig. 1).

Within one 10 km long linear city it is possible to place 20–60 'Habitat Ecology' IM-192 communities, villages or small towns which are sustainable, even without a strong mutual link.

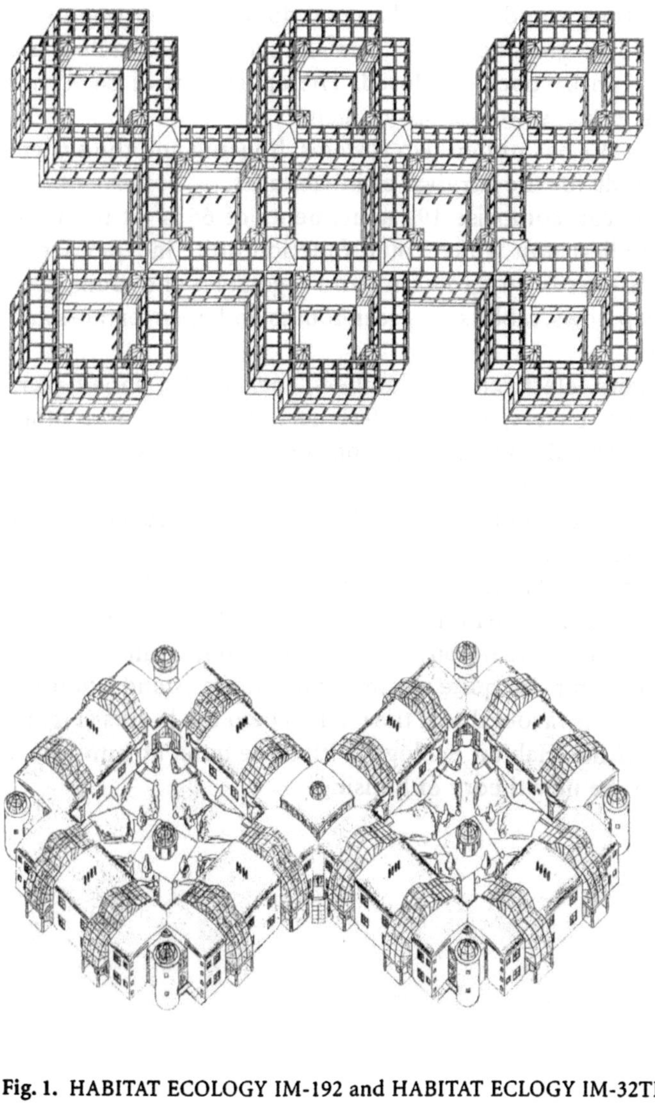

Fig. 1. HABITAT ECOLOGY IM-192 and HABITAT ECLOGY IM-32TP

The first 'Habitat Ecology' IM-192 is a clearly defined model of 192 (124) apartments of 66 m^2 to 196 m^2. It can be symmetrically or asymmetrically expanded on all sides. It comprises 8 structures-blocks with 9 completely closed atriums and another six atriums open on one side, i.e. 8 structures blocks use 15 atriums altogether.

The first variant is characterised by:

1. Flat roofs with open or closed areas for greenhouses, ateliers, workshops, etc.
2. Mini towers which serve for storing agricultural products, tools, etc.

The apartments are on the ground, first and second floors. All apartments on the second floor have closed greenhouses of 33 m^2. Above the second floor are gardens. These spaces can be open or closed. The apartment of 66 m^2 has another 66 m^2 with the gardens, workshops or ateliers.

'Habitat Ecology' IM-192 has united two systems: housing and the system of activities which provide for the survival of inhabitants.

The displayed model can comprise 192 apartments of 66 m^2 at the most and is probably the smallest unit which could function quite independently as a village or town. It is possible to grow fruit and vegetables and grain, keep domestic animals (sheep, cows, rabbits) in greenhouses and open terraces and keep fish in the ponds.

Joiners/cabinet makers, electricians, plumbers, locksmiths, shoemakers hairdressers, beauticians, bakers, fashion designers, tinsmiths and other craftsmen will have their small kiosks-workshops on the terraces which will be mutually interlinked. Of course, they will all help those people who live on the ground floor, first and second floor. It would be ideal if craftsmen themselves lived under the terraces.

The second variant form 'Habitat Ecology' IM-32TP (co-author is Tibor Palankai from Hungary) is considerably different from the first. Here each apartment has its own closed greenhouse. Big spacious halls run along the staircases. The halls can comprise quite a number of kiosks (stands, booths) with different functions, it is favourable if the tenants own the kiosks (Fig. 1).

Both variant forms of the 'Habitat Ecology' with large quiet atriums are, I hope, a contribution to the new theory of housing.

Urban Poverty and the Environment in Botswana

Aloysius Clemence Mosha

1 Introduction

Over the years, Botswana, in southern Africa, has recorded good and impressive economic growth through its production of minerals and the cattle industry. Its successes have surpassed those of the neighbouring countries. Urban areas have been the main beneficiaries of economic success, and urbanisation has been extremely rapid. While population in designated urban areas now comprises 24% of the total population, more rural areas have acquired an urban character, with 45% of the population fulfilling the relevant demographic criteria. However, not all the urban people have been beneficiaries of the rapid economic growth, and urban poverty is slowly and noticeably creeping in. The latest Income and Expenditure Survey carried out in 1993/94 shows that for the country as a whole 59% of households live below the Poverty Datum Line(PDL) and in the urban areas the index was 30% (Central Statistical Office, Gaborone 1995).

The primary cause of this urban poverty is the phenomenon of rural-urban migration, natural urban growth, lack of employment opportunities, and low incomes. As long as these problems remain unsolved, the plight of the poor gets worse by the day.

2 Environmental Problems Caused by Urban Poverty

Poverty has led to social as well as environmental impacts. The poor have limited resources in the urban centres; they depend on wood for fuel and, already, wood cutting for cooking and lighting is depleting woodlots near urban areas. In the absence of conscious measures to preserve and replenish these resources, the combination of poverty and population pressure will result in resource depletion and further degradation of the natural environment.

University of Botswana, P. O. Box 00 22, Gaborone, Botswana

J. Breuste, H. Feldmann, O. Uhlmann (Eds.)
Urban Ecology

The other impact of poverty is the invasion of open spaces by the poor which creates havoc. Since the poor cannot obtain access to commercial land in the urban areas, it is now common to find that they have moved into any open space they find in the urban area, be it a recreational park, land zoned as a botanical garden, playgrounds, tot-lots in residential neighbourhoods, right-of-way for infrastructure like roads, electricity, water or sewerage, or flower gardens. Here they establish all sorts of businesses to eke out some sort of living. Vendors erect wooden, plastic or discarded metal shacks or place caravans on open spaces where they offer canned drinks or sell cooked tripe, chicken, papa, mealies, pirated music cassettes etc. Patrons who visit such places discard much litter and this is left lying around on these open spaces. The worst sites are bus or railway termini. Repeated rounding up of these vendors has not yielded any solution.

In recognising the importance of an informal sector, the government needs to give more attention through its physical planning activities to make suitable land available for informal sector activities. These should be located in appropriate areas where it is profitable for business.

Poverty has also had an impact on urban ecosystems. As a result of shortages of appropriate land (in spite of the Accelerated Land Servicing Programme) or the high economic rents on land, poor urban people in Botswana have been forced to settle in less hospitable areas such as flood plains, as in the case of the squatter areas of Masimanyenga, in Francistown, poorly constructed shacks or huts in places like Naledi in Gaborone, on the peri-urban areas of the big towns like Tlokweng or Mogoditshane, or on vacant land adjacent to dangerous industries like in Selebi Phikwe. Squatter areas or unplanned settlements are slowly growing in and around the cities and towns. The result is that land is being used haphazardly; plots are changing land uses without permission; infrastructure is hardly being provided in such areas ostensibly because these areas have no approved land use plans guiding their development as is the case in planned areas. Recent cases of squatter encroachment have been reported from Sowa Town, Jwaneng and even near the new military airport site. People are allocating tribal land near urban areas themselves if they cannot get it legally, as reported by the Kgabo Commission of 1992.

The physically precarious conditions typical of these urban squatter settlements and the rapidly growing concentration of poor people in them combine to magnify the vulnerability of the urban poor to a broad array of ecological crises, both natural and man made.

As we also know, cities and towns create enormous amounts of waste, the disposal of which is one of the leading challenges to developing sustainable urban environments. Some of the poor urban families and their children earn their livelihood in the waste removal business, which can involve the disposal of highly toxic drums or other hazardous wastes that carry bacteria and disease. Again the leachate from the dump site can endanger people (and their environment) living downstream of the site as the poisons seep into the soils and contaminate the ground water supply.

Other environmental consequences of poverty include: urban agriculture practised by the poor, which in some places is carried out in poorly located sites; small-scale industrial activities carried out in residential areas and in open spaces, thereby causing a lot of inconvenience to residents; car repairs and other activities carried out everywhere, and on-street and pavement vending which causes traffic hazards in almost every town.

3
Conclusion: The Policy Framework for Urban Poverty Alleviation

From the above facts it is obvious that urban poverty is manifest in Botswana and is causing environmental problems. What needs to be done is to come up with policies to alleviate the plight of the poor. To-date several poverty reduction strategies have been adopted both by the government, the local authorities, NGOs and the general public themselves. These efforts are already bearing some fruit and should be encouraged. These need improving to achieve better success in the future.

Green Areas in the City – Acceptance of Changes. Halle (Saale) as an Example

Renate Patz and Ilona Kuhpfahl

1
Introduction

Natural processes and human activities continually change green and open spaces. We focus on such *changes in public spaces* which are planned and carried out by municipal agencies or by private initiatives. The success of changes depends on acceptance by the residents and the users of the green areas in general. The *acceptance* depends on the concrete situation (current situation in the green spaces in the city or neighbourhood, special situation in former east Germany etc.). It depends on the specific space and time as well as on specific persons and social situations, and on the degree of public participation in planning and transformation processes. All people can use green areas but the users do not usually plan or transform these areas.

The purpose of our investigations was to determine public acceptance of changes in various green areas in the city of Halle. *The large new housing estate Halle-Silberhöhe* is one of the districts in which various problems are concentrated. The housing in Halle-Silberhöhe was built at the beginning of the 1980's and is dominated by 5, 6, 7 and 11-storey apartment houses. 36000 inhabitants live in an area of merely 2 square kilometres. The *key problems* are:

- quantitative and qualitative deficits in green and open spaces: a great part of the open space is wild undeveloped land.
- Few playgrounds: only 7% of the required number of playgrounds exist.
- The traffic problem: over 6000 parking spaces are lacking, so that many cars park on green areas.

One objective of district development is to alleviate these problems. One of the *current projects* is the realisation of a *new green belt*. This municipal project started in 1994, and the first sections were realised in 1995. Another project is the *building playground*, a kind of adventure playground, which

University of Applied Science Merseburg, Project Group Technology Assessment, Geusaer Strasse, D-06217 Merseburg, Germany,

J. Breuste, H. Feldmann, O. Uhlmann (Eds.)
Urban Ecology

offers alternative recreation. Children and youth transform the playground according to their own wishes so that its appearance changes continually. This project was initiated in 1995 and counselled and supervised by students of the University of Halle. (It should be mentioned that the acceptance of this project decreased in the course of time. There were cases of vandalism and fights between children and youths so the project ended in 1996).

In our investigations we surveyed all 860 households next to the new green belt and around the building playground about those two projects. 219 adults and 96 children answered a standardised questionnaire.

2 The Answers and Some of the Results

- Green areas are rated as a significant factor of quality of housing and recreation.
- The assessment shows that city children need a place within the urban housing environment where they can play in the open air and experience nature.
- Other problems are rated to be much more acute, for example unemployment, crime or not enough places for vocational training.
- Residents want to have many parking lots, more trees and playgrounds, more noise prevention and more information.
- The residents are uncomfortable with the dirt in the streets and green areas.
- The building playground was rated as an interesting project with a wide spectrum of good ideas and activities. Nevertheless it was assessed to be a rather chaotic and noisy.
- Many residents are willing to do something to their surroundings in the neighbourhood without payment, for instances to care for their "own" tree.

3 Some Conclusions and Remarks

Renewal of large, new housing estates is a long term ecological and social task. Regarding alterations in spaces with as many deficits as those in Halle-Silberhöhe, nearly every change improves the situation and their acceptance is good. If residents are integrated into the renewal process and the transformation of their surroundings then acceptance, as well as satisfaction, with changes may increase.

(Sustainable) urban development is to a large extent dependant on social experiments and on "social spaces" where there is room for new experiences such as the building playground. However, such initiatives are under tremendous pressure because they take place in areas with social problems. Future urban development requires not only a variety of green areas in the city but also has to take into account a plurality of lifestyles in all facets of society. In our opinion, the alternative should not be to decide between a normal playground or a building playground, but between a parking lot and a playground.

Land Use and Environment of Mendoza, Argentina

Claudia Marcela Polimeni

1
The Environment

Mendoza is situated close to the Andes in the proximity of the west border of the precordillera, at 800 m above sea level.

The mountain streams are true uadis that appear during the strong summer rainfalls, causing serious floods.

At 32° 53 south latitude, Mendoza has a dry climate, with only 220 mm annual rainfall, mainly between October and March with a mean temperature of above 20°C but with great thermic amplitude, the absolute maximum being 42°C (January) and absolute minimum 6°C (July). The winds are sometimes strong and they carry the dust of the dry bed-rivers. The Andes constitute an impassable barrier for the winds of the South Pacific anticyclone. The dryness is the most characteristic feature of the city, so water and fertile land are invaluable resources for the inhabitants.

2
Land Use and Different Growing Stages of the City

Mendoza has not escaped the phenomenal global urban growth. Its growth commenced in the 1960's due to the industrial expansion and the immigration of rural inhabitants to the cities.

The accelerated development of the city trapped the industries in urban areas thus mingling administrative, commercial, cultural, financial and industrial functions with the resulting risk for the population and its surroundings.

The urbanisation of Mendoza reveals an unplanned development. Up to 1945 only secondary centres existed, far away from the capital city and isolated among the vineyards.

Between 1945–1962 the cropping area was reduced and Godoy Cruz, Las Heras, Guaymallén and Maipú were consolidated with a resulting

Universidad Nacional de Cuyo, Instituto de Geografía, Fac. de Filosofía y Letras, Corrientes 122, 2° piso, Dpto A, 5500, Mendoza, Argentina

J. Breuste, H. Feldmann, O. Uhlmann (Eds.)
Urban Ecology

population of 500,000. From 1962 onwards all the empty spaces were urbanised.

The expansion has been carried out at the expense of the crops and following the axis: Av General San Martin y Ruta Nacional n° 7.

At this stage the occupation of the glacis began with the building of the University Campus, a high class and middle class neighbourhood amongst the existing shanty towns. During this urban period the modern motorways Acceso Este and Acceso Sur were built and attracted the creation of new neighbourhoods.

There is only one well defined industrial district in the south of Guaymallén. The wineries which have been absorbed by the growing city were originally in rural areas.

The natural vegetation belongs to the fitogeographic "monte" of the province and has been degraded by the excess of pasture (piedmont).

In short, the urban expansion has been carried out at the expense of the cropping land and is moving westwards, occupying unsuitable land such as the piedmont.

EUROPEAN RURAL AGENDA 21 towards Environmental Protection and Sustainable Development in Rural Europe

Heuitieh Seul and Christiane Wellensiek

1 The Problem

Rural Europe is experiencing a deep crisis. The main driving forces of spatial, economic, ecological and social development face severe structural problems. The exodus of rural residents is due to constraints in infrastructure, a severe rural job crisis and few future perspectives. These developments could have strong negative impacts on the ecological balance of the landscapes and natural resources in the European Union. A coherent programmatic instrument for environmental protection and sustainable rural development is needed to influence positively this situation and meet the targets of sustainable development. However, at present local sustainability initiatives in Europe predominantly originate from an urban background. Instruments and strategies developed here may not be appropriate for rural situations in all cases. This paper calls for the development of an agenda towards environmental protection and sustainable development in rural Europe, an „*European Rural Agenda 21*".

2 The Task

Rural sustainability requires the integration of interests and new perspectives on linking profitable economic development, sustainable management of natural resources and local participation in an European context. Common visions on rural sustainability, common policies for sustainable development in rural Europe, practical cross-sectoral strategies to integrate the manifold conflicting rural interests, and a solid and transferable methodology are needed. In commercial terms, a political and administrative framework is needed to make practical contributions to rural sustainability more competitive and thus more profitable. Rural interest groups have to be motivated and guided to start shaping their sustainability targets and

CREAM Consultants, Gut Fahrenbach, D-37216 Witzenhausen, Germany

J. Breuste, H. Feldmann, O. Uhlmann (Eds.)
Urban Ecology

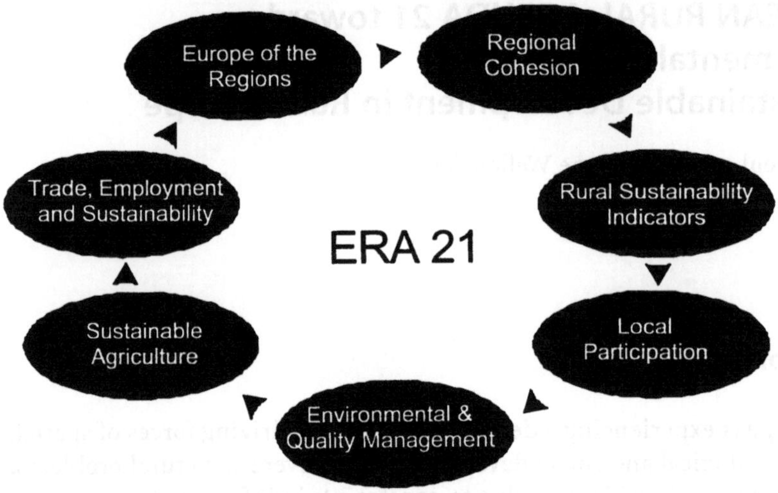

Fig. 1. Key Elements for a „European Rural Agenda 21"

processes. Main rural actors include local authorities, business sector and rural communities.

3 The Approach

In practical terms, a methodological solution to multi-sectoral integration is needed which combines sustainability goals with competitive market advantages. A *European Rural Agenda 21* should address and integrate seven key elements (Fig. 1).

4 Practical Goals

ERA 21 should help rural interest groups in their need to: react to the growing importance of regions for the European process; enhance regional cohesion and co-operation between urban centres and the rural periphery; develop specific, objective, applicable and measurable indicators for rural sustainability; include all local actors in ownership for the process and decision making; apply environmental and quality management systems on all levels; develop concrete goals and an action plan for sustainable key industries; create and market the positive synergism between trade, employment and sustainable development.

References

European Commission and European Parliament (1997) Environment and Employment. Conference Proceedings, Brussels

Organisation for Economic Development and Co-operation (1997) Sustainable Development - OECD Policy Approaches for the 21st Century. OECD, Paris

Organisation for Economic Development and Co-operation (1997) Environmental Benefits from Agriculture. OECD, Paris

United Nations (1992) Environment and Sustainable Development, Conference Proceedings, Rio de Janeiro

3 Land Use as a Controlling Factor in Terms of Urban Ecology

Compensation in Land-Use Planning – a Framework for Sustainable Land-Use within an Hierarchical System of Spatial Planning

Ortwin Peithmann

My starting point within 'Urban Ecology' is the view of a planner who has to find out which spatial planning methods have to be created to satisfy the needs of 'sustainable development'.

The discussion of the term 'sustainability' in the report of the Brundtlandt-Commission (1987) and in the AGENDA 21 has created the idea of the 'Environment-Space'. The concept of the 'environment-space' means to allocate natural resources to national territories in such a way that nations are able to function as materially closed systems – as far as possible. In the model of the 'Environment-Space' there are enormous demands for changes of life-style. The Friends of the Earth in the Netherlands (ISOE 1992) and the Wuppertal Institute in Germany (BUND/MISEREOR 1995) have calculated that it would be necessary to have several times as much area as is available in Europe for realising our life-style in closed natural circuits.

What are the consequences for future land-use? A system for rationing land-use has to be built up, which is based on the fact that land is finite and regions have optimal carrying-capacities. If this rationing is to be oriented toward sustainability then it has to be carried out in a very rigid and unfamiliar manner. The Wuppertal Institute for example recommends an immediate stop of all further extension of areas for settlement and traffic.

How are we going to influence this drastic decrease of space consumption? The traditional instrument for controlling land-use is 'spatial planning'.

1 Spatial Planning

The system of spatial planning has continuously been developing since the turn of the century and has shaped the current settlement structure. There could be two reasons why, at present, the actual land-use does not meet the demands of 'sustainable development':

Hochschule Vechta, Fakultät Raumplanung, Post-Box 15 53, D-49364 Vechta, Germany

J. Breuste, H. Feldmann, O. Uhlmann (Eds.)
Urban Ecology

1. Spatial planning has followed the wrong aims up to now; – conclusion: we have to change the objectives in order to obtain better results.
2. Spatial planning is not up to the demands for a reduction of space consumption, i.e. it cannot manage it on its own and has to be assisted by additional instruments.

My judgement in advance: the second reason is correct, spatial planning has structural deficiencies.

The deficiencies are not a result of the goals of spatial planning; these are acceptable. I refrain from citing the relevant passages in the German law for town and country planning. The actual problem is the degree of precision in which land-use is prescribed in spatial plans. Their contents only say approximately which actual use can be permitted for a specific area. They do not say at which degree of settlement the maximum capacity of an area is exhausted. Why does spatial planning not prescribe the use of land as precisely as it would be necessary for sustainable land-use?

The answer is that this is due to its liberal or democratic character. Spatial planning is drafted as an hierarchic framework of plans, fully integrated into our democratic system with a vertical division of powers. It is based on subsidiarity. This means that each level of the hierarchy only regulates to the extent that the next lower planning-level is not, itself, able to regulate. A rather loose or slack framework of binding contents of the plans has emerged within this field of overlapping competencies on the national, regional, local and private level. A suitable picture for the coherence within this system would be a loose pile of cardboard boxes or a slipping clutch.

In practice this means that there is a structural lack of possibilities to regulate the use of space in detail in a top-down-approach. The region or state has no right to strangle the development of a local municipality. The hierarchical system of spatial plans has been invented to regulate the functioning of growing settlement and infrastructure. It has not yet been adapted to the need for reducing additional land-use.

On the other hand we cannot trust in the capability of municipalities or private enterprises to self-restrict their land-use facilities. The wise piece of advice 'think globally, act locally' has not yet reached the sphere of local action. Local municipalities still live in rivalry with their neighbours. They want more inhabitants, more jobs and more taxes. Space serves as a factor of production and must be available.

2 Rationing

To fit the model of the sustainable spaceship, spatial planning would have to regulate land-use by absolute limits for the strain of soils. This sort of instrument is called 'rationing'.

We are familiar with rationing as a means of regulating maximum emissions in the environmental media.

The 'bubble-concept' is the example from the field of air pollution. In some regions of the USA emissions are restricted by absolute measures. The air-space is regarded as a limited system with a restricted emission-capacity. Emissions are licensed and to keep the system flexible, the licenses can be traded, but cannot be increased.

In Germany there is a similar method of regulation for the management of river-systems: there are limits for certain pollutants and for the water-temperature, additional pollution or heating of the water is only allowed if the limiting values are respected.

A third example is the rationing of fertilisers on agricultural soils, specifically for liquid manure, to protect the quality of groundwater.

Rationing is quite a normal method of regulation for the protection of our environment, with respect to both spaces and substances. For the system of spatial planning this would mean new and very concrete contents: regional planning would have to define a maximum carrying-capacity for each of the mentioned land-uses. This capacity would have to be distributed top down to the municipalities in the region, but, what is valid for air, water and fertiliser is not necessarily valid for land-use. This concept is impracticable due to the totality of the method of regulation. That which can be justified for single environmental media becomes impracticable for the entirety of claims to space. Any method of regulation which cuts off flexibility in decision-making for local planning has no chance today to be observed.

The question therefore must be: how can we combine a strict preservation of natural functions and resources in space with sufficient flexibility for local development?

3 Compensation

The preferential means in my opinion is 'compensation'. According to the polluter-pays-principle (PPP) compensation means the obligation to anybody who affects the functions of our natural environment to compensate for the losses incurred. In other words: compensation means nature's right to reparation from man.

Such an instrument of compensation has been introduced in the German law in 1976 as part of the right of natural conservation, it is named 'Eingriffsregelung', which can be translated as 'regulation in case of interference and impact resp.'.

The idea of compensation can be described by a simple formula:

Performance of Natural Asset = Performance of Natural Asset
(before impact) (after impact)

From the demand for constant worth and functioning of the natural asset it follows that any change for the worse has to be compensated for by an equivalent change for the better. When applying compensation the question arises almost automatically, whether a planned project in question is needed at all and whether it can be avoided. This part of the compensation-rule is called the 'requirement to avoid'. This does not have any severe effect, but it sets the right starting point for the course of the compensation rule.

The compensation rule includes all elements of the natural asset. Its application in practice therefore demands comprehensive assessment, as is known from the Environmental Assessment Analysis (EIA).

The main methodological problem is how to deal with impacts which cannot be compensated for in the same way as the project affects nature. Two examples may illustrate this:

1. When old trees have to be cut down they cannot be replanted at the same age elsewhere. How does one decide? Normally, in the practice of compensation, a multiple number of young trees have to be planted on an accordingly larger area.
2. When a site has to be built on, on which no buildings have been, the surface cannot be enlarged by a new unused site. In all cases of sealing soils another piece of soil has to be improved. If for example farmland is turned into settlement, other farmland can be improved in its ecological value by developing a meadow with multiple species of plants to become a habitat for insects and birds.

The following formula explains this coherence:

$$\underset{\text{(before Impact)}}{\boldsymbol{\Sigma}(\textbf{Area * spec. ecol. worth})} = \underset{\text{(after Impact)}}{\boldsymbol{\Sigma}(\textbf{Area * spec. ecol. worth})}$$

In practice planners work with similar formulas. The parameters are adapted to the special conditions of regions. To facilitate the job there are lists of biotopes with assigned values per acre. I must confess that these lists are just expert-guesses for planning uses, not the results of research. As soon as one leaves the sphere of compensation by identical functions and places one leaves behind the sphere of science and enters the sphere of compromise.

The practice of the compensation rule has been severely criticised by ecologists, but compromise is quite natural in the field of compensation. One can look at the case of insurance: to recover physical damages certain percentages of the sum insured are assigned for certain injuries: 100% for the loss of both eyes, 10% for the loss of a thumb. From the scientific point of view the worth of such features is intangible but in the practice of jurisdiction the pain of slander can be covered by a certain amount of smart-money. If this is carried out for human beings, why not do the same for nature to benefit both nature and mankind.

I have chosen the plan for the test-plant for vehicles of the famous make Mercedes-Benz to show a concrete example for the effectiveness of the compensation-rule. This circuit of 6 km length, and many additional small testing areas, is a comparatively large project with numerous impacts on the nature of a municipality in the west of Lower-Saxony. It has been planned in an area which had been used by agriculture and where peat has been exploited in recent years. Despite this initial deterioration of the area, the compensation for the planned test-plant led to a considerable amount of ecological improvement around the project site (Fig. 1). This example implies a sort of double compensation effect. It shows that large companies with well known products on the consumer-market agree to realise the necessary compensation, and to compensate for their non-ecological image.

The compensation-rule has only been applied to single projects since 1976. Since 1993 it has been extended to the plans of municipalities, that means to the legally binding plan and the preparatory land-use-plan. The topical amendment to the Federal planning code proposes a further extension of the compensation-rule to the level of regional planning. These extensions are steps toward its complete integration in the system of spatial planning.

4
Environmental Impact Assessment

I shall not finish without a look at the link between the compensation rule and EIA where there is a strong symbiosis. The advantage of EIA is that it forces authorities to analyse the impacts of projects in a comprehensive and thorough way under public control. The disadvantage of EIA is that it does not create any material standard for the protection of the environment itself. In the worst case, all the analyses produced on the basis of EIA-regulations may have no effect in the project.

Compensation, on the other hand, supplies decision-making with a strong material claim. The lack of precision in analysis which is often criticised by experts with regard to the practice of the compensation rule, can be reduced by the aid of EIA. The new guideline of the European Commission, which proposes the extension of EIA to plans and programmes, will ensure a good teamwork of EIA and the compensation rule in future planning-processes.

5
Résumé

Land-use plays an important role as a controlling factor in meeting the requirements of sustainable development. The pressure of problems has not yet become so high, that our liberal political system, which is based on subsidiarity, controls the detailed use of land. Land-use continues to be regulated

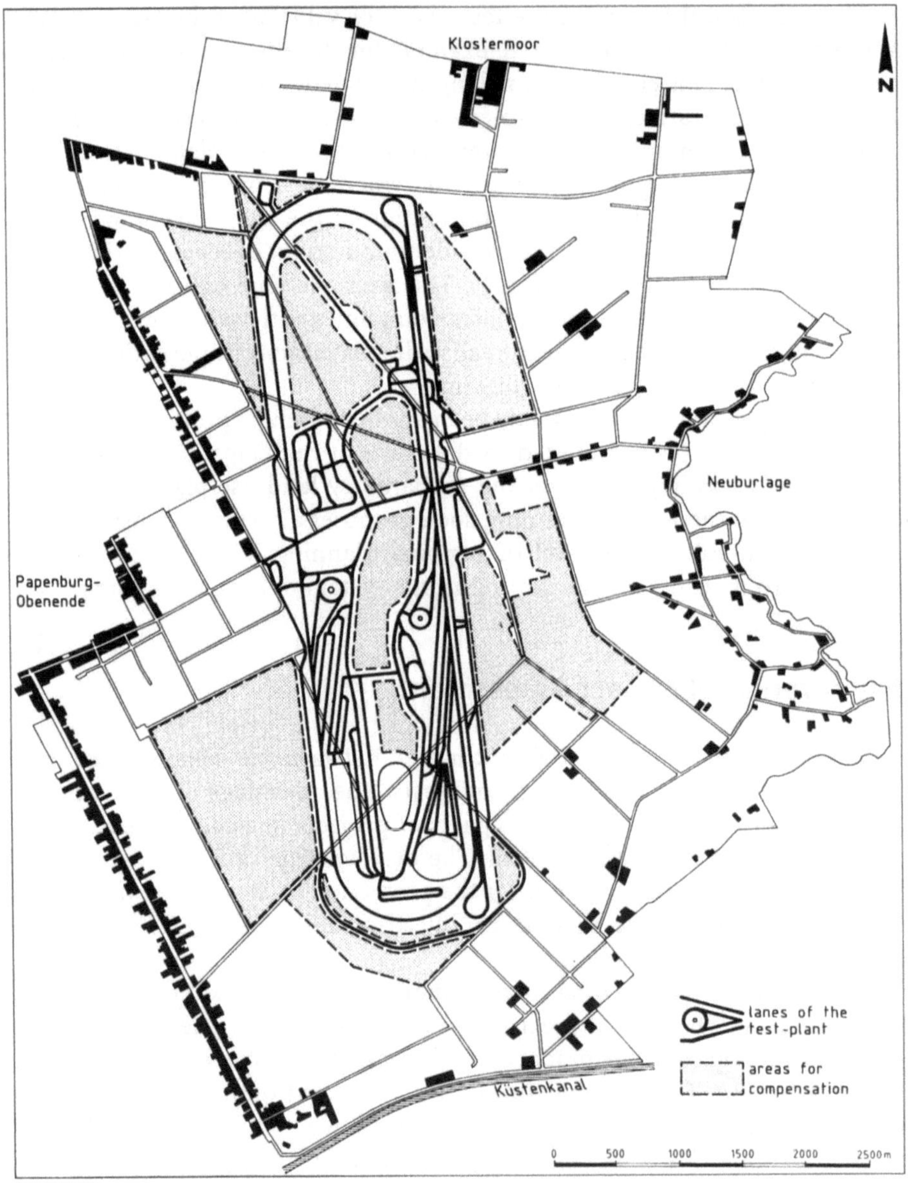

Fig. 1. Compensation for the test-plant of Mercedes Benz in Papenburg (Germany)

by a slack framework within an hierarchy of spatial plans. Under these circumstances compensation can solve part of the problems on our way towards greater sustainability.

Which are the main advantages of compensation?

1. Compensation maintains flexibility in land-use. It does not strangle development, but demands a maintenance of the level of ecological qualities.
2. The compensation constraint leads to higher costs for settlement projects on unspoilt grounds. This provides chances for the renewal and the recycling of already existing settlement areas. As the costs for compensation have a tendency to increase, even the chances of redevelopment of abandoned plants and dumps multiply.
3. The need for compensation is very convincing. All those people who accept compensation as a means of regulation among mankind to provide justice will easily accept this principle in the context of nature conservation and sustainable development.
4. The fact that the compensation rule can be easily understood, adjusted to any region and applied to easily accessible and controllable land-use-data make it an ideal item of international consensus. I hope that it will become the underlying principle for the discussion of land-use-planning on the basis of AGENDA 21.

References

BUND/MISEREOR (Hrsg) (Oktober 1995) Zukunftsfähiges Deutschland. Ein Beitrag zu einer global nachhaltigen Entwicklung, eine Studie des Wuppertal-Instituts im Auftrag von BUND und MISEREOR, Bonn und Aachen

Institut für sozial-ökologische Forschung (ISOE) (Hrsg)/Milieu defensie (Friends of the Earth Netherlands) (1992) Sustainable Netherlands - Aktionsplan für eine nachhaltige Entwicklung der Niederlande, Niederlande, übersetzung Frankfurt ca. 1994

A Procedure of Compositional Data Analysis for Land Use Planning

Tomohiro Ichinose[1] and Kazuhiro Katoh[2]

1 Introduction

Ecological analysis is essential for environmental assessment and planning. Based on ecological analysis we can recognize the ecological situation of the study area, and understand how environmental factors should be controlled to improve the ecological situation. Biological indicators have often been referred to in case of environmental evaluation. Various kinds of indicator organisms and methods have been proposed and practically used. For example, the simplest indicator is the number of species. But species richness can not always evaluate the quality of an area as a habitat appropriately. Especially in urban and suburban areas, species rich habitats are usually boundary habitats.

We should also keep in mind that ecological communities are influenced by various environmental factors. Major factors influencing communities should be detected and controlled in environmental assessment and planning. Among various ecological analyses, we focused on species compositional analysis, which has been the most popular way in community ecology. In species compositional analysis spatial and/or temporal variation patterns are detected and related to environmental factors to understand how community composition is determined.

2 Study Objective and Procedure

Among the ways of species compositional analysis, multivariate analysis is a quite popular one and has been used increasingly. It is said that multivariate analysis is useful for detecting spatial pattern of ecological changes and

[1] Munich Technical University, Chair of Landscape Ecology, Am Hochanger 6, D-85350 Freising-Weihenstephan, Germany
[2] The University of Tokyo, Experimental Station for Landscape Plants, Hata-machi 1051, Hanamigawa-ku, Chiba 262-0018, Japan

J. Breuste, H. Feldmann, O. Uhlmann (Eds.)
Urban Ecology

factors behind them (Gauch 1982) and available for selecting important habitats, monitoring species and controlling measures (Whitehead et al. 1992; Sætersdal and Birks 1993). Recently multivariate analysis has been used in many studies.

Several studies showed that multivariate analysis could illustrate spatial pattern of community variation and point out important environmental factors without detailed knowledge of species biology (ex. Kremen 1992; Debinski and Brussard 1994). Though most of such studies have been carried out in large natural areas, the methodology appears to be effective also in urban and suburban areas when appropriate environmental variables, especially variables indicating human activities and impacts, are included in the analysis.

Our study objective is to propose a procedure of ecological data analysis for environmental assessment and planning by using multivariate analysis of species compositional data. Recently, environmental or geographical information systems have rapidly become popular. Data in such systems are usually prepared as grid-cell data, which are quite suitable to numerical analysis. In Japan several kind of such data systems have been constructed, the practical application of such data to environmental planning is still few. In the case study carried out in the present study we tried to analyze species distributions based on grid-cell data system.

Our procedure can be divided into five steps. At first, we should prepare a set of species compositional data within the study area. The way of field survey depends on what kind of organism is studied. Though both quantitative and qualitative data can be used, we would explain the case in which quantitative data are provided. Anyway, field survey should cover the whole range of the study area. Based on the result we can prepare a set of grid-cell data indicating species composition of target organisms.

After the preparation of the species compositional data as grid-cell data, the cells are classified into several or more groups based on the compositional similarity. Multivariate classification technique is suitable for this stage. Among classification methods, cluster analysis and TWINSPAN (Two Way INdicator SPecies ANalysis) (Hill 1979) are popularly used.

The result of the classification in the previous step indicates community types in the study area. Average species composition within a group indicates typical figure of each community type. In case that each community type has strong relationship with its corresponding habitat type, habitat classification can also be obtained from the result of the classification. Spatial distribution of each habitat is also estimated based on the result.

In the forth step the environmental factors causing spatial variation of species composition are detected. Discriminant analysis using environmental parameters as independent variables is recommended to operate this step. Variables that are included in the discriminant model are regarded as indicators of the environmental factors that influence ecological communities in the study area.

Finally, the places where more detail surveys and analyses are desirable are selected based on the discrepancy of actual status and estimated status from the discriminant model. In such places factors other than those included in the model would be influential. These factors are usually local ones, so they can not be analyzed and evaluated in this kind on macro scale analysis.

3 Case Study

3.1 Study Areas

To test the validity of the procedure, we carried out a case study. Birds were selected as target organisms, because avian distribution clearly reflects landscape and/or habitat changes. Furthermore, birds are regularly accepted by citizens as symbols of natural landscapes. It is an important characteristic as indicator organisms in environmental management and planning.

Our study area was Tokorozawa City, Saitama Prefecture, Central Japan. It lies near Tokyo. Before the World War II most part of the city was agricultural area, however after the War residential and industrial areas have been developed all over the city, so habitats have become fragmented and isolated year by year. In spite of this situation there are not a few remnant habitats remaining (Inui 1992).

Before the analysis we provided a species compositional data in grid-cell data format for the whole city. The grid-cell has a length and height of about 500 m. Tokorozawa City consist 293. The number of observed individuals of each avian species was recorded for each grid-cell. In total 64 species in the breeding season and 83 species in the wintering season were observed (Tokorozawa City 1987).

We also prepared land use data for interpretation of the spatial variation of avian communities. Land use categories were integrated into 10 types that were woodland & forest, grassland, golf links, moist grassland, rice paddy, cultivated field, orchard, residential area with many trees, open water and residential with a few trees. To explain the land use situation within a grid-cell, a cell was equally divided into 25 sub grid-cells, of which size was about 100 m × 100 m. The number of sub grid-cells of each land use type was recorded for each grid-cell. We prepared the land use data based on vegetation maps provided by the Environment Agency, Government of Japan.

3.2 Case Study

The distribution of woods and open waters in Tokorozawa City is shown in Fig. 1. There are two small lakes and a large woodland area in the southwest part of the city. This part has been conserved as a prefectural natural park. There are some isolated woodlots in northern part of the city. These woodlots are deciduous ones having typical composition and structure of secondary forests in agricultural regions around Tokyo. Recently, they have been fragmented by various human activities. Other parts are covered with agricultural area and urban area. Tokorozawa City lies on upland, so there are few rice paddies.

Bird individuals were censused in a breeding season from May to June 1985 and in a wintering season from December 1985 to February 1986 (Tokorozawa City 1987). The grid-cells representing Tokorozawa City were divided into three groups by TWINSPAN based on avian species composition in the breeding season. Fig. 2 shows typical bird species in each group. For example, bird species listed in the top box were observed all over the city.

In group A, forest species like Short-tailed Bush Warbler (*Cettia squameiceps*) and Varied Tit (*Parus varius*) were uniquely observed. In group B species that mostly live in woodlands like Japanese Pygmy Woodpecker (*Dendrocopos kizuki*), Japanese White-eye (*Zosterops japonicus*) and Masked Hawfinch (*Eophona personatas*) appeared, while in group C species that live in agricultural or urban area like Barn Swallow (*Hirundo rustica*) and Grey Starling (*Sturnus cineraceus*) were dominantly observed.

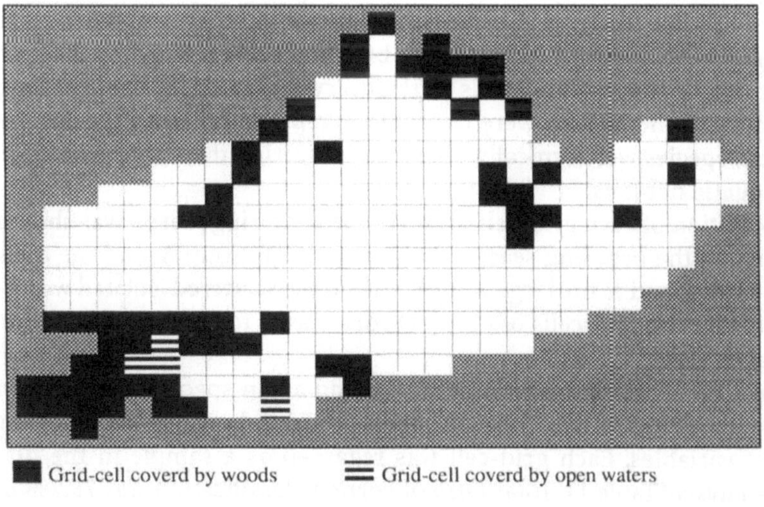

Fig. 1. Distribution of woods and open waters in Tokorozawa City

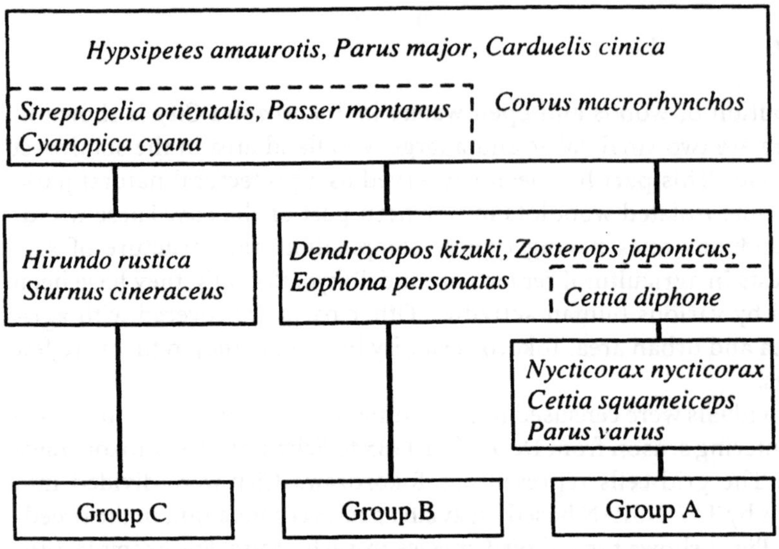

Fig. 2. Schematic diagram of the avian species compositional pattern (breeding season) The species shown in the broken line boxes were observed with more individuals in the corresponding group than in the other groups

The distribution of each grid-cell group was shown on Fig. 3. The grid-cells of group A almost corresponded to the southwest large woodland while the grid-cells of group B almost corresponded to the isolated woodlots located in the northern part of the city.

We also analyzed bird composition in the wintering season. All grid-cells representing Tokorozawa City were divided into 4 groups (Fig. 4). In group A some forest species such as Goshawk (*Accipiter gentlis*) and Goldcrest (*Regulus regulus*) were observed like in the breeding season. In group B some species that mostly live in woodlands were observed also like in the breeding season. In group C no species were observed except for common species. In group D Anas species were typically observed indicating that those grid-cells represent aquatic habitats.

The distribution of each grid-cell group in the wintering season was shown on Fig. 5. Like in the breeding season the grid-cells of group A, and the ones of group B, correspond to the large woodland and the scattered isolated woodlots, respectively. The grid-cells of group D correspond to the two lakes in the southwest area.

To find the relationship between land use and avian species composition we carried out canonical discriminant analysis using land use variables as independent variables. Each grid-cell was regarded as a sample in the discriminant analysis (Table 1). Total ratio of correct classification was 78.2% in the breeding season and 80.5% in the wintering season. These values were

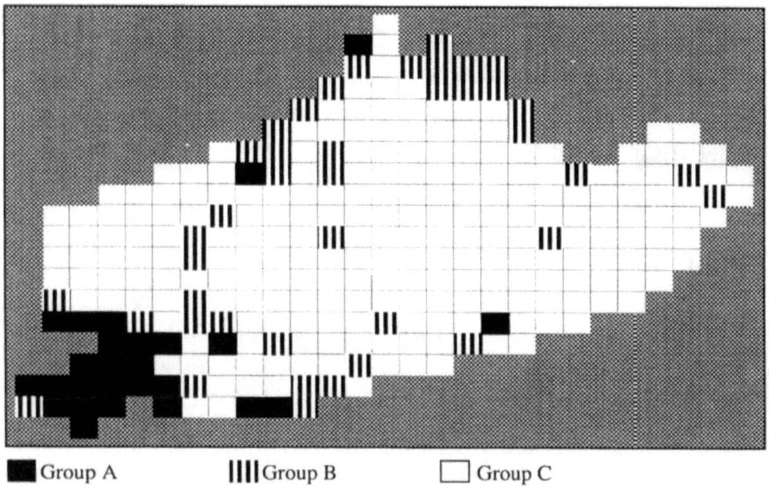

Fig. 3. Distribution of the three avian community groups (breeding season) in Tokorozawa City

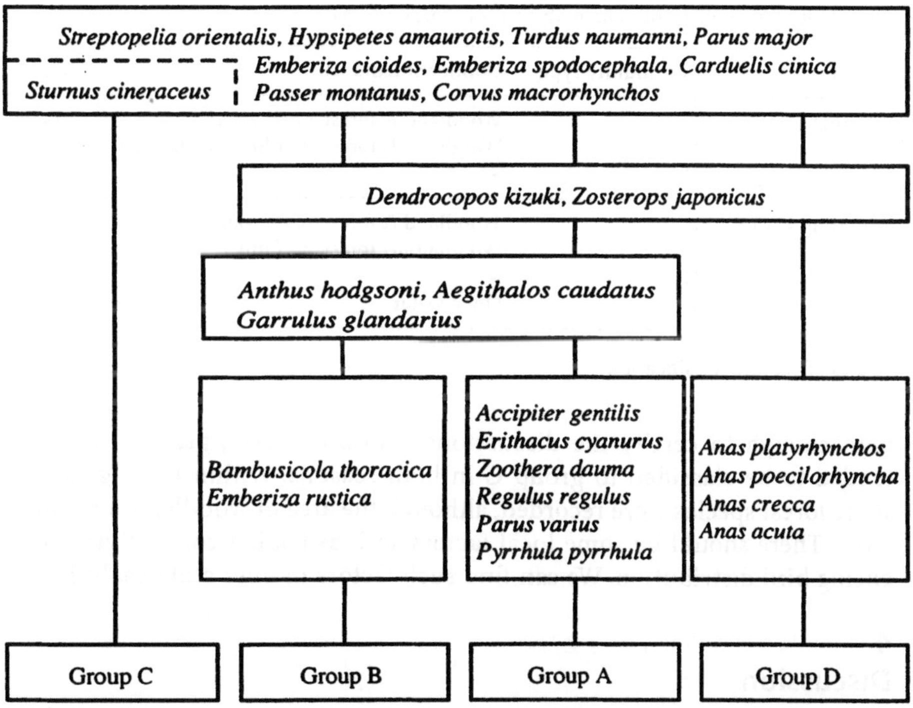

Fig. 4. Schematic diagram of the avian species compositional pattern (wintering season)

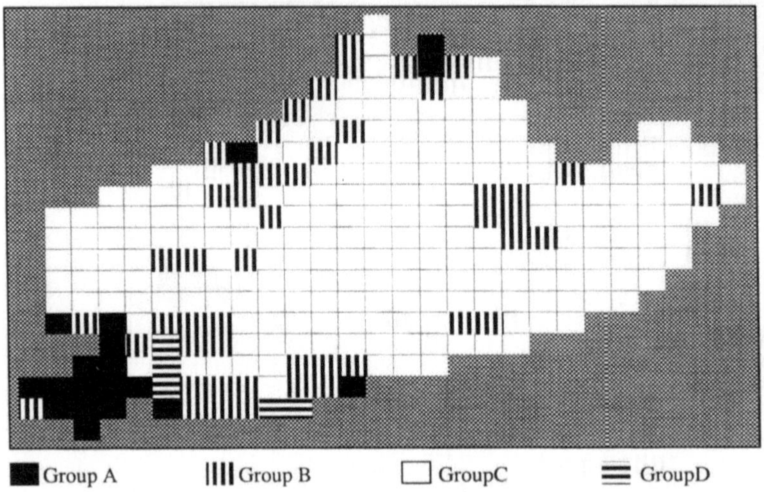

Fig. 5. Distribution of the four avian community groups (wintering season) in Tokorozawa City

Table 1. Results of discriminant analysis (Tokorozawa City)

Season	Community type	Dominant Land use	Ratio*
Breeding season	A	Woodland & forest, open water	23/26
	B	Woodland & forest, golf links, open water	21/41
	C	-	185/226
Wintering season	A	Woodland & forest, open water	15/19
	B	Woodland & forest, golf links	22/51
	C	-	195/218
	D	Open water	4/5

* Ratio of correct classification

high enough to certify the discussions above. However, grid-cell group B tended to be classified to group C in both seasons. Within these grid-cells some forest species were recorded, although the area of woodland was not so large. There should be some local factors such as habitat connectivity influencing bird distribution. We can find such factors in finer scale analysis.

4 Discussion

The result of the case study indicated that not only the largest woodland but also some isolated woodlots were important as avian habitats. The largest woodland have been and will be conserved as a natural park. However, many isolated woodlots are hardly protected, regardless of their values as avian habitats. Finally, several sites could be selected for detail surveys and analyses on finer scales.

We would like to conclude that multivariate analysis is effective to find spatial variation patterns of ecological communities, and to estimate factors behinds them. We proposed a procedure of ecological analysis using multivariate methods for environmental evaluation, planning and monitoring. Our procedure could point out important land uses influencing avian distribution.

Acknowledgments. We are grateful to Prof. Dr. Hisato Ide and Prof. Dr. Kazuhiko Takeuchi for their comments on earlier drafts of this paper. This project was supported with a grant from the Japan Society for the Promotion of Science.

References

Debinski DM, Brussard PF (1994) Using biodiversity data to assess species-habitat relationships in Glacier National Park, Montana. Ecological Applications 4: 833–843

Gauch HGJ (1982) Multivariate analysis in community ecology. Cambridge University Press, Cambridge

Hill MO (1979) TWINSPAN – A FORTRAN program for arranging multivariate data in an ordered two-way table by classification of the individuals and attributes. Cornell University, Ithaca

Inui T (1992) A geographical study on the use of the plain forests in the Kanto Plain. Kokonshoin, Tokyo

Kremen C (1992) Assessing the indicator properties of species assemblages for natural areas monitoring. Ecological Applications 2: 203–217

Sætersdal M, Birks HJB (1993) Assessing the representativeness of nature reserves using multivariate analysis: vascular plants and breeding birds in deciduous forests, western Norway. Biological Conservation 65: 121–135

Tokorozawa City (1987) Avifauna of Tokorozawa. Tokorozawa City, Saitama

Whitehead PJ, Bowman DMJS, Tidemann SC (1992) Biogeographic patterns, environmental correlates and conservation of avifauna in the Northern Territory, Australia. Journal of Biogeography 19: 151–161

Workshop: Land Use Planning and Management 3.1

The Assessment of Cumulative Environmental Impacts in EIA and Land-Use Planning

Karsten Runge

1 Introduction

Cumulative environmental impacts result from many different, often individually insignificant, effects. They are usually neither measured nor accounted for before they cause significant damage through accumulation. Although the existence of cumulative environmental impacts is known to be omnipresent, they are often overseen in environmental impact assessments (EIA) or in land-use-planning. The decrease of ground water quality and quantity, the deposition of toxic substances in aquatic sediments, the mobilisation of persistent or bioaccumulative substances, fragmentation and damage to habitats, loss of soil quality and the 'greenhouse effect' are all well known examples of cumulative environmental impacts. Many cumulative impacts are an issue of international environmental policy, but there is no reason why local EIA and land-use planning should not participate in finding solutions to these large scale questions. As a first step we consider how existing instruments of environmental planning could help in finding 'sustainable' approaches. 'Die deutsche Forschungsgemeinschaft' (DFG, German Research Association) is financing the research-project referred to in this chapter on the consideration of cumulative impacts within environmental impact assessment and land-use-planning.

Within the last ten years there has been a lively debate on the implementation of cumulative impact assessment within environmental assessment and land use planning in the USA and Canada. These approaches may provide a valuable basis for similar considerations within European planning systems. A comprehensive, and still most frequently mentioned, typology of cumulative impacts was developed during a combined workshop of the Canadian and the US environmental administrations (Beanlands and Eckermann 1986). It names the following types of cumulative impacts:

- *Time crowding*: frequent and repetitive impacts on a single environmental medium, for example wastes sequentially discharged into lakes or rivers.

Hamburg Technical University, Kasernenstr. 10, D-21073 Hamburg, Germany

J. Breuste, H. Feldmann, O. Uhlmann (Eds.)
Urban Ecology

- *Space crowding*: high spatial density of impact on a single environmental medium, for example habitat fragmentation in forests or estuaries.
- *Compounding effects*: synergistic effects arising from multiple sources on a single environmental medium, for example gaseous emissions into the atmosphere.
- *Time lags*: long delays in experiencing impacts, for example carcinogenic effects.
- *Extended boundaries*: impacts resulting from something some distance from its source, as is the case with major dams.
- *Triggers and thresholds*: disruptions to ecological processes that fundamentally change system behaviour, an example is the effect of the rising level of CO_2 on the global climate.
- *Indirect effects*: secondary impacts resulting from a primary activity, for example new road developments opening frontier areas
- *Patchiness effect*: fragmentation of ecosystems, examples are forest harvesting or the effect of port development on coastal wetlands.

A quick look at about 60 studies, articles and books of mainly US-American and Canadian origin shows the growing concern on cumulative environmental impacts. While between 1975 and 1985 just 11 papers on this issue were found, in the next five years 16 more titles were noticed. Of 31 titles, more than the half were published between 1991 and 1995. By scanning through these papers we identified eight methodological approaches to cumulative effect analysis, which will be detailed in the next pages.

2 Spatial Boundaries and Time Frames

Cumulative environmental effects commonly appear to be inter-regional and therefore exceed the boundaries of local plans and singular EIA-studies. In accordance with this, the concept of using inter-regional plans and large scale instruments for cumulative environmental assessment is very common. Many authors stress the importance of strategic environmental assessments, considering not just projects but also plans, policies and programmes. After years of preparation, an EU-guideline concerning EIA at the strategic level was published in 1996. The chances are good that the identification of cumulative effects within strategic EIA will soon obtain a legislative foundation. Investigation of cumulative environmental effects at the strategic level could be understood as a starting point for further tiered or stepped investigations on more detailed planning levels. Other types of planning, which focus on landscape structure and functions such as landscapes planning, could be valuable in detecting cumulative impacts.

Environmental impacts often accumulate over time, therefore it is necessary to have a clear focus on time frames within cumulative impact assessment. Historical considerations can have many different qualities. First of all one should look at the impacts of the past and the foreseeable future that will accumulate with present impacts. This accumulation may happen in the very distant future. Historical data could help to define trends of accumulation that could not be recognised from present data alone. This data could consist of measurements, historical maps or aerial photographs, or may at times even be collected by interviewing senior local people. Another quality of historical considerations is the consideration of planning projects. Projects have life-cycle phases that should be considered and, more importantly, may have an outflow of products that again have their own history in incoming resources. All these historical considerations improve the understanding of accumulation of environmental impacts over a period of time.

3 Matrices, Subjective Assessments and Networks

There are many helpful techniques to structure cumulative impact assessments, the most common being the use of matrices. The matrices used for the assessment of cumulative impact assessments differ from the interaction-matrices (which are well known in environmental impact assessments) since they do not only pair project elements and environmental indicators, but also focus on interactions between environmental indicators and between project elements. Networks, showing the complex causal relations between primary, secondary and tertiary effects, are often used in cumulative impact assessments, but in many cases the overview becomes obscured with these methods. Subjective assessments such as statements of individuals and Delphi-variants are common approaches to shorten lengthy investigations. Nevertheless the prediction capability for cumulative effects just on the basis of all these tools is very uncertain.

4 Simulation and Scenarios

There are different natural scientific works relating to cumulative environmental impacts. The best results of quantitative simulation of cumulative impacts have been achieved when examined in narrowly defined limnologic and maritime ecosystems. In open systems the value of quantitative simulation is limited, under these circumstances scenario-writing is a better concept. In 1994 the expert committee on ecological questions (in Germany) called for the inclusion of 'critical structural changes' in its environmental report, like the ones connected to the loss of soil quality and drainage into the critical load/critical level concept. The scientific council of the German government on global environmental changes developed a syndrome-concept in the latest annual report (WBBGU 1994) that is very similar to the understanding of cumulative impacts. Since there are structural limits of certainty when predicting or investigating very specific interactions between indirect effects, a plausible solution is to stick to the more generic term 'syndrome'.

5 Juridical and Organisational Issues

In some cases there are juridical or organisational boundaries that hinder the avoidance of cumulative environmental effects. The concept of emission-thresholds often does not take into account the accumulation of effects on the emission side. A potential solution to this problem may be the so called 'bubble policy' that is implemented within the US-American Clean Air Act. It limits emissions on the basis of larger units (bubbles); the way to comply with this generic standard is not fixed in this concept. Such juridical problems can only be stated, they are at this point not a question of spatial planning.

In practice cumulative effects often develop in niches between shared responsibilities, areas for which no institution feels particular responsibility. A realistic help in reducing the threat of unforeseen cumulative impacts not only has to deal with planning issues, but has to cope at an early stage with shared responsibilities on specific areas of investigation. An important research issue is the consideration of the significance of this aspect. The incremental character of present-day environmental planning seems to be one of the most important factors of cumulative environmental impacts. This aspect is sometimes called 'the tyranny of small decisions'. The German Council of Experts for Environmental Questions (SRU) stated in 1987 that environmental political and regional planning has not been able to handle fundamental complex interactions of effects in an adequate manner because of their limitation to very narrowly defined fields. Phenomena such as cumulative environmental impacts, which are not limited to a single administrative responsibility and demand interdisciplinary treatment, are usually left without due consideration.

6 GIS-based Accounting of Spatial Changes Substances and Environmental Effects

The usage of Geographic Information Systems (GIS) combines different functions that were noted as being valuable in cumulative impact assessment. The ability to zoom in and out of different scales improves the ability for spatial analysis. The ability of GIS to build up scenarios is well known. In cumulative impact assessment GIS based approaches were used several times for the overlay of historical maps and aerial photographs to uncover historical trends. We see the new generation of the so called Desktop-GIS as a most promising tool not only to fulfil these functions but also to improve co-operation within environmental planning, and thus to close gaps of responsibility. The fast development of 'intra-nets' allows us to think of a futuristic organisation of environmental planning; this would have different sectoral planning offices continuously submitting information to, and receiving information from, a central GIS-Server. The actuality of spatial information would allow historical trends to be considered and a compliance-check of spatial changes, substance-flows and environmental effects to be carried out within regular time frames. One of the major objectives of our research work is the definition of indicators for cumulative environmental impacts that fit into such a planning system.

References

Beanlands G, Eckerman W, Orians GH, O'Riordan J, Policansky D, Sadar MH, Sadler B (1986) Cumulative Environmental Effects – A Binational Perspektive. Canadian Environmental Assessment Research Council and US Research Council Hull; Washington

SRU (Der Rat von Sachverständigen für Umweltfragen) (1987) Umweltgutachten 1987. Kohlhammer Stuttgart u. Mainz

WBBGU (Wissenschaftlicher Beirat der Bundesregierung Globale Umweltveränderungen) (1994) Welt im Wandel – Die Gefährdung der Böden. Economica Verlag, Bonn

Environmental Impact Assessment *(EIA)* for Types of Land Use

Ulrike Weiland

1 Political and Legal Changes in Spatial Planning

Major changes in politics and law can be considered to be the leading motive for the search for improved information bases and appraisal instruments so that environmental requirements can be taken into account more effectively.

On signing the Agenda 21 the FRG (Federal Republic of Germany) committed itself to promoting sustainable development. The new quality of the approach consists of connecting economic permanence with ecological carrying capacity and social justice. The living conditions of future generations in a long term perspective are considered to be factors relevant to the political decisions of today. Sustainable development is both a guideline and a process for political actions; it has to be realised through participation of all actors and persons affected.

Guidelines for a sustainable human settlement development, as agreed in Agenda 21, serve to improve social, economic and environmental conditions in urban and rural settlements. When the Federal government signed the 'Global Action Plan' of the Habitat Agenda in Istanbul in 1996 it pledged to execute policies for sustainable housing, for a change of lifestyles and for a change in economic production. Implementation at the national, regional, and local level is essential for the realisation of sustainable development (Weiland 1997).

The Regional Planning Law and the Federal Building Code has been amended by the beginning of 1998. Sustainable spatial development shall serve as the main target of both the hitherto existing as well as the new targets of regional planning. The new targets are mostly socio-economic targets, but the 'protection of natural resources' and the 'long-term preservation of the options to develop patterns of the future usage of space' continue to be significant.

Technical University Berlin, Inst. for Management in Environm. Planning (FR 2-7), Franklinstr. 28/29, D-10587 Berlin, Germany

J. Breuste, H. Feldmann, O. Uhlmann (Eds.)
Urban Ecology

On a planning level, the region shall gain more importance because a regional plan shall be set up for the intertwined area of each 'central location of the first degree' and impacts on nature and landscape will be compensated for on a regional scale. Legally binding landuse plans, that are developed from the 'Preparatory Landuse Plan', do not have to be approved of any longer by the higher administrative authority. Consequently, the 'Preparatory Landuse Plan' will become outstandingly significant in landuse planning. At the same time, planning authorities have smaller staff capacities and smaller time budgets for planning projects (due to the continuing fiscal crisis). It can be assumed that short-term economic considerations will obtain priority more frequently than is the case today.

Considering the flexibility of the building and planning law, the weighing of interests gathers more importance which means that authorities and private investors need information on potential effects caused by their planning projects. Despite the future absence of the approval authority, a lack of consideration of environmental requirements can be regarded as a serious deficit and can lead to the invalidity of the plan.

The acceleration and flexibility of spatial planning on the one hand, and the new guideline for 'sustainable spatial planning' on the other, require further development of an ecological framework for comprehensive spatial planning. The existing planning instruments have to supplement each other, but they must be developed further. Landscape framework plans and landscape plans serve as information bases for spatial planning, yet there is a lack of an instrument which permits the evaluation of the contribution of regional and landuse plans to a sustainable comprehensive spatial planning.

2
Experiences in the Field of *EIA*

Many recent and older experiences exist at the regional and local level that can be used for an *EIA*; in the seventies the 'Analysis of Ecological Risks' was developed. In other European countries, much experience exists with 'Strategic Environmental Assessments for Plans and Programmes' (European Commission 1997). Several German research projects also developed methodologies for the assessment of environmental impacts at the city level (Hübler et al. 1995, Jacoby 1996). Several methods and tools are already used in landscape and spatial planning to analyse, represent and assess the environmental impacts of land-use types. Methods in question are scenario-writing, catalogues of test criteria, checklists, matrices, or reaction chains (Weiland 1994, Runge 1997). A combination of several methods will probably be relevant for *EIA*.

Accounting of flows of matter and energy is a new field of research. It is based on the assumption that the industrialised countries have to reduce their resource use by a factor of 10 if they want to create sustainable economies and sustainable ways of living. The results found so far have not been made applicable to regional planning and its procedures (Enquête-Kommission 1994, Baccini, Bader 1996).

One tenet of socio-ecological time-research is that the environmental crisis is partly created by the de-synchronisation of natural and cultural times and rhythms. The use of fossil energy can be interpreted as a rapidly accelerated use of the time accumulated in the fossil resources (Held, Geissler 1995). These experiences must be put into operation to achieve comprehensive sustainable and ecologically sound spatial planning.

3
The Methodology of the Proposed *EIA*

An *EIA* comprises the evaluation and presentation of potential environmental impacts of spatial plans. The notion 'environmental impact' includes potential effects and risks of types of land use for the environment. Not only the particular use in question shall be subject to an *EIA* (for example residential areas), but also resulting additional uses (e.g. infrastructure projects), and subsequent effects (e.g. a rise of the demand for transport). The *EIA* (ideally) should be prepared to consider a timespan of more than a generation, i.e. longer than thirty years.

The *EIA* has both describing and evaluating sections. A *target system* must be developed as a system of reference and criteria, which allow for auditing the contribution of regional development activities to a sustainable spatial development (SRU 1994, Enquête Kommission 1997).

In order to develop an *EIA* for spatial planning, one has to abstract from single projects and to define (proto-) types of uses; as a result, the *EIA* is methodically situated between 'Technology Assessment' and 'Environmental Impact Assessment'. An *EIA* can improve the outcome of the planning process only in cases where additional studies normally are not carried out. This concerns smaller, but more frequent planning projects whose effects are summarised.

Furthermore, sufficient information on environmental impacts is necessary. Because the field of 'building and housing' has been, and is being, investigated intensively (Enquête Kommission 1997), *General Residential Areas* shall serve as an example for a land-use prototype.

The description of the type of land use can be carried out in a similar manner to the 'Environmental Impact Assessment' for specific projects, as well as for types and duration of impacts, but they have to be amended to the scales of regional preparatory land-use planning. For the sake of a systems approach to the environment, the integration of material and energy inputs and outputs and the reduction/optimisation of the input and output have to be considered.

Solutions will have to be found to the generalisation of environmental impacts on the small scales of regional and preparatory land-use planning to build up an *EIA*-methodology. In addition, a method will have to be developed to deal with the inaccuracy of forecasts of land use trends. At the regional level, economic and social developments and their environmental impacts have to be anticipated on scales of 1:200,000, 1:100,000 and sometimes 1:50,000.

The timespan of more than one generation, which has to be taken into account in order to make the concept of sustainable spatial planning operational, has so far not been investigated, neither its perspectives nor its methods. Uncertainty grows exponentially if cumulative and indirect effects are considered; cumulative effects can probably only be predicted in a very simple way.

Finally, the results of the ecological contribution shall be compiled in an 'Environmental Impact Manual' (EIM). The EIM shall be easily available, so that affected citizens and Agenda 21 initiatives will be able to access the compiled information.

References

Baccini P, Bader HP (1996) Regionaler Stoffhaushalt. Erfassung, Bewertung und Steuerung, Heidelberg

European Commission, DG XI (1997) Case Studies on Strategic Environmental Assessment. Final Report, Brussels

Enquête Kommission 'Schutz des Menschen und der Umwelt' (1994) Die Industriegesellschaft gestalten. Economica, Bonn

Enquête-Kommission 'Schutz des Menschen und der Umwelt' (1997) Konzept Nachhaltigkeit. Fundamente für die Gesellschaft von morgen. In: Dt. Bundestag (ed) Zur Sache 1/97, Bonn

Held M, Geissler KA (ed) (1995) Von Rhythmen und Eigenzeiten. Perspektiven einer Ökologie der Zeit. S. Hirzel, Stuttgart

Hübler K-H, Riehl C, Winkler-Kühlken B (1995) Umweltverträglichkeitsprüfung in der Bauleitplanung. Band 2 UBA-texte 69/95. Erich Schmidt Verlag, Berlin

Jacoby C ed (1996) Strategische Umweltvorsorge in der Flächennutzungsplanung. ZAU-Sonderheft 7/1996. Analytica, Berlin

Runge K (1997) Umweltverträglichkeitsuntersuchung. Internationale Entwicklungstendenzen und Planungspraxis. Springer, Berlin

SRU (Sachverständigenrat für Umweltfragen) (1994) Umweltgutachten 1994. Für eine dauerhaft-umweltgerechte Entwicklung. Metzler-Poeschel, Stuttgart

Weiland U (1994) Strukturierte Bewertung in der Bauleitplan-UVP. UVPspezial 9. Dortmunder Vertrieb für Bau- und Planungsliteratur, Dortmund

Weiland U (1997) Nachhaltige Stadtentwicklung – Versuch einer Systematisierung und offene Fragen. In: Hübler KH, Weiland U (eds): Bausteine für eine nachhaltige Raumentwicklung in Brandenburg und Berlin, VWF, Berlin pp 105–138

Urban Structural Units as a Key Indicator for Monitoring and Optimising the Urban Environment

Peter Böhm

1 The Concept of Urban Structural Units

For urban planning purposes information at different scales is needed, ranging from the management of small habitats to entire urban agglomerations. Spatial typologies can help to structure data and provide an overview. In this chapter an urban ecological typology is presented, which is based on homogeneous units of the physical structure – Urban Structural Units (USU).

Urban structural units are physiognomically uniform urban areas with a characteristic formation of buildings and free spaces. They are more or less homogeneous with regard to the type and density of buildings and the type of free spaces (degree of sealing or the type and quantity of vegetation). The concept is based on the premise that there are causal links between the physical structure of an urban area and ecological and societal characteristics and functions (Duhme and Pauleit 1994).

26 structural units (Table 1) have been developed for the city of Leipzig (Böhm et al. 1996). Structural units can be identified and mapped using physical structures visible in aerial and satellite photographs or can be derived from available land-use data. The proposed typology USU is an aggregation of data, simplifying detailed information. The typology structures the real complexity and promotes recognition and a spatial overview. It should be used at a scale of 1 : 10,000 to 1 : 25,000. Most of the characteristics of USU are modified by numerous individual or systematic effects like the size of the area of an USU, homogeneity, distances to dominant natural structures, location or function in the city, type and intensity of relations to the vicinity, distance to main traffic axis, quality of the buildings, peculiarities, structure and dynamics of the population.

UFZ Centre for Environmental Research Leipzig-Halle, Department of Urban Landscapes, Permoserstr. 15, D-04318 Leipzig, Germany

J. Breuste, H. Feldmann, O. Uhlmann (Eds.)
Urban Ecology

Table 1. Urban structural units of Leipzig

Residential land and land of mixed use	City Center, Detached and Terraced curb-close apartment buildings with built-up courtyard, Detached and Terraced curb-close apartment buildings with open courtyard, Free-standing blocks of flats in rows, Large new prefabricated housing estates, Detached and semi-detached houses, Villas, Former village centres
Industrial areas	New, medium density industrial and commercial areas, Old, high density industrial and commercial areas
Special Estates	Large public facilities, Shopping Centers, Public utilities and waste management plants
Infrastructural facilities	Infrastructural facilities
Recreational areas	Parks and green space, Allotments, Cemeteries, Sports fields, playgrounds
Open spaces	Agricultural areas, Woodlands, Standing water, Rivers, Waste ground and changing areas, Quarries, pits, mines and related facilities

2 Characteristics of USU

A homogeneous physical structure and a typical arrangement of buildings provides both a general idea of the physical features and information for modelling of physical processes (arrangement of buildings and free spaces in an area, degree of sealing, average building and vegetation height).

The analysis of surface temperatures of urban structures in Leipzig showed characteristic differences between USUs in maximum and minimum temperatures (Spacetec 1993). The differences mainly occurred due to the volume of the buildings, degree of sealing and the type and quantity of vegetation. Thus, the information on land-use and structure can be used to identify climatic functions (areas of bioclimatic risk, cold air generation in open spaces), to give recommendations to solve problems (i.e. on the distribution of local discomfort areas, improvement of the ventilation, deficiency of urban greenery) and for the zoning of cities into areas with differing degrees of climatic characteristics. The type and arrangement of buildings and open space affected the transport of fresh and polluted air in the streets of Leipzig. When traffic emissions in small street canyons are high and the transport of fresh air into the area is low (because of reduced ventilation) no dilution can take place and dangerous concentrations of air pollutants occur. The benzole concentration in Eisenbahnstrasse (terraced apartment buildings close to the curb, street canyon) was much higher than the concentration in Gördeler Ring (wide, open area). The results indicate that rebuilding vacant sites in structures close to the street curb would seriously worsen the air quality in street canyons (Freyer et al. 1996).

Characteristic differences in the degree of soil sealing could be identified for Leipzig USUs. The degree of sealing can be used to calculate the surface runoff after rainfall events or the groundwater recharge (see calculations in Duhme and Pauleit 1994).

The characteristics of USUs – degree of sealing, type and quantity of vegetation, intensity of land-use and cultivation – affect the type and quality of biotopes/habitats. The evaluation of biotopes in Leipzig showed typical differences between intensively used, dense housing areas (low value), detached housing structures (medium value) and open spaces (medium to high value) (ökonzept 1994).

Residential USUs showed specific population densities. The population density can be used as an indirect indicator for the intensity and efficiency of land-use, traffic density, requirements for retail, service and public facilities.

3 Examples of Application in Planning and Research

USUs provide a possibility for processing physical data using GIS (for example using building height for wind-modelling, surface sealing for runoff-modelling, calculation of emissions). One of the maps in Berlin's Environmental Information System (the Environmental Atlas Berlin) is called "City Structure" and contains 17 urban structural units. The map was used as one piece of basic information to produce the following maps- soils, surface sealing, basic soil characteristics, groundwater recharge, ground level wind conditions, climatic functions, vegetation, supply of public green spaces, building- and vegetation heights, emission diffusion (in preparation).

The experience in processing the information of the city structure showed that mainly the physical characteristics were important for the creation of new data (Mrs. B. Fellmer, verbal communication).

One basis of the work in joint projects of the *Department of Urban Landscapes* is to encourage researchers to work with spatial typologies. We propose the USU typology in order to integrate spatially the works of different scientific disciplines. This encourages researchers to work at the same locations or with examples of similar urban structures. Using GIS, numerous maps of the city region (hypothetical maps of the microclimate, air pollution with sulphur and benzole) were produced. Hence, according to our experience in joint projects, the use of USUs make it possible to integrate various disciplines in urban research. We are also co-operating with the city council of Leipzig in order to test how the USU can support the city development (see E. Wickop in this volume)

4
Conclusion

USUs are able to describe the integration of some natural, physical and structural components of urban areas and are useful as spatial reference units in urban research and medium-scale urban development planning. They are a tool for modelling and processing physical data for the creation of maps for entire city regions and can therefore help to provide planning data by using mainly data already available. USUs can be both a stimulus for spatial research in various scientific disciplines and a platform for the interdisciplinary integration of numerous scientific fields.

References

Böhm P, Breuste J, Heyn C, Wickop E (1996) Map: Urban Structural Units of Leipzig, UFZ

Duhme F, Pauleit S (1994) Strukturtypenkartierung als Instrument der räumlich-integrativen Analyse und Bewertung der Umweltbedingungen in München.

Freyer K, Popp P, Treutler H-C, Wagler D, Schuhmann G (1996) Untersuchungen zu Wechselbeziehungen zwischen Immissionen und Flächennutzung auf strukturtypischen Testflächen in Leipzig. UFZ Bericht Nr. 10/1996

ökonzept Arbeitsgemeinschaft für Landschaftsökologie (1994) Stadtbiotopkartierung Leipzig, Bd. 1-3

Spacetec (1993) Stadtklimatische Untersuchung Leipzig - Abschlußbericht

Contaminated Sites and Urban Ecology in São Paulo, Brazil

Andreas Marker[1], Marco Antonio Gunther[2], and Rodrigo C. Cunha[2]

1 Introduction

A recent UNEP IE world-wide consultation revealed that land contamination is considered a serious issue by 78% of the countries consulted (Butler, 1996). However 61% stated that political awareness of the problem is low or non existent and only 28% of the countries have regulations dealing with contaminated land. The study shows clearly that there is a wide gap between consciousness of the problem and political action.

In the industrialised countries, specific environmental laws and problem-orientated institutional sectors are geared to the contaminated land issue. This is not surprising because the main assets in jeopardy are drinking water resources and public health (for example Germany takes more than 60% of its drinking water from groundwater aquifers).

The contaminated land issue is mainly restricted to urban industrialised regions with a long history of industrialisation and soil occupation. In this context, São Paulo Metropolitan Region (RMSP) is comparable with the large metropolitan areas of the so-called industrialised nations. With a population of more than 20 million, a daily generation of about 15,000 tonnes of domestic waste, and about 66,000 industries which account for about 30% of the Brazilian GNP, contaminated sites definitely can be considered a potential environmental issue in São Paulo.

During the last four years, the Technical Co-operation Project between CETESB and GTZ (Deutsche Gesellschaft für Technische Zusammenarbeit) assisted CETESB in establishing methods for site investigation and supporting the creation of a legal and institutional framework geared to the contaminated site issue. The initial situation and state of the art will be described below.

[1] Dr. Krätzig Eng. Ltd, Süsterfeldstr. 81, 52072 Aachen, Germany
[2] CETESB - Avenida Prof. Frederico Hermann Jr., 345, 05489-900 São Paulo, SP, Brazil

J. Breuste, H. Feldmann, O. Uhlmann (Eds.)
Urban Ecology

2
Contaminated Site Management in São Paulo

The first steps towards contaminated site management consist of a regional analysis of the issue through the identification of contaminated sites. The contaminated sites register comprises all data available concerning the sites, their vicinity and environment; it thus represents an „inventory of pollution sources". The inventory assists with the execution of the following tasks:

- overall analysis of the environmental situation related to the existence of contaminated land;
- institutional registration and control of the sites;
- support to urban and environmental planners;
- selection of priority sites with high risk potential in order to direct human and financial resources to the investigation and rehabilitation of these sites.

The potentially contaminated sites are locations where a possibility of soil and groundwater contamination resulting from industrial processes, generation of toxic waste, probability of leakage etc. exists. The sites included in the Sao Paulo inventory comprise 116 dump sites and landfills (considered to be contaminated sites) and about 2600 gasoline stations, fuel stocks and chemical warehouses, of which about 80 already have spillage and leakage problems. As far as the industrial sites are concerned, the selection of active and abandoned industries with a potential for soil/groundwater contamination are revealed to be extremely complicated. Out of 66,264 active industries and commercial enterprises and 4238 deactivated industries, about 24,000 and 2073 respectively have to be considered (theoretically) as potential pollution risks for soil/groundwater. In RMSP the main potential contamination of soil and groundwater can be expected to come from primary and fabricated metal production, electric/electronic equipment manufacturing and chemical industries. The results obtained permit the following initial conclusions to be made with respect to the potential contamination sources and the environmental risk (Fig. 1).

- The majority of the potentially contaminated sites related to active industries (849 priority locations) are located in densely populated urban areas of the city of São Paulo where contaminants and inhabitants are in close proximity. The careless and unrestricted use of the upper aquifers can bring risks to the population in the proximity of these sites.
- The industries and waste dumps in the catchment basins of the drinking water reservoirs Guarapiranga and Billings, situated in the southern water protection zones of RMSP, are likely to have an indirect influence on the

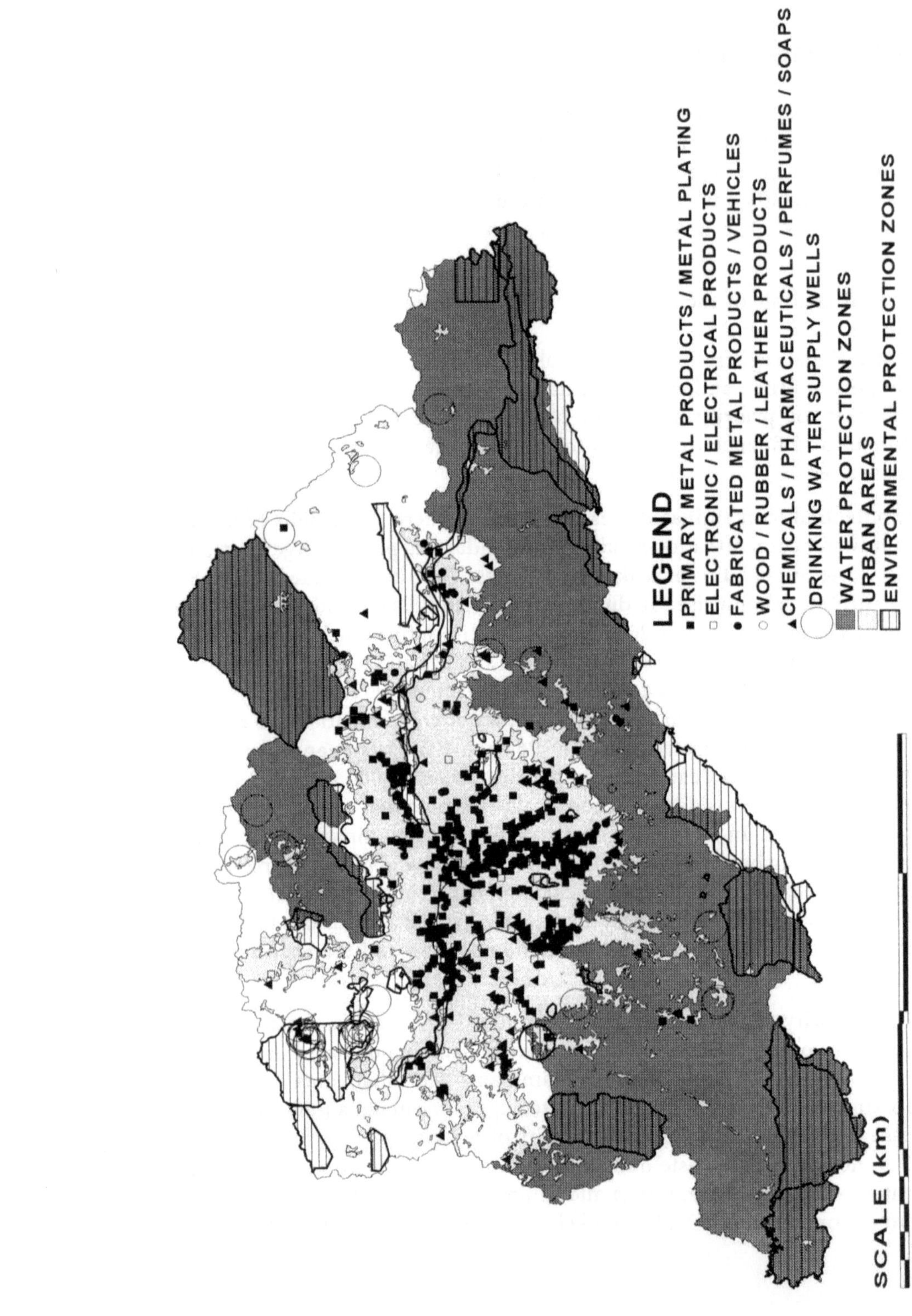

Fig. 1. Potentially contaminated sites – RMSP active industries

water quality of these reservoirs, since the upper aquifers are linked to the water bodies.

- The water quality of deep aquifers is likely to be influenced by contaminated sites. Halogenated solvents can penetrate various groundwater storeys because they are heavier than water and thus contaminate the drinking water supply wells, located mainly in the north eastern part of RMSP.
- A major potential problem created by contaminated sites is the redevelopment of land. Abandoned industrial sites represent future construction and redevelopment areas of high economic value due to their privileged location in the central urban regions of RMSP.

In order to implement contaminated site management, technicians of CETESB have been qualified in site identification and investigation. One of the present main concerns is the transfer of know-how to local authorities and private firms in order to decentralise the tasks and responsibilities. The institutionalisation and regulatory control of contaminated land through specific legal instruments is an essential step towards solving this issue. The contaminated land issue will be specified in a project for the revision of the decree 8468 to the state pollution law 997 which prohibits soil and groundwater pollution.

3 Perspectives

The issue of contaminated land requires very specific political and economic solutions. The local authorities and the real estate market have to be involved as property is concerned in the matter. Once an imminent risk for the population and/or ecosystems exists (arising from a contaminated property) it must be dealt with by the environmental control agencies, in their position as the competent authorities. Conflicts of interest are predictable and political solutions have to be discussed and translated into action. It cannot be expected that this process will happen from one day to the other, as the experience from European countries shows. Furthermore, the financing of the measures, especially when untitled or public sites are concerned, requires further analysis of the present and future resources and possibilities in order to assure the sustainability of the local development. More so than the "traditional" environmental issues like water and air pollution, contaminated land requires an inter-institutional and inter-sectoral approach involving state and local administrations as well as private sectors (none of whom are exempt of responsibility).

In this context CETESB assumes a protagonist position in Brazil and the rest of Latin America as being the only environmental control agency which makes the issue of contaminated land part of its list of priority goals in environmental policies, thus taking the first step towards an integrated management of contaminated land.

Reference

Butler BE (1996) Consultation with national experts: managing contaminated land. UNEP Industry and Development, April June, 55–56

Identifying Urban Soil Sealing by High Resolution Remote Sensing Methods

Maik Netzband and Gotthard Meinel

1 Introduction

Sealing of urban surfaces has negative ecological consequences. Effects on hydrology (water budget) and on micro-scale and meso-scale climate are well known and precisely described in literature. Planners need to have survey data and maps with information about scales of sealing in urban areas, for example to judge planned building projects and, if necessary, to prevent such projects. Conventional investigations use terrestrial test sites of representative urban structure types and extrapolate their sealing values to the whole urban area based on a structure type mapping. Such a method, however, results in an uncertain precision and is difficult to repeat at regular intervals to actualise sealing data. For that reason modern investigations use remote sensing techniques to identify urban surface sealing. For planners a map scale of 1:25,000 is sufficient to obtain an overview of town quarters, whereas for more detailed planning a scale of 1:5000 is needed. Such maps can be produced by innovative and efficient methods using remotely sensed data and will be introduced in the following pages (Netzband and Meinel 1996).

2 Acquisition, Analysis, and Evaluation of the Collected Data

2.1 Flight Mission and Data Pre-processing

The following investigations are based on a flight mission over parts of the city of Dresden, Germany. The chosen flight areas cover 57 test sites with 173.1 ha of a terrestrial investigation for which sealing degrees were calculated (Heber, Lehmann 1993, 1996). Most surface cover materials that are typical for urban agglomerations were recorded in the flight area. The aircraft was equipped with a DAEDALUS-ATM-Scanner AADS 1268 (Amann 1986). With

Institut für ökologische Raumentwicklung e.V.(IöR), Weberplatz 1, D-01217 Dresden, Germany

J. Breuste, H. Feldmann, O. Uhlmann (Eds.)
Urban Ecology

a flight altitude of about 500 m the geometric resolution was 1.1 m and flight paths possess a lateral overlap of 40%. In total, three flights were carried out, one flight at high noon, one flight after dusk and one flight before dawn. During the evening and morning flights only data from thermal infrared bands were recorded. During the day time all bands were registered. Aerial colour-infrared photos of the flight area were simultaneously recorded using scanner recording.

The scanner data have to be converted into true reflection values and into temperature values using the recorded flight parameters. For the rectification about 30 ground control points per km^2 are marked in the scanner data and in the scanned topographic maps at the same time (scale 1 : 10,000). Data are resampled with a resolution of 2 m. The mean rectification error therefore is 2.3–2.8 m pixels (4.6–5.6 m) with a maximum value of 5 pixels (10 m). As the rectified flight stripes need to be compiled into complete pictures, a mosaic process is then carried out.

2.2 Instruments to Ascertain Soil Sealing

To evaluate the collected scanner data the following three analysis methods have been developed and tested.

The *vegetation index* gives a very precise overview of earth surfaces covered with vegetation. In shaded areas a visual interpretation of vegetation cover is hardly possible. In this approach the vegetation index separates sealed from unsealed areas in detail, although the index is slightly raised in shadow zones. The vegetation index value can be transformed by visually identifying a threshold operator and can be used to transform the vegetation index data set into a binary picture. Water and unsealed areas without vegetation have a low vegetation index and consequently a high degree of sealing, because of this mistake it is necessary to exclude them before carrying out the transformation. For a multi-step analysis strategy the vegetation index 'Normalised Difference Vegetation Index' (NDVI) should only be used to classify areas which are covered by vegetation and therefore unsealed.

Thermal infrared data give information about surface temperatures. Sealed surfaces in urban areas are characterised by a significantly higher heat storage capacity than natural, unsealed surfaces. Therefore, they cause higher mean temperatures and overheating in sealed urban areas. The described phenomenon is called urban heat island. Generally speaking, received temperatures are mostly dependent on surface materials, but are also influenced by relief, daylight and wind exposure, as well as air and soil humidity (Goßmann 1984). Mean surface temperatures of different materials are determined for each of the three flight dates. The most important conclusions are:

- Traffic areas are characterised by high temperatures in the evening and morning situations with little standard deviation. On average these temperatures are 5°C higher than temperatures of other, not built-up areas, therefore it is possible to identify these mainly sealed areas by their temperatures.
- It is typical for building complexes to possess a high heating value during day time, and a fast and intensive cooling at night, dependent on their roof materials. Even roofs with the same materials vary considerably in temperature because of different exposure and age of the material.

An exact and distinct separation of sealed and unsealed areas is not reached because the temperature ranges are too similar and temperatures are influenced by external factors, i.e. relief, exposure, and local climate. It is quite possible to determine traffic areas in the evening as well as in the morning data set, because of their significantly higher temperature.

Both methods, that based on the vegetation index and the other on the cooling index, show limitations in precision caused by methodological disadvantages. Therefore it seems useful to filter out only the piece of information which can be identified clearly and integrate it in an *hierarchical classification* scheme. The initial data set for the classification is a bitemporal data set containing all bands from the midday flight, the thermal band from the morning flight as well as the vegetation index. As a first step, water areas are classified by a parallelepiped classifier to exclude them from the following classification process. In the next step, traffic areas which are not built-up, but sealed are separated using a threshold operation on the morning temperature data set. In a third step, areas covered with vegetation are determined by applying a threshold operator upon the vegetation index data set. Afterwards the identified vegetation areas are subdivided into grass, meadow and trees/shrubs with an unsupervised classification and are also masked out. Finally, the remaining class 'vegetationless' is subclassified by a Maximum-Likelihood-Classification into the classes 'buildings' and 'not sealed open areas'.

2.3 Accuracy Assessment and Terrestrial Reference Test Sites to Evaluate Classified Data

To evaluate the classification results with the *accuracy assessment* 150 independent control points are identified for each class by a random order. These points are controlled in the multispectral band composit and in CIR-pictures to certify their class membership. The overall classification accuracy is 89.1%. Compared with a monotemporal classification (accuracy: 76.3%; Kleinberg 1996) there is a significant increase in accuracy caused by the integration of the thermal data and the differentiated classification scheme. Problems mostly occur when separating the classes 'tiled roofs' and 'open soils' without any vegetation cover because their spectral characteristics are very similar and lead to misclassifications.

The comparison between the sealing values obtained by scanner classification and the *terrestrial takings for each test site* shows a strong correlation ($r = 0.9$) and results in a mean deviation of 13%. In total it was found that sealing values calculated by remote sensing methods are lower than the values from terrestrial investigations. This fact is mainly caused by the inclusion of adjacent traffic areas which are mostly sealed. Along the streets trees often cover the sealed ground which cannot be recognised by remote sensing data. The degree of sealing in remote sensing data can also be overestimated which can affect the outcome e.g. the incorrect interpretation of bare soil and dried out grass taken as sealed areas. It has to be kept in mind that the scanner flight took place in 1995, whereas the terrestrial taking was carried out in 1992 and 1993, so there are some land use changes which have not been identified but are responsible for different values.

3 Conclusions

The results of the different methods of analysis allow the following conclusions to be made:

- The transformation and calibration from NDVI sealing degree value is a less costly method for the production of sealing maps and delivers sufficiently good results.
- The evaluation of soil sealing causes some unexpected problems if using the temperature characteristics of urban surface materials. In general, sealed areas are characterised by undergoing an enormous gathering of heat during the day and a strong cooling off at night time. Surface temperatures are highly dependent on material and local climatic conditions, so that temperature values of sealed areas are partly similar to unsealed areas. The complete separation of sealed and unsealed areas by transforming data into an index was not accomplished, although separation of traffic areas is possible.
- The land use classification is a more time-consuming method but differentiates surface cover types, which is also useful for other urban investigations. Difficulties appear when classifying buildings because of their immense variation in roofing materials. This often causes a spectral overlap of classes with unsealed surfaces, especially with bare soil.
- The calculated sealing values are merged with urban boundaries (vector data) and can produce maps of mean sealing values for urban quarters or

even urban blocks which is a very useful instrument for urban planning not obtained by any other method.

References

Amann V (1986) Multipektraler Linescanner Daedalus AADS 1268 (ATM), Leitfaden für den Datennutzer, DLR, Oberpfaffenhofen

Goßmann H (1984) Satelliten-Thermalbilder, Bundesforschungsanstalt für Landeskunde und Raumordnung, Heft 16, Bonn

Heber B, Lehmann I (1993) Stadtstrukturelle Orientierungswerte für die Bodenversiegelung in Wohngebieten, IöR-SCHRIFTEN 05, Dresden

Heber B, Lehmann I (1996) Beschreibung und Bewertung der Bodenversiegelung in Städten, IöR-SCHRIFTEN 15, Dresden

Kleinberg A (1996) Studie zu Möglichkeiten der Landnutzungsklassifikation auf Basis von hochauflösenden ATM-Scannerdaten, unveröff. Diplomarbeit, TU Dresden

Netzband M, Meinel G (1996) Erarbeitung von übersichtskarten zur Versiegelungsintensität mit verschiedenen Fernerkundungsmethoden, UFZ-Berichte 12/1996, Leipzig

Integration of Remote Sensing Data and GIS to Facilitate Change Detection in Urban Areas

Florian P. Kressler and Klaus T. Steinnocher

A number of institutions and government agencies have the duty of monitoring the environment and detecting changes. Landscape and city planners face the problem of having to collect up-to-date information at acceptable costs and in the intervals required. Due to their comparatively low costs, their spatial, spectral and temporal resolution, satellite images are an attractive data source. As satellite sensors supply large quantities of data, the techniques used for the analysis of the satellite images should not only be reliable but also efficient. The strategy proposed in this chapter applies a spectral mixture analysis to satellite data and uses the results to detect changes in an urban area. A graphical interface for ArcInfo was developed to ease the comparison of this data with ancillary data and to help update existing databases.

1
Introduction

Monitoring changes in urban areas is typically carried out by photograph interpretation. Small-scale photographs (e.g. 1:10,000 and larger) and vertical stereopairs are used for urban studies on a general scale. Aerial photographs have the disadvantage of being very expensive which makes regular repeat coverages difficult to finance (Richards 1986). In addition much time and personnel are needed for the analysis of the analogue photographs. Compared to aerial photography, satellites offer a number of advantages that are necessary for a change detection system. These are: regular repeat coverage, recording data from the same geographic area at the same time of day, maintaining the same scale and look-angle, recording reflected radiant flux in consistent and useful spectral regions, and lower costs than aerial photographs (Jensen 1986). The major drawback of satellite images is the low geometric resolution compared to aerial photographs, but since high resolution sensors are becoming available (Jürgens 1996) the utility of satellite images for urban studies will be significantly improved. The procedure proposed here for

Austrian Research Centers, Department for Environmental Planning, A - 2444 Seibersdorf, Austria

J. Breuste, H. Feldmann, O. Uhlmann (Eds.)
Urban Ecology

detecting building activities in an urban area uses high spectral, low spatial resolution images to determine where building activities have taken place. These changes are then compared to a high spatial resolution panchromatic satellite image and an existing land use database. The database is then updated according to the results of the comparison. A graphical user interface for the geographic information system ArcInfo was developed specifically for this purpose. It allows fast comparisons of different data layers and user friendly updating of the existing database.

2
Basic Concept

The approach proposed here for the updating of land use databases in urban areas covers three steps (Fig. 1). The first step is the generation of a change detection map to highlight those areas where building activities have taken place. This is done by analysing multispectral satellite images recorded at different dates. Due to the relatively low resolution of multispectral imagery, nothing definite about individual buildings or the exact type of change that has taken place can be said. In the second step the change detection map is combined with ancillary data. A high resolution panchromatic satellite image may be used for the visual verification of the changes detected. This image is used as a background image, with the change detection map and land use

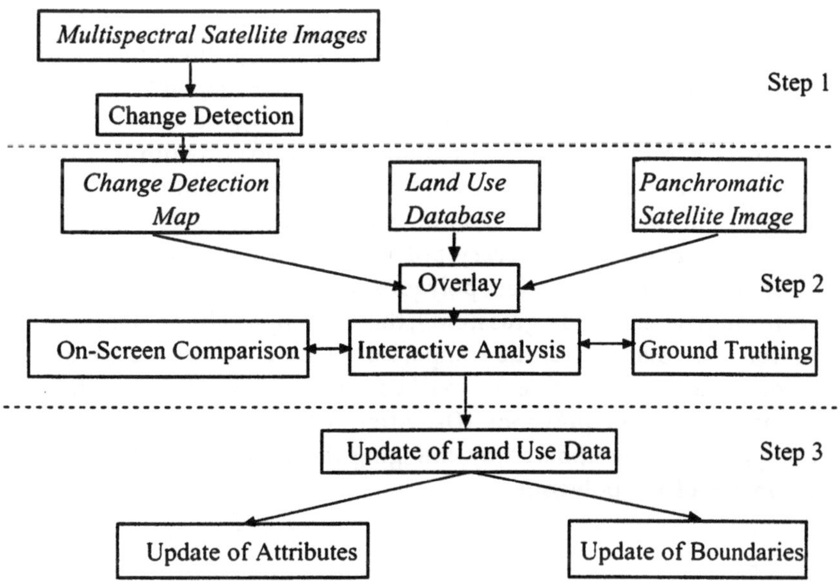

Fig. 1. Update of land use database using satellite images

database being overlaid. All those areas where changes have taken place can be marked for ground truthing. After the ground truthing (or directly if the type of change is immediately apparent in the high resolution image) the land use database can be updated and the attributes and boundaries of its polygons can be adjusted. The comparison of the different data sets and the update of the land use database is carried out using a newly developed graphical interface for ArcInfo, which facilitates speedy comparisons and updates.

3 User Interface

Steps two and three of the procedure described above were implemented in the geographic information system ArcInfo in the form of a graphical user interface. It has a modular structure, offering different display and edit options. Each module is represented by a separate display window. The top level comprises data selection and visualisation. This allows the user to display the different data sets (e.g. high resolution image, change detection map, land use data) within one display window. At this level polygons of the land use data set may be selected for further analysis. All the information connected with the selected polygon (e.g. attribute, area, identification number) is displayed separately. The modules for editing the attributes and editing the geometry may be selected or the polygons may be marked for ground-truthing.

Using this interface, the update of land use data is carried out in three basic steps: on-screen comparison, ground-truthing, and update. For the comparison the high-resolution satellite image is used as a background image and as a reference. The change detection map and the land use map are overlaid on the high resolution image. The user selects those polygons where changes have taken place and compares the attribute of the polygon with that of the high resolution image. If the new class cannot be determined by examining the high resolution image, the polygon may be marked for ground-truthing. These steps are carried out for every polygon where a change might have occurred. Another option is to select polygons with land use classes which are very likely to change (e.g. fields, grass, etc.); these again may be compared with the high resolution image. After the comparisons have been carried out a map can be printed showing all polygons marked for ground-truthing with their respective attributes and polygons-IDs. After the ground-truthing all the selected polygons can be updated by changing the attributes and/or changing the geometry of the polygons.

4 Application

Two Landsat TM quarter scenes covering the city of Vienna were available for the multispectral analysis. The images were recorded on June 5, 1986 and July 1, 1991. The images have a spatial resolution of 30 × 30 m. The sensor records reflected light in 6 different bands which cover the visible light bands and parts of the near and middle infrared; except for geocoding the images were not pre-processed in any way.

At the time of the study no digital high resolution satellite data was available. An analogue high resolution image recorded with the KFA-3000 camera system on the MIR Space Station on August 23, 1993 was used instead. A copy of this image with a resolution of approximately 2–3 m was available. It was digitised using a Zeiss PS1 scanner and afterwards geocoded.

The digital land use database of the city of Vienna comes in a vector format in which areas of identical land use are united into polygons. Each polygon has an attribute of one of 42 possible classes. The database was set up in the late 1970's using aerial photographs and has been updated every three years. The update is undertaken by comparing the analogue photographs with a corresponding analogue copy of the land use database.

For the change detection a spectral mixture analysis (SMA) is applied to both multispectral images. The aim of the SMA is to estimate how each ground pixel's area is divided among different cover types. The result is a series of images, each the size of the original image and each giving a map of the concentration of a different cover type across the scene (Settle, Drake 1993). The method used here to detect changes is to subtract fraction images, which represent the same land cover type, from one another. Changes should be clearly seen, as the fraction image values must be higher or lower for pixels where the land cover type has changed. The result of this subtraction is a new image, showing those areas, where changes are supposed to have taken place (Kressler, Steinnocher 1996).

The next step is to compare these results with ancillary data and the land use data to verify the changes and to update the land use database. This was carried out using the interface described in section 3. The high resolution image was used as a background image with the change detection map and the land use database as an overlay. The polygons where changes have occurred (according to the change detection map) were selected and examined. Through a comparison with the high resolution image, 39 of the 40 detected changes could be confirmed. In the next step polygons belonging to the classes field, building site, park, unproductive area, grass, vineyard, gardening, sport field, yard and forest were selected and again compared with the background image. In this way another 16 changes could be determined. As the high resolution image was recorded two years after the second multispectral image these additional changes could only be expected. After a ground-truthing exercise was conducted and the changes again confirmed, the database was updated and polygons, where necessary, adjusted.

5 Conclusion

The procedure presented here for change detection offers a powerful tool for city planners as it not only gives accurate information as to where changes have taken place, but can also be carried out in a very short time. Costs are not only saved in the collection of the data but also in the analysis, compared to other methods. Further analysis for the update of land use data may be limited to the areas indicated by this method. By combining the benefits of the high spectral resolution of the Landsat TM and the high spatial resolution of newer systems, the work necessary for updating existing databases can be significantly reduced. The use of the fraction images is not limited to detecting building activities but may be used for different applications in urban areas and other environments.

References

Jensen JR (1986) Digital Image Processing. Prentice Hall, New Jersey

Jürgens C (1996) Neue Erdbeobachtungs-Satelliten liefern hochauflösende Bilddaten für GIS-Anwendungen. Geo-Informations-Systeme, 9(6): 8–11

Kressler FP, Steinnocher KT (1996) Change detection in Urban Areas using Satellite Images and Spectral Mixture Analysis. International Archives of Photogrammetry and Remote Sensing, Vol. XXXI, Part B7, 379–383

Richards JA (1986) Remote Sensing Digital Image Analysis. Springer, Berlin Heidelberg New York

Settle JJ and Drake NA (1993): Linear Mixing and the Estimation of Ground Cover Proportions. International Journal of Remote Sensing, 14(6): 1159–1177

Strategies for Preventing Suburbanisation (1997)

Christoph Hümmeler

With some 480,000 residents, Leipzig is, after Berlin, the second-largest city in the new Federal German states and the centre of the conurbation Halle-Leipzig containing 1.6 million residents. Although Leipzig was among the largest and most influential of German cities before the Second World War, it stagnated during the forty years of the GDR. A population that has been continually decreasing since 1933 and the anachronistic economic legacy of the previous regime have contributed to Leipzig's loss of status among German cities.

Leipzig has witnessed a dizzying phase of development following the German reunification. Processes which have taken decades to occur elsewhere have been concentrated here in a very short period of time. The economic and political transformation following reunification has presented Leipzig with new opportunities, but has also posed a range of economic, ecological, planning, infrastructure and social problems. One of the main planning themes is the rapid progress of suburbanisation.

1 Suburbanisation

1.1 Retail Development in "Greenbelt" Locations

Leipzig lacked, as did all East German cities, adequate consumption opportunities in 1989 resulting in the erection of shopping centres, which played a large role in the first wave of investment in the city. Most of these shopping centres were disproportionately large and were built in the immediate suburban ring of the city, in locations, where, for a variety of reasons, large-scale projects were most easily completed. Perhaps the best-known example of this trend is the Saalepark, which contains over 100,000 square meters of retail space. As a result of these large-scale projects, inner-city retail development, in the sense of mixed use and a multi-functionality, lagged far behind. This

Stadt Leipzig, Stadtplanungsamt, Postfach 780, D-04007 Leipzig, Germany

J. Breuste, H. Feldmann, O. Uhlmann (Eds.)
Urban Ecology

imbalance between the inner-city and suburban retail industrial building complexes has hindered both the development of smaller, neighbourhood-oriented retail centres and the promise of an attractive city-centre. In 1993, for example, 50% of all Leipzig's purchasing power was spent in suburban retailers.

1.2 Housing

The development of suburban housing also began after 1990. In the ring of municipalities bordering Leipzig, 35,000 housing units have been planned. 22,000 of these have already been approved or built, offering housing for some 60,000 people. Suburban housing development is in no way limited to single-family houses, rather, high-density housing, often in the form of multi-story apartment houses, have dominated.

1.3 Development of Commercial Space in the Leipzig Suburbs

Suburban municipalities of Leipzig have dedicated large portions of their land to the development of commercial parks since 1990. More than 1230 hectares of commercial space have already been developed and a further 340 hectares are under construction. On completion of all of these projects, more commercial space will have been planned and built in Leipzig's environs than Leipzig itself has set aside in its 1994 Master Plan.

Commercial parks in the suburbs are dominated not by manufacturing industries, but by wholesalers, logistic and construction-firms, contrary to their original plans. All those could also have settled in Leipzig. Furthermore, the suburban location of the commercial parks makes them unreachable by public transport.

1.4 Population Trends

The development of the suburbs has led to a decrease in the population in Leipzig. This trend has reached critical proportions. The original population migration to West Germany immediately following reunification has been replaced by a migration to the surrounding suburban municipalities. Leipzig has lost 70,000 residents since 1989 and a change in the trend is nowhere in sight.

2
Leipzig's Planning Instruments

One of the most important planning tasks for the city is to find several methods to slow further suburbanisation and develop Leipzig in the tradition of the "European city", with its high level of urbanity. The terms "European city" and "urbanity" combine a variety of ideas. Residents should have access to workplaces, shopping opportunities, schools and free-time activities, all within close range of the flat. The city should be designed to minimise distances and to offer adequate public transport.

Contrary to the large-scale use of land needed for suburbanisation, Leipzig has focused on the development of the inner-city. Existing structures in the city offer adequate potential for housing construction, commercial and retail development. The development of new land has been limited to uses that cannot be integrated successfully into existing structures, such as the development of single-family housing.

2.1
Master Plan

The most fundamental instrument for development planning in Germany is the Master plan. Leipzig's Master Plan has been valid since 1995, but the general planning principles are more thoroughly described in the form of "Urban Development Plans".

2.2
Neighbourhood Centre Concept

A first sectoral development concept was the "Neighbourhood Center Concept", which was completed in 1993. The goal of concept is to protect the development of neighbourhood centres.

2.3
Housing Development Plan

The 1996 Housing Development Plan indicated, that the specified zones for housing use in the Master Plan would adequately cover housing needs for the city. In the housing land register there are 334 potential housing sites with an area of more than 2,000 square meters each. The construction of 62,000 housing units would be possible on these sites. The potential for a further 10,000 housing units has been estimated on what are now small parcels free of buildings between existing structures. A total potential of 72,000 housing units has been estimated for the city as a whole. Of these,

46,000 would be built on land containing existing structures and 26,000 on expansion land.

The Housing Development Plan paid special attention to single-family and semi-detached houses within the city, as the demand for this form of housing has played a large role in the trend toward suburbanisation. It came to the conclusion, that Leipzig has a large land potential for single-family and semi-detached housing. Finally, the Housing Development Plan mandated that priority should not be given to the expansion of housing development land for multi-story apartment houses, but should put emphasis on the preparation of land for concentrated single-family housing.

2.4
Commercial Space Development Plan

Following reunification, much of the basic industry in eastern Germany collapsed because it was not competitive in the newly opened markets. Leipzig had approximately 100,000 workers in the manufacturing sector in 1989. Now this number is just 12,000. The situation is quite dramatic compared to west German cities.

As a result of these trends, land formerly used for manufacturing industries now lies unused (approximately 180 hectares). These plots of land are generally well connected to city and regional transport systems and are in dire need of conversion.

The city has also developed four new, large-scale commercial parks which house a variety of new users. Within these 35 hectares are still available and possessed by the city.

As already mentioned, the Leipzig suburbs also offer large plots of commercial development land. The result of suburban commercial development is a drastic decrease in commercial development within the city. This represents not only an economic loss for the city, but has a negative effect on the structure of the city as well. The many abandoned industrial plots in the city are destined to remain undeveloped, as commercial re-use seems ever more distant.

In order to steer the direction of commercial development in Leipzig, a Commercial Space Development Plan has been drafted. The Development Plan contains guidelines intended to place priority on the development of central-city locations over the development of new commercial land, and the provision of jobs with the lowest possible commuting distances. The Development Plan shows that Leipzig already possesses enough commercial development land in the form of older commercial parks. Only in exceptional cases, such as industries requiring large plots of land and investors in large-scale manufacturing projects, should new land be developed for commercial purposes. The activation of small parcels of land for commercial development may, however, be necessary, as the development of older commercial space is a difficult challenge. However, the development of existing commercial land remains the city's first priority.

The Commercial Space Development Plan explains the strengths and problems of particular commercial plots. Provision of this information to investors increases the chances in competition for commercial plots.

In the conclusion of the Commercial Space Development Plan, uses for the various commercial land plots are suggested and the instruments with which the plots can be reactivated are explained. The Commercial Space Development Plan is, then, along with the already mentioned Housing Development and Neighbourhood Centre Development Plans, an important instrument in the city's battle to improve its status as a site for commercial development. The status of the city is, in turn, an important factor in conserving Leipzig's role as the main regional centre, thus hindering further over-development of suburban areas.

3 Summary

The City of Leipzig is attempting to stop a trend toward suburbanisation through planning for a high level of urbanity. City planning, in this sense, achieves its aim of stopping population loss through increasing the attractiveness of the city.

The conversion of plans into concrete projects, however, takes place under more difficult circumstances than in suburban locations due to higher land prices and the complicated legal proceedings surrounding the ownership of many of the properties. Many of the trends that first appeared after 1990, including the trend toward suburbanisation, have become hard facts in reality. The goal of development planning must be to stop these trends and to reorientate development toward a more ecological model.

References

Stadt Leipzig (1994) Flächennutzungsplan
Stadt Leipzig (1994) Statistisches Jahrbuch der Stadt Leipzig
Stadt Leipzig (1995) Statistisches Jahrbuch der Stadt Leipzig
Stadt Leipzig (1996) Statistisches Jahrbuch der Stadt Leipzig
Stadtplanungsamt (1993) Stadtteilzentrenkonzept der Stadt Leipzig
Stadtplanungsamt (1996) Stadtentwicklungsplan "Wohnbauflächen" der Stadt Leipzig
Stadtplanungsamt (Stand Ende 1996) Entwurf des Stadtentwicklungsplanes "Gewerbliche Bauflächen" der Stadt Leipzig
Stadtplanungsamt (Stand Ende 1996) Raumbeobachtungssysteme der Stadt Leipzig

Impacts of the Development of Belo Horizonte on the Serra do Curral Mountain Range

Cristina Menezes da Costa Croce[1], Maria Cristina Ramos de Carvalho[2], and Maria Rita Franco Gomes[3]

1 Introduction

Nowadays, the whole world is talking about the quality of life and the ecology of the earth and what is absolutely necessary for us to do in order to survive.

The main attention of governments is on the natural wealth of the world, especially because of the world-wide demographic density.

It is vital to be concerned about protecting the earth, its soil, water and air, and all the universe's other components for a high quality of life on the planet.

Belo Horizonte offers a high quality of life and it is considered to be Brazil's second best city to live in. Belo Horizonte is the capital of one of the most important states in Brazil: Minas Gerais. It is located in the south-eastern part of the country. The city was originally planned and constructed by the engineer Aarão Reis to hold 30.000 inhabitants. On December 12th, 1997 the city is going to celebrate its centenary.

The name Belo Horizonte was chosen because it is a pleasant place with a beautiful horizon framed by a mountain range called the Serra do Curral. Its name pays homage to the old village that was initially on the city's site and formed the basis of the new city, Belo Horizonte.

The Serra do Curral is the most important geographic point of the area. It extends in a 70 km line along the south-west and the south-east areas of the city; in order to preserve the Serra's profile a Federal and a municipal law were passed.

Since the 18th century, the Serra do Curral has attracted human occupation with its pleasant climate; it is now suffering the consequences of the development of urban expansion, mineral exploration and industrial activities.

[1] Avenida Contorno, 8.000/608, Santo Agostinho 30.110-120, Belo Horizonte, Minas Gerais, Brazil

J. Breuste, H. Feldmann, O. Uhlmann (Eds.)
Urban Ecology

The Belo Horizonte's original aim was to preserve the landscape of Serra do Curral. Nowadays, this landscape has been spoilt by the waves of development. The original urban plan of the city identified only an urban central area far from the Serra's foot. With an increase in development, occupation of land surrounding the Serra was unavoidable and the preservation of many cultural and historical features and mineral and natural reserves of the largest urban land mark were affected.

The region's vegetation cover is quite heterogeneous, vegetation types changing with altitude. It comprises the savannah forests, the iron hydroxide soil and tropical forests that show the characteristics of the Brazilian coastal Atlantic forests and the Brazilian river bank forests. This heterogeneous vegetation creates habitats that house a fauna composed of many species. The climate in this area is variable and ranges from humid to dry.

2 Main Causes and Consequences of the Impacts on the Mountain Range

In Belo Horizonte there is a law of 'Use and Land Occupation' which is not effective in protecting all of the Serra do Curral area; for instance it allowed residential districts to be built on the sloping area at the foot of the Serra. With its pleasant climate the Serra do Curral is attractive to human occupation. The Mangabeiras, a very wealthy residential district, was the first to be built in this area.

Another type of human occupation are the slums, which have sprung up at the foot of the Serra due to a disorganised and illegal occupation by the low class. This kind of occupation is a result of the lack of an areas for the lower class close to the city's central area, and this area was unusable due to the steep topography.

With these occupations, the development of urban infrastructure was inevitable in the area surrounding the Serra. Hospitals, schools, a large shopping centre, recreation squares and the Governmental Palace were built there. Around this area there are also some parks like the Mangabeiras Municipal Park, and forest reserves.

A new district called Belvedere III has been constructed at another part of the Serra's foot. The vertical form of the buildings there has been causing some dispute. Those buildings will work as a barrier to the normal flow of the wind, which will cause changes in the climate of the city. Also, new buildings mean a increase in the number of cars in the region. A new route is being planned to link Mangabeiras and Belvedere III districts, changing the heavy traffic flow from downtown to the foot of the Serra. In addition to the residential area, there is a large industrial complex that also causes a negative environmental impact.

With such a large concentration of people, the number of cars has increased leading to heavy traffic in the region and air and noise pollution which are changing the natural fauna, flora and climate of the area.

3 Mineral Extraction

The geological formation of Serra do Curral is typical of the Ferriferous Quadrilateral Area of Minas Gerais state. The geographic forms appear in a succession of complex geological events, forming iron oxide, quartz rocks, filito and dolomite soils. History can show us, clearly, how the impacts modified the ecosystem of the Serra region.

Some of those transformations began with the Saint John Mining Company's mineral extraction. This company obtained rights to explore all the Ferriferous Quadrilateral area in Minas Gerais. The Industrial Revolution increased the demand for iron minerals which led to the siting of new companies in the area; now a political and social conflict between the rights to exploration of the Serra and its preservation has arisen.

The MBR, a large mining company, created in 1966, inherited the rights of mineral extraction from the Saint John Mining Company. One huge mine of this company is the Aguas Claras, which started activities in 1973. It is located in the central part of the Serra do Curral, in the municipal area of Nova Lima city and overlaps the city limits of Belo Horizonte. It is the biggest focus of mineral exploration. Despite the MBR's intentions not to degrade the area, the mountain range in its central part has been reduced in size to about 100 meters.

Attracted by the mineral wealth of the Serra, some other mining companies are exploring in three different areas of Belo Horizonte: Lagoa Seca in the central area, Emitaq and Itanorte in the south-east and Inconfer and Sical in the south-west.

The mineral extraction in this area is about 27 million tonnes per year.

The companies cause irreversible damage to the natural appearance of the area, reduce the stability of the mountain range, change the hydrological cycle (the rivers courses and the springs) and change the vegetation cover, thus modifying the fauna.

4 Conclusion

To minimise these problems, many rehabilitation projects of degraded areas have been designed by the municipal and state agencies and mining companies involved in the Serra do Curral's complex. These projects include reforestation, hydro-planting, parks, hydro-geological surveys, and parcels of land in the degraded areas.

The community is trying to reintegrate the Serra do Curral with its natural ecosystem through some projects, such as new parks, natural reserves and ecotourism with scenic viewing.

Other public and private institutions, associations and organisations are working to preserve the monument of the Serra do Curral.

Ecologically Efficient Use of the Building Stock: A Concept of Agencies for Living Space

Wolf Engelbach

1
Model of the Use of a Building

One strategy to stop land use pressure and urban growth is to extend and intensify a building's use. In terms of material chain management either the speed of material flow should be slowed down *(extended use)* or the amount of materials should be reduced *(intensified use)* (Enquete 1995). My suggested strategy for the ecological improvement of buildings consists of the three components demand-side-management, facility management and maintenance, including reconstruction.

Demand-side-management is a strategy that was originally practised in the energy supply sector: instead of enlarging the supply capacity, the demand for energy was reduced, which turned out to be cheaper than developing new plants. When transferred to buildings, Demand-Side-Management aims at influencing the demand for space for dwelling, working or any other purpose. It deals with questions from the user's point of view: who needs buildings or parts of them at what time and where? Is it possible to fulfil these needs with less effort and fewer buildings?

Facility Management is an economically motivated strategy to deal with the resource "building" in the most efficient way. It integrates economic management, technical appliances and the organisation of services in a building. This sometimes has environmentally friendly side effects, for example the saving of energy. On the other hand it may cause a waste of materials if new products are cheaper than the repair of existing products.

Systematic inspection and maintenance allow the discovery and repair of damage in time and ensure the proper functioning of the building. The requirements for buildings change over time. It is therefore necessary to develop the building further in order to avoid its function becoming obsolete. There are successful examples of the conversion of old factories into offices, art galleries or apartments or turning castles into hotels; besides preserving

Fraunhofer IAO, Nobelstr. 12, D-70569 Stuttgart, Germany

J. Breuste, H. Feldmann, O. Uhlmann (Eds.)
Urban Ecology

the cultural heritage one saves grey energy that was used in the construction process.

2
The Agency for Living Space

One instrument within the strategy of demand-side-management is the sharing of living space. There is enough floor space in our society in general, it is a matter of re-organising its use to prevent the need to create new buildings. Today, the demand for more flexible use of apartments is increasing due to more rapidly changing social conditions and the insecure economic situations of households. Many households require an additional room, (e.g. if one more member joins the household or in order to work on a hobby or to store some (leisure) equipment). Whilst some households shrink in size or people cannot afford their apartment anymore, e.g. on account of unemployment or retirement (Eichener et al. 1994).

One problem is that moving house is always combined with efforts in terms of time and money; additionally one often leaves a well known neighbourhood. In Germany many cities support removal agencies for these reasons. Elderly people especially should leave their large apartment and move to a more suitable one, while a household with children should take their former rooms. The agencies support a removal not only financially, but they also take care of the whole organisation. I suggest a more flexible strategy of just renting another room, e.g. for working, hobby, storage or guests.

A modern understanding of privacy led to the rule of the indivisible apartment (Jenkis 1991). Historically, a house was not always restricted to only one family or person, it was used much more flexibly and interactively (Kühne-Büning, Heuer 1994). In today's situation of ecological and economical crises this thinking could provide a chance to re-define the responsibility of owners and users of buildings.

2.1
Objective Influences

The objective potential living space for reorganisation demonstrates the difference between a household's situation and its politically defined needs. One estimation, based on an opinion poll, is shown in the following tables. Table 1 contains a suggested definition of a "normal" supply of space and rooms for principal tenants in Germany. Working with these results, Table 2 shows the supply situation for the eastern and western part of Germany. Almost half of the households that live in a rented apartment either have too much or less than enough living space or rooms.

Table 1. Suggested normal supply of living space and rooms for principal tenants

Size of Household	Living space (m^2)	Number of Rooms
1 person	40–67	2
2 persons	55–82	2–3
3 persons	60–90	3–4
4 persons	65–97	3–5
5 and more persons	73–118	4–5

Source: Schwandt 1994, page 701. Modified results of interviews.

Table 2. Supply situation of principal tenants in Germany

	Under supply		Normal supply		Over supply	
Supply situation	East	West	East	West	East	West
Space and rooms	6,8	6,9			4,0	11,8
only space	20,0	7,2			4,6	8,6
only rooms	2,0	2,8			8,0	9,0
all together	28,8	16,8	54,6	53,8	16,6	29,4

Source: Schwandt 1994, page 702. Based on interviews in 1381 (East) and 1048 (West) households.

The chance of demand and supply being close together are better in socially and constructively mixed areas. Homogeneous structures often have similar households (e.g. single family housing) or social housing; dense areas have more on offer.

2.2 Subjective Influences

The subjective potential depends on the economic situation and the housing market, which is itself very much influenced by state legislation and financing. In times of crisis, shared apartments will be much more common than in times of prosperity; in this way they make the housing market more flexible. Besides the size and the price of rooms there are two important factors to be taken into consideration:

- The frequency of use: the more seldom a room is required, the longer is the accepted distance between the main apartment and the additionally rented room. Therefore it is easier to offer rooms to store ski or surf boards than

for heating coal. I assume that daily contacts would not be acceptable with more than 3 minutes walk, weekly contacts with a distance of more than 10 minutes.
- The type of use: the more private the proposed activities are, the more important is the closed situation of the supposed room. A separate entrance is more strongly demanded for a guest's apartment than for a music room. Thus, the acceptance of supplying a room in one's own apartment is even lower than that of using it, because the amount of privacy lost is larger.

Some household groups are most likely to participate in this flexible use of floor space. Expanding families often need more space, but cannot afford removal costs or large apartments. The well-educated middle-class understands the sense of having shared apartments and require extra space. On the supply side, poorer households are not able to finance their apartments anymore, unemployed people hope to keep their apartment by offering a part of it on a short-term basis. Some elderly people do not want to leave their large apartment, but enjoy regular visits.

In each apartment-sharing situation, it is possible that conflicts arise: can I trust the other party, or do I have to be afraid about burglary and personal security? Are children, music or television bothering me? Does one party want personal contact, but the other prefers just formal communication? For many problems the moderation and experience of an agency can be very supportive.

2.3
Possible Forms of Agencies

The potential supply and demand sometimes happens to be activated spontaneously through informal talks or advertising in a paper or in the Internet. For institutions, the arrangment of residential space calls for (1), a sufficient collection of objective facts about supply and demand, (2), the psychological skill to build up faith between two sides of a contract, and (3), the knowledge about legal solutions to avoid conflicts. This could be carried out by three different actors:

- Housing managers deal with many apartments which are often close together. They have the technical and organisational infrastructure to coordinate demand and supply, moreover they are used to reconstructing apartments if necessary. Housing managers are confidential persons that know some basic facts about the social situation and economic power of their tenants.
- Room pooling services nowadays only deal with whole rooms and apartments that are for rent for a certain period. These agencies provide the infrastructure for, and experience in, arranging supply and demand of

living space. They are not bound to specific apartments but are accessible to everyone.

- The housing office of a municipality is a contact for people who are looking for an apartment, but often do not find the right one on the free market. These offices do not depend on commission but make savings by avoiding investment in infrastructure and planning costs for newly developed areas.

2.4 Support by State and Society

Agencies for living space face problems of acceptance because any participating party will have to live with less privacy than in their own unshared apartment. An image campaign is important to show the individual and common advantages of shared apartments: They save money for both parties as well as for society, e.g. by avoiding people being in debt and loneliness.

The support of the government could be demonstrated by punitive taxes on over consumption of living space. Additionally local and regional governments should stress the importance of activities in the building stock in master plans or regional plans. This integrative accounting could justify investments in organisations instead of buildings, for instant the housing office could run the agency for living space.

References

Eichener V, Heinze RG, Bucksteeg M (1994) Neue Anforderungen an die Wohnungswirtschaft vor dem Hintergrund gesellschaftlicher Entwicklungen. Institut für Wohnungswesen, Immobilienwirtschaft, Stadt/Regionalentwicklung, Bochum

Enquete-Kommission "Schutz des Menschen und der Umwelt" des Deutschen Bundestages (ed) (1995) Studienprogramm Umweltverträgliches Stoffstrommanagement. Economica, Bonn

Jenkis H (ed) (1991) Kompendium der Wohnungswirtschaft. Oldenbourg, München Wien: 23

Kühne-Büning L, Heuer J (eds) (1994) Grundlagen der Wohnungs- und Immobilienwirtschaft. Fritz Knapp, Frankfurt a M: 190–198

Schwandt A (1994) Zu einigen Indikatoren der Wohnungsversorgung und deren Bewertung durch die Bewohner in den ersten Jahren nach der Vereinigung. Informationen zur Raumentwicklung IZR: 691–720

Poster 3.2

Environmental Care in Urban Land Use Planning – a Component of Sustainable Development

Veronika Mook

1 Introduction

In Germany the building code rules provide the basis for urban planning development. The zoning plan identifies the different types of land use which would result from the proposed urban development. The protection of the environment is one important local planning target identified in the building code as urban land use planning interferes with the environment and the scenery of the landscape.

§ 8 a of the German Nature Protection Act expounds the description of, and solution to, this interference in the zoning plan (Mook 1997).

In this chapter a concept will be described for the integration of the environmental aspects in urban planning for the municipal area of Jülich.

2 Methods

First of all the different environmental fields (in both the biotic and abiotic environment) are described and evaluated. The different types of land use and vegetation structure are ascertained by photo-interpretation and territory mapping. Land use and vegetation structure are classified into different biotope structures. Information about sealing, vegetation coverage or building types are also available from the data on land-use and vegetation structure. Existing maps, investigations or studies are analysed and integrated with this data.

The abiotic component – soil, ground water, surface water, climate, landscape scenery and recreational use – are analysed by using prior studies and maps; conservation areas and planning guidelines are also described.

Environment – Planungsgemeinschaft Stadt und Umwelt, Ruschenstraße 8, D-45133 Essen-Bredeney, Germany

J. Breuste, H. Feldmann, O. Uhlmann (Eds.)
Urban Ecology

Environmental quality targets have to be defined and tested for the municipal area. They provide the basis for the evaluation which is related to the functions of the ecological balance (Mook 1995, Pauels 1995).

3 Results

The evaluation of the environmental components show the deficits and the risks for further landscape development. The encroachments by planning projects are evaluated, advice for avoidance and minimisation of impacts are given, and the compensatory measures necessary are ascertained. These are part of a concept of landscape development for the urban area which itself is a further step that should be included in the urban land use plan.

4 Conclusions

The guidelines for sustainable development (Mook, Grauthoff 1997) in urban areas must take into account the aspect of avoiding environmental encroachments. For example:

- the creation of concentrated building areas,
- coverage types with minimal land use,
- land recycling,
- optimisation of existing settlement structures, and
- general reduction of resource use

are the aims for sustainable development. The concept described in this chapter is one aspect which shows the limitations of urban land use.

References

Mook V (1995) Umweltvorsorge in der Flächennutzungsplanung – Idealvorstellung oder Realität UVP-report 9, H. 4: 181–182

Mook V (1997) Die naturschutzrechtliche Eingriffsregelung in der Flächennutzungsplanung UVP – report 11, H. 3: 153–156

Mook V, Grauthoff M (1997) Nachhaltige Stadtentwicklung – Worthülse oder eine realisierbare Zukunftsvision? Naturschutz und Landschaftsplanung 29, H. 8: 247–253

Pauels J (1995) Der landschaftspflegerische Fachbeitrag – Wege zur Umweltleitplanung 105, H 11: 34–35

Developing an Urban Landscape Management Concept through the Integration of Environmental Quality Goals and Environmental Information Systems

Matthias Richter and Reinhard Böcker

1 Introduction

During the last ten to fifteen years the concepts of improving environmental quality in cities and environmental information systems have gained increasing significance (Böcker 1994, Fürst et al. 1996, Sukopp and Wittig 1993). How these two concepts relate to one another is, however, seldom discussed in detail.

2 Results

Environmental quality objectives which relate to various disciplines (e.g. nature conservation, climatology or soil science), are often represented by different city administrative departments. These disciplines must co-operate so as to reach a main objective that integrates and restores the balance of all that is environmentally important (see Figure 1). The desired state of the city's ecological system is determined by *environmental quality goals*. By comparing the *desired state* with the *present state*, *deficits* in the area of environmental quality become clear. From this information, specific *steps* working toward the desired state can be derived. Continual *monitoring* and assessment can be used to evaluate the success of these steps. The present condition can be assessed by an *environmental information system*. Environmental information systems require a spatial context. Such a context is available for Stuttgart from the city Department of Statistics. The hierarchy of districts, municipalities, subdivisions and city blocks lends itself to the effective structuring of bio-ecological data, despite the dissimilarity of the areas within a single spatial context. Part of this environmental information system is the *floristic vegetation data bank*. Data from this bank can be visualised through the use of a *Geographic Information System* and printed out in the form of a map.

University of Hohenheim, Institute for Landscape and Plant Ecology, D-70593 Stuttgart, Germany

J. Breuste, H. Feldmann, O. Uhlmann (Eds.)
Urban Ecology

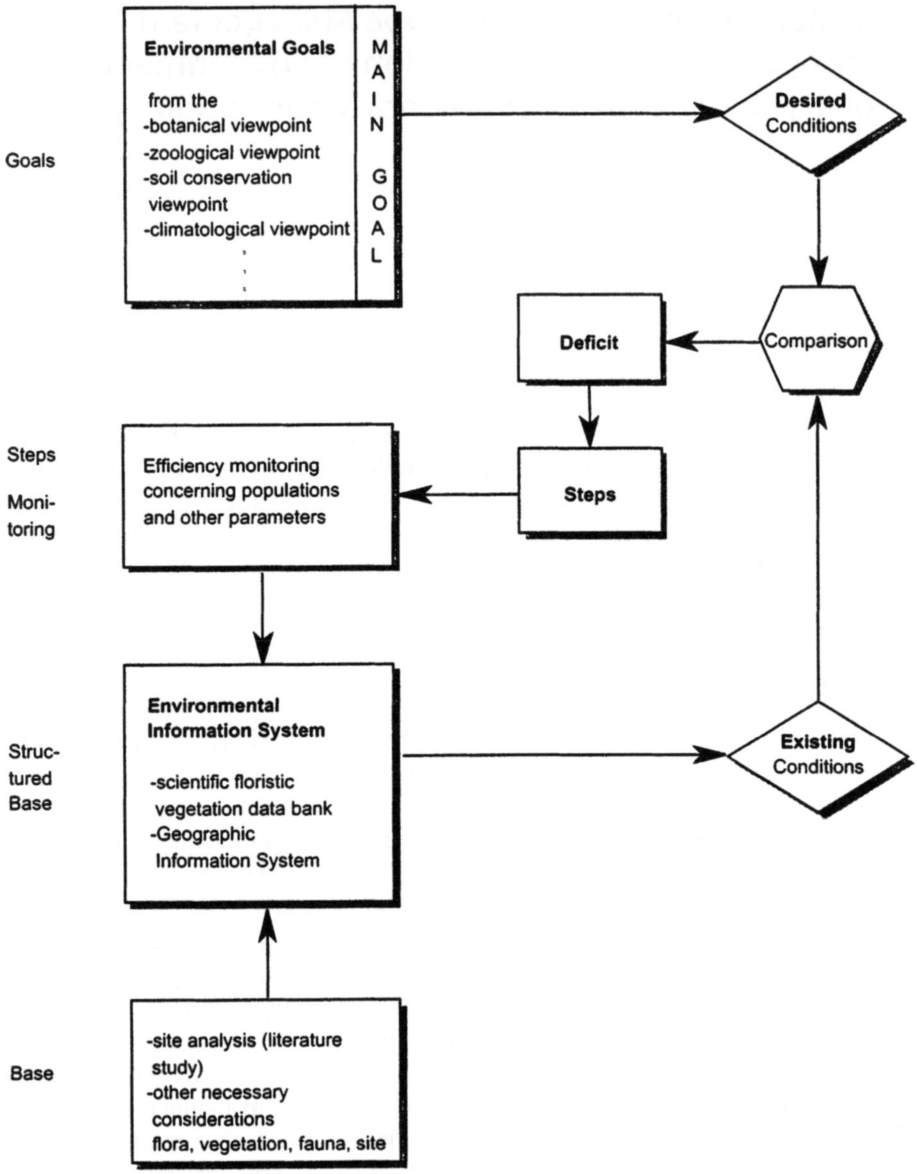

Fig. 1. Flow chart demonstrating implementation of integrated environmental quality goals and environmental information systems

Through the extrapolation of data from the environmental information system, patterns such as important population developments can be recognised. Significant parameters relevant to efficiency control are incorporated into the environmental information system as "current state" with respect to the *ecological data base*. Thus, a rational approach should separate the city into areas of similar natural and historical use.

3 Conclusion

It is extremely urgent that these goals are implemented in city areas. A comprehensive management concept is required which is designed using spatially oriented objectives that connect together relevant environmental disciplines. Within the framework of multi-faceted government regulations (UVP regulations, European Community Environmental Guidelines, regulations concerning construction, contaminated site clean-up, landscape planning, etc.) a quicker recourse to ecological data with an integrated spatial context is becoming increasingly important. When planning this concept it is of the utmost importance to consider the potential synergy between nature conservation, conservation of flora and fauna, soil conservation, valuable climatic functions, and the need of the population for recreation and relaxation in the vicinity of nature. Through the use of such an urban management concept, it would be possible to improve the integration of city planning and nature conservation.

References

Böcker R (1994) Stadt statt Landschaft. Der Bürger im Staat, 44 (1): 71–77

Fürst D, Roggendorf W, Scholles F, Stahl R (1996) Umweltinformationssysteme. In: Schriftenr. Fachb. Landschaftsarchitektur u. Umweltentwicklung d. Univ. Hannover, Contribution to spatial planning 46: p. 258, Appendix

Sukopp H, Wittig R (1993) ökologische Stadtplanung. In: Sukopp H, Wittig R (eds) Stadtökologie. G. Fischer, Stuttgart, pp 348–373

Ecological Planning in the UK and Suburbanisation in the Glasgow Conurbation

Bernd Stephan[1] and Olaf Kuhlke[2]

1 Introduction

The following study examines the role and importance of environmental planning in the UK, using the Glasgow conurbation as an example. Two major ecologically orientated counter policies to suburbanisation in the area are outlined.

2 Examples of Ecological Planning in the Glasgow Conurbation

2.1 Structural Changes and Industrial Decline

The structural change in the region was initiated by the decline of the main industries (shipbuilding, textile, mechanical engineering, mining and steel industry) after World War II and had severe long-term demographic, social economic and environmental effects. It went hand in hand with spatial effects of suburbanisation, urban blight and local population growth.

2.2 Suburbanisation of the Glasgow Conurbation

The suburbanisation process was initiated by the designation of the New Towns Cumbernauld and East Kilbride during the 1950s.

Four main government policies facilitated urban sprawl during the 1980s: privatisation measures for housing, investment politics for industry and trade, financial cutbacks (urban renewal and public transport) and a weakening of local planning by various central Government schemes (e.g. Enter-

[1] Kleine Ortenberggasse 1a, D-35039 Marburg, Germany
[2] Kent State University, Department of Geography, Kent, Ohio 44242-001, USA

J. Breuste, H. Feldmann, O. Uhlmann (Eds.)
Urban Ecology

prise zones, Unitary Development Plans (UDP) and Simplified Planning Zones (SPZ)). In 1996 a 2-tier political system was introduced for Scotland.

2.2.1 *Counter Policy I: Urban Renewal*

Urban renewal for vacant and derelict land provides one of the central pursuits of the region, as in other early industrialised areas of the UK (Strathclyde Regional Council 1995). Nonetheless, only minor results such as the set up of Action Areas could be achieved. Designated areas for environmental improvement were reduced due to cut-backs of central Government assistance.

2.2.2 *Counter Policy II: Environmental Planning*

The example of the "Central Scotland Forest Strategy" by the Central Scotland Countryside Trust" (CSCT) provides an insight into the planning process of a *quango* (quasi autonomous non-governmental organisation) which is supported by several woodland initiatives from the grass-roots level:

Objectives of the integrated strategy are the establishment of new woodlands and improved management of existing woodlands, realisation of productive potential of woodlands and wood industries, farm diversification, restructuring and woodland development, improved environment and image of the forest area, enhanced recreation, leisure and tourism opportunities in the countryside, community involvement and promotion of education and training, and, finally, the safeguarding of the valuable natural and cultural heritage. The partnership for the programme integrates 18 public and non-public groups and organisations through a formal partnership commitment, and another 16 bodies on a wider partnership agreement (CSCT 1995).

3 Conclusion

The British Planning system appears to be more action- and policy-oriented from both the grass-roots and local government level and also from the central government arena. This is supported by the British culture of debating and mediating which generates a higher degree of flexibility than the rather static system in mainland Europe. On the other hand, extensive changes were introduced by the former conservative government which supported planning as a tool for economic success. Privatisation schemes for public and public goods were popular. This was promoted by the generation of state competitive organisations such as Urban Development Companies and planning schemes (such as the SPZ) which have exclusive rights and statutory power beyond the "ordinary" planning system. On the other hand bottom-up planning from the local grass-roots level is more dominant within the UK, making planning decisions much more sustainable.

In terms of minimising ecological effects, there is still a lack of sophisticated and developed instruments with statutory power, such as the Environmental Impact Assessment in continental Europe. Whether or not New Labour will move forward to introduce extensive changes to the established system of the '80s has to be questioned. First signs, such as a general road building moratorium and proposals to strengthen funding for public transport, seem to indicate a shift towards a more ecology oriented policy. The bottom-up approach, which integrates various local initiatives on a grassroots level, is likely to gain in importance. A growth in participation of local groups could, in the long-term, create a higher degree of sustainability throughout the UK, compared to other parts of Europe.

References

Central Scotland Countryside Trust (CSCT) (ed) (1995) The Central Scotland Forest Strategy. Shotts (North Lanarkshire)

Strathclyde Regional Council, Department of Physical Planning (1995) Strathclyde Structure Plan 1995, Glasgow

Ecologically Responsible Mobility 4

Space Demand and Traffic Development – Ways and Means of Reduction (Sustainable Mobility and Urban Structure)

Dieter Apel and Dietrich Henckel

1 Introduction

The main purpose of the chapter is:

- to give some suggestions as to the meaning of ecologically sustainable mobility;
- to show the interdependence between mobility, traffic and space demand;
- to argue that the reduction of space consumption and traffic is a prerequisite to sustainability;
- to elaborate on ways and means to the reduction of space consumption and traffic;
- to make some sceptical reservations regarding the feasibility of achieving 'sustainable mobility'.

The chapter is based mainly on experiences in Germany, especially with regard to the possible instruments which are referred to in a German institutional and legal context.

2 What is Sustainable Development, what is Sustainable Mobility?

Human settlements and transport systems lead to consumption of space, air, water, energy and other resources. Settlements and transport systems that do not consume more than they replace may be considered ecologically sustainable; this is a high pretension on a very abstract level.

A broader definition of sustainability includes not only ecological, but also economic and social aspects (Camagni 1995). In our view sustainable mobility can only to be defined within a broader concept such as this. It is therefore not easy to define sustainable mobility in an operational way. One fact we can state is that the present traffic situation is definitely not sustainable

German Institute of Urban Affairs (DIfU), Straße des 17. Juni 112, D-10623 Berlin, Germany

J. Breuste, H. Feldmann, O. Uhlmann (Eds.)
Urban Ecology

regarding the heavy environmental, social, cultural and economic impacts, especially of motor car traffic. The simplest definition of a progression towards sustainable mobility (or ecologically responsible mobility) might be: the reduction of negative impacts. The impacts of motor car traffic on the environment and the urban structure are:

- consumption of resources (land, air, water, energy and others);
- accidents and danger to people and wildlife;
- destruction of urban mode of living (space demand, noise, danger and others);
- impairment of local economy (high speed traffic, longer distances travelled, out-of-town development).

It is well known that motor car traffic is responsible for a great part of the energy consumption and air pollution in the world and that in the future it will be both necessary and possible to reduce both to a great extent. What is less well known is that urban structures, which provide for a high degree of access and mobility by car, are inherently incompatible with structures served predominantly by other modes of transport. "This is because the car requires a lot more space than other passenger modes; in fact ten or twenty times more space per person travelling, not to mention the parking spaces required at homes and destinations. Where the car is accommodated, it will to some extent reduce the ability of the urban structure to support both the patronage and quality of public transport, and the possibility for journeys to be made on foot or by bicycle." "The other inescapable fact is that European Cities have been developing spatially in ways which are convenient for car use, and less convenient for other means of access. As people, jobs and other activities have moved further away from the city core, so the density and structure of built-up areas has loosened, and use of the car has increased. It is not the intention to explore the causal relationship here, but the trend of dispersal is inevitably linked to rising access to cars. Nor is the intention to challenge the undoubted benefits which people seek when they choose to live out of the city, such as more living space, more privacy and proximity to the countryside, but attempts to assess the scope and feasibility of traffic limitation strategies must inevitably take into account the powerful counter trend of city dispersal"(Pharoah, Apel 1996).

3 Traffic System, Space Demand and Urban Structure

Space demand is the one of the key problems of suburban development and traffic expansion. In the first place space demand is caused by tech-economic factors and by social changes. Today we consume for example much more living space (38 m^2 per person) than forty years ago (14 m^2). A similar expansion of specific space demands can also be seen for production, office space, retailing, leisure time developments and other facilities. Secondly the expansion of space consumption for settlements and traffic facilities is due, to a much greater extent, to car-based developments. The reasons are:

- car-orientation stimulates low density and dispersed development;
- car-orientation stimulates out-of-town shopping and similar developments with an extremely high space consumption;
- car-based developments require much more space for traffic facilities than structures based on public transport systems or pedestrian access.

In the old cities of German agglomerations 226 m^2 per person are consumed for settlements (including traffic facilities) whereas in the suburban rings around the cities the figure is over three times as high (824 m^2 per person).

The inter-relationships between the traffic system, demand for space, urban structure and traffic demand can also be shown by the following considerations, when comparing cities which have different structures.

- The "city type Delft" is characterised by high density, functional integration and is based on pedestrian, bicycle and public transport.
- The "city type Oldenburg" has only a medium density, less functional integration and is based on bicycle and car traffic.
- "The city type Denver" is characterised by a low density and a full car-orientation.

The comparison demonstrates the different space demands. The space consumed for housing areas, social facilities, industrial areas and traffic facilities is nearly four times larger for the "city type Denver" than for the "city type Delft".

This comparison demonstrates that in cities with a high density and low car ownership travel distances are relatively short and can mostly be made on foot, by bicycle or public transport. If car-orientation dominates, traffic problems and ecological problems arise: high space consumption, consumption of energy, pollution, noise, destruction of landscape and amenities, hampering of pedestrian, bicycle and public transport, destruction of urban mode of living etc. All these problems are much less important if we maintain and develop cities which have a high density, multifunctional land uses and a great deal of pedestrian and bicycle traffic and public transport.

4 Objectives, Strategies and Regulatory Instruments

The regulation of space consumption and traffic levels can be seen as a major components of strategies for a more sustainable urban development. Space consumption and growth of traffic are (as already shown) mutually dependent, therefore the only solution which seems conceivable should adopt a cohesive approach to both. Solutions in this field must also lend themselves to an integration of ecological restructuring of the economy within broader reform projects in the highly industrialised countries.

The first question is: which kind of urban structure is able to reduce the need to travel? Qualifying elements are as follows:

- generally an urban structure with a high density, mixed use (functional integration) and a good environment. (These elements are often found in old towns and city quarters with an historic character). For this purpose it is necessary to reduce car traffic and the number of parked cars;
- a good balance between the number of dwellings and places of work in towns and in quarters of larger cities;
- the concentration of new developments in locations accessible by a wide range of transport modes, not only by car;
- the concentration of traffic-generating activities (major centres of employment, leisure facilities etc.) at locations with a high accessibility by public transport and bicycle;
- higher densities in locations well served by public transport;
- the discouragement of low density and car-based developments on peripheral sites, especially out-of-town shopping centres and small, free-standing new dwellings in the countryside.

The objectives, most important strategies and regulatory instruments which are necessary for the reduction of space consumption and the need to travel (and therefore traffic, especially motorised traffic) are summarised. These are the results of a Difu-study worked out by an interdisciplinary group. This broad overview does not show the whole number of suitable strategies and instruments. (The whole overview of strategies and regulatory instruments and supplementing charts and figures is available in internet by http://www.difu.de English/occasional papers/space demand and traffic development). We focus here on key regulations on a general level (national, state and regional level) which mark the "regulatory framework" for local policy. These are:

- the reduction of space consumption;
- the concentration of development, "compact city";
- the reduction of motorised traffic;
- the encouragement of alternative means of travel.

We cannot elaborate on the whole range of strategies and instruments here, instead we concentrate on the three most promising strategies and instruments.

1. Space consumption generates substantial external costs which need to be internalised. Land values, as the long-established traditions of debate illustrate, are the outcome of social processes (rather than the consequence of a single investor's actions). The potential has not been utilised by the fiscal system where public measures have led to increases in land value. The present land tax solves neither the question of increases in value nor the ecological problems posed by the urban consumption of space and its external costs. We therefore propose a combined tax based on both land value and the quantity of used space. The land value component is based on the market value (here the local benchmarks can be used for calculation) and the other component is based on the amount of land absorbed. Each component contributes in its own way towards an economic incentive for a lower rate of land take-up, denser development and functional integration.
2. An indispensable step is to internalise the external costs of transport. A long-term, continuous increase in energy prices is needed. For transport, a higher petrol tax (implemented in well-defined, rising steps) seems to be the most appropriate and simplest measure. (This measure affects different classes of traffic to varying degrees. Higher petrol tax would probably have the least impact on freight traffic and the highest on leisure traffic.)
3. The high degree and continuous increase in space consumption and road transport is considerably enhanced by a lack of co-ordination within the regions. It seems to be essential to achieve a more cohesive urban development and transport policy at a regional level. Our proposal is to constitute urban regions within a distinct local authority framework. This means that urban regions would draw their legitimacy from a directly elected assembly and assume responsibility for all matters which can only be solved at the regional level. They would also take over certain functions from the federal and state governments. The aim is not to create another tier of local government, as the regions would replace the districts or evolve from them.

The proposed fiscal instruments will generate markedly higher revenues from these taxes. Since, however, it is not the aim of an ecologically oriented reform to increase government expenditure as a whole, fiscal revenue can be adjusted by reducing other taxes and levies, such as income tax or employer contributions.

The three main instruments described could, and should, be complemented by a wide spectrum of other instruments, although these also require reform. Examples are as follows.

- Under the current system of housing subsidies (especially of owner occupied houses) subsidies are given to any kind of development. In the future they should be given only to housing under the conditions of a maximum space consumption or a minimum density.

- Subsidies for commercially used buildings should be given only for concentrated developments according to the regional development plan.
- The current regulations preventing higher density parking and favouring out-of-city developments should be altered.
- Instead of parking space construction, there should be fiscal promotion of car-free housing areas.
- The existing parking-charge systems in towns and cities should be completed with regard to the entire higher density built-up-area. Secondly parking charges to the inhabitants should be included, as in Amsterdam and Stockholm where it has been in practice for more than 10 years.
- Instead of the current global motorway toll for lorries which leads to more traffic on other roads in some cities and regions, a levy on heavy goods traffic (based on the kilometres driven per year on all kind of roads) should be introduced.
- The many different tax deductions to car-drivers (for example the "Kilometerpauschale" or kilometre flat rate) should be abolished.
- Enterprises and facilities should be sited so that they are accessible by the relevant mode of public transport such as the so-called ABC planning instrument ("the right company at the right location") in the Netherlands. This planning instrument should also be introduced in other countries.
- Last but not least: investors will not prefer locations near public transport facilities if the transport systems do not change. To break the dominance of car traffic and road transport, transport policy must alter considerably, for instance by increasing the petrol tax, as already mentioned. It is also necessary to alter the priorities within the national transport network plan (Bundesverkehrswegeplan) (i.e. less money for motorways, more money for rail transport.). Moreover, an improved financial base for the municipal public transport is urgently needed. (In Germany we have to bear in mind that new problems for financing local public transport will arise due to deregulation of public utilities like electricity as cross subsidisation will no longer be possible.)

5 Concluding Remarks

We gave only a rather weak and negative definition of sustainability; any progress in the reduction of space consumption and traffic can be regarded as a step towards more sustainability. Even with this limited perspective the aim is very ambitious, it is not a trivial task.

We are rather sceptical about a short term improvement with respect to a reduction of space demand and traffic development for various reasons.

- The proposed strategies are necessary, but there are insufficient conditions for an improvement of the current situation. We are not convinced that the political will to introduce these measures is present, or that the political majorities to attain this will come into being in the near future.
- The major trends run counter to an improvement. The high investments in high-speed transport modes (high-speed railways, motorways, airports) will "shrink" the world further, i.e. reduce time distances and therefore extend the travel distances. This will lead to more inter-regional traffic and generate additional traffic on the local and regional level.
- The acceleration of many processes in our society (e.g. product cycles, land use cycles) foster a functional disintegration or a functional specialisation; this in turn leads to more traffic.
- The world wide integration of telecommunication and information networks will – in the net effect – also increase material transport, thus leading to more space and energy consumption.
- As already mentioned the implementation of the proposed strategies is a necessary, but by no means sufficient condition for an improvement; it is definitely not a solution for a more rigorously defined sustainable mobility; at best we can reach some relief in comparison to the current situation.
- The conclusion of this sceptical view is that we have to take every measure which tends to lead to an improvement, and that despite the grim perspectives we have to do all we can to make mobility more sustainable.

References

Camagni R (1995) The form for sustainable city. Milan (unpublished)
Pharoah T, Apel D (1996) Transport Concepts in European Cities, Aldershot p 8

Least Cost Transportation Planning – An Information System for Economically Efficient Low-Impact Mobility in Cities

Tilman Bracher[1], Matthias Bergmann[2], and Angelika Uricher[2]

1 The Need for Cost Transparency

Lately, discussion on transportation cost has been very emotional, irrational and ideological. Very often, the cost argument is used to prevent innovative policies. At present, city authorities lack financial means and are not able to finance mobility concepts other than mainstream.

A major deficiency in the transportation sector is the lack of cost transparency. There are many public and private institutions as well as individuals within the transport sector, whose expenditures are neither statistically summarised nor, in some cases, known at all. Public bodies have a budget structure where, in relevant planning steps, only current income and expenditure rates are taken into account – leaving out overall costs and benefits. In addition to direct expenses, there are numerous indirect expenditures (and incomes) within the transport sector. Among these are police forces, environmental as well as judicial authorities.

To compensate for the deficiency of a transportation planning system which is not properly based on economic transportation effects, the concept of Least Cost Transportation Planning (LCTP) was adapted for the German environment. In the United States, Least Cost Transportation Planning was applied earlier, however under different circumstances and with other objectives.

Different means of passenger transportation can be looked upon as substitutes. Road investment and public transport investment are usually considered as being independent values. They are organised in different institutional frameworks (i.e. different public bodies, laws, and budgets) on the municipal, state and federal level.

In political decisions, there is an increasing emphasis on cycling and walking as modes of transport due to their positive impact on urban ecology.

[1] IVU – Gesellschaft für Informatik, Verkehrs- und Umweltplanung mbH (Company for Information Technology, Traffic and Environmental Planning Ltd.), Bundesallee 88, D-12161 Berlin, Germany

[2] Öko-Institut e.V. (Eco-Institute), Friedrichstraße 165, D-10117 Berlin, Germany

J. Breuste, H. Feldmann, O. Uhlmann (Eds.)
Urban Ecology

But these modes are not yet thoroughly integrated into transportation planning processes; they still lack institutional integration. Thus, these modes are usually not regarded as substitutes for motorised modes. Inter-modal integration of transportation is necessary if modal shifts are supposed to minimise car-use, car-ownership and car-parking.

2 Calculating the LCTP Table

2.1 How can we Learn from the Energy Sector (LCP)?

The main idea behind Least Cost Planning in the energy sector is, that there is no increasing demand in energy itself – only in certain energy services such as heating or light. Energy supply companies could therefore meet this demand by either selling electricity or the equipment for saving energy. Through this concept, the energy sector succeeded in reducing the demand for energy, despite economic growth and an increasing demand for energy services.

There are two major differences between the energy sector and transportation: Energy as a product is rather homogenous, transportation relatively heterogeneous. Transportation vehicles have a high intrinsic value or status function, and even consumers themselves can be producers of transportation (self-driving). All modes have special characteristics which determine their market sectort.

Unlike energy, transportation is not offered by a single enterprise. Roads, public transport services and vehicles are operated and financed by various private and public bodies. Among them are car- or bicycle holders, private companies as suppliers of parking space on private ground, private and public companies operating public transport services and municipal, state and federal bodies that operate and finance road networks.

Since a single transport company assumed to be as economic as an energy supplier, we developed the idea of a virtual "company" with different departments and products. Each institution or person involved in urban transportation is considered to be a department and every mode a product of the firm.

2.2 Application to Freiburg and Schwerin

Freiburg and Schwerin are the two cities examined by our research group City: *mobil*. Freiburg has about 180.000, Schwerin has 120.000 inhabitants.

Both cities are known for their innovative approaches towards transportation: Freiburg is known for its "pedestrianisation", traffic calming, improvement of the cycle-track system and its efforts towards improvement of public transport within urban and - recently - rural regions. Schwerin has a tramway-oriented structure, with a high housing density along major tramway lines.

To use the Least Cost Transportation Planning concept in Freiburg and Schwerin, four major assumptions were made:

- we defined a relevant market sector of transportation; the cities of Freiburg and Schwerin were defined within their political boundaries. Transportation data on modal split and passengers' kilometre coverage existed,
- we assumed that Freiburg and Schwerin were both served by a single virtual enterprise offering all private modes of transport (car, bicycle, walking) as well as tram and buses in the city area. In addition, this virtual company takes care of infrastructure and operations,
- the virtual enterprise's main objective is cost efficiency,
- our model included user preferences as passengers may not change transport modes, if recommendations are only "theoretical".

To identify local expenditure structures, various data sources, including the municipal budget, have been thoroughly monitored. In those cases where no specific data was available, national data was used. We always tried to be as precise as possible. From municipal budgets, direct cost of the transportation sector and indirect cost of other departments and resorts were included.

2.3
Transparency of Present Cost Structures

To identify present transportation cost within a city, input data from private users, budgets of public transport companies, and municipal budgets have been combined in a comprehensive table. The table is calculated for a given year and includes (money) income and expenditure of that year only. However, not all available data directly fit into the structure of the table, one of which is road costs, i.e. construction, maintenance and operation.

Roads are used for commercial traffic as well; due to their high axle weight and the wide lorry-space, a considerable share of road costs does not depend on the mobility of private households. According to their road performance, private cars are responsible for nearly 50% of all road costs, commercial traffic, including heavy lorries, for another 48%, public transport (buses) and cycle traffic for 1% each.

2.4 Research on User Costs

User costs have been calculated based on expenses of all private households in Germany. Data were adjusted for Freiburg and Schwerin by considering the actual distances travelled in both cities. Parking costs were also included. The user costs collected in our research derived from different federal and local statistics.

The largest single expenditure is "vehicles", followed by "petrol". Similar data were calculated for public transport and for the cycle traffic. Costs were either fixed or variable. In total, users spend much more money per kilometre when using private cars (0,35 DM) than on public transport (0,08 DM in Schwerin, 0,17 DM in Freiburg) or on bicycle (0,11 DM).

2.5 Public Budgets

From public budgets, the relevant expenses on municipal, state, federal and company level have been identified. Municipal public transport is not only subsidised by fares, but also by several state budgets. A major source of income is an investment scheme for communities on the federal level, and there were also subsidies from the community itself or cross-subsidies from other municipal companies.

Transportation cost within municipal budgets include direct and indirect expenditures.

Direct expenditure on transportation is identified directly from the budget-book. Examples are road investment, expenses on public transport companies, bridges, parking, traffic signalling, and more. Indirect expenses are those being related to transportation and not being directly included in the transportation budget. These are costs from other offices, such as financing environment and public order. Their shares have been calculated according to experts' discussion on the individual budget positions. This amounts to about 20% to 22% of direct expenditures.

2.6 Calculation Methods and Results

The completed table for the city of Freiburg shows, that in Germany, users finance nearly 70% of all transportation cost, while city and public transport companies on one side and state and federal agencies on the other side cover the remaining 30%.

Most money is spent on car traffic; followed by public transport, and only a small amount can be devoted to cycle traffic. Pedestrian figures are 0 as there

Table 1. Summary of Transportation Cost in Freiburg from the LCTP-table

Costs in million DM	User	Town and Pub.Transp.	Federal and state	"Enter-prise"	External costs
Private car	281,9	48,6	3,5	333,9	81,7
Public transport	41,4	36,2	54,5	132,2	5,7
Bicycle	11,4	0,6	0,1	12,0	0,0
Pedestrian	0,0	0,0	0,0	0,0	0,0
"Enterprise"	334,7	85,4	58,0	478,1	87,4

were no relevant expenditures within this field that were not caused by other means of transport (see table 1).

2.7 Considering External and Social Effects

The annual survey of expenditures and budget data is important, but cannot be regarded as sufficient information for the evaluation of policies. Therefore LCTP data have to be integrated into a more comprehensive approach.

Especially when considering car traffic, not only possible positive effects on the economy but also negative social and external effects of car use should be stated. However, it is even more difficult to apply cost figures to the latter effects.

In order to compare the relation of budget data with the external effects, estimates of the latter have been made for both cities. The effects were calculated according to a method suggested by Huckestein u. Verron (1995) from the German Federal Environment Office. Their estimates are lower than those from most other sources (careful estimate).

3 Political Implications of LCTP

The results show, that both cities spend more than 50% of their expenditures on subsidies for cars, while giving priority to the "green" non-car mode in their politics. In addition, external transportation costs exceed the total annual community expenditure for transportation.

In addition to absolute figures, the cost data related to the distances travelled are shown on the following table.

The car is the most expensive means of transport. As most of its costs are not considered within municipal budgets, urban administrations and politicians do not get the full effect of car-oriented policies.

Public transport cost is mainly covered by the public. Therefore its social, ecological, urban and mobility benefits must be identified and promoted to strengthen and maintain its position in urban policy schemes.

Walking and being outdoors are fundamental for urban life, creating only minimal costs. Thus, improving conditions for pedestrians is not just a transportation issue. The bicycle is far more economical than all motorised vehicles. Therefore its potential should be fully developed. Bicycle use could be supported by

- institutional frameworks (budgets, law, planning system, modelling)
- activity patterns (e.g. shopping and leisure activities) where the bicycle fits in,
- spatial and land use patterns to reduce distances and to improve bicycle access,
- an image campaign to change the model's "unattractiveness function" and
- attractive facilities, as 50% of all existing car trips are 5 km or shorter. (see figs. 1 and 2)

4 The Evaluation of Projects with LCTP

The LCTP approach may also be used for the evaluation of projects or strategies. LCTP may be applied to a wide range of concepts, as it can integrate investment costs and operating costs of any kind. This was shown in three test scenarios. As plans and strategies can only be successful if users (and politi-

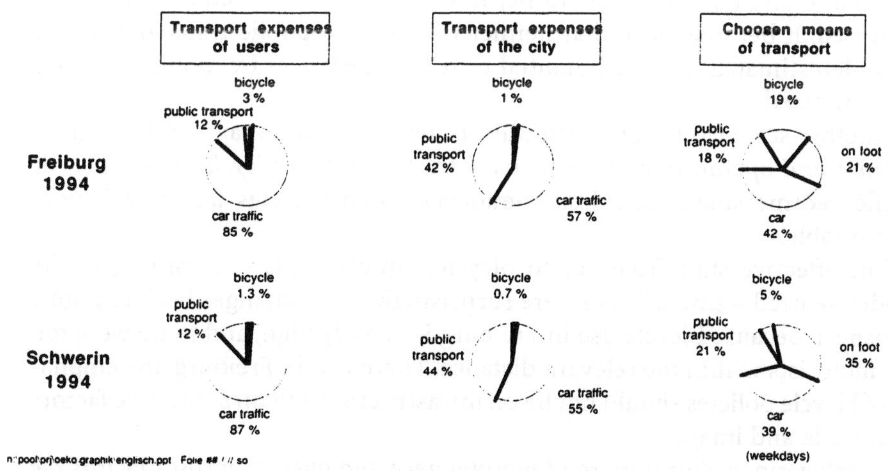

Fig. 1. LCTP Results from Freiburg and Schwerin

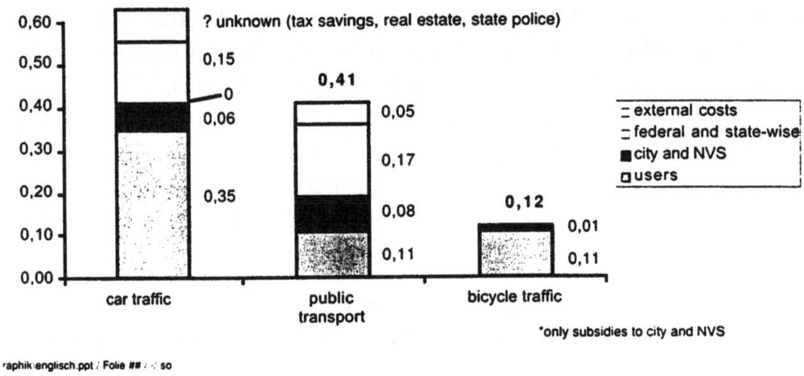

Fig. 2. Expenditure (and external costs) in DM per passengers' kilometre coverage regarding urban public transport in Schwerin 1994

cians) find themselves positively affected, a win-win-situation should be approached. To estimate user acceptance, we applied a Generalised Cost Model to include monetary, time and effort cost.

The first scenario was on *a shift of municipal subsidies for public transport operations in Freiburg to six new suburban mobility centres.* The aim was to achieve a shift from car to so-called "green modes", making suburban centres attractive for shifting some shopping trips away from the city centre. Demand effects were calculated on experiences and assumptions.

To fill the financial gap of public transport, a major price increase resulted in shifts towards car use, walking and cycling. To our surprise, most shifts happened with long distances – on trips too far for the bicycle or for walking. This led to an increase in car traffic, and thus the generalised cost term also increased. It became clear, that neither the value of public transport should be underestimated nor the potential to reduce distances by mobility centres overestimated.

Another scenario for Freiburg was on the effects of a major *shift of maintenance and operation costs from road traffic towards bicycle traffic*, i.e. car traffic became slower and less comfortable, and bicycles faster and more comfortable.

The effective shift from car to bicycle, however, was – according to the model we used – small. There were surprisingly many changes back to public transport, because bicycle use in Freiburg is already high, and there were too few motorists within the relevant distance. Therefore, in Freiburg, the emphasis of bicycle policies should not lie on infrastructure, but on subjective factors as attitude and image.

In Schwerin, *a shift from road maintenance and operation cost towards the improvement of local environment* was suggested so that road network quality would decline. By applying this scenario, attractiveness of the urban environment of Schwerin's major housing estate was increased. This should attract people to stay in the area more often for leisure purposes, shopping and other free-time activities.

According to our model, enough people might accept this strategy and stay in the area – thus reducing overall distances and transport cost.

5 Summary

Least Cost Transportation Planning for municipal transportation provides a thorough and comprehensive information system for communities. It can be used to identify all expenditures and incomes and the allocation of resources to different modes. It enables communities to analyse the present situation of transportation expenditure and to evaluate the economic side of planning projects. Cities will be able to evaluate various concepts to reduce traffic in its consequences for urban households, as shown in the three scenarios.

References

Huckestein B, Verron H (1995) Externe Effekte des Verkehrs in Deutschland. In: Mobilität um jeden Preis? Berichtsband eines Expertenworkshops, FGU Berlin. Im Auftrag des Umweltbundesamtes, Berlin

Workshop: Mobility Concepts and Strategies 4.1

Sustainable Mobility in Urban Regions Aims, Methods, Planning Strategies

Markus Hesse[1] and Harry Lehmann[2]

1 Project Information

The interdisciplinary research project on „Ecologically Acceptable Mobility in Urban Regions" (Ökologisch verträgliche Mobilität in Stadtregionen) was launched in May 1994. The project is supported by the German Federal Ministry of Education, Science, Research and Technology (Bundesministerium für Bildung und Wissenschaft, Forschung und Technologie, BMBF), and is situated within the ministry's Urban Ecology programme. The work is scientifically co-ordinated by the Wuppertal Institute for Climate, Environment and Energy (WI) and the Institute for Ecological Economic Research, Berlin (IÖW, Institute für ökologische Wirtschaftsforschung).

The goal of the five year project is to develop models, criteria and strategies for action towards a sustainable structuring of mobility in city regions. Beside the WI and IÖW, participants in the Research Group are: the universities of Bremen (T. Krämer-Badoni), Hamburg-Harburg (D. Läpple) and Kassel (H. von Winning), the Bureau for Integrated Urban and Transport Planning (BiS), Bonn, and the Pro Terra Team in Dortmund. In the cities of Bremen and Halle/Saale, local authority staff are involved from the Department of Women, Health, Youth, Social Affairs and Environment Protection and the Department of Construction, Transport and Urban Development, Bremen, and the Department of Planning and Environment, Halle.

This chapter shows alternative opportunities in modelling travel behaviour which reflect modifications in daily life mobility (change of experiences or „routines") and discusses practical options to deal with their environmental impact.

[1] IRS/Institut für Regionalentwicklung und Strukturplanung, Erkner (b. Berlin), Germany
[2] Wuppertal Institut für Klima, Umwelt, Energie GmbH Döppersberg 19, D-42103 Wuppertal, Germany

J. Breuste, H. Feldmann, O. Uhlmann (Eds.)
Urban Ecology

2
Background of the Problem

The idea of sustainable development is based on the recognition of the equal rights of both present and future generations in all continents and to the fulfilment of their basic needs and requirements. In order to meet these aspirations to a better life an environment must be maintained in which all human beings can live naturally. Science and planning disciplines today still provide little knowledge about the earth's system and have hardly even begun to investigate some parts of the ecosystem. We are aware of little information on the effect of human action, either with respect to the intensity of the ecosphere's reaction or regarding the time scale involved. As a precautionary measure we should therefore attempt to minimise anthropogenic effects on the earth's system and thus prevent as far as possible any negative consequences. Mobility, as a large contributor to environmental problems, needs to minimise all of its impacts.

When one considers the environmental significance of transport, specific shortcomings are obvious: the current situation in most of our cities is poor since air pollution is still severe and will not be reduced over the next few years. Long term problems that remain unsolved are present in the field of noise, CO_2-emissions and demand for space. Further environmental conflicts are visible if analyses reflect the environmental impact caused by the production of cars, trams etc. and transport infrastructure as well as emissions due to the use of vehicles. It is obvious, but widely underestimated, that buses and trams also contribute to the global warming potential, not as much as cars, but still significantly. Consequently, information on the modal split may not be sufficient to indicate ways of decreasing the pollution caused by transportation and need to be supplemented by absolute data on realised VMT (vehicle miles travelled). However, future forecasts predict a drastic long term reduction of NO_X-, CO^-, VOC^- and diesel particulate-emissions due to emission reduction technologies, improved materials and new combustion techniques.

3
Aims

Successful implementation of these measures and a positive impact on the environment require specific methodological adaptations within the planning system and planning culture. Traditional transportation planning is oriented towards the supply of services and an operational and organisational approach. The basic intention is to provide a given individual demand with transport options and infrastructure- thus neglecting environmental and capacity constraints and not asking the question whether mass mobility (e.g. by cars) is ecologically sustainable or not.

The basic aim of this project is to define steps towards more environmentally friendly mobility in urban regions. The project undertakes a critical review of existing (traditional and modernised) methodologies in transportation planning and political measures in this field. A second aim is to develop better planning strategies and political actions which are adjusted to the complex system of mutual dependencies in the urban transport sector. The intention is to influence the planning and the policy measures in terms of urban acceptance and environmental sustainability through improved understanding of the transport generation framework. The definition of a preliminary set of targets for minimising certain emissions is also part of the project.

4 Methodology

One of the main reasons for the policy gap in traditional planning procedures is that underlying macro models do not provide adequate knowledge about daily life and travel behaviour in modern society. They are based on poor data sources – using average and theoretical assumptions of individual choices – and offer incomplete descriptions of different transport modes. Finally, they do not consider changes in lifestyles or corporate policy that are basic requirements for an efficiently working transport policy.

The approach of the project was to analyse and conceptualise the system of urban mobility. This system has mutual dependencies and a network of interactions and interdependent cycles on different levels of society. A complete view of the urban transport system has to combine a more macroscopic view with microscopic perspectives of a multidimensional system.

Policy measures distinguish between top-down and bottom-up approaches. A typical top-down measure is the increase of transport prices or the introduction of pricing policies in general. A typical, frequently used bottom-up approach is providing information to actors. These approaches are reflected in model making by aggregated and disaggregated model theories. The different levels of society are also represented in the dissolution of the network of interactions of the different actors and groups of actors.

In this context, our understanding of urban transport consists of three elements:

- the transport infrastructure,
- the vehicles,
- the running system (VMT).

Empirical analysis in the cities of Bremen and Halle (eco-balance) offered a differentiated view of the environmental impacts of these three elements. Although the role of the diverse traffic emissions is well known in theory and practice, there has been little knowledge on the effects of infrastructure and material flows generated by the products (vehicles). Our comparative eco-balance of the global warming potential (carbon-dioxide-emissions) of the transport sector shows that almost 20% is related to the production of vehicles, 4% to the laying down of infrastructure and almost 75% to the use of vehicles (i.e. emissions) (Bremen). In Halle the contribution of vehicle production was more than 25% (based on per capita, Ankele, Jancke 1997).

One consequence of these findings is that methods of „sustainable“ mobility are investigated to find ways of reducing the impacts of all 3 elements, particularly by „lean“ infrastructure management, reducing the number of vehicles and a more efficient use of motorised transport. „Environmentally friendly“ urban transport policy requires a new insight into the role of the automobile in cities.

In trying to find appropriate views of the urban transport system the project first started with system dynamic analysis. At this point we realised that the behaviour of the individual is the central element of the urban transport system. Consequently, the process of generating a model concentrated on a „micro“-model, a model based on the simulation of an individual's activities. The sum of individual activities generates the urban transport situation (Brüggemann et al. 1997).

The simulation is organised in three phases:

- transport decision,
- transport enforcement,
- transport experience.

The simulation of the actor is based on three main ideas. Each actor has a subjective and individual view of the city (opportunities, transport connections, transport preferences: so-called „mental maps“). The actor makes decisions based on a set of individual rules and follows routines, i.e. if he/she has a solution for a certain transport problem, he/she will follow this solution even if it is not the best one. Through the use of such a model we hope to discover potentials and sensitive fields for changes in individual behaviour (e.g. routines, mental maps, decision rules).

5 (Planning) Strategies

Generating adequate planning strategies and policy measures necessitates the solution of a two-dimensional problem:

- the necessary individual and social changes (behaviour),
- the development of technical and planning issues and framework.

In practice this means that actors must not only change their routines and decision models, but individual changes have to be supported by structural and technological developments. For a significant reduction of environmental impacts to take place by changing individual's behaviour systematic strategies are needed. „Guard rails" for future development have to be defined to link environmental targets with the proposed changes in transport behaviour and structures (Hesse et al. 1997). These „guard rails" control the direction and constrain the problem, its potentials and priorities.

In this context we do not believe in comprehensive planning options which follow a top-down policy through a system dynamic approach (macro-view). Due to the complexity and instability of the entire urban mobility system we prefer searching for important „influence chains" and „mutual dependencies". These help us to identify „planning corridors" for urban mobility. These corridors could be:

- car free or car-reduced housing areas,
- the interrelationship between land use and transportation,
- transport efficient logistics/corporate policy.

References

Ankele K, Jancke G (1997) Ökologisch verträgliche Mobilität in Stadtregionen -Teilprojekt 2: Ökobilanz von Transportvorgängen in Stadtregionen, Stoff- und Energieflußbilanz, 3. Zwischenbericht April 1997, Berlin (unpublished)

Brüggemann U, Lehmann H, Reick C (1997) Handbuch zur Systemanalyse Mobilität. Zwischenbericht zum Teilprojekt 1: Systemanalyse. Wuppertal (unpublished)

Hesse M, Lehmann H, Reick C, Wilke G (1997) Thesen zur Systemanalyse im Forschungsverbund „Ökologisch verträgliche Mobilität in Stadtregionen". Berlin, Wuppertal (unpublished)

Transport and Environment: Principles and Guidelines Towards Sustainable Means of Transport

Udo J. Becker

1 Introduction

In this chapter, we intend to develop principles and guidelines for sustainable means of transport. It must be noted that such principles and guidelines – if implemented – imply far-reaching consequences for transportation, human beings and the way we live. The main question, therefore, is whether we should change our way of travelling and our lifestyle. This seems to be the global question that has to be addressed.

2 Transport and the Environment

The only university chair on Transport and the Environment ("Transportation Ecology") in the German speaking scientific community was established in 1994 in Dresden at the University of Technology. In the process of establishing this chair, it became quite obvious that a number of different approaches to the field of transport and the environment exist:

1. The first approach covers only two areas, i.e. transport itself and the environment; this approach tries to minimise the negative effects of transport.
2. The second approach covers three areas (people, transport and environment) and can be considered to include systematic aspects; feedback and connections between the transport-generating environment and between transport and the environment affected by transport are considered.
3. A third, rather innovative approach is based on "sustainability": limitations of the ecological systems, carrying capacities etc. are included, limiting the options available for the transport-generating environment, transport and nature.

At the end of this millennium, while approaching the 21st century, it becomes more and more obvious that the human way of life has to change. This is espe-

Dresden Technical University, Chair of Transport Ecology, D-01062 Dresden, Germany

J. Breuste, H. Feldmann, O. Uhlmann (Eds.)
Urban Ecology

cially obvious in the field of transportation: transport as it is known today is definitely not sustainable and is becoming more unsustainable, day by day, in all countries worldwide. For this reason, transport engineers and transport politicians have to consider changing transport systems.

3 Lessons and Measures

The lesson drawn from these considerations is obvious. It is not possible to discuss transportation as an isolated sector of human activity, and it is not possible to evaluate transportation measures only according to their effects on transport. "Time" and other affected areas influence our transportation behaviour and transport measures; these factors have to be taken into consideration.

A number of measures were developed to deal with transport as a system when considering carrying capacities or when trying to develop sustainable solutions for transport. These are relatively easily understood and are described as follows:

1. *Stable systems* are important. This means that systems, founded on continued growth assumptions are no longer tolerable.
2. When evaluating effects or "responses" of a system, it is not sufficient to base the evaluation of the current situation of the system on static responses only. *Dynamic effects*, i.e. changes within the system, must also be considered.
3. "*Positive*" *feedback* is dangerous for systems as the system may "explode".
4. *Irreversible processes* are to be avoided.
5. As discussed above, *long-term illnesses and incubation periods* complicate the evaluation process of transportation effects and may have dangerous effects.
6. "*Quick solutions*" are not always solutions at all.
7. *Comprehensive life cycle assessment* is required for all transportation decisions.
8. In transportation we should deal *with functions, not with products.* It is primarily not important how someone obtains his groceries or goods or travels to work if he can satisfy his needs in an appropriate way.
9. *Exergy*, i.e. "energy which can be used for productive work", should never be wasted.
10. When designing a system, we always have to build "*checks and balances*" into it. No part of a system should go unsupervised.
11. When designing a system, it is always helpful to use *existing forces and pressures* instead of trying to fight these forces.

12. Permanent solutions in transportation should be *decentralised*, based on the *principle of subsidiarity* and developed in a *democratic way*.

4 Rio de Janeiro and Vancouver

The UN Conference for Environment and Development in Rio de Janeiro in June 1992 was a milestone in the debate about sustainability. The Agenda 21 document with a list of the actions which need to be carried out in the next century was developed.

We do not agree with the name "Agenda 21". We think this name is misleading, since it is not an agenda for the next century; instead, this list of actions to be put in place describes Agenda II of humankind. While Agenda I dealt mainly with the problem of survival and fighting the "hostile" environment, Agenda II describes the task of integrating human societies into the system, i.e. into the limits of existing ecological systems. Agenda II, thus, is completely different from Agenda I and describes the tasks for human society not only in the next century, but well into the next millennia.

In the formal Agenda 21 document, transportation is dealt with in six points. To make transport more sustainable, our societies have to: (1) reduce transportation demand; (2) develop public transport; (3) promote non-motorised transport; (4) integrate planning with the maintenance of public infrastructure; (5) develop and communicate processes between the different countries and communities; and (6) change the patterns of consumption and production.

These principles were recently discussed worldwide. In 1996, at the Vancouver Conference, nine sustainable transportation principles were developed:

1. *Access.* People are entitled to reasonable access to other people, places, goods and services.
2. *Equity.* Social, inter-regional, and inter-generational equity must be ensured; developed and developing economies must co-operate.
3. *Individual and Community Responsibility*; Individuals and communities are responsible for acting as stewards of the natural environment by making sustainable choices.
4. *Health and Safety*; Transport systems should protect the health (physical, mental and social) and safety of all people.
5. *Education and Public*; People and communities need to be fully engaged in the decision-making process about sustainable transportation.
6. *Integrated Planning*: Transportation decision makers must integrate environmental, health, energy, financial, urban design and other approaches.

7. *Land and Resource Use*; Transport must use land and other natural resources efficiently while preserving habitats and maintaining biodiversity.
8. *Pollution Prevention*; Transport emissions should not threaten public health, global climate, biological diversity or essential ecological processes.
9. *Economic Well-Being*; Prices should relate to sustainable transport; full environmental, economic and social costs, present and future, are to be covered.

The Vancouver principles are primarily based on human needs. It becomes clear that transportation systems fulfilling the requirements described above will be completely different from those currently found in the world. Whether we look at developing or industrialised countries, there is a worldwide trend towards more frequent and lengthier travel as well as heavier motorised traffic. Today, products are becoming more similar. The concept of a factory serving the whole world is becoming increasingly attractive. These processes are based on very low transport prices: prices of transport are low because environmental effects are not included. Sustainable transport will thus have to reverse all of these current trends.

5
Criteria and Categorical Imperatives

Transport has to change, i.e. it has to become sustainable. What will sustainable transport look like? At present, it is not possible to describe sustainable transport in detail. It is not even helpful to describe the characteristics of sustainable transport in detail. The final *destination*, our *vision*, as well as the *aim of sustainable transport* should not be defined in detail, but kept open. We can, however, define key features of sustainable transportation and describe what needs to be done at present when we make a decision.

Sustainable transport will, for example, entail a reduction or elimination of CO_2. It is quite probable that we shall have to reduce fossil energy consumption and thus reduce fossil CO_2 emission of transport to zero. This holds true even for NOx emissions. While CO_2 emissions entail global challenges, the problem of NOx emissions is on a regional or continental scale. Transportation also has to be sustainable on a local scale, where noise may serve as an indicator.

In order to achieve these aims, transport has to be improved *technically*. Our industrialised world is based on technical solutions and approaches. We have to develop these approaches. The train called "technology", which we boarded a long time ago, is running and should be used to serve our purposes.

However, transportation in general has to be *shorter and slower.* Both aspects are associated with each other. It is only possible to make longer trips if – in the context of a given time frame – transportation modes become faster, i.e. we have to change a trend. Transportation will only become sustainable if we reverse the direction in which transport is currently heading worldwide.

It is not clear at this point which solutions will be developed. It is, however, very clear that two transport modes have to be part of the solution: *bicycling and walking* should be promoted whenever possible.

In order to reach these goals, we have to stop subsidising transport. At this point in time, transport is subsidised by non-transport users, foreign countries and future generations. Transport will only become sustainable if we stop these subsidies and internalise external costs. By our doing this, transport may approach system conditions which are *stable, global* (no differences between different parts of the world) and in which all humans are treated *equally.*

From these considerations, three *categorical imperatives* can be derived:

1. Improve transportation ecologically by technical means.
2. Reduce passenger/vehicle miles travelled.
3. Expand, or at least maintain, the options and possibilities of all other/all future humans to change or choose their own transport systems.

We have to find clever solutions to reduce these distances. The third principle addresses the options for future generations, all regions and all humans. We have to keep options open for other people, no matter where and when they live. These people have to be able to address their transportation needs in a way *they* decide upon.

The consequences of the path described above are quite drastic. While the nature of sustainable transportation is not yet clear in detail, it is obvious that steps towards sustainable transportation are necessary now. All of us may consider these principles and guidelines in our daily decisions and evaluations. The future (transportation) world will either be sustainable – or it will not. Of all the factors influencing our way of life, *time* is the most crucial resource. Thus, we have to start this process now. The sooner we start the process, the easier it will be to adopt; but the longer we wait, the more drastic, expensive, anti-social and catastrophic it will turn out to be.

Modern Society and Urban Transport – Understanding the Difficulties in Achieving Ecologically Responsible Mobility

Thomas Krämer-Badoni

The sociological observer is confronted with an ever present paradox in modern western societies. Although information on how much damage our behaviour does to the environment is becoming more widespread and the ecological and climatic harm caused by car pollution is a well-known fact, the number of cars is on the increase. The number of registered vehicles in Germany topped over 40 million in 1995, and there is no end in sight to this trend.

Many ecologically oriented academics tend to hold the opinion that the population of highly developed countries is just plain irresponsible, selfish and inconsiderate. If we do not want to share this same opinion then we have to make an attempt to try to understand the trends in social development which characterise modern society's relationship to space.

Apart from the basic trends which have promoted mobility among widespread sections of the population, certain sections have experienced an improvement in their income (the so-called elevator (upwardly mobile) effect) and an increase in spare time (the latter not being essential for an improvement in living standards). There are two other trends I would like to deal with: 'individualisation' and 'suburbanism'/suburbanisation. These two trends have developed parallel to one another, are epochal, and partially supplement and reinforce one another. I shall only be dealing with these two trends or processes although there are, of course, developments such as differentiation which have long-term effects; spatial impacts can be shown very clearly using the example of 'individualisation' and 'suburbanism'/ suburbanisation.

Individualisation means that the individual breaks away from traditional social ties such as family or class. It is the disintegration of a specific form of integration. At the same time, 'individualisation' entails liberation and an opportunity to be free. 'Individualisation' forces individuals to take their biography into their own hands. Nothing can be taken for granted. It is a real freedom (and emancipation); however, this liberation also involves many risks.

University of Bremen, KUA, ZWE "Work and Region", P. O. Box 33 04 40, D-28334 Bremen, Germany

J. Breuste, H. Feldmann, O. Uhlmann (Eds.)
Urban Ecology

The spatial impacts of the process of 'individualisation' can be seen to be the consequence of the disintegration of traditional integration mechanisms. One such impact is that households are becoming smaller. In 1994, 67.3% of households in Germany were single or two-person households. In 2015, this number will amount to about 70%, leading to a considerable increase in the number of households. At the same time, population growth is stagnating. Thus there was about a 4.2 million increase in households between the 1970 census and the microcensus of 1994 which was not based on immigration.

This has resulted in the constantly rising demand for floor space. In 1972 the average German used 26.4 m^2, in 1995 this increased to 38.9 m^2 in the core states. Inevitably this will lead to an expansion in the number of populated areas as fewer and fewer people inhabit the same residential areas. Cities are bursting at their seams, and distances between home and other functional places (shopping, work, cultural activities and hobbies) are forever increasing.

These trends in development are reinforced by the increasing number of single female households. This is an effect of 'individualisation' in which the person is employed and economically independent and thus subject to the same demands on mobility as their male colleagues. 'Individualisation' has an impact on mobility and space.

These development trends, which can be included under the process of 'individualisation', to a certain extent intensify the spatial mobility of the increasing number of households.

These processes are made possible and are reinforced through an imminent trend due to technological developments. Technological appliances can be used individually and are at the individual's disposal, spatially and socially, even temporally (Sachs 1984). A large number of such developments exist which have long lasting social effects: from cinema to television, and from trains to the car, to mention but a few. Mobility in our society can only really unfold in connection with this dynamic development. The car is, despite vehicle mobility/'automobilism', developing into a large technical system, a genial means of mobility and communication since it allows us to organise our own mobility according to social (individual), spatial (wherever) and temporal (whenever) criteria.

Suburbanisation goes hand in hand with the process of 'individualisation' but it is, however, not a new phenomenon. It continually marches on together with the process of 'individualisation' and reaches dimensions which were previously inconceivable, especially with respect to its effects on how we see our cities.

More and more land is being covered with urban life, leading to the decreasing importance of the role of urban centres. A typical example is Bremen, but it could just as well be any other city; between 1970 and 1987,

Bremen's population dropped by about 50,000 inhabitants. The population of the region (within a radius of 30 km) however increased by about 75,000. The suburbanisation of the population was followed by the suburbanisation of trade and industry (including area requirements and approach roads for the motorway). The expansion of the areas settled is accompanied by a simultaneous thinning out of the population, which also takes place in heavily built-up city centres- the built-up area has increased but even less people live there than before. This expansion and thinning out, coupled with a dispersed job distribution, has led to a new form of mobility which is no longer centred on inner city areas as was the case 20 years ago with residential areas on the periphery and jobs in the city centre. This new form of mobility is widespread, especially in areas away from town centres. City centres are no longer the focal point of this way of life. It is now the house or flat which is the centre of the traffic radii and from which vehicle mobility is organised. It is a mobility which is running wild. Fishman (1991) based his analysis of this process on examples from the USA.

Thus 'individualisation', suburbanisation and vehicle mobility are all interconnected. Speed and individual mobile potential have led to a dwindling of space. Distances have become immaterial, even within certain mobile parameters. However an increase in vehicle mobility does not necessarily lead to an obligatory use of the car (determined by distance). On the contrary, as Gerhard Bahrenberg and Thomas Krämer-Badoni (1995) showed in their analysis, an increase in distance only accounts for a small increase in vehicle mobility (measured by the increase in inner city mobility). How does the remaining increase in vehicle mobility come about?

Traffic is an autodynamic social process. Cars create their own terms for their further use. This process has recently been comprehensively examined in two detailed studies carried out by Kuhm (1995, 1997). The car is a technical appliance with whose help everyday activities can be separated from one another. Once it is there and already integrated into everyday life, an increasingly differentiated structure in everyday activities can be attained. We can choose where we want to do and/or buy something. This possibility of differentiating our daily structure implies, however, using a car and leads to its future use, leading to further segregation. Even though a car is a technical appliance, it is a part of society, it is more than just a mechanism or instrument; we cannot imagine our life without it. The technical system of automobilism is a social system which produces the material and psychological conditions for its continued use. For the individual, the car is a tie to modern society, it guarantees social participation in which mobility has become a stipulation for its own reproduction. When viewed in this way, car mobility cannot be controlled, reduced or prohibited. Car mobility can only really be replaced by a mobility system whose mobility functions better than a car, from a spatial and social point of view, as well as temporally, and perhaps even from an ecological point of view.

Although the number of cars is on the increase, our best hope is that motorisation will not manifest itself in an increased car use. In fact there is some empirical evidence which points to this, such as the figures for Munich where the number of cars has increased but car use has stagnated.

Our own empirical research on how cars are used does seem to show that people follow reasonable mobility patterns: people -at least those who live within cities- seem to use the car more and more only when it is absolutely necessary, and tend to organise their everyday life around use of the bicycle or public transport.

References

Bahrenberg G, Krämer-Bodini T (1995) Arbeit und Wohnen, Berufs- und Ausbildungsverkehr in der Stadt Bremen und Region Bremen. Eine Analyse der Entwicklung von 1970 bis 1987. Arbeitspapier No. 20/8 der ZWE 'Arbeit und Region', Bremen

Fishman R (1991) Die befreite Megalopolis: Amerikas neue Stadt. In: Arch + 109/110 pp 73–83

Kuhm K (1995) Das eilige Jahrhundert. Einblicke in die automobile Gesellschaft. Junius, Hamburg

Kuhm K (1997) Moderne und Asphalt. Pfaffenweiler: Centaurus (in press)

Sachs W (1984) Die Liebe zum Automobil. Ein Rückblick in die Geschichte unserer Wünsche. Rowohlt, Reinbek bei Hamburg

The Alternative Means of Transport for Protection of the Ecological Environment in Hanoi – Vietnam

Vu Thi Vinh

Hanoi is the capital of Vietnam and is the political, cultural, scientific, technological and economic centre of the entire country. In recent years, Hanoi and most of the big cities in Vietnam have developed rapidly, in line with the development of the whole country. Hanoi is important not only as the capital of Vietnam, but also as the capital of an ASEAN member country which involves it collaborating with other large cities in the region.

1 Background

At present Hanoi's total population is 2.3 million in an area of 921 km^2. The inner city covers 43 km^2 and has a population of 1.1 million and a population density of 25,600 people/km^2. During recent years the speed of urbanisation and the population growth rate in Hanoi have risen. The population increased by 10% between 1991 and 1996 with an accompanying increase in the birth rate of about 1.5% (Nyugen Lan 1997).

The above factors, as well as the slow infrastructural development, have caused Hanoi to face many problems, one of which is urban transport. As in other large cities in many developing countries, Hanoi's traffic-jams have caused serious pollution and affected the urban environment.

The traffic intensity in Hanoi is not high, but the traffic system is very inefficient, road surfaces are poor, and vehicle speed is retarded by the disorder of the system, causing traffic jams at many junctions (Nga Tu So, Kham Thien, Cua Nam etc.). Cars are forced to wait in queues for a long time thus polluting the air with their exhaust fumes: CO, NO_x, SO_2, Pb and mixtures of lead. Air in the Giai Phong, Nguyen Trai, Mai Dong and Lo Duc streets is heavily polluted, the pollution is high even in areas 40–50 m away from the streets. A survey showed that many factors exceed the allowed standards:

Hanoi Architectural University, Dal hoc Klen truc Hanoi, Thanh Xuan, Hanoi, Vietnam

J. Breuste, H. Feldmann, O. Uhlmann (Eds.)
Urban Ecology

- concentration of CO_2 is 1.5–1.7 times the standard,
- heavy dust: 43–60 times the standard,
- air-dust: 5–20 times the standard,
- NO_2: 2.5–2.9 times the standard,
- SO_2 (at Nga Tu So): 3–5 times the standard,
- Pb dust on the road: 6–14 times the standard.

The traffic jams in Hanoi are caused by a badly structured road network and poor public transportation system. The road network in Hanoi is divided into 3 areas, each with their own different features. While the average road density of the whole city is 2.5 km/km^2, that of the ancient quarter is 15 km/km^2, in the old quarter it is 10 km/km^2 and only 1 or 2, or even 0.4 km/km^2 in the new quarters (Thi Vinh 1997). In the inner city 88% of the streets have a width of less than 11 m, and among these 50% have two lanes.

In addition to the poor road network, the number of vehicles has increased rapidly, especially motorbikes. From 1991 to 1996 the annual increase in the number of cars was 8–10% and motorbikes from 15–20%; meanwhile, the number of bicycles fluctuated from 12–20% (Nguyen van Buc 1993). According to a 1995 survey in Hanoi, there was then 1 motorbike per 3 people and 11 cars per 1000 people. During rush hours bicycles and motorbikes make up 92 to 94% of all means of transportation. The continuous increase in the number of motorbikes is causing serious environmental pollution in Hanoi.

The shortage of public transport in Hanoi is one of the major causes of the increasing number of private vehicles. There are many reasons why people do not use public transport: it is uncomfortable, distances between bus stops are long and it takes a long time. A distance of 7 kilometres takes 45 minutes by bus (including the walking time from the house to the bus stop and the waiting time at the bus stop), 35 minutes by bicycle and only 14 minutes by motorbike (Vu Thi Vinh, 1997).

2
Solutions for Limiting Traffic Jams and Protecting the Environment in Hanoi

The master plan of Hanoi up to the year 2020 is currently being studied and altered. One of the major reasons for this is the transportation problem. Due to its limited sources of land and capital, Hanoi must choose an appropriate system of road development and urban transportation.

Planning for the development of the transport system in Hanoi to the year 2020 should pay attention to the following 2 main points for the reduction of traffic jams and protection of the urban environment.

2.1 Suitability of the Road Development

- Construction of ring roads should be completed,
- appropriate road density should be aimed at and planned for: increase the road density from 2.5 km/km^2 to 8 km/km^2, increase the land use for transportation form 5% to 14%, increase the land area per person for transport form 2 m^2 to 11 m^2,
- car parks: increase the land use for car parks from 0.5% to 4%,
- renovate important transportation nodes,
- separate mechanised and non-mechanised means of transport.

2.2 Appropriate Rates of Transport

The level of public transport should increase from 3% to 40% and the remaining 60% should be met by other means of transportation.

The reasonable division of transportation which Hanoi city must achieve to reduce traffic jams and protect the urban environment in the future is as follows:

- Large buses: 20–25%,
- Small buses, minibuses: 15–20%,
- Motorbike: 20%,
- Bicycle: 20%,
- Car and others: 5–15%.

Three actions should be carried out on the basis of this planning orientation:

1. Transportation solutions should be worked out for each quarter – ancient, old and new. In the ancient quarter, where the roads are only 6–7 m wide and very close together (50–100 m), certain types of transport should be prohibited; some areas should be pedestrianised. In the old quarter, where the road structure and width is relatively favourable, traffic lights should provide sufficient regulation of traffic and bicycles should have special lanes. A new road system for public transportation should be developed in the new quarters.
2. An alternative traffic system is necessary to avoid traffic jams in the 3 major street axes, where the passenger flow is 6000/hour. There have been disputes about the following alternatives: underground, monorail or tram; however, all of these are unfeasible in the current financial climate.
3. To promote the participation of private sectors in the urban transportation service.

(KC 11-06 1995, Le Hong Ke et al. 1995).

3 Conclusion

Through the open door policy of the Vietnamese government the Hanoi municipality is mobilising all financial sources and is ready to create favourable conditions for domestic economic sectors and foreign countries to invest in the development of the city, particularly in infrastructural development.

The combination of well thought-out policies, the master plan, and a belief in possible co-operation and assistance from international supporters, it is hoped that Hanoi's transport system will soon be improved contributing to the sustainable development of the city.

References

KC 11-06 (1995) Upgrading and improving the quality of transport in big cities. National Institute of Urban and Rural Planning (UIURP)

Le Hong Ke, Vu Thi Vinh, Tran Thi Huong, Xuan Hiep (1995) Policy analysis study on environment management and infrastructure. The Netherlands, Poland, Vietnam

Nguyen Lan (1997) Urban architecture of Hanoi, current situation and orientation

Nguyen van Buc (1993) Ways to solve traffic jams in Hanoi. Magazine of Transport no.7, 1993

Vu Thi Vinh (1997) Some basic issues in the development of the road system of Hanoi up to 2010. In: Construction Review 2 1997

A Multidimensional Concept of Mobility – A New Approach to Urban Transportation Research and Planning

Thomas Jahn, Peter Wehling

1 New Tasks of Transportation Research

Whenever one speaks about urban ecology and sustainable cities, it very quickly becomes clear that traffic is a central cause of ecological and social strain in cities, as well as of "urban unsustainability". This has resulted in the emergence of the idea of "sustainable transportation", or "sustainable mobility". In this chapter, we look at the social process of shaping a "more sustainable urban mobility"; we will concentrate on some aspects of this process rather than providing a normative description of "sustainable transportation". Initiating a process aimed at shaping a more sustainable mobility implies the need for a critical assessment of existing approaches and conventional methods of transportation research and planning. With respect to this conceptual reorientation, our first claim is that the ultimate failure of strategies and policies supporting more environmentally viable modes of transportation can, among others, be attributed to shortcomings in current transportation research.

If we turn to the field of behaviour oriented approaches to transportation research we see a characteristic division into three branches, more or less separated from one another:

- first, descriptive research into travel behaviour dealing primarily with the quantitative registration of trips, routes travelled, time taken, distances and modal choice;
- second, (social scientific) research into the attitudes and values assumed to influence travel behaviour and which, therefore, are generally regarded to be a major starting point for changing that behaviour, in particular the modal choice;
- third, psychological research into the latent motivations of car travel, focusing on the internal and individual dispositions, thus taking the motivational background of mobility and travel behaviour out of its normal context.

Institut for Social-ecological Research (ISOE), Hamburger Allee 45, D-60489 Frankfurt am Main, Germany

J. Breuste, H. Feldmann, O. Uhlmann (Eds.)
Urban Ecology

Given these disciplinary and sectoral divisions, we suggest that a more adequate research approach to urban mobility could be achieved by pursuing three tasks and objectives:

- Overcoming the divisions mentioned above among the branches and approaches within the field of transportation research. This means, above all, connecting the travel behaviour of individuals and groups (which can be empirically described) to research into the social contexts of action and the motivational backgrounds.
- Drawing a clear distinction between transportation and mobility, and focusing on mobility. "Mobility" refers to a complex social-ecological phenomenon (or what we call, "a societal relationship with nature") which includes the social causes of why and how people travel, while "transportation" refers more narrowly to the technical aspects. In addition, mobility refers not only to modal choice but also to particular activities and their goals.
- Putting the context of mobility back into an element of the everyday life and social practices of individuals, household-members and groups: this means understanding mobility as a form of social action, one that is both at the material level (the connection of activities located at different places) as well as at the symbolic level (linkage with cultural forms of expression) part of encompassing modes of living and life styles.

2
Three Dimensions of Mobility

In our project, "Viable Forms of Urban Mobility," conducted by City:mobil (a network of 5 institutes) and funded by the German Federal Ministry of Research, we have tried to achieve these goals, and thus a new conceptual point of view by developing an interdisciplinary and multidimensional concept of mobility. We distinguish analytically the following three perspectives, or dimensions, which are always connected as a social activity within mobility: spatial mobility, socio-spatial mobility, social mobility (Fig. 1).

First we would like to explain these three dimensions, then we shall turn to some practical perspectives for the social shaping of urban mobility and to the different ways this could be realised.

"*Spatial mobility*" means the movement of people (or things) within physical space. This dimension focuses on spatial distances, as well as on the amount of time required for getting from one place to another with the available means of transport.

The second dimension, "*socio-spatial mobility*", refers to the movement of people (or things) between social locations for the sake of given

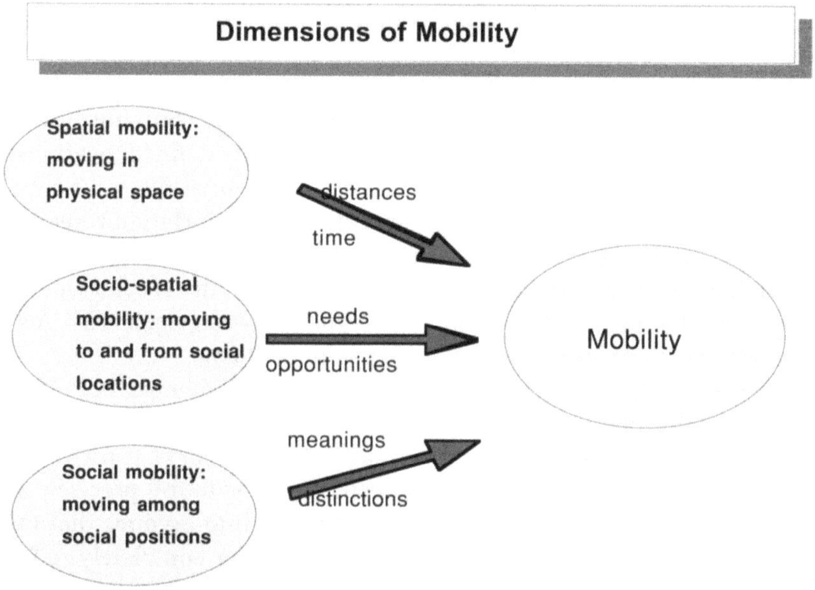

Fig. 1. Dimensions of mobility

opportunities (business, recreation, etc.), and, therefore, to the activities and the purposes of these activities which individuals pursue at these locations (work, shopping, etc.).

The third dimension, "*social mobility*", which one could also call "socio-cultural mobility", refers primarily to the social and cultural meaning of mobility in general, as well as to the meaning of certain means of transportation. This dimension encompasses a wide field of connected symbols that can be perceived, however, very differently by different social groups. For example, to demonstrate membership of a particular social milieu, or to display a particular life style, it may be deemed necessary to drive a certain kind of car, and to drive it in a particular fashion, or to travel often with a bike. Where and how often one travels can also be of great significance for a "positioning" in social space.

Here it must be stressed that such cultural meanings and modes of symbolisation cannot to be explained in terms of individual psychological dispositions, rather they are constructed and anchored socially. Certain kinds of mobility are linked with such attributes as modernity, efficiency, status, freedom or they symbolise social inclusion in a fundamental way. Of course, such attributes cannot be completely independent of the experience of the actual use of means of transportation when stabilising themselves in society, but they are at the same time not simply a result of everyday experience. They build a step towards practical experience. The social construction of such modes of symbolisation, however, implies that they can also be changed or shifted within and by social processes.

Our conceptual approach maintains that mobility always encompasses, and draws on, these three dimensions – although not always consciously and explicitly: people move in physical space to specific socially defined locations; they do so with specific goals in mind, with this movement at the same time providing a specific symbolic expression of mobility. Transportation research and planning has concentrated so far almost exclusively on the spatial and socio-spatial dimension of mobility, and has assumed that the crucial factors influencing travel behaviour can be found on these levels (e.g. spatial distances, accessibility, socio-demographic patterns of activity).

It is clear that the third dimension holds the greatest challenge, both theoretically and methodologically, for traditional models of traffic research (and for general behavioural theories, such as rational choice theory). It turns out that in many cases the decisive factor of influence is to be found precisely at this socio-cultural level. Furthermore, we have to take into account that (at the very least) the spatial dimension of mobility is not a completely independent factor; it is itself influenced by social meanings and cultural symbols. Spatial distances and time costs for travel are evaluated differently in different social contexts. In general, one of the most important tasks of future mobility research will be to investigate how the different dimensions, and their social perception, mutually influence one another.

Distinguishing the three dimensions of mobility opens new perspectives for developing effective strategies to achieve a sustainable mobility. They make it possible to analyse the barriers within all three dimensions and thus move towards a de-coupling of mobility and automobility, while at the same time not ignoring their interconnection. In contrast, strategies for sustainable mobility that do not consider all three dimensions are likely to fail, or to have only minimal success. (This applies, for instance, to the concept of "traffic avoidance", which refers almost exclusively to the first and second dimensions).

3 Strategic Mobility Management – a New Perspective

What we have presented so far cannot, in its present state, claim to be a theory of mobility, but it does represent an innovative perspective on research which can contribute new knowledge to the fields of mobility research and urban ecology (Götz et al. 1997). The multidimensional concept of mobility, however, is not only intended to be a new basic concept of (social science) research but also a stimulus to new approaches to urban (transportation) planning aimed at sustainable mobility. Thus, we would like to conclude by very briefly sketching some rough ideas for a new approach based on this broader concept of mobility. We call this approach, "strategic mobility management".

Mobility management has emerged in recent years both as a further development of, and as a contrast to, traditional transportation planning (Ferril 1997). While transportation planning, even in its most advanced form as "integrated transportation planning", deals mainly with optimising the technical infrastructure of transportation, mobility management focuses on social actors in their different social contexts (e.g. business firms, households, neighbourhoods, etc.). Within these contexts not only is a demand for transportation created but also symbolic meanings are constructed, or at least reproduced and stabilised. Thus, the aim of mobility management is to offer new opportunities and services to fulfil mobility needs without automobility, as well as to de-couple social positioning from auto-mobility.

For us "strategic mobility management" is a bottom-up approach which begins with the assumption of different contexts of action for social actors, and then goes on to link and integrate these in the perspective of an environmentally viable mobility. In this way, it will be possible not only to rebuild the systemic and infrastructural framework of mobility, but also to change the symbolic meanings and symbolic attributes which are so powerful in this area. The strategic key elements of such a process are, above all, the establishment of new action networks (or "social-ecological alliances") among different actors (e.g. local retail stores, car-sharing agencies, bicycle messenger services, sport clubs, recreational facilities, etc.), the expansion of new mobility services by means of such networks and, finally, the reduction of private ownership of automobiles in favour of their more flexible and conscious use in the form of a "mobility leasing".

References

Ferril S (1997) State of the art in mobility management in Europe. Momentum (Mobility Management for the Urban Environment), Issue 1

Götz K, Jahn Th, Schultz I (1997) Mobilitätsstile in Freiburg und Schwerin. Stadtwege 3: 10–19

Motor Vehicles and Sustainable Urban Transport – How Can this Conflict be Solved?

Rudolf Petersen

1 Introduction: Motor Vehicles and the Environment

The growth in the number of automobiles causes environmental as well as other burdens to citizens and nature. Although exhaust emission and noise standards for automobiles were already introduced at the beginning of the seventies and have been significantly tightened since then, technical improvements achieved at the individual level have been overridden by the overall increase in the number of cars (OECD/ECMT 1995). Conditions for non-motorised modes of transport and for public transport are worsening (Newman, Kenworthy 1989).

Motorised vehicles contribute about 20 to 25% of world-wide CO_2 emissions, most of which are produced by the rich and highly industrialised countries (OECD 1996). The developing countries demand more motorised transportation. Without the rich countries (especially the US) reducing energy consumption there is no chance to mitigate the greenhouse effect (MacKenzie, Walsh 1990).

To a high extent, mobility can be organised very effectively by basing it on public transportation and non-motorised modes of transport (IEA/OECD 1995), but in countries like the US these modes play only a minor role. In this chapter only technical perspectives for a more sustainable automobile based transport system are discussed. The findings are based upon a recent study (Diaz-Bone, Petersen 1997). For criteria of sustainable transportation we refer to an OECD country study (Verron et. al. 1997) in which reduction rates of 80% for CO_2 and 90% for NO_x and VOC until 2030 are mentioned.

Wuppertal Institute for Climate, Environment and Energy, POB 10 04 80, D-42004 Wuppertal, Germany

J. Breuste, H. Feldmann, O. Uhlmann (Eds.)
Urban Ecology

2
Technological Options for the Car of the Future

Three main areas of development to reduce energy consumption and emissions from cars are currently being researched by manufacturers (PNGV 1996):

- internal combustion engines (ICE) with high efficiency;
- energy storage systems (especially for alternatives to ICE);
- vehicle bodies with about 40 to 50% less weight.

In this paper we will focus on improvements of the conventional ICE (gasoline and diesel), on alternative fuels and on electric vehicles as well as on fuel cells. The vehicle weight problem will not be discussed.

2.1
Short-Term Option: Optimisation of Traditional Technology

Today's cars are not optimised for their typical uses. Their engines deliver unnecessarily high power. Engine displacement, size and weight of the cars are increased nearly every year (DeCicco 1995). A car designed to meet the real challenges is not yet on the market: it should weigh less than 450kg, have a 20-kW-engine (maximum) for up to 100km/h, with ultra low emissions (a factor of 5 less than today) and low noise output (5db (A) less). It should keep the passengers safe and enable the driver to act responsibly towards non-motorised traffic participants (cruise control for speeds of 30 or 50km/h within settlements). Even for somewhat larger models for the family weekend, the fuel efficiency should still be better than 2 or 3 litres of gasoline per 100km.

This kind of car should not be an additional "City Car", but substitute the common car. All elements are available so the car could be on the market soon.

2.1.1
Problems and Perspectives

For thermodynamic reasons the optimal effectiveness of an internal combustion engine is about 45 to 55%. The most energy efficient ICE is the direct-injection diesel engine. In the current US research initiative "Partnership for a New Generation of Vehicles (PNGV)" DI Diesel engines are seen as a good choice for the 80mpg-goal (2.8 litre per 100km for a US medium sized 6 seat family car), although toxic emissions are still high (PNGV 1996). A critical point is that diesel particulates cause respiratory cancer. Currently, no technical solution is at hand to eliminate diesel particulate from vehicle exhaust.

2.2 Alternative Fuels

Since the oil crisis of the early seventies, numerous projects have been funded to become independent from OPEC via alternative fuels. Nowadays, the focus is on lower exhaust emissions (DeCicco, deLaski 1995, IEA/OECD 1996).

Two groups of fuels are reviewed: alcohols produced from natural gas or from biomass. Alcohols burn more efficiently than gasoline and produce fewer toxic emissions. Methanol derived from natural gas is not expected to produce less CO_2 than gasoline; methanol from burning coal brings even more (DeLuchi 1991).

Ethanol can be produced from nearly all plants. The energy balance has to include cultivation, harvest and chemical processing. When this energy input comes from biomass itself, the energy benefit would be about 10 to 20% for plants such as sugarbeet or corn (maize). The ecological aspects of large scale production should be discussed carefully. Brazil's ethanol programme has caused serious problems.

The second group of alternatives is scheduled to replace diesel fuels. In Germany the use of rapeseed oil has been proposed. Non-modified rapeseed oil in diesel engines does not fulfil durability-needs because of problems with plugging valves and pistons. Chemical processing of rape oil forming methyl ester solves this problem but demands further energy input. When energy demand for cultivation (also fertilisers and herbicides), harvest and processing is included a net CO_2-benefit of about 20 to 25% compared to diesel fuel is expected. In Europe, only a small share of the diesel market could be replaced, and this alternative is very costly.

Gaseous fuels, either Compressed or Liquefied Natural Gas (CNG resp. LNG) and LPG (Liquefied Petrol Gas) can be used in conventional combustion engines, but need costly storage equipment. Energy demand for compressing or liquifying gas worsens the CO_2 balance. Nevertheless, the use of these fuels will bring some benefits. Toxic exhaust emissions of catalyst-equipped automobiles burning gaseous fuels are lower than from gasoline engines.

2.2.1 *Problems and Perspectives*

The production cost of biomass-based fuels will give higher market prices than fossil fuels. Prices only would be competitive if crude oil prices were doubled.

2.3
Electric Battery Vehicles (EVs)

A nickname of EVs is "energy crisis on wheels". This describes the main problem: their limited energy storage capacity. Energy losses in the supply chain are high: for every kWh electricity that comes out of a net plug in Germany, two kWh primary energy are lost, another 15 to 18% loss occurs when charging the battery (DeLuchi (1991). The limited durability of the battery makes EV driving very costly. Further development is unclear (DeCicco, deLaski 1995).

2.3.1
Problems and Perspectives

Depending on the primary energy mix and emission control system EVs produce more or less pollution than ICE cars. EVs with electricity produced by coal power plants have higher toxic emissions than ICE-vehicles (DeLuchi 1991), depending on desulphurisation or denitrification. Under current German conditions, battery EVs produce more GHG. Benefits could be obtained if electricity was produced in combined heat-power plants based on natural gas, or from renewable sources.

2.4
On-Board Electricity Generation by Fuel Cells (FC)

In FCs, fuels react with oxygen without a combustion flame to produce electricity. When hydrogen is used as a fuel, water vapour is the only emission. Thermal efficiency is quite good (50 to 60%), but lowering of manufacturing costs is needed. FCs are not expected to enter the vehicle market in the next few decades (PNGV 1996).

The high hydrogen storage cost aboard vehicles cause manufacturers to plan the use of methanol or gasoline to serve as hydrogen-containing fuels. On-board-reformers would produce hydrogen from it; but this reduces the system efficiency figure to about 40 to 45% (Daimler Benz 1996). Production of methanol from natural gas causes additional losses. Total efficiency of the system then equals IC engines.

2.4.1
Problems and Perspectives

The conversion efficiency is poor if the whole fuel chain is calculated. With hydrogen from vapour reformed natural gas, a passenger car would emit about 65g CO_2 per km; via onboard reformed hydrogen from methanol/

natural gas about 90 g. Conventional ICE vehicles emit about 110–120 g. Thus, CO_2 reduction is only about 25 to 50%, which shows no improvement compared to optimising ICE, see section 2.1.

3 Conclusions

Concepts have been evaluated to solve the environmental problems of today's motor vehicles. The main result is that even with very optimistic assumptions of technical progress, sustainability objectives cannot be reached.

Optimisation of the traditional IC engine, alternative fuels and other developments will be able to improve some aspects of the situation, but a more sustainable transport development demands significant shifts in transport modes and a reduction (or at least a mitigation) of the demand for motorised transport (OECD/ECMT 1995).

References

DeCicco J, deLaski A (1995) The Green Machine Challenge: A Concept for Promoting Ultra-Efficient Vehicles. Washington D.C. Berkeley

Daimler-Benz (1996) Fakten zu Brennstoffzellen. Presseinformation

Diaz-Bone H, Petersen R (1997) Passenger Car Technology for the next Decade. Report on behalf of WWW-Germany (in prep)

DeLuchi M (1991) Emissions of Greenhouse Gases from the Use of Transportation Fuels and Electricity. Final Report. Davis, Princeton

IEA/OECD (1995) Reconciling Transportation, Energy and Environmental Issues: The Role of Public Transport. Paris

IEA/OECD (1996) Towards Clean Transport. Fuel Efficient and Clean Motor Vehicles. Paris

MacKenzie J, Walsh M (1990) Driving Forces: Motor Vehicle Trends and their Implications for Global Warming, Energy Strategies, and Transportation Planning. World Resources Institute Washington, Washington

Newman P, Kenworthy J (1989) Cities and Automobile Dependence. An International Sourcebook. Aldershot

OECD (1996) Sustainable Transport Policies. CO_2 Emissions from Road Vehicles, Policies and Measures for Common Actions. Working Paper 1, Paris

OECD/ECMT (1995) Urban Travel and Sustainable Development. Paris

Modelling the Ecological Effects of Traffic on a Regional Scale Using GIS

Carola Stauch

1 Introduction

Planning is a decision making process. A digital planning environment can be set up by expanding a GIS using decision support components. Multiple criteria evaluation is a toolbox consisting of different approaches to combining spatial data according to their importance in decision making. A major goal of this research is to eliminate, or at least reduce, the subjective part of the planning process by making every part of the decision transparent and comprehensible. The distinction between scientific results and values introduced by legal, social or ecological guidelines is stressed. The decision making process is elucidated by taking the example of modelling the impact of road fragmentation on birds. This paper emphasises the step of criteria selection.

2 From GIS to Spatial Decision Support

GIS in landscape planning is often used in the same way as analogue data were previously dealt with: different layers of information are overlayed, hereby identifying areas where particular combinations of conditions exist. The question which is not supported by GIS is: which of the remaining sites offer the most promising characteristics for development? A mechanism for representing choice and priority when evaluating conflicting criteria and objectives is missing in a GIS, i.e. a tool for decision support. The decision between one site or the other is no longer a scientific question but depends on values of the society, and ecological, moral or political concepts, etc.

A spatial decision support system integrates the spatial modelling capabilities of a GIS with the decision optimising methods of a decision support system (Heywood et al. 1994). After defining the goal the planner has to select

Institute for Landscape Planning and Ecology, University of Stuttgart, Keplerstr. 11, D-70174 Stuttgart, Germany

J. Breuste, H. Feldmann, O. Uhlmann (Eds.)
Urban Ecology

the criteria for evaluating a site depending on the ecological or political guideline for the spatial unit in question. This determines which criteria are considered to be relevant. Spatial evaluation is carried out in a GIS; this could be a spatial search of potential sites given certain limitations. These sites are then evaluated by applying different multiple-criteria evaluation techniques (MCE techniques), i.e. different techniques of criteria combination and/or different weighting schemes to account for different views. If several MCE techniques are used in conjunction with the same set of criteria then different lists of acceptable sites will be produced. A sensitivity analysis is required to come up with a single list of sites (Carver 1991). This process can be iterative, either for the whole process or for parts of it.

3 Fragmentation by Roads

The first three steps are elucidated now by taking the example of evaluating the detrimental effects of fragmentation by roads, taking birds as an interest group. The distinction between criteria selection and criteria evaluation is stressed so that changing ecological views can be easily integrated (Jessel 1996).

3.1 Identification of the Problem

Fragmentation is a detachment or separation of spatially connected landscape elements into spatially segmented tracts. Roads can be regarded as a chronic, anthropogenic disturbance source which produce divisive fragmentation.

3.2 Evaluating Divisive Fragmentation

In this chapter a method which was suggested for evaluating landscape elements on different scales (Plachter 1994) is altered to evaluate a different element such as a road. The unit of scale will be the regional scale, represented by the natural region. The process comprises two levels, the type level, which consists of one criteria value for the whole natural region, and the object level, which encompasses the aggregation of single objects, i.e. the road sections.

The type level specifies the character of all roads within the natural region compared with a larger spatial unit, i.e. the region of Stuttgart. This level requires data about distribution, abundance, etc. to evaluate the natural region; the output is one score for each criterion. The object level gets a summed score of individual scores for single objects. The road section which is defined by the homogeneous amount of traffic is compared to the general description of the natural region. For example one can calculate the amount of sensitive areas being cut by each road section and compare it to the average value for the natural region. Thereby a ranking of the road section can be made with respect to the affected sensitive areas.

3.2.1 *Criteria Selection*

At first criteria have to be selected according to the ecological view of the natural region, both for the type level and the object level. The ecological view again depends on the nature conservation priorities given for this region. Table 1 shows an example of criteria that can be used to characterise the spatial unit with respect to fragmentation, Table 2 contains possible criteria at the object level. The criteria were selected according to an ecological view which concentrates on birds (Steiof 1996), (Reijnen 1995). To account for a

Table 1. Criteria used on the type level

No.	Description	Evaluation (natural regions)
t1	Fragmentation	$frag_{(region)}$
t2	Abundance of Structures	$struct_{(region)}$
t3	Rate of Sensitive Areas (Biotopes, ...)	$sensitiv_{(region)}$
t4	Urbanity (urban, suburban, rural):	$urbanity_{(region)}$
...	...	$\ldots_{(region)}$

Table 2. Criteria used on the object level

No.	Description	Bird Species A	Bird Species B	...
o1	Speed	$speed_{(birdA)}$	$speed_{(birdB)}$	$speed_{(\ldots)}$
o2	Average daily traffic (DTV)	$DTV_{(birdA)}$	$DTV_{(birdB)}$	$DTV_{(\ldots)}$
o3	Noise	$noise_{(birdA)}$	$noise_{(birdB)}$	$noise_{(\ldots)}$
o4	Shape of a Road	$shape_{(birdA)}$	$shape_{(birdB)}$	$shape_{(\ldots)}$
o5	Road level	$level_{(birdA)}$	$level_{(birdB)}$	$level_{(\ldots)}$
o6	Soundproof barriers	$barrier_{(birdA)}$	$barrier_{(birdB)}$	$barrier_{(\ldots)}$
o7	Bridges, underpasses	$passage_{(birdA)}$	$passage_{(birdB)}$	$passage_{(\ldots)}$
o8	Direction of structures (e.g. hedgerows)	$struct_{(birdA)}$	$struct_{(birdB)}$	$struct_{(\ldots)}$
o9	rivers within buffer (crossings)	$river_{(birdA)}$	$river_{(birdB)}$	$river_{(\ldots)}$
o10	Adjacent biotopes	$biotope_{(birdA)}$	$biotope_{(birdB)}$	$biotope_{(\ldots)}$
o11	Sensitive area within buffer	$sensitiv_{(birdA)}$	$sensitiv_{(birdB)}$	$sensitiv_{(\ldots)}$
o12	Contamination	$contam_{(birdA)}$	$contam_{(birdB)}$	$contam_{(\ldots)}$
o13	Width of road	$width_{(birdA)}$	$width_{(birdB)}$	$width_{(\ldots)}$
...	...	$\ldots_{(birdA)}$	$\ldots_{(birdB)}$	$\ldots_{(\ldots)}$

specific bird species the criteria have to be specified in more detail and additional criteria may have to be added.

3.2.2
Multiple Criteria Evaluation

The next step is the combination of all criteria on both the object level and the type level into one score. MCE techniques consist of two elements, the criterion scores and the planners' preferences (Jankowski 1995). The output of single or multiple aggregation functions could be either one recommended alternative, as reduced decision space consisting of several 'good alternatives' or a ranking of alternatives from best to worst. A MCE Analysis starts with the construction of an evaluation or decision matrix where the impact of alternatives on every criterion is estimated.

The problem becomes three-dimensional if one looks at one or more alternatives from different ecological points of view. The dimensions of the evaluation process are (1) the view, (2) the set of alternatives if a new site is planned/ the set of measures if an existing site is evaluated/ evaluation only of the existing site and (3) the different MCE techniques.

The resulting equation which has to be solved for every alternative during the planning process, if one considers only one view, would therefore look like this:

$$Evaluation_v = f\left((t_{1(v)}, t_{2(v)}, \ldots, t_{n(v)})\sum_{s=1}^{n}(o_{1(v)}, o_{2(v)}, \ldots, o_{n(v)})\right)$$

t: a criterion at the type level *s: a road section*
o: a criterion at the object level *v: a specific view or guideline*

The MCE techniques used to solve this equation can be classified according to the cognitive processing level which can be compensatory (a low value of one criterion can be compensated for by a high criterion), or non compensatory (Jankowski 1995). For the evaluation of roads one probably has to use a combination. If one wants to evaluate the sensitive areas (criterion o11 in table 2) then it is not possible to use a compensatory approach. A disturbance of a highly sensitive groundwater area cannot be compensated for by the preservation of a nature conservation area, but it is conceivable that a very winding road, which birds cannot overlook, might be compensated for by a very low speed. These approaches will be tested with data coming from an ongoing project about ecological acceptable mobility in the region of Stuttgart (Stauch et al. 1997).

4 Conclusion

GIS is still a tool for experts. The purpose of planning support systems is to provide the non-expert user with spatial data and analytical tools as well as providing a graphical user interface with online-help to guide the user through the whole process of decision-making, from data selection to the final choice of a site. We are still far away from this vision but many pieces of the puzzle already exist which have to be put together.

References

Heywood, I, Oliver J, Tomlinson S (1994) Building an exploratory Multi Criteria Modeling Environment for Spatial Decision Support. In: Proceedings of the EGIS/MARI, volume 1. Paris pp 632–641

Jankowski P (1995) Integrating geographical information systems and multiple criteria decision-making methods. Int J Geographical Information Systems 9(3): 251–273

Jessel B (1996) Leitbilder und Wertungsfragen in der Naturschutz-und Umweltplanung. Naturschutz und Landschaftsplanung, 28(7): 211–216

Plachter H (1994) Methodische Rahmenbedingungen für synoptische Bewertungsverfahren im Naturschutz. Z Ökologie u Naturschutz, 3(2): 87–106

Reijnen R (1995) Disturbance by car Traffic as a Threat to Breeding Birds in the Netherlands. Dissertation Rijksuniversität Leiden

Stauch C, Roedel P, Scheuerer W, Kaule G (1997) Bewertung von Umweltwirkungen durch den Verkehr. In: Schrenk, M (ed) Computergestüzte Raumplanung – Beiträge zum Symposium CORP'97 vom 12.-14.2.97, Wien pp 265–273

Steiof, K (1996) Verkehrsbegleitendes Grün als Todesfalle für Vögel. Natur und Landschaft, 71(12): 527–532

Workshop: Traffic Behaviour and its Possible Changes 4.2

The Impact of Land Use Patterns on Mobility Structures

Andreas Kagermeier

1 Introduction

In the context of the discussion about sustainable cities in the last few years, the idea of a concentrated decentralisation –which means putting more emphasis on developing central places of medium order– has been fairly often debated as a possible solution for regions within the catchment area of greater conurbations (BfLR 1996, Empirica 1996). It is expected that the realisation of this settlement structure concept will reduce the need for motorised traffic and therefore minimise the use of energy. Until now there is little empirical evidence of the role which settlement structures play within the context of transport efficiency. Most of the time it is argued that it is common sense that a neighbourhood with different functions automatically induces functional interactions between the various areas. In this chapter I would like to discuss the question: what share of transport can be attributed to settlement structures? Southern Bavaria has been chosen as the area for a case study, a region characterised by its interactions with the city of Munich.

2 Transport Patterns and Volumes in Daily Traffic

Within this region (according to categories established by the *Bundesforschungsanstalt für Landeskunde und Raumordnung)* six types of communities (apart from centres of higher order) have been identified to cover the main types of settlement structures in the region.

Within the metropolitan area:

- centres which are officially denominated for intense settlement activities;
- communities with or without a low degree of centrality;
- central places of medium order.

Technical University of Munich, Department of Geography, Arcisstraße 21, D-80290 Munich, Germany

J. Breuste, H. Feldmann, O. Uhlmann (Eds.)
Urban Ecology

Outside the metropolitan area:

- central places of medium order;
- communities without a low degree of centrality with a railway station;
- communities with or without a low degree of centrality without a railway station.

The aim of the survey was to obtain an estimation of the total traffic volume for each type of community, based on the whole spectrum of activities, their frequencies and the choice of the transport mode (for more details on the methodology of the research and the empirical findings cf. Kagermeier 1997).

As far as commuter traffic is concerned, a high degree of orientation towards Munich has been found. This orientation towards the city of Munich decreases only very gradually with longer distances from the city. The various proportions of the orientations within the same community are of particular importance with respect to the traffic volume In all communities with a low degree of centrality, and in centres which are officially denominated for intensive settlement activities, the orientations within the same community represent less than one fifth of all orientations. Central places of medium order located near Munich have a binding quota of 40%, whereas in the communities with a medium degree of centrality outside the metropolitan area the quota barely reached 50%.

The amount of traffic caused by commuters is estimated by using the distance (as the crow flies) between source and target community. In total, the highest traffic volume is observed in communities with a low degree of centrality outside the metropolitan area- up to three times higher than the lowest value measured. In the communities within the metropolitan area the kilometre values are (in total) at a lower level, corresponding with their shorter distance from Munich. The differences between the communities with a low degree of centrality and the centres which are officially denominated for intensive settlement activities are relatively low because of the conformity concerning the remarkable outside orientation. This means that in these cases the distance to the city determines the traffic volume.

Daily orientations for the supply of convenience goods and services shall be outlined briefly as a second example. In most of the communities with a lower degree of centrality convenience goods are, to a great extent, bought outside the community. The situation for the purchase of shopping goods is quite similar. In communities with a low degree of centrality, including the settlement centres, high degrees of orientation to other communities are dominant.

In Table 1 the index values of the monthly amount of transport per household member are shown for various types of daily traffic. To eliminate the effects caused by different socio-demographic structures of the sample

Table 1. Index of the monthly motorised daily traffic volumes per household member. Source: own survey

Community Types	Index of Traffic Output (South Bavarian Average = 100)				
	Profession	Education	Convenience goods	Shopping goods	Services
Inside the metropolitan area					
Settlement centres	70	60	89	80	84
Communities of low centrality	89	98	132	91	127
Communities of medium centrality	66	90	72	73	64
Outside the metropolitan area					
Communities of medium centrality	95	114	48	107	73
Communities of low centrality with railway	132	117	149	114	118
Communities of low centrality without railway	111	108	110	105	118
Relation between maximum and minimum	2,00	1,95	3,10	1,56	1,98

communities, the responses of each probationer are weighted in such a way that each community has the same proportions of socio-demographic groups. It is evident that the communities with a low degree of centrality outside the metropolitan area have the highest amounts of transport. As far as the daily traffic in Southern Bavaria is concerned, specific relationships exist between the different types of settlements and the amount of traffic generated by them. On average, communities with a low degree of centrality, located far away from the city-centre, i.e. communities which recorded the highest population growth during the last few years, generated twice as much traffic as other communities. One reason for the steadily growing amount of private motorised traffic is that these communities recorded the highest population growth rates (c.f. Kagermeier 1997).

A considerable proportion of the amount of transport essentially depends on the interference between the factors 'distance to the city-centre' and 'part of the orientations outside the home community' (cf. Fig. 1). Low traffic volumes in daily traffic are found in places located close to Munich or in communities with a well developed infrastructure, therefore most needs can be satisfied at the place of residence. High traffic volumes for each journey made to the city-centre are compensated for by a low intensity of interactions in these communities.

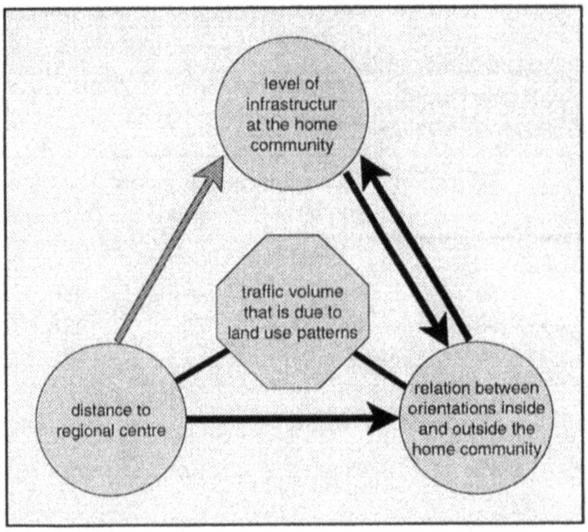

Fig. 1. Scheme of the relationship between the impact of land use patterns on traffic volume and specific factors of influence

3
Leisure Traffic Volumes

The total traffic volume is determined considerably by leisure traffic as almost half of the traffic is due to leisure activities. In communities with a low degree of centrality sub-average transport volumes are produced. Thus high traffic volumes in the daily traffic contrast with lower ones in leisure traffic. This means that the inter-relations are different from those in the daily traffic. The higher traffic volumes found in settlement centres and central places of medium order are not caused by longer distances per journey, but by higher journey frequencies for visits and trips.

The availability of private outdoor recreation space is the real grade value which can measure different frequencies between the various types of communities. Due to the higher proportion of households with direct access to green space around the house in communities with a low degree of centrality, the volumes of leisure traffic are significantly lower (similar results were found by Fuhrer, Kaiser 1994 and Holz-Rau, Kutter 1996).

Lower leisure transport output in communities with a low degree of centrality is generally not able to compensate for the high daily traffic volumes. Regarding the extrapolation of the total energy consumption (shown in Fig. 2), communities with a low degree of centrality outside the metropolitan area still show the highest values. The lowest values have been found for communities with a medium degree of centrality and for settlement centres. Through compensatory effects of leisure traffic, the insufficient implementation of

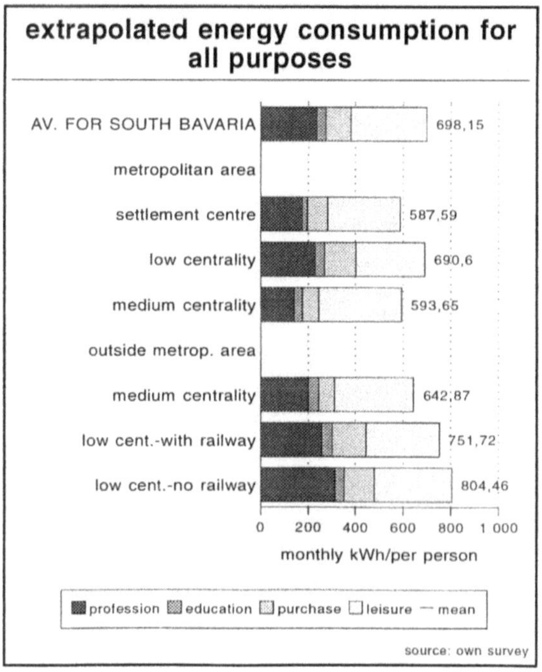

Fig. 2. Extrapolated monthly energy consumption for all traffic purposes

regional planning objectives for the development of settlements is responsible for a considerable part of the increase in traffic volumes.

In conclusion it should be mentioned that a measurable reduction of traffic volume can be achieved through the realisation of a concentrated, decentralised settlement structure. It should also be stressed that a concentrated settlement structure is a good precondition for other traffic policies because:

- requirements for an increasing proportion of public transport are granted due to the relatively concentrated flow of traffic;
- existing orientations within the same community allow for an effective promotion of non-motorised traffic;
- increasing the proportions of orientations within the same community is comparatively unproblematic and is possible without causing serious social friction due to rising mobility costs.

References

BfLR (Bundesforschungsanstalt für Landeskunde und Raumordnung) (1996) Nachhaltige Stadtentwicklung. Herausforderungen an einen ressourcenschonenden und umweltverträglichen Städtebau. Bonn

Empirica Gesellschaft für Struktur- und Stadtforschung (1996) Dezentrale Konzentration. Neue Perspektiven der Siedlungsentwicklung in den Stadtregionen? Schriftenreihe "Forschung" des Bundesministeriums für Raumordnung, Bauwesen und Städtebau 497, Bonn

Fuhrer U, Kaiser FG (1994) Multilokales Wohnen. Psychologische Aspekte der Freizeitmobilität. Bern

Holz-Rau C, Kutter E (1995) Verkehrsvermeidung. Siedlungsstrukturelle und organisatorische Konzepte. Materialien zur Raumentwicklung 73, Bonn

Kagermeier A (1997) Siedlungsstruktur und Verkehrsmobilität. Eine empirische Untersuchung am Beispiel von Südbayern. Verkehr spezial 3, Dortmund

Mobility Concepts in a Large New Housing Estate in Eastern Germany. The Example of Leipzig-Grünau

Thomas Keidel

1
Traffic Development in Eastern Germany in General

1.1
The Traffic Situation before 1989

The traffic situation in socialist states such as the German Democratic Republic (GDR) was a result of a state controlled policy. To encourage people to use public transportation tickets were very inexpensive. High passenger numbers were not the result of a high standard of comfort or high vehicle speed. Moreover, roads were mostly in a poor condition. Some people used private cars, but they were very expensive in relation to the average income, and there was a waiting period of up to 15 years for a new car; the number of cars increased very slowly. Only a few parking spaces were necessary for the resulting low number of private cars. Traffic jams were very seldom.

1.2
The Traffic Situation After 1989

As a result of the reunification of Germany, the population had the same opportunities to buy and use cars as in other capitalist countries, so a new car became less expensive in relative terms. Public transportation prices rose by as much as 20-fold but the comfort of the vehicles did not increase as fast as the prices. As a result the number of cars doubled and the use of public transportation was reduced to half (Stadt Leipzig 1994). A lack of parking spaces and a high incidence of traffic jams followed immediately. Until today, the old streets have not been completely repaired, and the number of parking spaces has not increased fast enough to cope with the increase in traffic.

UFZ Centre for Environmental Research, Department Urban Landscapes, Permoserstrasse 15, D-04318 Leipzig, Germany

J. Breuste, H. Feldmann, O. Uhlmann (Eds.)
Urban Ecology

2
The Traffic Situation in Large New Housing Estates

Large new housing estates built from prefabricated concrete elements are typical residential areas of all Eastern German cities. Most such areas are between 10 and 20 years old (Stadt Leipzig 1992). The buildings are made from prefabricated concrete elements. The high density of flats and the high numbers of inhabitants as compared to other parts of the cities are typical of these areas. Social infrastructure such as kindergartens and schools had been provided, but little other infrastructure (shops, pubs, meeting houses etc.) has existed up to now. As far as traffic infrastructure is concerned, only the connections to the city centre via public transportation have been completed; new roads and parking sites are mostly unfinished until this day.

The initial residents were typically young couples with small children of a mixed social background that included a relatively high percentage of university graduates. Today, the majority of the initial residents are still living in these areas. As a result of this situation, there is a higher number of cars per inhabitant than in other parts of the city.

The high density of the buildings causes extra traffic problems. The parking spaces around the buildings are not sufficient for the residents' cars.

3
The Example of the WK8 in Leipzig-Grünau

3.1
The Traffic Situation in the WK8 in Leipzig-Grünau

A typical example of such a structure is the WK8 in Leipzig-Grünau, part of a residential area for 80 000 people (one fifth of the entire population of Leipzig). In 1992 an investigation initiated by the Environmental Research Centre Leipzig-Halle was lauched in the area. One major aspect of the project was the analysis of the traffic situation (Keidel 1996).

The results concerning car ownership are not astonishing. The majority of households (94%) own cars, and 26% of these have two or more cars at their disposal (Keidel 1996). These high incidence of car ownership is the reason for the unsatisfactory parking situation in the residential areas (Table 1).

More than 20% of the cars counted were illegally parked. The most critical areas were in the inner areas of the WK8. Wrongly parked cars were found in courtyards, on pedestrian ways and on formerly green areas. In contrast some parking sites on the outskirts of the residential areas were nearly empty (Keidel 1996). It is therefore no wonder that the parking situation was considered by residents to be the most serious problem of the residential area (64% thought that the situation was chaotic; Keidel 1996).

Table 1. Results of car counts

	1. Count	2. Count
Cars on parking sites	3988	3988
Incorrectly parked cars	1167	1220
Other incorrectly parked vehicles	34	33
Incorrectly parked motorcycles	16	21
Car-wrecks	25	18

Sources are our own counts from 1993

The residents were then asked for ideas to solve the parking problem. The majority suggested new multi-storey parking systems (69%) and underground parking sites (41%). Further solutions suggested were an increase in police supervision (28%) and towing away of illegally parked cars (32%). The most un-ecological idea was to convert green areas into parking spaces (26%; Keidel 1996).

A consideration of the costs of residents' wishes proved that most wishes were not affordable. Only 10% of residents said that they would pay more than 200 DM per year for a parking site (Keidel 1996), but this would not be enough to finance underground or multi-storey car parks. On the one hand, the residents were not willing to pay high parking fees, on the other hand, flat owners or municipalities cannot afford to pay for the construction of expensive new parking spaces. As a result it is not possible to build enough new parking spaces.

However, is it really a good idea to build new parking spaces? What other possibilities are there?

3.2 Solution-Strategies for the Parking Problems

Our investigation showed that it was necessary to develop a completely new concept for such residential areas (Keidel 1996; see also Knoflacher 1993).

The point is that it is impossible to build enough low-cost parking spaces in the vicinity of the flats. The solution may be to have *no* parking spaces directly next to the flats. Instead, huge parking sites should be established at the borders of residential quarters on some less frequently used parking areas. After the nearby parking sites have been closed, the outer rebuilt sites will be frequented more often by pedestrians, making them as safe as the former inner parking sites. Following such a reorganisation, most inhabitants of the area would have to walk the same distance as is necessary to get to the next public transportation stop. This is one aspect of the concept of equal conditions for all transportation facilities. Of course, the calculation of the real costs of the different transport facilities should also be a serious aspect of a new traffic system. It is necessary that the fee for the using of *any* parking site in the residential area must be as high as the ticket price for public transportation.

To increase the use of public transportation and bicycles, the concept of a transportation office in the centre of residential quarters should be established. There, residents could obtain detailed information about public transportation timetables, tickets, bicycle repairs and rentals, and car sharing.

As a result of the demolition of inner parking spaces, the streets near the flats are frequented less often and the area becomes quieter and less polluted. Furthermore, there would be new space for other uses, for example playgrounds, sport areas, allotment gardens etc. Nature can be better integrated into the living space. The goal of such a concept is to rouse residents' consciousness for the environmental situation.

To make this concept a success, it is necessary that residents contribute to finding solutions. They must realise that it is impossible to solve the problematic situation by building new parking sites and that they have to pay for the luxury of a private car. If the residents accept this situation and actively seek solutions, the concept is suited to encourage them to take more responsibility for their residential quarter. The risk of vandalism in the reorganised open spaces is thus minimised.

References

Keidel T (1996) Untersuchungen zur Situation des Wohnumfeldes ostdeutscher Großsiedlungen am Beispiel von Leipzig-Grünau. Leipzig

Knoflacher H (1993) Zur Harmonie von Stadt und Verkehr. Freiheit vom Zwang zum Autofahren. Wien Köln Weimar

Stadt Leipzig (1992) Kommunale Gebietsgliederung der Stadt Leipzig. Leipzig

Stadt Leipzig (1994) Statistisches Jahrbuch 1994

Mobility Models and Traffic Behaviour – An Empirical Socio-Ecological Research Project

Konrad Götz and Thomas Jahn

City-Mobil is the name of a 3 year project supported by the German Ministry of Research involving interdisciplinary investigation of a 'viable form of urban mobility' (in German: stadtverträgliche Mobilität). This means: the mobility requirements of the urban population need to be satisfied in an ecologically and socially agreeable, but also economically efficient way. The network is composed of five institutes: Contract, Karlsruhe; ISOE, Frankfurt; IVU, Berlin; Öko-Institut, Freiburg; Österreichisches Ökologieinstitut, Vienna. In the empirical phase of the integrated project the Institute for Social-Ecological Research investigated mobility orientations and traffic behaviour in two cities – Freiburg in the south west and Schwerin in the north east of Germany. The starting point for this project was that measures to bring about a 'viable form of urban mobility' can only succeed by taking into account the everyday behaviour and the motivational background of road users. We worked on the assumption of the differentiation between mobility and traffic. Besides moving things and persons in the physical space, mobility involves a positioning within the "social sphere"/"social space". Social positioning is coupled with the symbolic dimensions of products and actions. For example: driving a certain make of car can signal membership of a particular milieu, but also detachment (distance) from another, thus the symbolic dimension was taken into consideration when planning the empirical design. All phases of research, both qualitative and quantitative, were comprised of the main areas: style and way of life; mobility orientations (including the symbolic dimension); traffic behaviour. The analysis should lead us to typical mobility patterns which we call "mobility styles". They also can be seen as target groups for practical intervention. The goal of this intervention will be the partial uncoupling of both physical and social mobility from auto-mobility.

Institut für sozial-ökologische Forschung GmbH, Hamburger Allee 45, D-60486 Frankfurt, Germany

J. Breuste, H. Feldmann, O. Uhlmann (Eds.)
Urban Ecology

1 Design of the Project

We decided on a research design with qualitative and quantitative phases:

100 open face-to-face explorations, lasting up to two hours; a traffic diary recording all trips undertaken in a fortnight; a subsequent representative study with 1000 persons in both cities (randomly selected).

2 The Representative Study

From the extensive material obtained from the qualitative research phase hypotheses were made for subsequent approval by quantification. Secondly, the data served as 'raw material' for putting the quantitative questionnaire into operation. In practice this means that all important attitude dimensions (auto-mobility, public transport, bicycling, pedestrian) were presented in statement form. Besides the mobility orientations, important dimensions of lifestyle were also included in the form of items: – Work orientation; – Leisure orientation; – Way of life; – Basic orientations. Data on the movements of the previous day and the previous weekend were collected as an indicator of traffic behaviour: – Starting time; – Purposes of trips (17 purposes); – Mode of locomotion/means of transport; – Distance.

3 Multivariate Analysis: Factors

After analysing the "whole count" we carried out a factor analysis; this means the statements on different attitude dimensions were scanned for inner correlation. This analysis was carried out for the topics of lifestyle as well as mobility orientation. An example is given to enable better understanding: in Schwerin the 18 items of the topic area "bicycle" came down to the following factors: 1st factor: *enthusiasm for the bicycle*; 2nd factor: *antipathy and prejudices against cyclists*; 3rd factor: *danger and fear while cycling*; 4th factor: *commitment to risky cycling*. Factor models of this kind were also calculated for the topics auto-mobility, walking, use of public transport.

4 Multivariate Analysis: Cluster

The cluster analysis represents a further step in the compression of data. The goal is the differentiation of groups of persons (types) whose members all have similar orientations.

4.1
Freiburg: The Traditional Domestics (24%)

A type oriented towards family and security, mainly characterised by the fact that he/she does not hold pointed views. Older people and women are over-represented; there is a focus on the lowest educational levels; there is an above-average number of pensioners, housewives.

4.2
Freiburg: The Reckless Car Fans (20%)

A type oriented towards career and achievement and is characterised by a very strong affinity to the car. Above-average figures for the following factors:

commitment to risky and aggressive driving; the car as a requirement for independence and escape but also: cycling as the experience of risk and technological enjoyment (during leisure time). No pleasure gained from walking; type with the highest proportion of men (90%!). Youngest and middle age-groups are over-represented as well as those employed full-time and those having above-average incomes.

4.3
Freiburg: The Status-Oriented Automobilists (15%)

A type oriented towards prestige with an affinity for the car and also characterised by a strong sense of insecurity concerning non-automobile forms of locomotion : above-average figures for the following factors:

danger when cycling; danger when walking in town at night; dislike for passengers in public transport; the car as a status symbol and object of prestige; set image of tram users; the car as an important leisure partner.

This is a socio-demographically inconspicuous type; women are slightly over-represented (65%).

4.4
Freiburg: The Traditional Nature Lovers (24%)

Basically traditional type of person. Preservation of nature is a top priority;

members of this group enjoy going places on foot and do so frequently; the tram is a favoured form of transport; travelling at night is found especially threatening. 67% of the respondents are women; older people are over-represented, as are unemployed persons.

4.5
Freiburg: The Ecologically Resolute (17%)

Rather young and technically minded group. They reject the car for ecological reasons. Whenever they do use a car, which is seldom the case, this type find themselves faced with a moral dilemma due to their strong principles of environmentally friendly behaviour. The main and most popular mode of transport is the bicycle.

4.6
Schwerin: The Insecure Status Seekers (38%)

This type is oriented towards authority and family. He/she is characterised by a moderate affinity towards the car but also by a strong feeling of insecurity in traffic. Important factors are: danger while driving; does not enjoy walking; in favour of a slowing down of traffic; dislike of public transport passengers.

This group legitimises the car as a rational necessity. The socio-demographic characteristics of this type are average in every respect.

4.7
Schwerin: The Mobile Sensation Seekers (12%)

This is an hedonistic type, longing for experience, enjoying all kinds of mobility which promises a change. It is possible that this type gets pleasure from all methods of locomotion (especially walking). Above-average figures especially for the factors: walking as relaxation and experience of nature; enthusiasm for the bicycle. In addition the following factors are noticed:

the car is a partner for experience and adventure but rational legitimisation of the car as a necessity for everyday life is of no relevance at all for the mobility orientation. Men are slightly over-represented (60% of respondents), as are the youngest age group. Lowest educational levels and the group of skilled workers are over-represented.

4.8
Schwerin: The Environmentally Anxious (32%)

This type corresponds in some respects with the nature-oriented type in Freiburg.

Important factors: ambivalent relation to the car (thoughts of parting with it are of importance); sense of being threatened when going for a walk in town at night; rejection of the car out of concern for the environment.

Situation in life/phase of life: women and older people are over-represented, as are unemployed people and male and female pensioners.

4.9
Schwerin: The Aggressive Drivers (18%)

A type oriented towards achievement and money, with a strong affinity to the car but differing from the 'risk-oriented' type in Freiburg in two respects: in showing a pronounced dislike for cyclists and cycling, and in the greater proportion of women. The most important factors are: the car as a requirement for feeling part of it all, feeling social integrated; commitment to risky and aggressive driving; antipathy and prejudices against cyclists; legitimisation of the car as rational necessity.

Below-average figures for the factors: going by tram as a pleasant experience; walking for relaxation and experiencing of nature; enthusiasm for the bicycle. Proportion of women: 33%, men: 67%. Over-representation of middle age group (31–49 years) as well as persons in full-time employment. Many members of the higher occupational groups (executives/self-employed persons).

5
Traffic Behaviour Of The Types (Mobility Styles)

The typology is based solely on mobility-orientations. The data for traffic behaviour was evaluated once the typology had been fixed. At the same time this turned out to be a test for the hypotheses of a relationship between mobility orientations and traffic behaviour. The differences in the choice of means of transport are remarkable. *The Reckless Car Fans* in Freiburg cover 56% of all trips by car. In contrast to that *The Ecologically Resolute* in Freiburg use the car for 10% of all trips. *The Aggressive Drivers* in Schwerin use the car for 65% of all trips – on the other hand *The Environmentally Anxious* in Schwerin for 19%.

These results show that there is a strong relationship between mobility orientation and traffic behaviour (if you take into account the driven distances, the differences are much bigger).

So our results are: when dealing with the problem of finding forms of sustainable mobility, one must think about the society's models of mobility and about the mobility orientations in different social groups. Thus these results show: the main problem is not the gap between ecological consciousness and behaviour – as very often claimed. The problem is: there are target groups which are extremely different – although similar in terms of consciousness and behaviour. First we have groups whose lifestyle is not very much connected with the way they move around (social mobility is not much connected with spatial mobility). Second we have a group who show something like a car-fixation. They will not accept measures to reduce car driving; the question is how to handle this group. Thirdly: we have groups whose members are prepared to be mobile without a car. We think the only problem is: full advantage has not yet been made of this orientation, thus it represents a resource, a capacity for the future.

Reference

Götz K, Jahn Th, Schultz I (1997) Mobilitätsstile – ein sozial-ökologischer Untersuchungsansatz. Arbeitsbericht

Car-Sharing – from the Point of View of a Municipality

Michael Glotz-Richter and Frauke Baumann

1
Car-Sharing – an Element of an Intermodal Transport System Leading towards More Independence from the Car

With the high basic and low user costs, the private car gives the illusion of inexpensive travel. Having a car in front of the house tempts one to use it. Why should a car-owner use other modes of transport ? Owning a car means using the car.

The availability of a car offers high flexibility as a supplement to public transport and the other environmentally friendly modes of transport – but it does not have to be a privately owned car. A good reliable car-pool may give the same flexibility as a private car with higher economic and ecological efficiency.

Urban traffic problems have to be seen not only in terms of pollution – which can be improved by technical innovations – but also in terms of the threat of accidents – which has direct impacts on the opportunity for children to play on the streets without being supervised – and of course in terms of the space required for moving and parked vehicles. With the aim of making the urban lifestyle more attractive – which we see in direct relationship to the process of the Agenda 21 – we have to find new solutions to the spatial problem. We see Car Sharing as one of the rare options to reduce the negative impacts of urban traffic without restricting the mobility.

2
How does Car-Sharing Work in Detail?

Organised Car-Sharing is a centralised network of user groups and vehicles for shared use. The aim is a more efficient use of resources. A privately owned car is typically in use for only one hour a day. 23 hours a day it requires a parking space – using valuable real estate even when not in use. Logically a

City of Bremen, c/o Senator for Environmental Protection, Section sustainable development of the urban environment, Hanseatenhof 5, D-28195 Bremen, Germany

J. Breuste, H. Feldmann, O. Uhlmann (Eds.)
Urban Ecology

shared car is a more efficient use of an expensive means of transportation and also has ecological advantages.

The question is however: can a shared automobile offer the same flexibility and reliability as a private car?

StadtAuto in Bremen has grown in size from 28 participants in November 1990 to more then 1300 participants (in a city of approximately 550,000 inhabitants) with a present pool of almost 100 cars.

Imagine, you are a StadtAuto member and you want to use a car, right now, very spontaneously. What do you have to do?

Booking: You have to be a member of the organisation to get access to the StadtAuto vehicles. The organisations have a pool of cars at their member's disposal. The cars can be booked by telephone 24 hours a day. The booking office works in a many cities in co-operation with the local taxi company which has a booking office available 24 hours a day; of course the booking is computerised.

Getting the car/Location: The cars are located at reserved parking spaces. The car keys are kept in a safe adjacent to the parking space. Each member has a key to the safe. These 'stations' are distributed throughout the city. In Bremen, for example, in there are about 37 locations. That means that the user usually has only a short walk or bike ride to the car. Car-sharers have access to a car at any time of the day. Spontaneous rides are possible – about one third of all booking is done within 1 hour before the ride.

The cars: All cars are high standard, low emission vehicles. They can be filled up at gas-stations, booking the costs by credit card (tank card) to the organisation's account. All maintenance is carried out by the Car-Sharing organisation. An additional advantage of Car-Sharing over car ownership is that Car-Sharing often owns a variety of vehicles – from the small „city-car" to a van thus offering the appropriate car for various purposes.

Several lockers also contain child-seats and roof-racks.

In 1998 the system will change to electronically controlled lockers using smart cards with an automatic billing system.

Payment: Before using the car you have to fill in the log-book with the data of your ride, time and mileage. After returning you fill in the log-book with the new mileage and time of return. You leave one copy in the locker's mail-box and keep one copy for yourself. Members are billed regularly. Each use is individually listed including the cars of partner organisations.

The system of organised Car-Sharing works as simply as booking a tennis-court: it is obviously more efficient to book a tennis court than to buy one.

3
The Economic Facts for the User

Car-Sharing is a modern form of use of the automobile. Many people do not use the cars very often and when they do they use the kind of car they need just for as long as they need. The average rate is 10–20 users per vehicle.

Car-Sharing can help save money. Except for an initial membership fee only the use of the car is paid for. The purchasing costs, insurance, parking, maintenance and fuel fees are borne by the organisation.

In comparison to the private car the organised Car-Sharing has very low basic costs and realistic running costs, linear depending on the mileage so there is an efficient incentive to reduce the mileage.

One of the ecological qualities are *linear* tariffs. You are charged for each kilometre driven. The more you drive, the more you have to pay. Thus there is a strong incentive to reduce the individual mileage; this has been proved to have positive impacts on the urban environment.

Experience shows that each Car-Sharing car replaces about five to ten private cars which demonstrates car-sharing's positive impact on the urban traffic development.

Organised Car-Sharing completes the network of environmentally friendly means of transport (public transport, walking and cycling). It works like a mobility insurance for all cases when public transportation, walking and cycling are not adequate (e.g. at night or when public transport is not available or inconvenient, when bulky or heavy objects have to be transported, etc.).

A cautious estimate is that 1.2 million private vehicles can be replaced in Germany (Baum, Pesch 1995). The European potential is much higher. We estimate that at least 6 million private cars could be easily replaced without any restriction on personal mobility (Fig. 1).

4
Car-Sharing Reduces Car Traffic

According to a study carried out in Switzerland by the Federal Ministry of Energy in 1992 and to a study undertaken by the German Ministry of Transport in February 1995 (Baum, Pesch 1994), people who have joined a Car-Sharing organisation and who previously owned a car have reduced their car mileage by 50%. For Germany the potential is estimated by the Federal Ministry for Transportation to be about 7000 million fewer kilometres per year. The European perspective is much higher – an annual reduction of about 30000 million kilometres (Fig. 2).

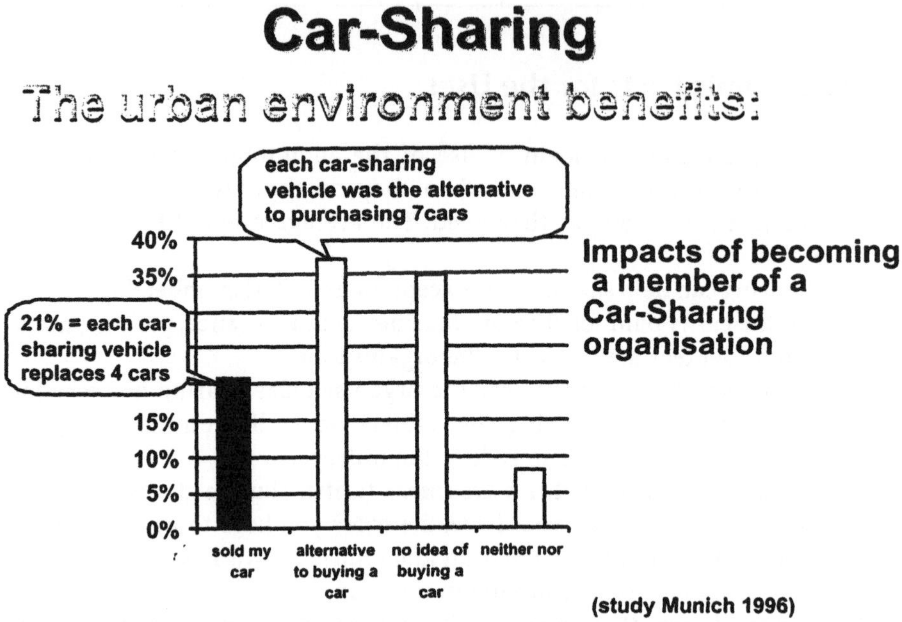

Fig. 1 Impacts of car-sharing on the number of vehicles (MVV München 1996)

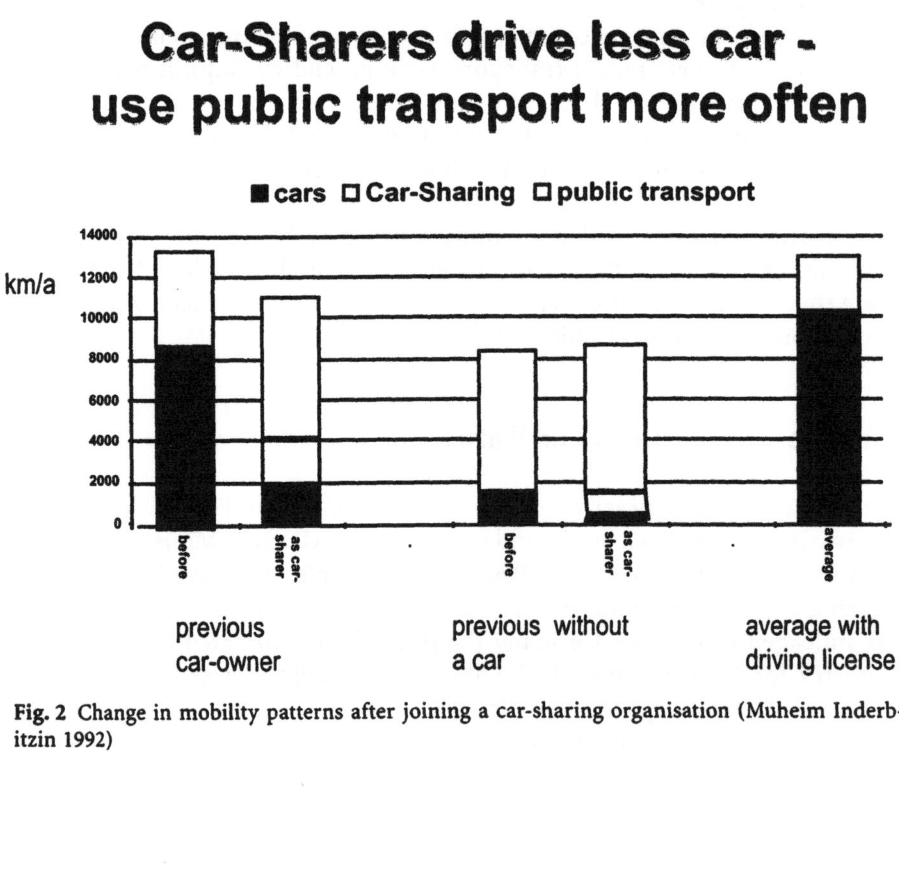

Fig. 2 Change in mobility patterns after joining a car-sharing organisation (Muheim Inderbitzin 1992)

5 Car-Sharing Promotes Public Transport

Car-Sharers do not reduce their mobility, but are found to use public transport about twice as often as car-owners. As a result, public transport also profits from Car-Sharing.

6 Car-Sharing Uses Less Energy

According to the Swiss Study, former car owners now use about 50% less energy for transport - just by changing their mobility behaviour and using shared cars with higher technical standards.

The authors

Michael Glotz-Richter is head of the section 'sustainable development of the urban environment' in the Bremen city-state Ministry for Environmental Protection, and chairman of the working group 'practical alternatives to the car' within the Car Free Cities Network for new Mobility Culture.

Frauke Baumann is working in the section 'sustainable development of the urban environment' in the Bremen city-state Ministry for Environmental Protection focusing on the Agenda 21 process.

A brochure on car-sharing published by the Car Free Cities Network is available in English, German and French available from: Mr. Glotz-Richter or Mrs. Baumann.

References

Baum H, Pesch S (1994) Untersuchung der Eignung von Car-Sharing im Hinblick auf Reduzierung von Stadtverkehrsproblemen. for: German Ministry of Transport. Institute for Traffic Science, University of Cologne

Baum H, Pesch S (1995) Car-Sharing als Lösungskonzept städtischer Verkehrsprobleme. Institute for Traffic Science, University of Cologne

Bundesamt für Energiewirtschaft (Federal Ministry of Energy) (1992) Untersuchungsergebnisse am Beispiel der ATG AutoTeilet Genossenschaft (Schweiz) in: Muheim, Peter u. Partner (eds) Das Energiesparpotential des gemeinschaftlichen Gebrauchs von Motorfahrzeugen als Alternative zum Besitz eines eigenen Autos. Bern

Muheim P, Inderbitzin J (1992), Das Energiesparpotential des gemeinschaftlichen Gebrauchs von Motorfahrzeugen als Alternative zum Besitz eines eigenen Autos. Luzern

MVV München (1996) Repräsentative Mitgliederbefragung des MVV München und StattAuto München, März/April 1996, results published in: Daten Analysen Persepktiven Band 2, Schriftenreihe der Münchener Verkehrs- und Tarifverbund GmbH, Munich

Motivation and Conditions for Mobility without a Car

Leonhard Kasek and Helga Gantz

1 Introduction

Reducing the environmental impact of road traffic presents one of the major challenges of the present time. Air pollution, noise, consumption of land and congestion caused by motorised traffic constitute serious problems in almost all German cities. These effects of motorised traffic are leading to a social devastation of city centres; many people are moving to the outskirts to avoid noise and air pollution caused by car traffic.

Cars are seen as lifestyle statements by most people in Germany, indeed almost as a basic human right. Few motorists really understand the ways in which road transport now causes impacts on health or the environment. For many people the automobile is a symbol of our civilisation and for these people the use of a car is a sign of being part of the society. "What, exactly, is the problem with transport?" asked Kim Loughran, Managing Editor of the Journal Tomorrow (3, 97, page 7). His answer was: "There's too much of the wrong kind; it's too tied in with old-fashioned consumerism, marketed as a way to make us happy rather than a way to get us from A to B. . . ."

Without understanding the social motives for car possession and its use it is impossible to improve the situation. It is not possible to stop the deterioration of the quality of life caused by motorised traffic only through technological progress, therefore we have analysed the social conditions and motivations for the use of cars.

2 The Examples of Leipzig and Dresden

This chapter is based on several surveys on mobility behaviour and motivation in two cities in Saxony, Germany: Leipzig and Dresden. Altogether 76% of the surveyed residents in Dresden have the opportunity to use a car, 53%

SAS Regional, Transport and Environmental Research, Cunnersdorfer Straße 25, D-01189 Dresden, Germany

J. Breuste, H. Feldmann, O. Uhlmann (Eds.)
Urban Ecology

use the car as their main means of transport. These constitute, on the whole, the male members of households, whereas the women are more inclined to use public transport.

The share of people using a car as the main means of transport is growing. Without a fundamental change in the framework conditions at least three quarters of the adult residents will use a car as their main means of transport. About half of the car-users have a very strong commitment to their cars so that they are not prepared to use public transport or a bicycle more often. About one quarter are prepared to use public transport or a bicycle in the case of an emergency. In practice it is difficult to convince these people to reduce the use of a car. About a fifth can be more easily persuaded to change from using a car to public transport or bike and only 3% to 5% are ready to change immediately if they have an opportunity to reach their destination without a car.

Our findings show that, in Dresden, the users of cars require the same amount of time for their journeys (to go to work, to go shopping or for leisure purposes) as the users of public transport or bicycles. The small gain in time for car users is lost in the search for a parking space and the fact that the distance to shopping areas is then longer. In Leipzig one third of all journeys by car have a distance of only 3 km and a further 22% from 3 km to 5 km. Over these small distances car drivers have almost no opportunity to gain time in the city in comparison to users of public transport or bicycles. The car user loses from 5 to 10 minutes during every journey to the city centre in Leipzig through searching for a place to park. This is a reason why express couriers use bicycles in the city centre. About 2/3 of the people researched would be able to live without a car, calculated on the time required for their journeys and the accessibility of their destinations. Only about 30% depend strongly on the car (mostly for work purposes).

Our first question is therefore: for what is the use of a car intended? or, through which conditions is the use of a car enforced?

We obtained the following findings by our empirical research in Dresden (725 adult residents of Dresden were interviewed by SAS GmbH in March 1994, the sample is representative for all adult residents of Dresden and the findings in Leipzig show a similar situation): independence 67%; car as means for the transportation of goods (shopping etc.) 59%; convenience 59%; faster than going by bus or tram 49%; easier to cope with everyday life in the family-life 42%; going by car is fun 36%; possible to shop cheaply 36%; a must for the job 34%; more possibilities for leisure time 32%; to fulfil the daily duties 29%; to avoid disadvantages 27%; going by car is cheaper 27%; some important destinations cannot be reached otherwise 25%; protection against criminals 23%; habit 20%; caused by illness 9%; high prestige 6%.

These findings can be summarised into three main factors: social norms, the accessibility of destinations, and attitudes.

1. Social norms: going by car is a symbol which shows: I am a normal, fully integrated member of society. Doing without a car symbolises a step away from normality. Only people with a very stable social position or people with a low social position that have no hope of rising up the social ladder are able to do this. For most people the car is firstly an attribute of normal social behaviour and only secondly a means of transport. Freelancers and entrepreneurs are heavy car-users: it is very important for them to have the image of being a serious businessman. For them the car has the same importance as a suit and tie for the bank employee. A strong reason for conforming to these norms is the habituation of using a car as the normal mobility method. For every third person car-driving is simply fun and they do not ask whether there is a better (faster, cheaper, not so detrimental to the environment and so on) means of reaching their destination as the car.
2. Accessibility of destinations: above all, migration of residents to the outskirts (see Herfert 1996) and the pressure to take jobs far from the living place are leading to a growing number of car-users. If people own a car it is more economical for them to go shopping in the large supermarkets on the outskirts of town. There the prices are lower and it is possible to buy a everything one needs at the same time. For a growing number of people the car has an increasing importance with respect to leisure pursuits: it is possible to take more equipment and to reach more destinations with the car than with public transport. On the other hand car-driving is becoming more and more a leisure pursuit in itself: for a growing number of people the driving-experience is the goal, not reaching a specific destination. The growing need to use a car in order to reach important destinations will lead to a fall in the number of users of public transport- 30% in the next few years, if policies do not change the framework for mobility in the cities. Above all, due to a combination of social pressure and accessibility, about half of car drivers are by no means ready to change to public transport or bicycling. This is mainly true for men from the age of 30 to about 50 and workers, self-employed persons and civil servants. It is interesting that most people in this group of heavy drivers overestimate the advantages of the use of cars and they are badly informed about public transport and underestimate its opportunities (Brög 1992). Their ties to car-driving are so strong that possible alternatives are no longer perceived.
3. Attitudes: the impact of attitudes on the choice of means of transport is often overestimated. In our analyses only 20% to 25% of the variability was explained by the different attitudes. The following findings reveal the attitudes having the greatest impact (statistically significant coefficients of regression): gender 0.27, to live in peace 0.16, to live healthily −0.13, to have many friends −0.12, to give help to other people 0.11, to protect the environment −0.07.

Habitation to car use is, in most cases, very strong and a change in conditions (for example: a higher fuel price, a new job near the flat or better bus and tram services) would not result in a great change in car use. The impact of ecological awareness on the choice of the means of transport is quite weak. This has much to do with the so called prisoner's dilemma: if only a small proportion of the population reduce their use of cars, there is no obvious benefit for the environment. Only if a greater part of population reduces their use of cars would the environment benefit substantially. Many people do not know much about the influence of car driving on the destruction of our environment and a large proportion of people also do not want to know more about this problem.

A greater reduction of the use of cars in cities would only result from a whole set of measures and a long-term traffic policy directed towards this aim. The following findings shows under which conditions people would use public transport in place of the car (source see above, survey in Dresden 1994): buses and trams are at least as fast as cars 47%; lower fares 45%; charges for using streets 31%; opportunity to go by bus or tram without changing 30%, high frequency of buses or trams 23%; alarming deterioration of the status of environment 22%, many prominent persons use buses or trams too 22%; ease of buying tickets everywhere 21%; easy possibilities to change from car to bus or tram 21%; employer would pay the fare 20%; rising of fuel price 18%; no crimes in buses or trams 18%, a lack of places to park 17%; sufficient seats in buses and trams 14%; cleanliness of buses and trams 11%, information on buses and trams 8%; short distance to next stop 7%.

3 Summary

These findings can also be summarised into three main factors: price comparison between car and public transport, reputation of public transport, and speed of public transport in comparison to cars in the town.

1. Price: a combination of higher fuel prices, higher prices for parking and a lack of places to park in the city-centres is probably the most effective means of promoting public transport and reducing car-driving in the towns, but this method would not be popular and the threshold for a change in behaviour is very high. A higher fuel price would lead, for example, to saving money in other spheres of everyday life and not to less use of the car. Only if the price were very high and there were no more compensation measures possible, would people start to reduce car-use. If the fuel price were higher than 3 DM/litre a greater proportion of people would fundamentally reduce car-driving and if the price were as high as 5 DM/litre a large proportion of people would probably sell their cars. These

findings are confirmed by the reaction to a rise in fuel prices in 1994 (almost 20%): 72% of the residents did not react and only 3% had strongly reduced the use of the car.

2. The reputation of public transport is also important. If prominent figures like artists, sportsmen, leading managers and businessmen, as well as politicians travel by bike or public transport in the town, then 'Joe Public' will use public transport and bike too. These findings confirm that car use is a kind of social behaviour.
3. The time required to reach a certain destination by means of public transport is also important. Short routes to the bus/tram stops, and from the stop to the destination, a high speed of buses and trams, a short time between two vehicles of public transport and as few changes as possible between the start and destination would make public transport more attractive. Attractiveness of public transport is produced, above all, by higher speed and only in second place more and better seats and modern vehicles. There is currently a vicious circle in all cities of eastern Germany: the number of users of public transport is declining. To save money the prices will increase and some routes will be cancelled or buses and trams will go more seldom. This results in more people using their cars.

References

Brög W (1992) Mobilitätsverhalten beginnt im Kopf. In: Altner G et al. (eds) Jahrbuch Ökologie 1993. München, pp 174–190

Herfert G (1996) Wohnsuburbanisierung im Verdichtungsraum Leipzig. In: Stadt Leipzig, Amt für Statistik und Wahlen (ed) Statistischer Bericht I. Quartal 1996, pp 12–16

SAS Regional-, Verkehrs- und Umweltforschung GmbH (1995) Räumliche und monetäre Steuerungsmöglichkeiten des Verkehrsmittelwahlverhaltens in der Landeshauptstadt Dresden

Stadt Leipzig, Amt für Statistik und Wahlen (1995) Bürgerumfrage Frühjahr 1995, Leipzig

Woods J (1997) Cars Stuck in Second Gear. In: Tomorrow 3/1997, p 14

Workshop: Traffic Emissions and their Ecological Impact

4.3

Nitrogen Deposition from the Dennis Agreement and Terrestrial Ecosystems Impact Identification Using GIS

Eric Rapaport and Folke Snickars

1 Introduction

The Dennis Agreement is a transportation proposal for improvement of both the public and road transportation systems in Stockholm County. One of the aims is to decrease the external effects of road traffic, such as air pollution, by construction of two orbital systems, referred to as the Stockholm Ring Road and the Outer Bypass Route.

A case study was set-up to build a model for the determination of threshold limits for terrestrial ecosystems with regards to nitrogen deposition. This led to the development of a nitrogen critical loads model to compare the impacts of the new transportation plans.

It was found that although the Dennis Agreement may reduce emissions, it may not necessarily affect the regional nitrogen deposition pattern in Stockholm County. Based on this and the nitrogen critical load model it may also be concluded that the transportation plans will not necessarily bring about an improvement in the natural environment.

2 Background

In Sweden and other countries, atmospheric nitrogen pollution arises from farming practices, private industry emissions, the production of energy for heat and electricity, and from vehicle traffic emissions. Concern over the role of vehicle emissions led to the adoption of the Swedish Vehicle Emission Ordinance (1991:1481). The passing of strict emission standards might not have been enacted if trends in transportation were not so alarming.

In Sweden, there has been a 73% increase in the measured volume of road traffic in vehicle-kilometres between 1970 and 1989. The number of roads constructed has increased by 148%, and there was a 56% increase in passenger car usage. Of course, it is not possible to state flatly that these trends have

Dept. of Infrastructure and Planning, Royal Institute of Technology, S-100 44 Stockholm, Sweden

J. Breuste, H. Feldmann, O. Uhlmann (Eds.)
Urban Ecology

caused the increase in atmospheric nitrogen. New technology and emission requirements for vehicles take place at the same time. Increasing car ownership may, however, coincide with a net increase in emissions. (Button and Rothengatter 1993.)

Road traffic contributes to 62% of the total nitrogen emissions in Stockholm county (Ekström et al. 1995). Consequently any new significant transportation infrastructure project which proposes to change the amount or flow of traffic within Stockholm County should consider the impacts on regional air pollution patterns.

In Sweden and other countries nitrogen pollution impact assessments are often defined by hourly concentration. Little effort has been made to measure the long term impacts such as soil acidification and eutrophication. A project was developed to determine how long-term nitrogen pollution could be assessed for large scale transportation plans.

3 Methods

The aim of the study described in this chapter is to link nitrogen deposition arising from infrastructure improvements to impacts on terrestrial ecosystems. This was carried out in two steps. First a road traffic model, EMME/2, was linked to a geographical information system to determine the amount of nitrogen emissions and deposition. EMME/2 was used to estimate traffic flow with and without the Dennis Agreement. The scenarios included trends related to economic growth which could affect the amount of traffic and the type of vehicle on the road. Results from EMME/2 included nitrogen and carbon dioxide emissions.

The emission information was transferred to a geographical information system in which a dispersion and deposition model was built. This model provided crude estimates of nitrogen deposition throughout the Stockholm region. The model was not calibrated and thus the results must be look at critically.

Second, a GIS-based nitrogen critical load model was created using a similar method applied by the Stockholm Environment Institute. Kuylenstierna and Chadwick (1989) created a relative acidification sensitivity map for Europe by using a classification of ecosystems according to relative sensitivity, and by subsequently assigning critical load values which had been predefined for a specific site or region. Four digital layers reflecting nitrogen sensitivity were weighted and combined using simple additive map algebra. Each indicator was scaled to reflect sensitivity to the deleterious impacts of nitrogen (Table 1).

To handle uncertainty about the selection of factors, four different scenarios were created based on the inclusion and exclusion of factors and iteration

Table 1. Example of some of the controlling factors used to reflect nitrogen sensitivity in the study

Class Sensitivity	Soil Texture	C/N Ratio	Bedrock	Rainfall (mm/yr)
1 Very high	Rock outcrop	>20	Young Granite	>1220
2 High	Gravel/sands	20–25	Granite/Gneiss	800–1220
3 Medium	Tills	25–30	Gneiss	600–800
4 Low	Silts	30–40	Leptite	400–600
5 Very low	Clays, others	<40	Gabbro	<400

Table 2. Example of the allocation of empirical critical loads over the nitrogen sensitivity classes for coniferous forest

Nitrogen sensitivity class	1	2	3	4	5
N critical load (kg/ha/yr)	3	6	9	12	15

(Information on critical load values comes from Nilsson 1989).

of weights. The scaling of nitrogen sensitivity was used to extrapolate empirical nitrogen critical load values (Table 2). Forests situated on relatively more nitrogen sensitive sites were assigned lower critical values compared to those forests located on sites relatively less sensitive to nitrogen. The nitrogen critical load maps were compared with nitrogen deposition maps. The resulting maps showed which ecosystems are over or below the critical load limit. From the results it was possible to determine whether or not the deleterious impacts linked to nitrogen would continue to occur; it did not, however, identify the specific effects.

4 Results

The critical load model was applied to a study area in the south western part of Stockholm County. The entire area covers 49,056 hectares and falls within the boundaries of six municipalities all situated outside what may be considered the urban core, inner-city Stockholm. The selection for the study site was based on available digital information.

The reduction in nitrogen deposition from the transportation plans did not decrease enough to cause any significant changes regarding the critical load (Table 3 and Table 4).

About 95% of ecosystems in the study area showed no change. The positive changes range from a low of 1.3 to 9%. The high economic growth results

Table 3. The results form the critical load models using all factors and weighting

Mean economic growth			Fast economic growth		
Impact	Total	Change	Impact	Total area	Change
Negative	0 (ha)	0.0 (%)	Negative	5 (ha)	0 (%)
No change	2696 (ha)	98.7 (%)	No change	2533 (ha)	91 (%)
Positive	36 (ha)	1.3 (%)	Positive	255 (ha)	9 (%)
Total	2732 (ha)	100.0 (%)	**Total**	2793 (ha)	100.0 (%)

Table 4. The results from the critical load models using all factors and No weighting

Mean economic growth			Fast economic growth		
Impact	Total	Change	Impact	Total area	Change
Negative	0 (ha)	0.0 (%)	Negative	18 (ha)	0.6 (%)
No change	2577 (ha)	98.7 (%)	No change	2572 (ha)	97.0 (%)
Positive	34 (ha)	1.3 (%)	Positive	63 (ha)	2.4 (%)
Total	2611 (ha)	100.0 (%)	**Total**	2653 (ha)	100.0 (%)

in the greatest positive changes, 2.4 and 9%. This is mainly due to the fact that with increased economic growth, the purchasing of new cars increases leading to more fuel efficient cars with catalytic converters on the roads, which results in less emissions.

The dominate ecosystem found in the study area, as well as in Stockholm County, is coniferous forest; this showed little improvement. At the most, 5 to 6% of coniferous forest will alter from being above to below the critical load level, critical load scenario 3. Critical load scenario number 4 resulted in no improvements.

5 Conclusion

The project demonstrates an ability to link nitrogen deposition from transportation plans to terrestrial impact damage by using critical loads to set thresholds. Thus future large scale transportation projects could consider the impact not just in terms of emissions related to concentrations, but also in terms of deposition and its related consequences. This seems to be an area for future research.

The rate of economic growth affects the amount of nitrogen deposition. Fast economic growth resulted in greater reductions in deposition. It is often thought that technology improvements result in greater emission reduction compared to transportation infrastructure improvements. The Dennis Agreement is likely to reduce the amount of nitrogen deposited, but this will not result in dramatic improvements of certain terrestrial ecosystems located in the south-western corner of the County of Stockholm. This means that acidification, nitrogen leakage and other negative consequences will probably continue in ecosystems in the study area.

References

Button K, Rothengatter W (1993) 'Global Environmental Degradation: The Role of Transport.' In: David B, Kenneth B (eds) Transport, the Environment and Sustainable Development. E & FN Spon., London

Ekström A, Johansson C, Höglund P, Jonson T (1995) Air Pollution in Stockholm County 1994–1995. Stockholm Air Quality and Noise Analysis. Miljöförvaltningen, Stockholm. (In Swedish)

Kuylenstierna JCI, Chadwick MJ (1989) The Relative Sensitivity of Ecosystems in Europe to the Indirect Effects of Acidic Depositions. The Stockholm Environmental Institute at York, University of York, UK

Nilsson J (1989) Critical Loads for Nitrogen and Sulphur. In: Malanchuk J, Nilsson J (eds) The Role of Nitrogen in the Acidification of Soils and Surface Waters. Report 1989:10. Nordic Council of Ministers

Habitat Fragmentation and Roads: Strategy, Objectives and Practical Measures for Mitigation and Compensation

Hein D. van Bohemen

1 Strategy of the Second Transport Structure Plan

The Dutch Second Transport Structure Plan (Ministry of Transport 1990) is an example of an approach to integrate environmental aspects into a policy document for which an environmental impact assessment was not compulsory.

The strategy of the Second Transport Structure Plan consists of the following five elements:

1. Tackling problems at their source (e.g. vehicles must be as clean as possible).
2. Managing and restricting mobility (e.g. reducing the travelling distances between home and work by means of location policy).
3. Improving the alternatives to the private car (e.g. improving facilities for cyclists, raising the standards of public transport).
4. Selective accessibility on the roads (e.g. special corridors for each mode of transportation).
5. Strengthening the foundations with support measures (e.g. better communication, financial planning, and research).

The five elements of the strategy have been translated into 35 policy areas (Fig. 1): e.g. air pollution, energy-saving, noise nuisance, etc. grouped into 4 categories: environment and amenity, managing and restricting mobility, accessibility, and foundations (means of support).

For all the policy areas qualitative and quantitative targets have been formulated.

By way of illustration I will discuss the aspects of roads and traffic and habitat fragmentation in more detail.

Road and Hydraulic Engineering Division, Ministry of Transport, Public Works and Water Management, P.O.Box 5044, 2600 GA Delft, The Netherlands

J. Breuste, H. Feldmann, O. Uhlmann (Eds.)
Urban Ecology

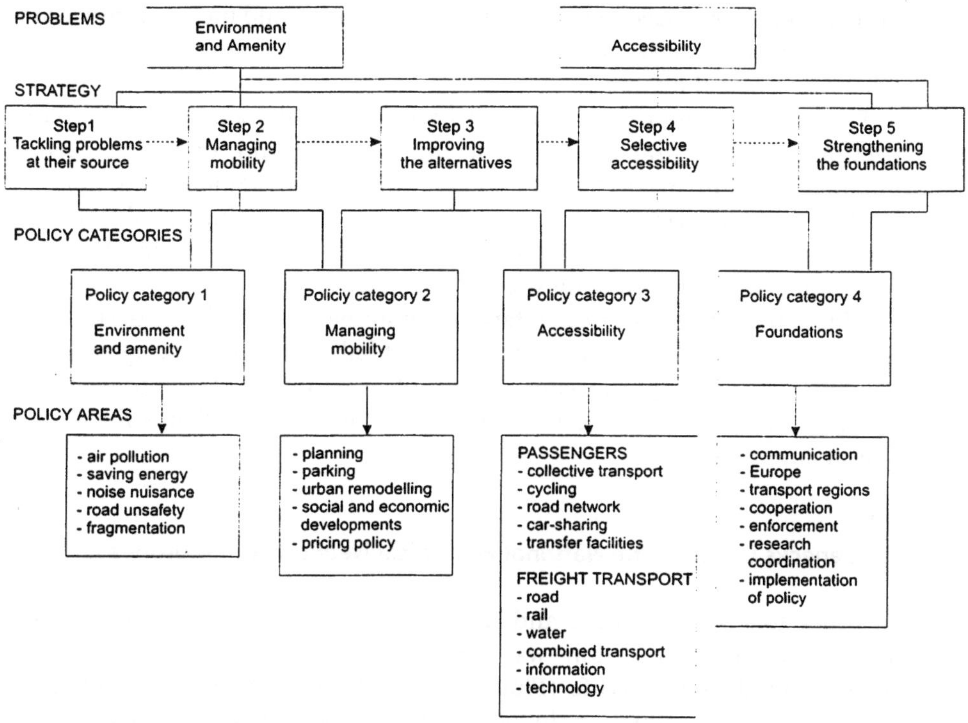

Fig. 1. Objectives of the Second Transport Structure Plan

2 Habitat Fragmentation

The overall effect of infrastructure on nature is called fragmentation: the partition of ecosystems and/or habitats of plant and animal populations into smaller, more isolated units. Infrastructure is one of the main causes of habitat fragmentation (Canters 1997).

Four aspects of fragmentation can be distinguished:

- destruction: absolute loss of a habitat area through the physical presence of the road and associated infrastructural elements,
- disturbance: deterioration of the habitat due to e.g. traffic noise, emission of xenobiotic substances and the effects of visual stimuli (artificial lighting),
- barrier action: separation of functional areas,
- collisions: injury or death of fauna due to motorised traffic.

In order to mitigate negative effects of roads on faunal elements, certain measures such as warning signs, wildlife warning reflectors, fences and badger tunnels were introduced in the seventies and eighties, and more recently ecoducts (overpasses for fauna), allowing animals to move more safely between otherwise isolated habitats.

First of all an attempt is made to prevent habitat fragmentation by restricting infrastructural development and by integrating the infrastructure into the landscape. Secondly, the aim is to counteract fragmentation. This can be done by mitigating existing situations (by constructing suitable verges, providing barrier fencing and tunnels for badgers and amphibians, by building ecoducts, and by reducing speeds and noise levels).

Where the effects of fragmentation cannot be mitigated it is desirable to take compensatory measures (replacing lost habitats or enhancing marginal habitats through appropriate forms of environmental improvement). Recently, compensation became compulsory in the Netherlands for certain conservation areas.

2.1 Bottlenecks between the National Ecological Network and Motorways

In order to find out where and how to implement measures in concrete situations, an investigation was carried out on the main bottlenecks between roads and wildlife (Fluit et al. 1990). In this study a bottleneck was defined as a situation in which a road is situated in an actual or potential habitat of a species which is vulnerable to fragmentation, in a priority area from a viewpoint of conservation, or which shows large numbers of dead animals due to traffic collisions. This was followed by an investigation of all the locations of intersections between the National Ecological Network and the existing and planned main infrastructure (Specken 1992).

Later more concrete targets concerning reducing the effects of habitat fragmentation due to roads were formulated:

- In 2000 40% of the bottlenecks will be resolved
- In 2010 90% of the bottlenecks will be resolved
- In situations where mitigation is not possible compensation measures will be implemented.

3 Mitigation Measures

A definition of mitigation is: measures used to stop or reduce negative effects on ecological values caused by building and use of infrastructure.

3.1 Badger Tunnels

In the Netherlands, yearly 15% (approximately 400–450 individuals) of the total badger population is killed by road accidents. The main bottlenecks between badger and motorways have been solved by implementing mitigating measures such as badger tunnels and barrier fencing along the road to realise safe connections between habitats. During the last few years the reduced mortality brought about an increase in population size.

3.2 Tunnels for Amphibians

Besides experiences with the temporary closing of roads and the moving of caught amphibians in buckets, nowadays special tunnels for amphibians together with prefabricated conducting elements are a successful solution.

3.3 Ecopipes

All kinds of species can use the stream culverts to cross a road, if the culvert is completed with a dry path beside the water. Different forms of culverts (ecopipes) are available which can be used by different groups of species.

3.4 Ecoducts

Together with the construction of a new highway (A50) in 1988, two overpasses (ecoducts) were built to enable the red deer to migrate from one side of the highway to the other.

The ecoducts are protected against the visual and acoustic influences of the highway by means of a wall and trees.

Recently a new ecoduct was built over the highway (A1) near the Dutch town Almelo (Boerskotten) and more ecoducts are in a planning stage.

3.5 Joint Use of Constructions

Bridges, viaducts and aqueducts can also serve as crossing places for animals. Isolation of parts of these constructions from traffic increases the possibilities for animals to cross the roads.

4
Compensation

Compensation entails that ecological damage, expected to result from human intervention which cannot be mitigated, is compensated for by replacing lost habitats or by enhancing marginal habitats.

In the case of mitigating measures such as barrier fencing and badger tunnels, a connection can be established between the effect of the intervention and the measure that is implemented. In the case of compensation, however, it is more difficult to establish such a connection.

A logical order of priority might be to apply compensation (1) *in*, (2) *adjacent to* and (3) *outside* the intervention area. The product of the area and the quality of nature in the intervention area should remain equal before and after intervention.

If identical replacement of ecological functions and natural values proves unfeasible, the best approach is to aim for similar ecological functions and natural values and, if this is not feasible, for dissimilar ecological functions and nature values.

5
Conclusions

It is important to realise that mitigation and compensation are elements of an integrated landscape conservation strategy to maintain values of natural ecosystems. Moreover, an ecological network across the landscape can help to achieve the goals of biodiversity conservation. Both are invaluable components of the overall conservation strategy.

References

Canters KJ (ed) (1997) Habitat Fragmentation and Infrastructure. In: Proceedings of the international conference on habitat fragmentation and infrastructure and the role of ecological engineering. Ministry of Transport, Public Works and Water management, RWS/DWW, Delft

Fluit N van der, Cuperus R, Canters KJ (1990) Mitigerende en compenserende maatregelen aan het hoofdwegennet voor het bevorderen van natuurwaarden, CML/DWW, Leiden/Delft

Ministry of Transport (1990) The Second Transport Structure Plan, The Hague

Poster 4.4

Analysis of Transport Measures for Traffic Avoidance through Reduction of Motorised Private Transport (MPT) in Leipzig

Erhard Geisler

Basic aims of a socially tolerable and ecologically orientated traffic planning are the traffic avoidance, the moving towards more environmental friendly modes of conveyance and compatible processing of the traffic. Avoidance of MPT consists of reduction of distances traveled through a altered spatial network and/ or the change to more environmental friendly types of traffic. The development and aims of the Leipzig traffic are marked in Table 1.

Internal and external factors of the traffic sector are interdependent. Use of motorised transport and car use increase every year, while the budget of time for traveling on average remains the same. Faster traveling speeds of the MPT are compensated for to an extent by the increase in distance traveled (Table 2). Regarding the purpose of mobility there is no significant increase in the statistical average. On the contrary, the activities out of the house are reduced because of mass unemployment and an increase in the percentage of old people in the population (Table 3). A primary decisive factor for the increasing MPT is the *ownership of cars*. From this arise

- the acceptance of larger and increasing distances,
- the endeavour to use the car as often as possible, i.e. for journeys of short distances as well as long ones,
- necessary analyses select instruments for the minimisation of the MPT in the FRG are,
- analyses of the elasticity of prices (costs of the maintenance of cars, particularly for fuel, taxes and disposal, costs for the management of park areas),
- effects on the building and the sustaining of technical and social infrastructure,
- criteria for the accessibility and acceptance of the distances, particularly for the non-motorised forms of transport,
- criteria for an indispensable MPT,

University of Leipzig, Interdisciplinary Institute of Nature Conservation and Environmental Protection, Augustusplatz 10/11, D-04109 Leipzig, Germany

J. Breuste, H. Feldmann, O. Uhlmann (Eds.)
Urban Ecology

- criteria for minimal equipment of an area with technical and social infrastructure,
- criteria for the public transport alternative compared with the MPT,
- building legislation (public transport as an integral component for the development of new building areas).

Table 1. Leipzig modal split: development and future aims

Mode of travel	1987	1991	1994	2010*
afoot	36,0	39,5	38,2	35,0
bicycle	5,2	5,8	5,8	12,0
motorcycle	1,2	1,0	0,1	1,0
car driver	16,6	22,2	27,4	20,0
public transport	35,1	22,8	20,1	23,0

*aims of the Leipzig traffic concept

Table 2. Development of the average time pro journey, distances and speed in Leipzig

Mode of travel	1987			1991			1994		
	T	D	S	T	D	S	T	D	S
afoot	12	–	–	14	0,9	3,8	14	1,0	4,0
bicycle	13	–	–	16	3,1	11,7	14	2,5	10,3
motorcycle	15	–	–	16	6,3	23,8	–	–	–
car driver	17	–	–	22	8,0	22,4	30	13,8	27,7
public transport	36	–	–	41	11,9	15,2	35	8,1	12,9

T = average time pro journey (min./journey), D = distances (km/journey), S = speed (km/h)

Table 3. Selected reference mobility numbers in Leipzig (all data fundations: Traffic Planning Dept., Leipzig)

Mobility number	1991	1993	1994	1995
increase of the average days traffic (%)		5	5	6
cars/1000 citizens		378	396	398
number of the journeys pro person and day				
- mobile persons	3,9		3,6	
- all persons of a household	3,4		3,2	
time pro journey (min.)	26,3		30,0	
Cars become predominant (share of the traffic kinds) used:				
- journey to work or to the training place (%)		46		52
- shopping/services (%)		52		58
- everyday life recreational (%)		69		51

Transport in the Municipality of Iasi – a Challenge for Sustainability

Peiu Nicolae

1 Statements of Objectives

Iasi, an historical and cultural centre (with about 400,000 inhabitants), is the first city in Romania to have joined the European Sustainable Cities Campaign. The main objective of our study is the attempt to present:

- The assessment of the impact of the presently changing mode of transport towards more use of private cars on the environment.
- Measures to control the environmental impact of transport.

2 Methods

The following methods were used to assess the impact of the presently changing mode of transport towards more use of private cars:

- Monitoring of air quality with the help of a mobile laboratory in a campaign in the period 25 November – 3 December 1996. Pollutants: NO_2 (chemiluminescence), SO_2 (flame-photometry) and O_3 (chemiluminescence)
- A dispersion study carried out on the basis of an inventory of urban mobility, EU Corine air emission factors for cars, CALINE 3 and CAL3QHC air quality dispersion models from US EPA (1995)
- Monitoring of emissions from cars. Pollutants: CO, CO_2 and VOC.

3 Results

Despite the high humidity during the monitoring period a maximum level concentration for O_3 was registered (Table 1) at the cross-road Podu-Rosu

11a/3, MACAZULUI Street, 6600 Iasi, Romania

J. Breuste, H. Feldmann, O. Uhlmann (Eds.)
Urban Ecology

Table 1. Results of air quality measurements (μg/m³)

Cross-road	Copou	Central B.S.	Central R.S.	Mitropolie	Podu-Rosu
NO2 max. level	20	103	14	88	42
NO2 average con	13	47	8	36	15
SO2 max. level	55	58	38	21	55
SO2 average con.	27	29	14	11	33
O3 max. level	36	23	12	5	102
O3 average conc.	16	14	6	3	22

(one the busiest in the city) which exceeds the Romanian Air Quality Standard ('100 μg/m³ maximum concentration for short time −30')

The results of the dispersion study show that many receptors receive levels of concentration of CO that exceed the Romanian Air Quality Standard (19 mg/m³ in the cross-road Independentei, 10 mg/m³ in the cross-road Podu-Rosu and 7 mg/m³ in the cross-road Podu de Piatra, in comparison with a standard value of 6 mg/m³ maximum concentration for short term −30'). Concerning the monitoring of emissions from cars, 182 cars representing 42.6% of cars monitored, failed to observe the Romanian Emission Standard for cars.

4 Proposal and Conclusions

We propose the following measures to control the environmental impact generated by the urban mobility that already threatens the population's health and architectural heritage of the city,

- the decrease in the need for mobility (services in place);
- the extension and good maintenance of public transport;
- the rehabilitation and extension of cycle tracks;
- construction of noise protection walls along some railways through the city.

Other measures include regular emission tests for motor vehicles, encouraging efficient driving behaviour, educational campaigns etc. All these proposed measures and others, that will be included and enforced in a Local Agenda 21, can represent steps towards the achievement of a sustainable development of the transport in our city.

References

US EPA (1995) User's Guide CAL3QHC Version 2.0: A Modelling Methodology for Predicting Pollutant Concentrations Near Roadway Intersections

5 The Integration of Nature and Landscape into Urban Development

Integration of Nature in Urban Development

David Goode

1 Introduction

Concern for nature in the city is a relatively recent phenomenon which has developed spontaneously in many places during the past 20 years. In some towns and cities, particularly in Europe and North America, new programmes based on ecological approaches have been developed for the protection and management of nature in urban green space. New initiatives, from small-scale neighbourhood wildlife gardens and ecology parks to more extensive networks of green corridors and urban fringe forests, are all part of this green revolution. There has undoubtedly been a change in attitude resulting in greater appreciation of the value of nature in the city. Such changes include greater awareness amongst the professions responsible for planning and management of urban open space.

Nature's ability to survive in cities is clearly demonstrated by many studies describing the wide range of habitats and species commonly occurring within urban areas (Sukopp and Werner 1987). Most cities contain examples of ancient habitats which survive as pieces of encapsulated countryside within the urban sprawl. Such places are often rich in biological diversity and can be regarded as natural gems within the urban fabric. In addition, there are other habitats which have developed in man-made situations through natural colonisation of derelict industrial sites, railway sides and even cemeteries. Such unintentionally wild places provide an important range of habitats in many towns and cities (Goode 1989).

Increasingly, it is being argued that such habitats should be preserved, not only for their wildlife per se, but to provide places where nature can be experienced by people at first hand. In the UK, many towns and cities have now identified the most important areas of this kind and are implementing programmes for their protection within local plans. Protecting such areas is, however, only one means of catering for nature in the city. After all, such habitats have developed only by chance, not by design. If we were to utilise

The London Ecology Unit, Bedford House, 125 Camden High Street, London NW1 7JR, Great Britain

J. Breuste, H. Feldmann, O. Uhlmann (Eds.)
Urban Ecology

ecological knowledge in the design and management of urban green space, then considerably greater provision could be made for nature, at the same time catering for people's enjoyment of nature within urban surroundings.

The first part of this chapter reviews approaches developed in the UK for ecological planning, with particular reference to London. The second part is concerned with designing for nature within the urban environment.

2 Planning for Nature in UK Towns and Cities

One of the most significant developments in the UK has been the adoption of strategies for nature conservation as part of town and city planning by many local authorities. Such strategies provide an overall framework for policy development, explaining why nature conservation is important and how it can be implemented. A typical strategy will contain a clear set of objectives, list the necessary policies to achieve these objectives and identify areas worthy of protection. It also specifies which areas of public open space are appropriate for ecological management. Recent examples of such strategies have also referred to the necessary mechanisms for implementation of policies, either through planning or by means of partnerships with local organisations. Some of the earlier examples such as Greater London (1984) and Manchester (1986) were not only significant locally, but also provided important models for use elsewhere. Such strategies have certain features in common. All recognise the need for protection of habitats of value, enhancement of existing open land for wildlife, and the creation of new habitats in areas of particular need. They all place considerable emphasis on the local value of nature to residents of urban areas. Over the past ten years numerous examples of such strategies have now been produced in the UK, a recent example being the detailed plan for people and wildlife in Peterborough (1992). The development of strategies for nature conservation in urban areas has been reviewed in detail elsewhere (Goode 1989, 1993).

The rapid progress made over the past 10 years is illustrated with reference to London. Nature conservation policies for the capital were first produced in a popular handbook published by the Greater London Council in 1984. Since that time, the London Ecology Unit has produced detailed strategies for about 25 individual London boroughs. This involves comprehensive survey and evaluation of all areas likely to be of value for nature conservation, including parks and public squares. Such areas are categorised according to their significance on a London-wide, borough or local level. Each strategy is published as a handbook which performs the dual role of providing both specialist input to planning and a popular account of the natural history for borough residents. The descriptions and management recommendations also provide a basis for drawing up management plans for implementation in the individual boroughs by parks' departments. It is expected that over 1300 'Areas of Nature Conservation Importance' will be identified within the 33 boroughs comprising Greater London. The resulting system has been widely adopted in the capital for protection of 'Sites of Nature Conservation Importance', including those of value to local neighbourhoods.

3
Designing for Nature in the Urban Environment

The integration of nature can be achieved on a number of different scales. It is possible to accommodate a great deal more naturalistic planting within the built-up environment, for example in courtyards or even through the creation of rooftop habitats. More common perhaps are neighbourhood wildlife gardens which are small scale projects which enable the enjoyment of nature within the local community. Such gardens include children's play areas as well as a pond and other wildlife habitats. Ecological parks are rather larger areas (1 to 2 ha) where new habitats are created for public enjoyment of nature and for environmental education. Most ecological parks in the UK have this dual role.

Many schemes have now been implemented in the UK to cater for nature within existing parks. Several stages are required in this process:

- Survey of present ecological features and existing uses of the park.
- Assessment of potential for ecological enhancement and for creation of entirely new habitats.
- Consultation with park users.
- Production of management plan with clear objectives for ecological management.
- Further consultation on proposals.
- Implementation of new habitats with adequate information for the public on what is being done.

When identifying existing features of nature conservation value, it is important to identify existing habitats such as areas of ancient grassland or long-established woodland. Individual old trees are often particularly important, especially for their dead and dying timber. Other features of value include natural irregularities in local terrain and, particularly, water bodies such as streams and ponds. Areas where the vegetation structure is undisturbed, with a good shrub layer, are also important.

Ecological design involves a number of factors which will differ from normal park management. These include the effective application of ecological knowledge, for instance in developing new herb-rich grasslands where knowledge of soil conditions and nutrient status is crucially important. It is also important to create a diversity in vegetation-structure with open glades, and shrub layers within park woodland. Waters-edge habitats will also be important, allowing a natural hydrosere to develop. In all these approaches it will be important to use native species where possible.

4
Key to Success

It is suggested that successful integration of nature within the urban environment will depend upon the following:

- Applying ecological knowledge effectively.
- Integration of approach between the professions.
- Adequate training in ecology for open space managers.
- Public interpretation of nature.
- Partnerships with the local community.

Improved professionalism in creating naturalistic landscapes within urban parks could result in a greatly increased variety of wildlife within our towns and cities, but this will only be achieved by bringing together the skills of several different professions including ecological science, horticulture, landscape design and park management. I believe that co-operation between these disciplines, together with involvement of the local community, will provide the key to success. There are enough examples of good practice available for us to be confident of succeeding. The challenge now is to produce new landscapes in the urban setting, equalling the quality of some of the park and garden landscapes of the past.

References

Goode DA (1989) Urban nature conservation in Britain. J Appl Ecol 26: 859–873

Goode DA (1993) Local authorities and urban conservation. In: Goldsmith FB, Warren A (eds) Conservation in progress. Wiley, London

Greater London Council (1984) Ecology and nature conservation in London. Ecology Handbook 1. GLC London

Greater Manchester Council (1986) A nature conservation strategy for Greater Manchester. Greater Manchester Council, Manchester

Peterborough City Council (1992) Peterboroughs strategy for people and wildlife. Peterborough City Council, Peterborough

Sukopp H, Werner P (1987) Development of flora and fauna in urban areas. Nature and Environment. Series No 366. Council of Europe, Strasbourg

Urban Development and the Integration of Nature: Reality or Fiction?

Rüdiger Wittig

1 Introduction

There can be no question that it is highly desirable to integrate nature into the development of cities. Many authors have enumerated the advantages of a city containing 'nature' in comparison to a city being only a desert of stones. Hence there is no need to list and discuss these advantages a second time.

'Nature in cities' is a lucrative business. Many jobs depend on it, a lot of money is involved, and there is strong competition. Competition often leads to exaggerated promises, therefore scientists have to carefully explore the question of whether, in which cases and to what degree, the integration of nature into urban development is reality or fiction. In addition, the statements and promises associated with the subject 'integration of nature into cities' have to be questioned.

Whether the integration of nature into urban development is regarded as reality or as fiction is not just a question of the definition of the two terms 'nature' and 'integration', it also depends upon the geographic and economic situation. For developing countries located in arid zones, in particular, the situation is quite different from that in developed countries in temperate and humid regions. We start with a short look at the situation in developing countries in arid zones, touch on the problems of definition of 'nature' and 'integration' and then come to the integration of nature into urban development in developed countries; finally some conclusions will be made.

2 The Situation in Developing Countries, Particularly in Arid and Semi-arid Zones

In arid zones, particularly in desert regions, no one has the idea of integrating nature (i.e. the desert) into urban development. That is why such a sharp

Johann Wolfgang Goethe-Universität, Institut of Botany, Department of Geobotany and Plant Ecology, Siesmayerstraße 72, D-60323 Frankfurt am Main, Germany

J. Breuste, H. Feldmann, O. Uhlmann (Eds.)
Urban Ecology

border is found between urbanised areas and nature. Of course, even in desert cities, there is some urban green, however, this is highly artificial and depends strongly on permanent watering. Considering that water is often a rare and valuable resource in these areas, one should pose the question of whether there is no better use for it than the irrigation of lawns and exotic trees and flowers.

In developing countries in arid and semi-arid areas, ensuring a supply of food and firewood is one of the greatest problems. In the vicinity of large cities, only fruit trees can survive. All arable land is used for the production of food and non-arable areas are destroyed by intensive grazing. Around large cities in savannah areas a broad belt of cleared landscape is found (Kéré and Wittig 1995). Large mammals avoid these areas and all edible birds, reptiles, small mammals etc. have no chance of survival because they are hunted intensively. Some green areas can be found in hotel grounds and in the quarters of the rich, particularly in the housing areas of ambassadors, ministers and representatives of European and North American companies but these areas are alien elements, rather than integrated parts of urban development. This leads us to the question of what is meant by 'integration'.

3
The Term 'Integration'

The term "integration" comes from the verb 'to integrate' which means to combine parts into a whole, to complete something by adding parts, or, in a figurative sense, to treat someone as an integral part of a community (e.g. integration of handicapped persons into the working community). Integration requires acceptance and not just tolerance combined with mental disapproval. Although spontaneously growing plants are no longer strongly disapproved of and attacked, a dandelion (*Taraxacum officinale*) growing between paving stones in the city centre is only tolerated, not accepted, and is consequently not integrated into the city environment. In upper-class neighbourhoods dominated by single-family dwellings, spontaneous vegetation is regarded as being a weed, and is consequently attacked or at least strongly disapproved of. The congruence of the vegetation map and the social map of a city, which was proved for Osnabrück by Hard (1985), is still true for all European cities.

On the other hand, there have been some successful attempts to integrate spontaneous vegetation into urban development, e.g. on abandoned industrial land in the Ruhr area (Reidl, 1995). However, these positive examples have not been able to change public opinion generally. It will take at least one or two generations, i.e. several decades, before spontaneous vegetation has the chance to become accepted as an integrated aspect of urban areas by the majority of the citizens.

There is a second type of natural area, which we cannot consider to be integrated into urban development: many cities have nature reserves which contain a biocenosis so sensitive to trampling, eutrophication, littering, etc. that city authorities are only able to protect it by erecting fences and prohibiting trespassing. By taking such measures it is sometimes possible to preserve rare and threatened species of the original landscape within urban areas (unfortunately not in all cases, due to other factors as will be explained below). However, in these cases we can not speak of the integration of nature into urban development. The areas referred to have not been integrated, but excluded, from urban development.

4
The Term 'Nature'

This paper shall omit highly sophisticated discussions about whether there is a borderline between nature and culture, or whether culture is a part of nature. If the latter were true, even city centres would have to be regarded as nature. In this case there would be no need for a discussion on 'the integration of nature into urban development', because urban development would indeed be a part of the development of nature.

The fact that 'integration of nature into urban development' was one of the key subjects of the Urban Ecology conference in Leipzig shows that ecologists do not view culture as a part of nature but see a border (albeit an blurred one) between nature and culture. Nature can be influenced by measurements of culture (in particular agriculture, silviculture, horticulture, and urban development), but it must have a strong spontaneous component. Without any doubt, all types of landscape originating without human (i.e. cultural) influences are natural (e.g. forests containing natural vegetation, moors, fens, waters, edaphic savannahs, alpine regions, deserts etc.). Traditional agricultural landscapes consist mainly of nature, in particular, heaths, dry meadows, wet meadows, and extensive pasture grounds that have developed spontaneously under the influence of mowing or grazing; even in a corn field there is a lot of spontaneity.

In an ornamental park, in contrast, everything is planned and prescribed by garden architects. Gardeners scrupulously watch over the maintenance of these anthropogenic plans. No plant is allowed to grow spontaneously and animals are welcomed or tolerated only if they do no harm to the park. Of course, even such a park consists of natural elements, i.e. plants, soil and some animals. However, the result (the ornamental park) should no longer be called nature. This can be illustrated with an example from the field of geology: there is no doubt that a geologist will regard rocks as nature. He/she will also regard geological profiles as nature, even if they are man-made (i.e. in a stone pit, or along streets or railway tracks leading through mountain areas), but he will never regard a building as nature, even though it consists of sand, stones and other geological material (i.e. of elements which obey the laws of nature and fall in the category of the natural sciences). In the same way that one would not call the skyline of San Francisco nature, I am not willing to call the gardens of the Royal Palace in Copenhagen or the gardens of the Vatican in Rome, nature, but I concede, of course, that they consist of natural elements.

A clear distinction should be made between areas containing nature and areas containing some natural elements within the scope of 'garden art' or of intensive production. Within the term nature one should differentiate between remnants of the natural or agricultural landscape and urban nature.

5
The Integration of Nature into Urban Development in Developed Countries

In the following we will at first deal with nature of the natural and agricultural landscape (5.1) and then turn to urban nature (5.2). In both cases only spontaneous nature will be considered.

5.1
Remnants of the Natural and Agricultural Landscape

Excluding the alpine region, where normally no urban development takes place, the natural landscape of most of the developed countries consists mainly of forests, accompanied by water bodies and, in some cases, moors (fens, bogs). All studies dealing with forests included in urban development have found a great difference between the species combinations in urban and in non-urban forests. In the Ruhr area, beech forests, which are believed to be the former natural vegetation of this area, are still widespread today. The herb layer of the Ruhr beech forests, however, has nothing in common with that found in beech forests in the areas surrounding the Ruhr area, although they grow on the same soil and under the same climatic conditions (Wittig 1991a).

In the herb layer of pine-oak forests in Berlin *Acer platanoides* and *Acer pseudoplatanus* are very common. Neither species belongs to natural pine-oak forests, nor do *Sambucus nigra*, *Urtica dioica*, *Euonymus europaeus* and *Ribes uva-crispa*. The latter species are not represented in all areas of the pine-oak forests of Berlin, but can frequently be found in some of the subtypes of these forests (Seidling 1990). We also observe that in Berlin urbanisation has caused a remarkable change in the species composition found in natural forests.

Urban development up to now has always been associated with a lowering of the ground water level (Wessolek and Renger 1998, Burghardt 1991). It may be possible to reduce this phenomenon, but it will impossible to avoid it. The integration of all wetland habitats into urban development will be associated at best with a dramatic change in species composition; often such integration will remain a fiction. For instance, some decades ago there was a large wetland area, mainly used as a hay meadow, in the north of Frankfurt covered with plant communities of the Calthion. Part of this was declared a nature reserve. Today this reserve is no longer situated near Frankfurt (as it was at the beginning of this century) but in Frankfurt itself, as the growing city has incorporated this area. This has resulted in a large part of the former wet hay meadows (communities of the Calthion) being changed into fresh meadows (Arrhenatherion).

All bodies of water integrated into urban development have changed the state of their banks or shores, their chemistry and, of course, their species composition. There is probably no urban body of water which has a submerged plant community as rich as is possible in 'natural' waters. Usually, submerged plants are totally missing and floating species are reduced; often only one or two species are left. The fact that reed communities in rivers are normally strongly reduced in cities was proved by the example of Berlin (Sukopp and Markstein 1981).

In the vicinity of cities and, of course, particularly within cities, traditional agriculture is declining. Consequently, it is very difficult to integrate parts of the agricultural landscape into the development of cities. It is only possible to maintain hay meadows in cities, or in the immediate surrounding of cities, when the mowing is paid for. However, at times of decreasing budgets, such a financial effort is no longer possible. In the Frankfurt area there are some attempts to change meadows into pastures to maintain the grassland character at least.

Some pioneer communities of dry extensive grass land, particularly of sand dunes, may survive in cities, as is proved by the example of the city of Mainz (Frey, in prep.). Nevertheless, in comparison with the former agriculture landscape, which is preserved in the nature reserve 'Mainzer Sand', the city area has lost the rarest communities.

5.2 Urban Nature

The characteristic urban nature, i.e. ruderal plant communities, is the only type of nature which can be easily integrated into urban development because it is adapted to the specific urban conditions (Wittig 1991b), and thus needs no pampering. However, as mentioned before, if 'integration' is real acceptance and not just tolerance (often combined with mental disapproval), we must say that up to now this type of nature has not been accepted by the majority of citizens. We still observe that ruderal vegetation is characteristic of neighbourhoods with low social status. This has been demonstrated by the example of Osnabrück (Hard and Kruckemeyer 1990) where the occurrence of the typical community of north-west European cities, the Hordeetum murini, is strongly related to the social status of a particular urban neighbourhood.

If we take the amount of litter as a measure of approval or disapproval, we find that this type of vegetation is definitely not approved of. In Frankfurt and Düsseldorf we counted the number of pieces of litter (tins, cigarette packets, bottles, newspapers etc.) greater than 1 cm^2 (cigarette stubs were not counted) in the vicinity of trees. In the areas around tree trunks covered with ruderal vegetation, we found an average of eight to nine pieces per tree. When the area around the tree was covered with ornamental vegetation there was normally no litter, but if it was occupied by ground-covering plants, there was also a remarkable amount of litter (four pieces per tree).

When the advantages of spontaneous urban vegetation are enumerated, one sometimes finds the term 'biodiversity' among them. In fact, at first glance it seems that the typical urban vegetation contributes strongly to biodiversity, because the species number per area in urban biotopes is often much higher than in biotopes in natural or agricultural landscapes. Nevertheless, it is a great error to believe that the specifically urban communities contribute to biodiversity. In contrast to the flora of natural habitats the spontaneous flora of North American and Central European urban habitats (for example) is almost identical; most of these species also occur in the cities of the temperate regions of East Asia. Hence diversity is only enlarged on a small scale as far as one single city is concerned. On a global level, however, there is reduction in diversity.

6 Conclusion

If 'integration' means acceptance, we must admit that the spontaneous city vegetation has not generally been integrated up to now. Integration of the natural and the agricultural landscape is possible, but always combined with a change in species composition. In particular, the rarest plant communities and species often disappear when their biotopes are integrated into urban development.

References

Burghardt W (1991) Wasserhaushalt von Stadtböden. In: Schuhmacher, H. & Thiesmeier, B. (eds) Urbane Gewässer. Westarp Wissenschaften, Essen: 395–412

Frey J (in prep.) Stadtbiotopkartierung - Erfassung, Beschreibung und Bewertung städtischer Raumstrukturen zwischen naturwissenschaftlicher Methodik und Naturschutz. Dargestellt an Beispielen aus der Stadtbiotopkartierung Mainz. - Diss. Univ. Mainz, Mainz

Hard G (1985) Vegetationsgeographie und Sozialökologie einer Stadt. Ein Vergleich zweier „Stadtpläne" am Beispiel von Osnabrück. Geogr Zeitschr 73: 125–144

Hard G, Kruckemeyer F (1990) Die Mäusegerste und ihre Gesellschaft in Osnabrück 1978–1990. Über den Zusammenhang von Stadt- und Vegetationsentwicklung. - Osnabrücker naturwiss. Mitt. 16: 133–156

Kéré U, Wittig R (1995) Die Gliederung von Landschaft und Vegetation in der Region Tenkodogo (Burkina Faso, Westafrika). - Verhandl. Ges. Ökol. 24, 17–18

Reidl K (1995) Emscher Landschaftspark - Wiederaufbau von Landschaft und Biosphärenreservat?.- Natur und Landschaft 70: 485–491

Seidling W (1990) Räumliche und zeitliche Differenzierungen der Krautschicht bodensaurer Kiefern-Traubeneichenwälder in Berlin (West). - Ber. des Forschungszentrums Waldökosystme, Reihe A, Bd. 61: 249 pp

Sukopp H, Markstein B (1981) Veränderungen von Röhrichtbeständen und -pflanzen als Indikatoren von Gewässernutzungen, dargestellt am Beispiel der Havel in Berlin (West). - Limnologica 13 (2): 459–417

Wessolek G, Renger M (1998) Bodenwasser- und Grundwasserhaushalt. In: Sukopp H, Wittig R (eds): Stadtökologie. 2nd ed., G. Fischer, Stuttgart, 188–202

Wittig R (1991a) Biozönotische Veränderungen in Wäldern von Industriegebieten.- Annali di Botanica 49: 175–183

Wittig R (1991b) Ökologie der Großstadtflora.- G. Fischer, Stuttgart, 261 S

Workshop: Analysis and Evaluation of Urban Nature 5.1

Relative Susceptibility of Plants in a Dry Tropical Urban Environment

Madhoolika Agrawal

1 Introduction

The tremendous increase in the size of the urban population and the consequent increase in the number of vehicles and extent of energy utilisation have led to the problem of urban air pollution throughout the world. A high rate of economic growth in India during last few decades has resulted in a large scale increase in urban pollution problems and in Indian cities, the toxic limit of air pollutants is often exceeded (Pandey et al. 1992). Air pollution has become the most important factor of plant growth inhibition in urban areas (Pandey and Agrawal 1994; Mansfield and Freer-Smith 1981) where, because of urban activities, plants in parks and along roadways are under greater stress than in rural areas. A study was conducted to quantify the effects of the ambient concentration of urban air pollutants on plants with the aim of screening the plants into sensitive and tolerant groups.

2 Materials and Methods

2.1 Study Area

The study was conducted in the urban environment of Varanasi with a population of about 1.03 million located in the eastern Gangetic plain of the Indian subcontinent (25°18′N, 83°01′E, 76.19 m above sea level). The study area has a tropical monsoon climate with an annual average temperature of 24°C, relative humidity of 60% and 1000 mm annual precipitation. Wind direction is predominantly westerly or north westerly from October to April and easterly or north westerly in the remaining months.

Banaras Hindu University, Department of Botany, Varanasi – 221 005, India

J. Breuste, H. Feldmann, O. Uhlmann (Eds.)
Urban Ecology

2.2 Study Sites

The city was divided into four zones with five micro-sites in each zone. Two of the zones (I and II) represented the area along major highways which receive pollutants from motor vehicles, small scale industries and railway tracks. The third zone is characterised by moderate traffic frequency but often experiences traffic congestion. Zone IV, situated at the south west of the city, represents residential localities with official buildings, gardens and cultivated land.

2.3 Air Monitoring

Continuous air quality monitoring was carried out at each zone for sulphur dioxide (SO_2) (West and Gaeke 1956), nitrogen dioxide (NO_2) (Merryman et al. 1973), ozone (O_3) (Byers and Saltzman 1958) and total suspended particulate matter (TSP) (Pandey et al. 1992).

2.4 Plant Analyses

Seven plant species common at all sites were selected for the study. These include two fruit trees, *Mangifera indica* L and *Psidium guajava* L, two leguminous avenue trees, *Delonix regia* Rafin and *Peltophorum pterocarpum* DC Baker ex. K. Hyne, one deciduous timber tree, *Dalbergia sissoo* Roxb, one ornamental shrub, *Bouganvillea spectabilis* Willd and a spiny shrub, *Carissa carandas* L.

Five equal aged individuals of each species were marked at each micro-site. Branches of the same age group were tagged for each individual in every direction. Leaf samples were collected three times in a season from the tagged branches in the morning hours and safely brought to the laboratory. Dust deposition, leaf area and specific leaf area (SLA) were measured, as described by Pandey and Agrawal (1994). Samples were thoroughly washed and stored in a refrigerator at 4°C and later used for analysis of chlorophyll (Maclachan and Zalik 1963) and ascorbic acid (Keller and Schwager 1977) contents. The data presented are averaged annually over a period of two years. Analysis of variance was carried out to test the effect of site and plant species on recorded data using the SPSS/PC$^+$ programme for microcomputers.

3 Results and Discussion

On the basis of air quality data, the zones in the urban area of Varanasi can be ranked from maximum pollution load to a minimum as I > II > III > IV (Table 1). Air quality varied significantly among sites ($p < 0.05$). Since the data for zone IV showed the lowest values, this was used as a reference for comparing the changes in plant parameters obtained in other zones.

Dust deposition varied significantly among sites ($p < 0.05$) and plant species ($p < 0.01$), the value being greatest for *M. indica* followed by *B. spectabilis, P. guajava, D. regia, P. pterocarpum, C. carandas* and *D. sissoo. Dalbergia sissoo, D. regia* and *P. pterocarpum* showed less dust deposition due to their smooth leaf surface and smaller leaf size. In contrast, plant species having higher dust deposition had larger leaf size, were pubescent and had a leathery texture of leaves. Dust deposition corresponded directly to the TSP concentration.

Plants growing at zones I, II and III showed significant reductions in leaf area (Table 2) and chlorophyll and ascorbic acid contents (Table 3) compared to those at zone IV. Reduction in these parameters is a direct reflection of the reduced photosynthetic capacity of plants. Reduction in photosynthesis due to reduced leaf area has been observed in several studies (Pandey and Agrawal 1994). Gaseous pollutants like SO_2 and O_3 have been shown to be involved in the formation of cytotoxic free radicals leading to chlorophyll oxidation. Ascorbic acid (AA) has been reported to prevent or reverse the damaging effects of pollutants. It is possible that a reduction in AA content at polluted zones might have reduced the detoxifying capacity of plants leading to a high reduction of chlorophyll. A sufficiently high level of AA in *B. spectabilis* and *P. guajava* rendered them less sensitive to urban air pollutants.

SLA indicates dry matter accumulation per unit leaf area thus a reduction in SLA at polluted zones further indicates a poor physiological and biochemical status of plants (Table 2). A reduction in SLA was found to be greatest in *C. carandas* and least in *B. spectabilis*.

Table 1. Annual 24h average air pollutant concentration ($\mu g\,m^{-3}$) in different zones of Varanasi city

Pollutant	Zones			
	I	II	III	IV
SO_2	71	65	48	15
NO_2	55	50	45	18
O_3	47	46	30	15
TSP	315	300	220	130

Table 2. Variations in selected parameters of plant species growing in different zones of Varanasi city

Plant species	Zones											
	I			II			III			IV		
	DCC	LA	SLA	DCC	LA	SLA	DCC	LA	SLA	DCC	LA	SLA
M. indica	8.3	34.3	13.0	8.1	40.2	13.4	6.3	38.5	15.3	2.2	48.9	17.8
D. sissoo	2.4	3.8	8.1	2.4	3.9	8.6	2.0	4.2	8.6	1.0	4.4	11.0
D. regia	4.4	11.2	4.2	4.2	12.1	7.4	3.3	13.9	8.1	1.6	16.1	9.2
B. spectabilis	6.6	7.5	7.6	6.2	7.8	7.7	4.1	7.9	7.8	2.2	9.2	8.1
P. guajava	5.2	38.2	10.4	4.8	40.2	10.8	3.8	42.7	11.0	1.7	46.3	11.6
C. carandas	3.2	2.6	8.1	3.1	2.6	8.3	2.3	3.0	9.0	1.1	3.9	13.0
P. pterocarpum	3.7	17.5	9.0	3.6	18.0	9.2	2.6	20.0	9.6	1.8	23.1	11.0

DCC: Dust capturing capacity ($mg\,cm^{-2}$); LA: Leaf area (cm^2); SLA: Specific leaf area ($mg\,cm^{-2}$)

Table 3. Variations in selected parameters of plant species growing in different zones of Varanasi city

Plant speices	Zones							
	I		II		III		IV	
	Chl	AA	Chl	AA	Chl	AA	Chl	AA
M. indica	2.5	1.7	2.6	1.8	3.2	2.0	4.5	2.6
D. sissoo	3.0	1.5	3.2	1.8	4.7	2.1	6.0	2.4
D. regia	1.4	0.9	1.4	0.9	2.0	1.3	3.8	1.8
B. spectabilis	2.6	1.3	2.6	1.3	3.1	1.6	4.2	2.1
P. guajava	3.0	1.6	3.0	1.6	3.4	1.8	5.1	2.2
C carandas	1.4	0.7	1.4	0.9	2.5	1.1	3.2	1.7
P. pterocarpum	3.8	1.6	4.0	1.7	4.8	1.9	6.1	2.7

Chl: Total chlorophyll ($mg\,g^{-1}$ dry leaf); AA: Ascorbic acid ($mg\,g^{-1}$ fresh leaf)

The results demonstrated that there is a wide range of variability in the response of plants to urban air pollution as the environmental conditions are more or less similar between the sites. After a critical observation of all the parameters *C. carandas* appeared the most sensitive to urban air pollution followed by *D. regia, D. sissoo, M. indica, P. pterocarpum, P. guajava* and then *B. spectabilis*. The differential susceptibility of species is likely to be due to the differences in plants' ability to avoid the entry of the pollutants and/or the physiological and biochemical differences between species. This study further suggests that *B. spectabilis*, *P. guajava* and *P. pterocarpum* should be grown in parks and green belt areas of the city and *C. carandas* should be used as a bio-indicator for monitoring pollution status.

References

Byers HD, Saltzman BE (1958) Determination of ozone in air by neutral and alkaline iodine procedure. J Am Ind Hyg Ass 19: 251–257

Keller T, Schwager H (1977) Air pollution and ascorbic acid. Eur J for Pathol 7: 338–350

Maclachlan S, Zalik S (1963) Plastid structure, chlorophyll concentration and free amino acid composition of a chlorophyll mutant of barley. Can J Bot 41: 1053–1062

Mansfield TA, Freer-Smith PH (1981) Effects of urban air pollution on plant growth. Biol Review 56: 343–368

Merryman EL, Spicer CW, Levy A (1973) Evaluation of arsenite modified Jacobs Hochheiser procedure. Environ Sci Technol 7: 1056–1059

Pandey J, Agrawal M (1994) Evaluation of air pollution phytotoxicity in a seasonally dry tropical urban environment using three woody perennials. New Phytol 126: 53–61

Pandey J, Agrawal M, Khanam N, Narayan D, Rao DN (1992) Air pollutant concentrations in Varanasi, India. Atmos Environ 26B: 91–98

West PW, Gaeke GC (1956) Fixation of SO_2 as sulfitomercurate (II) and subsequent colorimetric estimation. Analyt Chem 28: 1816–1819

An Analysis of the Condition of Peatlands in the Urban Landscape

Tamara I. Kucharchic and Sergey V. Kakareka

The inventory, transformation and use of peatlands which are located within the boundaries of the city of Minsk were considered in this study. Peatlands that had been preserved in their natural or semi-natural conditions were described and the main factors of the transformation of peatland areas were summarised. A number of variants were suggested with regards to the integration of preserved peatlands into the urban landscape structure.

The formation of a favourable environment on urbanised territories is impossible without the optimisation of city landscapes. City development tends to try to preserve environmental diversity and increase the natural component fraction of city landscapes. The most frequent places for the increase in the variety of urban landscapes is suggested as being the preserved parts of forest, meadows, deserted old gardens and different kinds of waste grounds.

The main objectives of our research were to reveal the condition of peatlands on city territories in Belarus and to investigate the possibilities of the integration of peatlands (preserved in their natural or near-natural condition) into the urban landscape.

1 Materials and Methods

The following sources of data were used in the investigation: official statistical data (Land Survey of cities; Peat Fund); branch materials (district geological and soil maps, soil maps of forestry, prospecting reports for peat deposits); topographic maps of different years of publication. Field investigations were carried out to obtain more precise information about the situation and also to evaluate the modern state of peatland areas. A description of the vegetation, sampling of soils, subsoil, surface waters and plants were made at each site as well as the characteristic of the location of the peatland in the landscape, the usage of the peatland and its adjacent territories and possible sources of pollutant entry.

Institute for Problems of Use of Natural Resources & Ecology, 10 Staroborysovski tract, Minsk, 220114 Belarus

J. Breuste, H. Feldmann, O. Uhlmann (Eds.)
Urban Ecology

The main laboratory methods of analyses were general hydrochemical methods, microscopy, atomic-absorption and emission spectral methods.

2 Results

According to the official statistical data, peatlands usually account for less then 1% of the territory in cities (while they occupy about 12% of the territory of Belarus). The area of peatland in absolute quantities is marginal, and only in some cases exceeds 100 ha. This is significantly low in comparison with forests.

We investigated the present condition of peatlands in Minsk – the capital and the largest city in Belarus – and revealed their basic problems connected with their use and preservation.

The account of urban peatlands in Minsk is less than perfect. The information in the Land Survey relates only their presence and provides an incomplete reflection about the real situation; Peat Fund reference materials are found to be useful mainly for the retrospective valuation of peatland distribution within city borders. Reference book data is renewed very slowly and thus is unable to cope with the changing situations in cities, besides which, only those peatlands which are of interest as peat fields are shown in reference-books.

The results enabled us to draw up the distribution of peatlands in current and prospective borders of Minsk. In addition we detected 10 naturally existing peatlands within the city circular road limits; there are also some renaturalised plots on peatlands which have been drained and peat extraction has been carried out.

In our estimation, there were initially about 20 peatlands in the city, with an area of approximately 1000 hectares (4–5% of the city territory). Low moors predominated (50%), with transition bogs and high bogs encompassing similar sized areas (25%). The area of a peatland varied from 1 to 220 ha.

City peatlands have been considerably transformed. The present use of peatlands in Minsk is as follows: under reservoirs −10%; under buildings −36%; as agricultural land (arable lands, hay-makings, pastures) −25%; as park zones – 10%; as protected territories −2%; as land-fills – 5%; without any special purposes – 12%.

On the whole, about half of all peatlands which previously existed (more than 400 ha) were drained and occupied with buildings or submerged by reservoirs. Common areas of natural and restored peatlands and separate plots now cover less than 110 ha; a combination of different kinds of activity is typical on large peatland territories.

The area of the majority of peatlands ranges from 1 to 3 ha, larger peatlands (from 6 to 14 ha) are situated on the flood-plain of the Svisloch river.

The remaining peatlands are rather diverse in structure. This depends on how they were formed (flood peatlands, peatlands of hollows, lake basins and narrows); on the type of nutrition: eutrophic (fens, mires), mesotrophic (transition bogs) and olygotrophic (high bogs); on the amount of peat (from 0.3 up to 2 m); and on vegetation (Oxycocco-Sphagnetea, Vaccinietea uliginosi, Alnetea glutinosae, Salicetea, Phragmitetea etc.); the geochemical situation differs depending on the load intensity.

The characteristics of the most well preserved and interesting peatlands of Minsk is given below.

2.1 Drazhnya

This has an area of about 1 ha and is located in the industrial zone, 500 m from the Tractor Plant. It has been preserved within its initial limits, the main influences being mechanical (trampling down) and atmospheric fall-out. It is a high bog (Oxycocco-Sphagnetea), with typical oligotrophic vegetation. The main plant species are: *Eriophorum vaginatum, Vaccinium uliginosum, Andromeda polypholia, Oxycoccus palustris, Sphagnum apiculatum*, and rarely – shortish *Pinus sylvestris*, on the edge of the peatland – *Phragmites communis*. The level of subsoil waters is 0.5 m and in spring water is present on the surface. The increased ash content of the upper peat horizons shows the introduction of mineral substances into the peatland. Peat depth is more than 2 m and is found to be moss in the upper horizons and 0.15 cm below these horizons, cotton-grass is found.

Sampling of peatland waters showed that subsoil waters are acid (pH 2.45), with a low mineral content (ion sum 154 mg/l), with the composition – HCO_3-Ca-Mg. High concentrations of lead (26.24 μg/l) and zinc (483.0 μg/l) were detected in the water and the content of zinc also increased (56.3 mg/kg of dry substance) in the surface peat sediments.

2.2 Kountsevschina

This peatland has an area of about 1.5 ha. It is located in the western part of the city and borders a residential area on one side, and an arable land on the other side. Part of the peatland was destroyed for the building of a new road. The peatland has a mosaic vegetation structure: the southern part forms a tussocky fen with *Alnus glutinosa*, the western part is much wetter with floating *Sphagnum* and some *Typha latifolia, Carex spp.* and gramineous plants. *Sphagnum* moss makes up 80–90% of the vegetation, with a prevalence of *Sphagnum angustifolium; angustifolium* and *hypnum peat* were noted in the vegetation survey. The depth of the floating bog is more than 1.2 m and the waters are acid (pH – 4.05). They contain insignificant quantities of mineral elements (126.3 mg/l), the formula of composition is HCO_3-Cl-Ca-Na-Mg; the content of heavy metals in the peat is on the same level as background values.

2.3 Loshitsa

This is a flood-plain type peatland, located in the southern part of the city in the valley of the river Svisloch and has an area of about 30 ha. Peat extraction was carried out in the past and it the area been partially filled in for road construction. The main threats are grazing and the deposition of pollutants from the flood waters of the river. There is a terrace about 1 m in height on the peatland; the lower part of the area, near the river, is often flooded and floating sites are typical here. In the higher part of the peatland the level of subsoil waters was found to be 0.5 m. The vegetation is determined by the level of subsoil waters. Along the banks of the river *Typha latifolia*, *Phragmites communis* and *Alnus glutinosa* prevail and the main area of peatland is occupied by horse-tail and sedge associations. The peat is mainly of sedge type, with reed-type in the deep layers and the peat capacity varies from 0.5 to 1 m. The waters are neutral (pH – 6.8), with increased mineral content (517 mg/l), and are classified as HCO_3-Ca-Mg. High concentrations of zinc and cadmium were noted in the peat deposits (respectively 42 and 2.5 mg/kg).

3 Discussion

We suggest that natural peatlands, or their remaining areas, should be seen as potential areas for nature-protection purposes. They can be form the kernel of park systems, especially in new districts of the city, where green cover is almost absent and the surrounding land is mainly agricultural. Remaining small peatlands are relatively stable ecosystems containing certain plants, soil fauna, small mammals and birds.

Some peatlands comprise small lakes whose protection is very important both from the point of their natural functions as well as for improving the landscape diversity of the city.

Some peatlands can become nature reserves with local value. An example of such a protected territory is the biological reserve Lebjagy, which was created on the bog Masjukovskoe (where peat was formerly extracted) and includes a restored area of bogs and reservoirs.

4 Conclusions

- Official statistics about the availability of peatlands in cities do not show the real situation. All categories of peatlands should be taken into consideration: those still in their natural condition, peatlands used for extraction, neglected peatlands, those with protected plots, restored peatlands and others. It is necessary to distinguish the most interesting peatlands with respect to biological, hydrological and landscape values.
- Every year the area of urban peatlands in Belarus and their quantity are reduced because of drainage and their use for different purposes; in some cases they end up becoming rubbish dumps! There is a danger that these natural ecosystems will disappear from urbanised territories before society realises their importance.
- The main influences on the preserved peatlands of Belarusian cities are recreation, alteration of water conditions, influence of fires, non-regulated storage of wastes, atmospheric fall-out, and fluxes of pollutants with surface drainage. Non-regulated recreation activity, littering, fires made for appearance on the peatlands, ruderal and synantrophic plant species (their composition depends on type of peatland and on a water regime) are also features associated with peatlands.
- General tendencies of the spontaneous changes on natural and semi-natural urban peatland are difficult to define at this time since the period of investigation is too short. Peatlands are clearly at great risk in built-up areas, although in the absence of direct influence, they are found to be stable ecosystems. However, intensive chemical loads causes the increase of mineral nutrition to lead to specific changes- e.g. in high bogs olygotrophic plants (cranberry, great bilberry, sphagnum mosses etc.) disappear.
- The most valuable peatlands should be protected to ensure the preservation of landscapes and biological diversity preservation as well as ensuring the continuation of the important ecological functions of peatlands. Some peatlands can become reserves, or others can be incorporated into the green zone around the city.

Documentation of Genetic Differences in a Volunteer Grass, *Poa annua* (Annual Meadowgrass), under Different Conditions of Golf Course Turf, and Implications for Urban Landscape Plant Selection and Management

Peter H. Velguth[1] and Donald B. White[2]

1 Introduction

Poa annua, commonly known as annual meadowgrass or annual bluegrass, is worldwide in distribution (Beard et al. 1978), and is one of the ten most abundant weeds in city allotment gardens (Gilbert 1989). Plants range from an annual bunching habit to a prostrate creeping plant that roots at nodes and tends to be perennial (Beard et al. 1978; Gibeault and Goetze 1972). Particular forms have been found in association with specific habitats and several studies have suggested that differences in populations have a genetic basis.

Warwick and Briggs (1978a) examined *P. annua* from adjacent environments which differed in clipping height and found a mix of erect and prostrate forms from flower bed environments and nearly exclusively prostrate forms in closely mown adjacent bowling greens at Cambridge University, UK. Plants grown from seed of these adults indicated consistency between erect versus prostrate habit in progeny compared to parent plants. The authors concluded that the habit of growth and flowering are under genetic control. In a companion report, these authors provided evidence that divergent selection occurred in these populations (Warwick and Briggs 1978b). McNeilly (1981) succinctly summarized this work: "... when regularly clipped, dry-matter production of prostrate lawn forms of *P. annua* is less drastically reduced than in adjacent erect flowerbed forms. These differences were shown to be genetically determined, and the product of considerable selection pressures, and would undoubtedly affect performance under competitive stress."

Gibeault and Goetze (1972) surveyed *P. annua* subspecies in Oregon and Washington State, USA and related the plant types present to irrigation

[1] University of Wisconsin, Department of Horticulture, 1575 Linden Dr., Madison, WI, U.S.A.
[2] University of Minnesota, Department of Horticulture, 1970 Folwell Ave., St. Paul, MN, U.S.A.

J. Breuste, H. Feldmann, O. Uhlmann (Eds.)
Urban Ecology

regimen. The presence of a higher frequency of perennial types (*P. annua reptans*) in the greens was related to frequent irrigation and "syringing" of greens during periods of moisture stress. The presence of annual types (*P. annua annua*) in the rough, in contrast, was consistent with infrequent rains where *P. annua* survives as seed during dry periods.

Differences also occur in growth rate in different populations of *P. annua*, and these differences may, in part, relate to the DNA content of the nucleus. Mowforth and Grime (1989) found a correlation between growth rate and nuclear DNA content in *P. annua*. They found a positive correlation between DNA content and cell size, and negative correlation with growth. This relationship is encompassed in the nucleotype hypothesis. Coined by Bennett (1971), the "Nucleotype describes those conditions of the nucleus that affect the phenotype independently of the informational content of the DNA." This effect is thought to occur, for example, through the effect of the mass and volume of DNA. With more DNA in the nucleus, more time is required to complete mitosis and meiosis. Longer cell generation times translate into slower growth rates.

DNA flow cytometry, a technique used to measure the DNA amount in the nucleus, was used to study ploidy level variation in *P. annua* from a sampling of green, fairway, and rough environments as well as to examine DNA content variation of tetraploid plants in these environments. Earlier research documented the occurrence of diploid ($2n = 2x = 14$) and tetraploid ($2n = 4x = 28$) *P. annua* in plants collected from golf courses, and that nuclear DNA content correlates highly with ploidy level (Velguth and White 1992). The smaller stature of diploid *P. annua* together with different cultural practices of golf turf, particularly close mowing height of the greens compared to fairway and rough, prompted the hypothesis that diploids may be favored on greens.

2
Materials and Methods

Samples were collected from greens, fairways, and roughs from two golf courses near Minneapolis, Minnesota, USA. Holes to be sampled were selected at random and plants at each location were selected by placing a transect line between two known landmarks, such as sprinkler heads and distance markers. Plants were first chosen at a predetermined point along the line and then at 2 foot intervals (61 cm) until 5 samples of *P. annua* were collected. Three holes at one course and seven holes at the other were sampled providing a total of 50 plants from each of the three environments.

Plant material was prepared for flow cytometric analysis of DNA content largely according to Keeler et al. (1987). Differences from this method included the use of *Pisum sativum* cv. Little Marvel as an internal standard and use of propidium iodide stain solution containing boiled ribonuclease A (Sigma) in chopping buffer. A computer program allowed the selection of data peaks in histograms, and calculation of mean channel number representing sample and *Pisum* DNA for each sample.

3 Results and Discussion

Results indicate that 1/4 of the 50 randomly collected plants from the green were diploid, while no diploids occurred in the 100 plants collected from the fairway or rough. These results also showed that tetraploid plants from the green had statistically significantly less DNA (6 to 7%) than tetraploids of the fairway or rough. The genome size was different in two components, in ploidy level and in DNA content of the tetraploid. Additional details are presented by Velguth and White (1998).

What is responsible for the differences in genome size? These findings are consistent with the hypothesis that different environmental conditions selected for or allowed plants of different ploidy level to survive on the green and also selected or induced changes in nuclear DNA content of a fraction of the tetraploid plants of the green. The environment of the golf course green typically differs in management from the fairway and rough environments. Differences include lower clipping height, more frequent irrigation, fertilization practices, pest management, and soil conditions. Although it is impossible to discern the specific factor or factors responsible for selecting or inducing the differences observed, one possibility is that clipping height could select for the typically smaller-statured diploid plants. Further, regarding the tetraploid nuclear DNA content differences observed, Wakamiya et al. (1993) reported finding a negative correlation between nuclear DNA content variation in the genus *Pinus* and two climate factors, lowest mean monthly and annual precipitation, and the highest mean monthly air temperature. They suggest this illustrates that variation in genome size may be an adaptive response to climatic conditions. Similarly, these microclimatic factors, or others, may be involved in selecting or inducing differences in tetraploid nuclear DNA variation in *P. annua*. Aneuploidy was considered unlikely as an explanation of DNA content differences, as these would have to be of independent origin on each golf course and on each of the greens analyzed.

This study illustrates that the human-influenced conditions of cities have the potential to change plant populations genetically. Other examples exist illustrating genetic change in plant populations due to human activity. Genetic differences have been shown in plant populations due to heavy metal toxicity; from lead deposited by automobile traffic along a roadside in *Plantago lanceolata* L. (Wu and Antonovics 1976), or in *Agrostis tenuis* Sibth.

on an abandoned copper mine (McNeilly 1968), for example. In addition, Snaydon (1970) reported on selection for differences in *Anthoxanthum odoratum* L. due to varied soil conditions present for 40 years as part of the Rothamsted Park Grass experiment. Common to the present experiment, as well as these examples, are conditions in which environmental factors are stable over time but vary in closely adjacent space. These conditions are noted by Snaydon (1970) as being more likely than others to show population genetic differentiation in plants.

When able to undergo sexual recombination, flowering, and setting seed, plant species have potential to undergo genetic change in response to selection due to urban environmental conditions. Given that ecological conditions of cities differ from the natural environment, the idea of selected genetic change has implications for plant material used in the designed or naturalized landscape. The mode of reproduction of a given species, sexual or asexual, in nature versus under urban conditions, is one important consideration. Selection for adaptation to urban conditions through directed breeding programs or through natural regeneration in the urban environment will assist in the establishment of species with greater adaptation to the urban landscape.

References

Beard JB, Rieke PE, Turgeon AJ, Vargas JM (1978) Annual Bluegrass (*Poa annua* L.) Description, adaptation, culture and control. Research Report #352. The Michigan State University Agricultural Experiment Station

Bennett MD (1971) The duration of meiosis. Proc R Soc Lond B 178: 277–299

Gibeault VA, Goetze NR (1972) Annual meadow-grass. Sports Turf Research Institute 48: 9–19

Gilbert OL (1989) The Ecology of Urban Habitats. Chapman and Hall, London

Keeler KH, Kwankin B, Barnes PW, Galbraith DW (1987) Polyploid polymorphism in *Andropogon gerardii*. Genome 29: 374–379

McNeilly T (1968) Evolution in closely adjacent plant populations. III. *Agrostis tenuis* on a small copper mine. Heredity 23: 99–108

McNeilly T (1981) Ecotypic differentiation in *Poa annua*: Interpopulation differences in response to competition and cutting. New Phytologist 88: 539–547

Mowforth MA, Grime JP (1989) Intra-population variation in nuclear DNA amount, cell size and growth rate in *Poa annua* L. Functional Ecology 3: 289–295

Snaydon RW (1970) Rapid population differentiation in a mosaic environment. I. The response of *Anthoxanthum odoratum* populations to soils. Evolution 24: 257–269

Velguth PH, White DB (1998) Frequency of diploid and tetraploid plants and tetraploid DNA content variation in *Poa annua* L. and as related to golf turf environments. Crop Science (in review)

Velguth PH, White DB (1992) Ploidy level variation in breeding materials of *Poa annua* L. (Annual bluegrass). American Society of Agronomy. Agronomy Abstracts p 177

Wakamiya I, Newton RJ, Johnston JS, Price HJ (1993) Genome size and environmental factors in the genus *Pinus*. American Journal of Botany 80: 1235–1241

Warwick SI, Briggs D (1978a) The genecology of lawn weeds I. Population differentiation in *Poa annua* L. in a mosaic environment of bowling green lawns and flower beds. New Phytologist 81: 711–723

Warwick SI, Briggs D (1978b) The genecology of lawn weeds. II. Evidence for disruptive selection in *Poa annua* L. in a mosaic environment of bowling green lawns and flower beds. New Phytologist 81: 725–737

Wu L, Antonovics J (1976) Experimental ecological genetics in *Plantago* II. Lead tolerance in *Plantago lanceolata* and *Cynodon dactylon* from a roadside. Ecology 57: 205–208

Long-term Observation of the Fauna of Two Manmade Nature Habitats in Cities of the Ruhr-Valley Area

Frank Bergen, Axel Schwerk, and Michael Abs

1 Introduction

Long term studies on faunal succession are urgently needed as they provide a basis for prognosis. In the last few decades the reduction in coal mining and steel production produced new types of manmade habitats (Rebele and Dettmar 1996). We studied the phenomenon of the effect of vegetation succession on animal communities in two different habitats and with two different animal groups: the birds of an urban park and the carabid beetles of a colliery spoil heap. In both habitats the faunal succession is measured by species turnover rate. In western central Europe, however, only a few monitoring studies of the avifauna of urban parks have been published (Nowakowski 1996), and the faunal succession of the carabid beetles of an urban colliery spoil heap has not yet been researched (before now). The *Westpark* covers an area of 10 ha and is situated 1.4 km from the city centre of Dortmund (51°30′N, 7°28′E) (NRW, FRG). The park was founded in 1811 as a cemetery and transformed after 1945 into a park; it now represents a piece of encapsulated nature surrounded by an extensive housing area. The dominant tree is ash (26% of the total trees), maple and birch are also present in abundance and, less frequently, oak, poplar, plane tree and horse chestnut can be found; some of these trees are about 100 years old. 18–20% of the vegetation is covered by shrubs, but most of the area is covered with intensively cultivated lawns. In the fifties the tree layer was less dense and during this period 30 nestboxes were installed, only one of which now remains.

The *colliery spoil heap* lies at the border of the small town of Waltrop 5 km north of Dortmund. The heap covers an area of 10 ha and can be divided into three parts. One part is the oldest section of the spoil heap which has existed for 90 years. This area was planted with birch 40 years ago (W5, successional stage IV: forest like stand). A newer area of the spoil heap was created in 1986; the east part being seeded in alternate plots by Jochimsen (1989) with seeds

Ruhr University Bochum, Chair for General Zoology & Neurobiology, Faculty for Biology, D-44780 Bochum, Germany

J. Breuste, H. Feldmann, O. Uhlmann (Eds.)
Urban Ecology

of the association of the Dauco-Melilotion (Daucus carota & Melilotus albus) (W2). These plots developed from the vegetational stage I: Ruderal pioneer vegetation of annual and biennial species to a mixture of stages II (Ruderal tall herbaceous perennials) and stage III (growing young birch) in 1996 (Dettmar 1991). A young forest like stand with young birch and ash was used as a control as it was situated just at the foot of the waste heap (W7).

2 Methods

The breeding density of the birds were monitored by one of the authors in the nineties (Bergen 1996) using Oelke's method (1974); a similar method was used by Erz (1964) who monitored the birds in the fifties and sixties. In the past 43 years 8 monitoring records were produced (Table 1).

We calculated the species turnover rate using the formula: $T = (J + E) \times 100/(S_I + S_{II})$ (modified from Mühlenberg (1993)) where J is the number of additional species in season II, E the number of species which have disappeared and S_I the number of species in season I and S_{II} those in season II. Statistical analysis of differences between the data obtained for each decade was carried out using the Mann-Whitney-U-test.

Our carabid studies covered the vegetation periods from 1989 to 1996. We set pitfall traps in 8 sampling plots, three traps situated in each plot. Only three selected plots were analysed. The carabids caught were identified to the species level. We compared the live biomass of single individuals to a formula used to find the mean individual biomass (MIB). This formula determines the biomass (y) from the body length (x) after Szyszko (1983): $\ln y = -8.92804283 + 2.5554921 \times \ln x$. The species turnover rate was also calculated. A non-parametric one-way analysis of variance (Kruskal-Wallis-test) was used to gain information about differences concerning the MIB of the sampling plots.

Table 1. The avifauna of the Westpark (Dortmund)

Years	1954	1955	1956	1961	1962	1994	1995	1997
species	18	17	16	21	21	21	23	20
density	100.1	98.4	100.2	132.4	132.4	138.0	180.0	125.0
Years compared								
	54/55	55/56	61/62	94/95	95/97	56/61	62/94	54/97
Turnover	8.6	9.1	4.8	4.5	16.3	18.9	23.8	42.1

Density = territories/10ha (Erz 1964; Bergen 1996)

3 Results and Discussion

3.1 The Birds of the Westpark (Dortmund) in the Last 4 Decades

The number of bird species increases significantly (U-test, $p < 0.05$) from 17 (median) in the fifties to 21 (median) in the sixties and nineties (Table 1), but the species composition changed between the observation periods. The species turnover rate was found to be 42.1% when the years 1954 and 1997 were compared. Only 10 species were regularly observed, of which the spotted flycatcher is an indicator species of ecologically well equipped parks (Flade 1994). 7 species, including whitethroat and redstart, disappeared but 8 species (e.g. nuthatch and short-toed treecreeper) settled after the monitoring in 1962. In the nineties further woodland species (e.g. great spotted woodpecker and green woodpecker) invaded, but are still not permanently established. The breeding density rose from 100 territories/10 ha to 131.5 territories/10 ha between the monitoring in the fifties and the sixties (U-test, $p < 0.05$). In the fifties only 3 species were dominant (>10%) (Bick 1989): blackbird 20.2%, house sparrow 16.6% and chaffinch 12.5%. These occupied 49.3% of the territories. In the nineties, however, 5 species (blackbird 36.5%, wood pigeon 15.5%, blue tit 12.5%, great tit 11.75% and chaffinch 11%) occupied 87 territories/10 ha comprising 66% of the total territories (Table 1). The disappearance of the nest boxes explains why the house sparrow and tree sparrow vanished. The tits, however, increased in density.

The data presented here are very valuable because little is known on the avifaunal succession of urban parks in western Germany over such a long study period. From this monitoring study we concluded that generalist species increase in number and also in breeding density at the expense of specialists. The species turnover rate of the birds of the Westpark is higher than in deciduous forest succession far away from settlements (1976: 27.5%) or pine forest succession during a comparable period (Dierschke 1973: 25%). The faunal succession seems to favour woodland type species.

3.2 The Carabid Fauna of the Colliery Spoil Heap in Waltrop over 8 Years of Observation

The MIB values of the plots are given in Table 2. The MIB values of the carabids reveals significant differences between the plots using the Kruskal-Wallis-test ($p < 0.01$). The succession of the carabid fauna of the colliery spoil heap Waltrop seems to be delayed in spite of the high species turnover (Table 3). Considering the high constancy of the MIB values (Table 2) it seems

Table 2. Mean Individual Biomass (MIB) of the sampling plots of the colliery spoil heap Waltrop

Plot	Year								Statistics	
	1989	1990	1991	1992	1993	1994	1995	1996	mean	s
W2	49.1	44.8	43.9	46.2	46.6	53.3	-	49.6	47.6	3.2
W5	-	-	52.4	126.0	134.3	198.7	155.9	103.1	128.4	49.3
W7	89.1	-	60.2	88.1	79.0	79.6	103.0	132.0	90.1	22.6

s = standard deviation. W2: ruderal vegetation; W5: birch copse; W7 young ash and birch stand. (Schwerk 1996)

Table 3. Species Turnover of carabids of the colliery spoil heap Waltrop

Plot	Years compared					
	91/92	93/94	94/95	89/94	91/96	mean/year
W2	22%	27%	-	37%		8%
W5	33%	35%	24%	-	35%	7%
W7	20.3%	29.7%	33.3%		35.9%	7%

Plots: see Table 2

that a period of 8 years is too short to observe changes in the carabid fauna. On the other hand, the increase of the MIB at the birch copse (W5) in comparison to the ruderal vegetation (W2) indicates the process of a succession. This result is strengthened by the statistical significance of differences between the sampling plots. The results are in agreement with those of Dunger (1968) and Vogel and Dunger (1991) who also demonstrated the slow succession on industrial fallow ground produced by brown coal mining.

Acknowledgements. The financial support of the research work at the waste heap Waltrop by the RAG Bergbau is much appreciated. Sincere thanks to an anonymous reviewer for his valuable suggestions.

References

Bergen F (1996) Ökologische Studien an ausgewählten, urbanisierten Vogelarten – unter besonderer Berücksichtigung des Gesangsverhaltens. Dissertation submitted for a diploma, Fakulty for Biology, RUB Bochum

Bick H (1989) Ökologie. G. Fischer, Stuttgart New York

Dierschke F (1973) Die Sommervogelbestände nordwestdeutscher Kiefernforsten. Die Vogelwelt 94: 201–212

Dettmar J (1991) Industriebrachen – vergiftete Wüsten oder lebendige Oasen. Internationale Bauaustellung Emscherpark, Gelsenkirchen

Dunger W (1968) Die Entwicklung der Bodenfauna auf rekultivierten Kippen und Halden des Braunkohletagebaus. Ein Beitrag zur pedobiologischen Standortdiagnose. Abh Ber Naturkundemus Görlitz 43: 1–256

Erz W (1964) Populationsökologische Untersuchungen an der Avifauna zweier nordwestdeutscher Großstädte. Zeitschr f wiss Zool 170: 1–111

Flade M (1994) Die Brutvogelgemeinschaften Mittel- und Norddeutschlands. IHW-Verlag, Eching

Jochimsen M (1989) Begrünung von Bergehalden auf der Grundlage der natürlichen Sukzession. Mitt Dt Bodenkundl Ges 58: 226-232

Mühlenberg M (1993) Freilandökologie. UTB Wiesbaden

Nowakowski JJ (1996) Changes in the breeding avifauna of Olstyn (NE Poland) in the years 1968-1993. Acta Ornithologica 31: 39–44

Oelke H (1974) Siedlungsdichte. In: Berthold P, Bezzel E, Thielcke G (eds) Praktische Vogelkunde. Kilda, Greven

Rebele F, Dettmar J (1996) Industriebrachen. Ulmer, Stuttgart

Schwerk A (1996) Charakterisierung von Laufkäferzönosen (Coleoptera: Carabidae) auf städtischen Industriebrachen anhand freilandökologischer und populationsgenetischer Untersuchungen. Dissertation, Faculty for Biology, RUB Bochum

Szyszko J (1983) Methods of macrofauna investigatioms. In: The process of forest soil macrofauna formation after afforestation of farmland. Warsaw Agricultural University Press, Warsaw

Tischler W (1976) In: Czihak G, Langer H, Ziegler H (eds) Biologie. Springer, Berlin Heidelberg New York

Vogel J, Dunger W (1991) Carabiden und Staphyliniden als Besiedler rekultivierter Tagebau-Halden in Ostdeutschland. Abh Ber Naturkundemus Görlitz 65: 1–31

Impacts of Site Disturbance on the Small Mammal Fauna of Urban Woodlands

Charles Nilon and Scott Huckstep

1 Introduction

Woodlands are an important component of urban and suburban residential areas in North American cities, making up 30% of open space land in a study of the open spaces in the Chicago, Ill. region (Nilon and Huckstep, in press). Previous researchers have pointed to the educational and social importance of wildlife habitats in and around residential neighborhoods (Burgess et al. 1988), and have considered the potential of these sites in protecting rare species (Dickman and Doncaster 1987).

Previous studies of small mammals have suggested that site disturbance associated with human activities can influence small mammal habitat and species composition and abundance (Nilon 1987). Residential neighborhoods vary in their socioeconomic, ethnic/racial, and demographic characteristics. These differences are may be associated with human activities that influence woodlands such as preferences in types of nearby vegetation and ownership of pets that may be predators on small mammals (Grove et al. 1995).

We studied small woodlands in neighborhoods that reflected major differences in North American cities. Our study asked three questions relevant to the management of these woodlands: Does human-related disturbance differ between small woodlands that are adjacent to neighborhoods that have different social, economic, and demographic characteristics? Does small mammal species composition and abundance differ between small woodlands that are adjacent to neighborhoods with different social, economic, and demographic characteristics? Is there a correlation between measures of human-related disturbance and small mammal species composition and abundance?

University of Missouri-Columbia, School of Natural Resources, 112 Stephens Hall, Columbia, MO 65211 U.S.A.

J. Breuste, H. Feldmann, O. Uhlmann (Eds.)
Urban Ecology

2 Methods

2.1 Study Area

Our study was conducted in Lake and Cook Counties, Illinois, U.S.A., part of the second largest metropolitan area in the U.S. Our study sites were associated with the *ChicagoRivers* Demonstration Project, and within 0.8 km of three branches of the Chicago River system: the Middle Fork of the North Branch; the Skokie Lagoons/Skokie River; and the Cal-Sag Channel (Nilon and Huckstep, in press).

2.2 Neighborhood Classification

We used six variables from the 1990 U.S. census of population and housing to classify census tract block groups next to 193 woodlands in our study sites. Block groups are small homogeneous units that correspond to neighborhoods in U.S. cities (Grove et al. 1995). Block groups in neighborhood type 1 (NT1) were characterized by housing units with a high mean home value, percentage of owner-occupancy, and access to vehicles, and a low percentage of black residents and single parent households. Block groups in neighborhood type 2 (NT2) were characterized by housing units with a lower mean home value, a high percentage of renter-occupied units, a more moderate percentage of access to vehicles, and a higher percentage of black residents and single-parent households (Huckstep 1996).

2.3 Selection of Study Sites

We studied woodlands that were between 1 and 3 ha in size and directly adjacent to block groups belonging to one of the two neighborhood types. Six of 193 woodlands, three associated with each neighborhood type, met these criteria and were included in our study. The woodlands were similar in plant species composition, basal area, shrub stem density, and canopy closure (Huckstep 1996).

2.4 Woodland Disturbance and Trap Station Characteristics

We used three measures of human impact to evaluate each woodland: the percentage of woodland covered by trash and yard waste; small mammal microhabitat characteristics associated with each trap station; and the occurrence of *Felis catus* and *Canis familiaris*. We used a t-test to make comparisons between woodland types for each variable (Huckstep 1996).

The percentage of the woodland covered by trash and yard waste was determined by measuring the total area of each woodland covered by trash, brush piles, and lawn clippings. Microhabitat variables were measured in plots associated with small mammal trapping stations. We established one sampling transect along the long axis of each woodland. Trap stations (10 × 10 m) were set up at 30 m intervals along the transect. The number of trap stations varied according to the size of the woodland microhabitat. Five random 1 m^2 quadrats were in placed in each trap station. We measured the percent cover of eight microhabitat variables (ground cover by herbaceous vegetation, woody vegetation, woody litter, herbaceous litter, leaf litter, bare soil, rock/rubble, and trash) in each quadrat (Huckstep 1996). We used scent post surveys to detect the occurrence of *Felis catus* and *Canis domesticus* in each woodland (Linhart, Knowlton 1975).

2.5 Small Mammal Species Composition and Abundance

One Sherman live trap (23 cm × 8 cm × 9 cm) was placed at each corner of the 10 m × 10 m trap station. All traps were baited with peanut butter and rolled oats, and checked daily within 3 hours after sunrise. Traps were checked at 24 hour intervals for four consecutive days during separate trapping periods in June and July 1995. Each animal captured was marked with a No. 1 Monel ear tag and released at the point of capture. We recorded the species, date of capture, trap station number, weight, and sex of each animal captured. Because there were no differences in mean number of captures/trap station between trapping periods, the capture data were pooled. We used a Mann-Whitney U-test to determine differences ($P < 0.05$) between neighborhood types in mean captures per trap station for each species (Huckstep 1996).

3 Results

3.1 Woodland Disturbance and Trap Station Characteristics

There were no differences ($P < 0.05$) in the percentage of trash and yard waste covering woodlands adjacent to the two neighborhood types. Woodlands adjacent to NT1 block groups had more ground covered by herbaceous vegetation ($P = 0.01$), and less covered by woody vegetation ($P = 0.01$), and woody litter ($P = 0.02$) than those adjacent to NT2 block groups. Trap stations in woodlands adjacent to NT2 block groups had more ground covered by bare soil ($P = 0.01$), trash ($P = 0.02$), and rock/rubble ($P = 0.04$) than those adjacent to NT1 block groups. We found no differences in mean detections/scent post station of *Felis catus* and *Canis domesticus* between woodlands adjacent to the two neighborhood types.

3.2
Small Mammal Species Composition and Abundance

Three species (*Peromyscus spp., Tamias straitus, Microtus pennsylvanicus*) and 283 individuals were captured in woodlands adjacent to NT1 block groups. Five species (*Peromyscus spp., Tamias striatus, Microtus pennsylvanicus, Mus musculus, Rattus norvegicus*) and 83 individuals were captured in woodlands adjacent to NT2 block groups. We found no differences in mean captures/trap station between woodlands adjacent to the two neighborhood types for four species. *Peromyscus spp.* was more abundant ($P = 0.01$) in woodlands adjacent to NT1 block groups.

There were significant correlations ($P < 0.05$) between six variables used to describe woodlands and mean captures/trap station of small mammal species. The percent of woodland covered by yard waste was correlated with captures of *Peromyscus* spp. ($r = 0.5$). The percent of a woodland covered by trash was correlated with captures of *Rattus norvegicus* ($r = 0.9$) and *Microtus pennsylvanicus* ($r = 0.9$). Trap station cover by herbaceous vegetation was associated with the capture of *Peromyscus spp.* ($r = 0.3$), and trap station cover by rock and rubble correlated with captures of *Mus musculus* ($r = 0.4$). Mean detections of *Canis domesticus* were correlated with capture of *Rattus norvegicus* ($r = 0.9$), and detections of *Felis catus* correlated with *Mus musculus* (0.4).

4
Discussion

We found that woodlands adjacent to block groups that are characterized by a high percentage of low income, predominately black renters, had greater amounts of disturbance than woodlands adjacent to block groups with higher income, predominately white home-owners. *Mus musculus* and *Rattus norvegicus*, non-native species typically associated with disturbed sites, were found only on woodlands next to NT2 block groups. Mean captures/trap station of *Peromyscus spp.*, a species associated with woodland habitats in North America, were higher in woodlands next to NT1 block groups. We also found significant correlations between measures of human-related disturbance and the abundance of four species. These results suggest a potential relationship between small mammal species composition and abundance and human-related disturbance. Determining the actual relationship between site disturbance and species composition and abundance will require long-term study. However our findings do suggest that managing woodlands to reduce human-related disturbance may be beneficial for native species.

Acknowledgments. This paper is a contribution of the Missouri Cooperative Fish and Wildlife Research Unit. This project was funded through a cooperative research agreement with the USDA Forest Service North Central Forest Experiment Station.

References

Burgess J, Harrison CM, Limb M (1988) People, Parks, and the Urban Green: A Study of Popular Meanings and Values for Open Spaces in the City. Urban Stud. 25: 455–473

Dickman CR, Doncaster P (1987) The Ecology of Small Mammals in Urban Habitats. I. Populations in a Patchy Environment. J Anim Ecol 56: 629–640

Grove JM (1995) Social Indicators and Classification For Urban Ecosystem Management. In: Kollin C, Barratt M (eds) Proceedings of the Seventh National Urban Forest Conference. American Forests, Washington, DC

Huckstep GS (1996) Effects of Adjacent Residential Land Use on Small Mammals Inhabiting Woodlands. M.S. Thesis. University of Missouri-Columbia, Columbia, Mo

Linhart SB, Knowlton FF (1975) Determining the Relative Abundance of Coyotes by Scent Station Lines. Wildl Soc Bull 3: 119–124

Nilon CH, VanDruff LW (1987) Analysis of Small Mammal Community data and Applications to Management of Urban Greenspaces. In: Adams LW, Leedy D (eds) Integrating Man and Nature in the Metropolitan Environment. National Institute for Urban Wildlife, Columbia, Md

Nilon CH, Huckstep GS (In press) Analysis of Chicago River recreation habitats. In: Gobster PH, Westphal LM (eds) People and the River. *ChicagoRivers* Technical Report No. 2. USDI National Park Service and Friends of the Chicago River, Chicago

human-related disturbance. Determining the actual relationship between site disturbance and species composition and abundance will require long-term study. However our findings do suggest that managing woodlands to reduce human-related disturbance may be beneficial for native species.

[illegible]

References

[illegible]

Dickman CR, Doncaster CP (1987) The ecology of small mammals in urban habitats. I. Populations in a patchy environment. J Anim Ecol 56:629–640

[illegible]

Workshop: Developing of Urban Green 5.2

Assessment of Habitats for Nature Conservation in Japanese Cities – Procedure of a Pilot Study on Biotope Mapping in the Urban Agglomeration of Tokyo

Norbert Müller

1
Introduction

Nature conservation in cities has advanced in the last decades in many industrial countries in connection with the rapidly increasing urban population (Müller 1997, Sukopp 1990). Nature conservation in cities focuses on the conservation of at least a minimum of plants and animals and their habitats (biotopes) as a basis for direct contact between urban dwellers and the natural elements of their surroundings. It also aims to preserve biodiversity and improve the quality of the environmental media air, water and soil.

The growing concern about urban nature is illustrated by the "Green Paper on the Urban Environment" published by the European Community (1990) which stresses the importance of nature conservation in cities. In Germany, for example, a special law for nature conservation demands that nature is conserved, maintained and developed in populated areas as well as in unpopulated areas.

During the 1970's investigations commenced of plants and animals in German cities, their habitats (biotopes) and their assessment for nature conservation (Müller 1997, Starfinger, Sukopp 1994). The investigations, which are known and respected as "biotope mapping", have been completed in most of the large German cities. In other European cities similar investigations have started and in recent times the European method was also tested in Brazil (Weber, Bede 1997).

Although there are already many applied plant sociological studies in Japanese cities (Miyawaki et al. 1987) similar standardised investigations do not yet exist there. A pilot study was started in the summer of 1996 in the urban agglomeration of Tokyo to test this method in Japan (compare Müller 1997, here also further references).

University of Applied Sciences Erfurt, Faculty of Landscape Architecture, Leipziger Str. 77, 99085 Erfurt, Germany

J. Breuste, H. Feldmann, O. Uhlmann (Eds.)
Urban Ecology

2
General Features of Japanese Cities

Japan is one of the most densely populated countries of the world. In general the archipelago is characterised by a high concentration of the population in cities, which are mainly situated on flat areas along the coastline.

Due to the fast growth in this century and the high frequency of disasters such as earthquakes and the ensuing fires, there is a low number of old settlements and old green spaces in Japanese cities compared to European cities. Particular exceptions are Shinto shrines and temples where the woody vegetation has remained untouched since ancient times thus relics of old forests can often be found. Another feature of Japanese cities is reclaimed land along the coast. The sea is filled up along the coast with soil and rubble to satisfy the high demand for settlements, as has taken place along the Tokyo Bay.

3
Test Areas

Two test areas in the urban agglomeration of Tokyo were selected to test the European method for biotope mapping within a short time period. The pilot study was carried out in Yokohama City which, with three million people, forms the southern part of the Tokyo agglomeration.

Many investigations in cities have shown that the landuse types constitute homogeneous habitats for flora, vegetation and wildlife (Gilbert 1989). The test areas were selected so that various typical landuse types of the urban agglomeration of Tokyo would be represented. Investigations were carried out in:

1. test area 1 – a hilly area which was mainly settled from the middle of this century – (around the campus of the Yokohama National University),
2. test area 2 – along the flat coast with settlements of different age (around Minato Mirai). One part comprises the oldest settlements of Yokohama City which were founded prior to this century (Yokohama port). The other part lies on reclaimed land and was founded in the last two decades.

4
Methods and Procedure of the Pilot Study

In principle there are two methods for biotope mapping in urban areas (Müller 1997, Starfinger, Sukopp 1994):

1. Selective mapping which researches areas worthy of protection; this presupposes a framework of mapping and evaluation

2. Comprehensive mapping which investigates biological data (flora, fauna, vegetation) of all landuse types in the city. Normally it is not possible to investigate the whole city area in such a detailed way so examples of all landuse types are mapped and the results transferred to all areas with the same use (comprehensive-representative mapping). The evaluation of areas which are important for nature conservation is not directly linked and is carried out in a separate step.

In the pilot study a modified method of the comprehensive-representative mapping was used in connection with an evaluation for nature conservation. In the first step actual biological data were researched. Detailed data of flora and vegetation already exist for parts of test area 1. Further basic information sources were old and actual vegetation maps of Yokohama City.

In the next step the landuse types were marked and controlled by field investigations on the basis of aerial photos on the scale of 1:4000. This enabled the biologically rich areas to be marked, i.e. areas with a high amount of spontaneous vegetation and low human impact, for example green spaces, waste grounds, forests and plantations and extensively used agricultural areas. They were investigated in more detail in the next phase through field research of representative samples of flora and vegetation of the biologically rich areas. Afterwards all areas worthy of protection were selected. For the evaluation the fundamental parameters were:

- *Reproducibility and age of the site* – areas which are replaceable in the long or medium term (at least 3–4 years) were mapped.
- *Size* – the minimum size of a selected site was in general 0.2 ha.

Additionally, at least one of the following parameters was necessary for selection of the site as a biotope worthy of protection:

- *Spontaneous occurrence of the site* – areas with a typical spectrum of spontaneous species;
- *Structural variety of the site* – areas with a high variety of vegetation types and age (e.g. green spaces with different vegetation types – lawns, shrubs and plantations);
- *Rarity of a site and of the spontaneous species* – areas which are rare or contain rare species in settlements (e.g. wetlands and natural forests).

Through a field investigation all biotopes worthy of protection were described with standardised forms. These comprise a short description of the biotope, a list of the characteristic and dominant plants and suggestions for further management. After the field investigations a map was drawn of the landuse types as well as a map of the biotopes worthy of protection (on a scale of 1:4500).

Finally, a list was made with a short description of the biotopes worthy of protection.

5 Further Investigations

For a final evaluation for nature conservation the selection of the biotopes worthy of protection is not enough. For example special landuse types as agricultural areas may be important as corridors in the city. Due to the limited test areas an evaluation is not possible. Some areas may be important for urban wildlife but were not marked at all because the investigations only focused on floral and phytosociological features.

To obtain information about all areas which are important for nature conservation the pilot study in Yokohama should be continued in the following way:

1. Mapping of all landuse types and biologically rich areas in the city (using aerial photos).
2. Representative research of flora, vegetation and fauna of the landuse types and biologically rich areas in selected parts of the city, which will entail the registration of spontaneous vascular plants, planted trees and shrubs, plant communities and selected animal populations (especially species with a high indicator value for urban habitats). The following animals should be researched: birds, selected groups of mammals (bats, shrews and hedgehogs), amphibians and reptiles, grasshoppers, butterflies (*Macrolepidoptera*), beetles (*Carabidae*), dragonflies, spiders (*Araneae*) and wild bees (*Apoidea*) (Arbeitsgruppe "Methodik der Biotopkartierung im besiedelten Bereich" 1993).
3. Development of specific mapping guidelines for biotopes worthy of protection and other landuse types important for nature conservation in the city (priority areas for nature conservation).
4. Investigation of priority areas for nature conservation in the whole city area
   - description of the biotopes worthy of protection using relevant forms
   - description of the landuse types important for nature conservation in the city (e.g. such as those important as corridors)
5. Summarising the results and demonstration in maps
   - map of landuse types and map of biotopes worthy of protection and landuse types important for nature conservation in the city.
6. Development of general recommendations for management and green planning for all landuse types from the viewpoint of nature conservation.

After finishing the pilot study, the mapping guidelines can be used for further investigations on biotope mapping in the agglomeration of Tokyo as well as in other Japanese cities. This will provide an overview of the situation of nature in the different Japanese cities, give information on existing vegetation in Japan (Miyawaki et al. 1987), National Parks etc., and will be of use for the development of further nature conservation strategies. Biotope mapping will be organised and conducted by the government so that comparable data can be obtained, as it is done for example in Germany.

References

Arbeitsgruppe "Methodik der Biotopkartierung im besiedelten Bereich" (1993) Comprehensive biotope mapping in urban areas as basis for nature conservation planing - Natur & Landschaft 68: 491–526 (in German with English summary)

Gilbert OL (1989) The ecology of urban habitats. Chapman & Hall, London, 369 pp

Miyawaki A, Fujiwara K, Box EO (1987) Toward harmonious green urban environments in Japan and other countries - Bull. Inst. Environmental Technology Yokohama National Univ., Japan, 14/1: 67–82

Müller N (1997) Biotope mapping and nature conservation in cities - Part 1 - Bull. Inst. Envir. Science & Technology, Yokohama Nat. Univ. 23/1: 47–62

Starfinger U, Sukopp H (1994) Assessment of urban biotopes for nature conservation - in: Cook E, v Lier H (eds) Landscape planning and ecological networks. Elsevier, Amsterdam, 89–115

Sukopp H (1990) Urban ecology and its application in Europe. In: Sukopp H, Hejny S, Kowarik I (eds): Urban Ecology. Plants and plant communities in urban environments - SPB Academic Publishing, The Hague: 1–22

Weber M, Bede LC (1997) Comprehensive approach on the environmental status in very discontinuous landscapes and strategies for implementing basic environmental improvement guidelines in Brazil. In: UFZ (ed) Abstracts of the International Conference "Urban Ecology" Leipzig 1997. UFZ Centre for Environmental Research Leipzig-Halle, Leipzig pp 89–90

Comprehensive Approach to the Urban Environmental Status in Brazil Using the Biotope Mapping Methodology

Markus Weber and Lúcio Cadaval Bedê

1 Introduction

Without doubt, the urban environment in Latin America has been a large arena for the environmental, social and economic problems experienced by the component countries for a long period in history. Taking Brazil as an example, about 75% of the population is now concentrated in urban areas. These are precisely the landscapes where the natural resources are vanishing rapidly and impoverishment of environmental quality is taking place.

The joint action of factors such as urban planning errors, poverty, accelerated industrial development and the lack of an agricultural policy capable of keeping the rural population in the countryside led some cities to face critical environmental situations. The results are (amongst others) housing deficits, lack of basic infrastructure, air, soil, water and visual pollution, chaotic traffic and sanitation problems.

The lack of a planning culture in countries with such tropical wealth has led to arbitrary or non-existent decision making. It is typical for any Brazilian city to have at least one zone or neighbourhood of entirely spontaneous (organic) development, with a lack of sanitation and facilities. In contrast, other areas rapidly develop into stylish shopping centres and modern commercial and industrial zones, smart residential areas and other development models imported from the traditionally industrialised countries. The urban areas in Brazil are thus exceptionally dynamic, with highly heterogeneous landscapes interwoven with the un-developed and natural surrounding rural spaces.

These conflicts, however, do not seem to have a short term solution; their origin is also historical and cultural. Assuming a creative perspective to these conflicts has become the main pathway to the so-called modernity.

Brandt Meio Ambiente, Rua Timbiras 270, 30.140-060 – Belo Horizonte – MG, Brazil

J. Breuste, H. Feldmann, O. Uhlmann (Eds.)
Urban Ecology

2
The Traditional Diagnosis

The starting point for efficient environmental planning is a good diagnosis. The traditional environmental diagnosis in Brazil, however, has laid an emphasis on the identification and the characterisation of specific and isolated environmental problems.

Environmental planning is traditionally managed under the "fire fighting" concept i.e. in most cases it attends exclusively to specific and urgent demands. This means that vitally important benefits are routinely wiped out, giving place to other development priorities. The usually scarce urban green spaces and leisure parks are seen as the only link between the citizens and the natural environment. At the weekends in Belo Horizonte (as in the majority of the largest Brazilian cities) one can observe a rush of people for alternative leisure options on more preserved environments which, in turn, pose threats and pressurise the peri-urban landscape and neighbouring towns.

Under these conditions, new environmental diagnosis methodologies have been identified and adapted to the Brazilian reality, aiming at an efficient, rapid and low-cost approach.

3
Brief Overview of the Biotope Mapping in Brazil

3.1
About the Methodology

In Brazil the biotope mapping methodology was adopted through co-operative work between Brazilian and German Institutions. These include the Ministry for Science and Technology (CNPq), the Minas Gerais State Science and Technology Secretary, the Federal Agency of Nature Conservation (BfN – Bonn, Germany which is linked to the Bundesministerium für Umwelt, Naturschutz und Reaktorsicherheit), the private consultancy companies Biola (from Hamburg) and Brandt Meio Ambiente and the Alexander Brandt Foundation (both from Belo Horizonte, Brazil).

The co-operative work started in 1991 in the form of workshops undertaken by members of these institutions and the UFMG (Federal University of Minas Gerais, Brazil). A pilot project was carried out in a section of Belo Horizonte city in 1993 and a handbook containing the fundamental method and the preliminary Brazilian experience was published in 1994.

The fundamental concepts are those developed by the German Federal and State working group for biotope mapping in urban areas (Bund-Länder Arbeitsgruppe Biotopkartierung im besiedelten Bereich). Until 1996 this was co-ordinated by Prof. Dr. Herbert Sukopp (Berlin University) and is currently co-ordinated by Dr. Wolfgang Schulte (BfN, Bonn).

It refers to dividing a landscape (be it urban or rural) into environmental units with similar structure, management and usage forms and associated biota. The basic result is a mosaic map of the several environmental unit types (or biotopes), extensively described and overlaid with varied thematic maps. The thematic maps are used as tools for understanding the environmental role of each unit and for inferring functionality patterns at the landscape level, including ecological indicator plant and animal species.

The evaluation procedure is carried out under a descriptive rather than a quantitative approach based on the work of several German and Brazilian authors.

3.2
The Brazilian Experience

The scope of the biotope mapping studies in Brazil have so far emphasised the understanding of the urban dynamics based on social and economic parameters as a way of investigating the current land usage patterns and trends.

Environmental quality data are obtained through obtaining data on a selection of physical and chemical parameters (water and air quality, for example) and ecological indicator groups (flora, birds, amphibians, odonates and others, depending on the situation and on the availability of specialised personnel). In many cases precise cartography of the urban mesh and other themes are lacking and mapping is carried out using the material available.

In relation to the physical environment, the major emphasis in urban areas so far has been the study of the relationship between land usage patterns and pedologic, hydrologic and geomorphic factors related to risk areas (e.g. settlements in landslide prone areas).

A major emphasis, however, has been the specification of actions for integrated environmental reclamation, i.e. applied guidelines for local improvements and impact mitigation, sometimes as subsidiaries for Environmental Director Plans.

3.3
Acceptance of the Biotope Mapping Scope

We soon found out how hard it is to convince the public planners to request basic environmental diagnosis, despite the general acceptance of the proposed scope of the biotope mapping studies. The following topics were identified as possible explanations for this problem:

- the lack of public funds for diagnosis and planning;
- high costs for environmental improvements, conflicting with urgent demands on the social area;
- time required for the diagnosis and implementation of environmental improvement actions compared to the short period of the political mandates (4 years);
- the emphasis on the biological-ecological aspects in the diagnosis, as opposed to the traditional purely physical, chemical and sanitary aspects;
- the perspective of an increase of the awareness and consequent demand of urban environmental improvements by the population;
- little knowledge about the biotope mapping methodology in Brazil.

Integrated diagnoses are costly and the municipalities normally have no funds to support them. The main financial support for biotope mapping projects in Brazil so far has come from Federal sources or from private companies.

In general, demands related to biotope mapping originate from the need to deal with specific environmental problems which result in the justification of multidisciplinary and integrated approaches. For example, some of the biotope mapping projects were initiated as pilot projects funded by the Federal government in response to popular requests for the solution of specific problems (e.g. mitigation of mining impacts).

When funded by private companies the scope of the studies are limited to the area of influence of a specific enterprise. In one case, this approach was used by a private urban development company as a tool for the environmental planning and enhancement of a future urban space (BMA/Lagoa Grande 1996).

4 Possible Solutions and Perspectives

Despite the relative success of the integrated environmental approach in urban areas through the biotope mapping, the success of the use and the spread of the methodology in Brazil will depend closely on the ability to match the obtained results with the immediate needs of the planners. This has to take place under the severe restrictions imposed by limited budgets, lack of basic data, lack of specialised personnel and, the most difficult of all, by the short term range and lack of continuity of the public policies and programmes, usually restricted to the four years of mandate.

As stressed before, convincing the decision makers is by far the hardest task and breaking this barrier depends on showing concrete perspectives. Our experience show that any such studies should follow these guidelines:

- work on comprehensive scales with a minimum degree of refinement on the definition of biotope types and subtypes;

- reduce the scope of multidisciplinary (thematic) studies, allowing an integrated approach to environmental quality covering the most urgent environmental issues;
- emphasise social and economic approaches for detecting current land use patterns and trends, each case being analysed according to the local social and environmental peculiarities;
- emphasise the perspectives for providing assistance to urban planning and policies in the shaping of new developments and on broad scale actions on developed areas (master plans);
- emphasise perspectives/subsidies for environmental education programmes;
- consider planning and management perspectives for urban conservation units;
- in the case of sites with relevant historical and natural patrimony, ensure enlightenment of perspectives of economic growth and diversification through a "sustainable development" approach (e.g. tourism).

References

BMA/Lagoa Grande - Brandt Meio Ambiente Ltda & Lagoa Grande Estudos e Projetos, Planejamento e Empreendimentos Ltda (1996) Empreendimento Urbanístico Lagoa dos Ingleses. Estudo de Impacto ambiental. Relatório não publicado p 235

Comprehensive Biotope Mapping in the City of Mainz – a Tool for Integrated Nature Conservation and Sustainable Urban Planning

Jochen Frey

1 Introduction

In 1993, the working group 'Methods of biotope mapping in populated areas' (set up by the Federal Nature Conservancy and the State Offices for Nature Conservation and Landscape Management) updated its so-called 'basic programme' for gathering, organising and analysing data relating to urban biotopes (Schulte et al. 1993). The programme's goal is to provide uniform research methods at State level to allow the scientific comparison of the results of different urban biotope mappings.

The same year, a research project called 'Biotope mapping of Mainz' was commissioned by the City Council of Mainz and carried out by an interdisciplinary group of geographers, biologists and environmental planners at the University of Mainz (Arbeitsgruppe Stadtbiotopkartierung Mainz 1996/1997). The project's work is based entirely on the strategy of integrated nature conservation, i.e. simultaneous protection of biotic, abiotic and aesthetic resources (Pfadenhauer 1991), which is considered as a fundamental basis for sustainable development in populated areas. One of the scientific goals of the 'Biotope mapping of Mainz' was to transfer the objectives and recommendations of the 'basic programme' to the local situation and to remove shortcomings within its methodological framework. An outline of the project is given here, focusing on introduced methodological aspects.

2 Some New Aspects of Biotope Mapping in Mainz

Table 1 provides a synopsis of two recent biotope mapping schemes in Germany. While the left column depicts the methodology of biotope mapping recommended in the 'basic programme', the opposite column gives an

FÖJ Rheinland-Pfalz, P. O. Box 1951, D-55009 Mainz, Germany

J. Breuste, H. Feldmann, O. Uhlmann (Eds.)
Urban Ecology

Table 1. Biotope mapping schemes in urban areas: the 'basic programme' and its applications and modifications in the 'Biotope mapping of Mainz'

Recommendations of the 'Basic Programme'	Applications (✓) and Modifications (• . . .) in the Research Project 'Biotope mapping of Mainz'
1. Delimitation of the area of investigation	• City limits of Mainz, incl. adjacent nature reserves
2. Analysis of available biotope data	✓
3. Comprehensive mapping of biotope-types (scale of 1:5000) with a given biotope-type reference key	• (partly) New systematization of the biotope-type reference key • Very detailed mapping of biotope-types, subtypes and their variations • Compilation of additional field notes
4. Representative mapping of sample biotopes	
Domains:	
• Flora/vegetation and fauna	✓
• Small-scale biotope features	• Mapping of small-scale biotopes with a special small-scale biotope-type key
• Landscape aesthetics (scenery etc.)/Human contact with nature / Experience of nature	• Mapping of natural phenomena and recreational activities
• 'Threats'	• Mapping of environmental impacts on biotopes
5. Evaluation of the data	
Description of biotope-types with regard to	
• their location / spatial distribution in the city	✓
• abiotic factors (geology, soil, climate)	✓
• small-scale biotope features	✓
• plant and animal species to be found	• Characteristic vegetation (plant communities, specialized and endangered plants, lichens)
• their 'significance for the experience of nature'	• Variety of natural phenomena/recreational activities
• 'the reasons and causes of threats'	• Environmental impacts
Description of the biotope-types' specific value for nature conservation	*! Topics not to be presented in a descriptive, but in a normative context (see 6.)*
Delimitation of 'important areas' for nature conservation	
Recommendations for protection, development and management of 'important' biotope-types	
Zoning of the cityscape according to geographical and ecological criteria	• Zoning of the cityscape • Delimitation and description of 'urban landscape units' with regard to the items mentioned in the biotope-types' description as well as to faunistic aspects

Table 1. (*Continued*)

Recommendations of the 'Basic Programme'	Applications (✓) and Modifications (• . . .) in the Research Project 'Biotope mapping of Mainz'
6. Assessment of biotope-types with regard to species protection, human contact with and experience of nature and landscape aesthetics Explanation of assessment procedures (targets, criteria, gradation)	Confronting the guidelines for environmental development with the findings of the biotope mapping • Development of environmental quality targets and standards, individually specified for 'urban landscape units' in the domains of flora, fauna, 'landscape and recreation', soil and climate • Comprehensive assessment of each unit's biotope-types with reference to the established environmental quality targets and standards
7. Handling of the project's results Recommendations for protection, development, management etc.	• Synoptic tables of targets, strategies and measures for protection, development or management of biotopes within the 'urban landscape units' of Mainz

overview of the items completed or modified in the 'Biotope mapping of Mainz'.

An important task of nature conservation in cities is the preservation of plants, animals and their habitats in order to make 'human contact with nature' possible (Sukopp et al. 1980, Gilbert 1989). Since this is mainly achieved by perception and/or recognition of natural phenomena or through the recreational use of natural areas, the mapping was focused on commonly known natural phenomena, everyday activities for leisure and recreation and their traces being found within biotope-types (Frey 1997). Table 2 gives a brief selection of these features, which were mapped in about 50 sample areas in Mainz.

To obtain a reasonable basis for spatial description and assessment of biotope-types in cities, the landscape within the city limits should be divided into different environmental zones. Three zones are determined from the basic principles of natural Geography and Geobotany (see e.g. Meynen, Schmithüsen 1962, Kowarik 1992),

1. *Area of remnants of virgin landscape*: area characterised by ecological processes typical of virgin landscapes (in Mainz: sand dunes, floodplains),
2. *Area of diversified rural landscape*: area characterised by agriculture and viticulture, with landscape elements like hedgerows, Loess slopes, rough stone walls etc.,

Table 2a/b. Natural phenomena (left table), recreational activities and activity traces (right table) related to biotopes (selected lists, after Frey 1997)

Natural Phenomena related to Biotopes
Scenic phenomena: sand dunes, pine forest, alleys, Rhine river, . . .
Structural phenomena: big old trees, stone walls, hedgerows, slopes, . . .
Aquatic phenomena: springs, water levels, waves, shorelines, aquatic wildlife, . . .
Animal phenomena: songbirds, nests, rabbits, foxholes, dragonflies, glow-worms, . . .
Seasonal phenomena: blossom, fruit aspects, colouring of the leaves, . . .
Sound phenomena: woodpeckers, birds' chant, rustling of leaves, croaking of frogs
Olfactory phenomena: scent of flowers, hay, mushrooms etc., stench of sewage etc., . . .

Recreational Activities and Activity Traces related to Biotopes	
Activities	Activity Traces
walking, strolling/jogging	footpaths
biking	bike traces
playing: ball games, scouting games, . . .	broken twigs/ *Clematis* liana
flower- or fruit-picking	cherry stones, nutshells etc.
picknicking	campfires
relaxing	lugged around sand and leaves
flirting . . .	shacks . . .

3. *Area of urban-industrial landscape*: built-up area (city centre, former village centres, industrial and commercial areas etc.) with different types of urban greenspace (parks, cemeteries, wasteland etc.) and characteristic small-scale biotope features (cobble gaps, solitary trees etc.).

Yet, the delimitation of such zones does not fulfil the needs of a detailed spatial evaluation of biotope-type data. According to geological, geographical and land use criteria, the three environmental zones can be subdivided into a variety of smaller landscape units, each of them having an individual environmental character within the city. In Mainz, 54 'urban landscape units' were differentiated and described in detail (Arbeitsgruppe Stadtbiotopkartierung Mainz 1996/1997).

Relating to the 'concept of environmental quality targets', which was officially recommended after the *Rio de Janeiro* Conference in 1992 (BMU o. J., RSU 1994), the 'Biotope mapping of Mainz' attempted to assess biotope-types by comparing their biotic, abiotic and aesthetic inventory with criteria and standards required in environmental quality targets. Previously, specified principles, guidelines, quality targets and quality criteria were established for each 'urban landscape unit' in the domains of flora, fauna, landscape aesthetics, soil and climate (see details and examples in Table 3). All propositions

Table 3. Development of environmental quality targets for urban landscape units and assessment of the landscape unit's biotope-types

Principal Ideas for the Sustainable Use of Land in the City of Mainz

The environment within the city limits of Mainz, as characterised by remnants of virgin landscape, diversified rural landscape and different types of urban-industrial landscape, is to be regarded as an *interdependent system* whose components have to be treated according to the principles of *integrated nature conservation* and the *sustainable use of land*. An important aspect in this case is the preservation of the landscape's individuality, diversity and beauty as a basis for quality of life.

↓

Guidelines for the Sustainable Use of Land in the Zones of Urban Landscape in Mainz

↓	↓	↓
Remnants of virgin landscape *Domains: Climate/Soil/ Fauna/Flora/'Landscape & Recreation'*	Rural landscape *Domains: Climate/Soil/ Fauna/Flora/'Landscape & Recreation'*	Urban-industrial landscape *Domains: Climate/Soil/ Fauna /Flora/'Landscape & Recreation'*
↓	↓	↓
...	...	*Example:*

Urban-Industrial Areas of Mainz: Guidelines in the Domain of 'Landscape & Recreation'

- Preservation of areas with relevance to urban nature as well as to cultural history, i.e. as documents of historic-cultural and historic-natural interdependence in the urban-industrial landscape of Mainz.
- Preservation of (historic-)natural areas for recreation, enjoyment, inspiration, observation of urban wildlife & environmental education.
- Preservation of relics of rural or virgin landscape.

↓

Environmental Quality Targets for 'Urban Landscape Units' in Mainz

Example:

Urban Landscape Unit # 53 (Housing estates on sand dunes, northwestern suburbs of Mainz): Quality Targets for 'Landscape & Recreation'

- Presence of *dune landscape elements* as documents of natural history and individuality, as well as for inspiration, enjoyment, education and recreation in harmony with nature.

Quality standards / Quality criteria

Example:

Urban Landscape Unit # 53: Quality Criteria for 'Landscape & Recreation'

Presence of

- relics of calcareous sand dunes, sand mounds and sand flats
- relics of sparse pine wood and other indigenous tree species
- relics of orchards with thermophilous fruit trees, e.g. apricots
- dry grassland on autochthonous (sandy) soils
- dry lawn on autochthonous (sandy) soils
- wasteland on autochthonous (sandy) soils
- sandy slopes
- sandpits
- sandy paths and trails
- sandy embankments and verges

↕

Assessment of the landscape unit's biotope-types

Assessment of the conformity of each biotope's structures with the required quality standards, i.e. with the required environmental quality targets

➢ „How does the biotope meet the specified environmental quality target ?" (4 steps; depicted in coloured maps)

Table 4. Targets and measures of biotope-related environmental management in urban landscape units (as an example: 'Housing estates on sand dunes')

Targets of Environmental Management	Suggested Measures	Biotope-types
Preservation or regeneration of dune landscape elements (see Tab. 3)	Drawing up of a 'Landscape management plan for housing estates and common greenspace on calcareous sand dunes in the north-western suburbs of Mainz' comprising • the support of sandflora species through considerate and moderate cutting of dry lawns, embankments and verges • the planting of thermophilous, indigenous tree species like *Pinus sylvestris, Quercus robur*, thermophilous shrubs, tall fruit trees like apricots, plums, cherries, almonds etc. • the avoidance of deposits with allochthonous soil material and humus • the avoidance of fertilizers and pesticides • the support of small-scale biotope features (see Tab. 3) • public information	• all biotope-types associated with buildings, esp. housing estates, churchyards, former military area 'Alte Kaserne' • biotope-type # 4100: playgrounds, verges of sportsfields • biotope-type # 6250: verges of parking lots (e.g. open air pool „Am Großen Sand")

endeavoured to meet the idea of sustainable urban planning (DRL 1992, Stadt Mainz 1995, Diamond, Noonan 1996).

As a practical outcome of the mapping's detailed spatial description and assessment, some advice for appropriate measures concerning biotope-type development in urban landscape units was supplemented (see Table 4).

3 Conclusions

Through the adoption of the currently discussed ideas of integrated nature conservation and sustainable use of land for the means of biotope mapping in populated areas, the research project 'Biotope mapping of Mainz' attempted to make its scientific outcome more accessible and understandable. It is hoped that local planning authorities will apply the project's findings and give a feedback of the results.

References

Arbeitsgruppe Stadtbiotopkartierung Mainz (1996/1997) Stadtbiotopkartierung Mainz. Endbericht. 8 Bände. (CD-ROM in pr.) University of Mainz, Mainz

Bundesministerium für Umwelt, Naturschutz und Reaktorsicherheit (BMU) (ed) (o J) Konferenz der Vereinten Nationen für Umwelt und Entwicklung im Juni 1992 in Rio de Janeiro. - Dokumente - Agenda 21. Bonn

Deutscher Rat für Landespflege (DRL) (1992) Natur in der Stadt - Beitrag der Landespflege zur Stadtentwicklung. SchrReihe des Dt Rates f Landespfl 61

Diamond HL, Noonan PF (eds) (1996) Land Use in America. The Report of the Sustainable Use of Land Project. Island Press, Washington D.C.

Frey J (1997) Die Stadtbiotopkartierung Mainz - ein Versuch zur Umsetzung und Weiterentwicklung des 'Grundprogramms Biotopkartierung im besiedelten Bereich'. Geobot. Kolloq. 13: 3–12

Gilbert O (1989) The Ecology of Urban Habitats. Chapman & Hall, London

Kowarik I (1992) Das Besondere der städtischen Flora und Vegetation. SchrReihe des Dt Rates f Landespfl 61: 33–47

Meynen E, Schmithüsen J (eds) (1962) Handbuch der naturräumlichen Gliederung Deutschlands. Bd.1. Bad Godesberg

Pfadenhauer J (1991) Integrierter Naturschutz. Garten und Landschaft 101(2): 13–17

Rat von Sachverständigen für Umweltfragen (RSU) (1994) Umweltgutachten 1994. Verlag Metzler-Poeschel, Stuttgart

Schulte W, Sukopp H, Werner P (eds) (1993) Flächendeckende Biotopkartierung im besiedelten Bereich als Grundlage einer am Naturschutz orientierten Planung. Programm für die Bestandsaufnahme, Gliederung und Bewertung des besiedelten Bereichs und dessen Randzonen. Natur und Landschaft 68: 492–526

Stadt Mainz (1995) Stadtentwicklung und Umwelt. Grundlagen und Ziele für eine umweltverträgliche Stadtentwicklung. Amt für Stadtentwicklung und Statistik, Mainz

Sukopp H, Kunick W, Schneider C (1980) Biotopkartierung im besiedelten Bereich von Berlin (West). Garten und Landschaft 90(7): 560–564

Derived Vegetation Map of Berlin

Walter Seidling

1 Introduction

The landuse (with its specific patterns of sealing and disturbance) is the main factor determining the presence and type of vegetation in built up areas. On the edges of cities remnants of forests and agricultural land intermingle with urban areas resulting in a mixture of man-made and natural factors which is reflected by the vegetation.

Each landuse type also has a more or less characteristic set of "planted vegetation". It interacts not only with the spontaneous vegetation and fauna but also fulfils psycho-sociological functions. For most citizens it is the true and esteemed vegetation of a city and should not be ignored in vegetation mapping of cities.

Within the framework of the Environmental Atlas (Umweltatlas) of Berlin a map of vegetation types (1:50,000) was drawn up for the whole city (891 km^2, almost 3.5 mil. inhabitants). The detailed knowledge on vegetation collected over decades in Berlin (West) (Sukopp 1990) was used in a city-wide approach with high spatial resolution.

2 Land-Use and Vegetation Categories

Landuse produces the mosaic of habitats for plants in urbanised landscapes. Since information about the prevailing landuse type (altogether 94 types) existed for each city block, it was used to derive a map of the *most probable actual vegetation*. Information on the degree of sealing or area extent of each city-block or on soil type etc. was also available as well as aerial photographs. For many landuse types sub-divisions were made according to those additional features.

Institute for Ecology and Biology, Technical University of Berlin, Schmidt-Ott-Str. 1, D-12165 Berlin, Germany

J. Breuste, H. Feldmann, O. Uhlmann (Eds.)
Urban Ecology

The dominant vegetation types were quantitatively determined for each landuse type and its sub-divisions using information from former surveys. The spontaneous vegetation was compiled after the Braun-Blanquet approach. Categories of low-ranking associations were pooled according to the phytosociological hierarchy to attain categories of sufficient spatial extent.

For the planted vegetation a classification mainly based on structural or expedient features (e.g. stock of street trees, stands of ornamental shrubs, plantations of ornamental annuals, cf. Heinrich, Rohner 1992) was applied. Usually composite categories (communities) such as "ornamental annuals with Chenopodietalia-communities" or "shrubs with Alliarion-communities" were defined. Estimates of spatial proportion (threshold: 5%) were used as a universal measure of abundance of each final vegetation category (pure or composite).

A cluster analysis was performed to reduce the number of cartographic units of vegetation for the built-up area to a manageable extent; this used the set of vegetation categories with estimates of their spatial proportion. Landuse types with similar vegetation below a certain fusion level were pooled resulting in 16 vegetation types (cartographic units) with 22 characterising vegetation categories (communities) (Seidling 1994).

For areas with semi-natural landuses the pure Braun-Blanquet approach was used. Much information for the western part of Berlin could be obtained from the preceding vegetation map (Böcker, Sukopp 1985). These types of vegetation were sub-divided into blocks with the help of additional information. For instance: water surfaces were subdivided into those with and without floating vegetation by means of aerial photographs.

Within forests, monocultures and old mixed stands differ significantly in floristic and ecological respects. "Composite communities" could be constructed but traditionally woodland communities are dealt with in a manner which integrates the managed tree layer and its spontaneous undergrowth in terms of the Braun-Blanquet approach. For this reason a series of forest types were used which were derived from one natural forest association differentially altered by foresters ("Gesellschaftsreihe", Scamoni 1963). The human influence was additionally expressed by 3 degrees of hemeroby (Sukopp 1972), mainly based on age and species composition of the stands (Seidling 1993).

3 The Resulting Map

The map shows the *most probable real vegetation types* of Berlin. It reflects the well known gradient from the fringe to inner city with distinct "extra zonal" elements such as parks, traffic axes etc. If we compare this map (Fig. 1a) with the preceding one (Fig. 1b), the vegetation of the built-up area

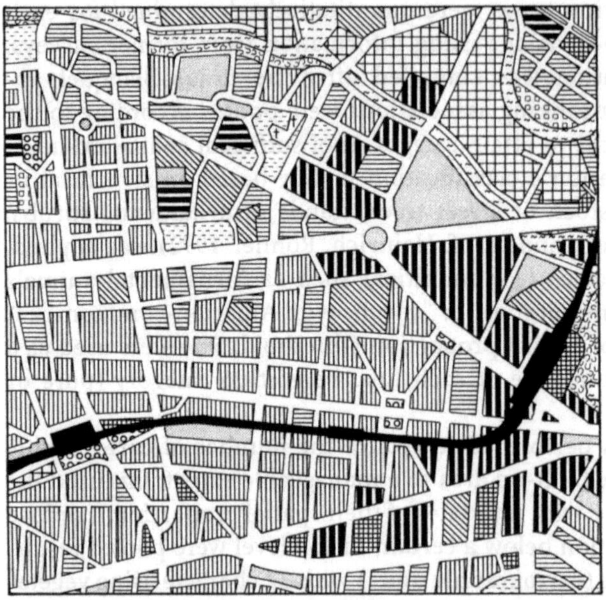

(a)

(b)

Fig. 1. Example drawn from a densely built-up area (Berlin-Charlottenburg) of the actual map of vegetation types (1:50,000, SenSUT in prep.) (a), and the same section from the previous map of vegetation types (Böcker, Sukopp 1985) (b)

is spatially much more varied. For the construction of the preceding map conclusions were transferred from mapped example areas to larger sections; in the new map each block was dealt with separately according to the attachment rules.

4 Discussion of Outstanding Issues

Biotope maps are a common tool for recording biota and their environment within cities (>160 projects in Germany, Schulte et al. 1993), but according to Baader et al. (1991) only 4 vegetation maps of cities and towns exist. Two methods (sigma-syntaxonomy, "vegetation types" derived from biotope mapping) and different scales (1 : 5000 to 1 : 50,000) are used. There are several reasons why maps of vegetation types of cities are rare in comparison with those of open landscapes: *1.* Vegetation occurs within highly fragmented habitats ranging from sq. cms (joints in pavements) to sq. kms (parks) with vast spaces in between without vegetation. *2.* There is a large amount of planted vegetation which is not integrated within the typological system of Braun-Blanquet (Klotz 1992). *3.* Even the spontaneous vegetation is often not in line with phyto-sociology because of high floristic and site dynamics (see Kopecky 1992).

The presented map is based on rules of attachments. In this respect it is similar to the vegetation map of Bavaria constructed by Seibert (1968) and differs from simulated maps based on conditioned probability functions (Brzeziecki et al. 1993). Sources of statistical noise within the resulting map are: *1.* Systematic and stochastic noise from phyto-sociological and structural typification which is inherent in any classification (Feoli, Zuccarello 1994). *2.* Each assignment such as: real city block → main use, land-use → vegetation type, has a certain probability of error and can be viewed as fuzzy interference (Feoli, Zuccarello 1994). *3.* High rates of changes of landuses are a general feature of cities cause thematic maps to become quickly out of date.

The integration of planted vegetation is still under discussion: Kowarik (1995) proposed a terminological framework for tree stands. Klotz (1992) designated a partly spontaneous vegetation (Halbspontanvegetation) composed of planted shrubs/trees and their spontaneous undergrowth. The combination of all kinds of planted vegetation with the co-existing spontaneous vegetation as practised here may not lead to a new hierarchical system of typification but, in combination with multivariate methods, it can be considered as an appropriate tool to solve classification and typologic problems, at least on city-oriented scales.

Acknowledgements. Thanks to the staff of the 'Umweltatlas' esp. Betina Fellmer at 'Senatsverwaltung für Stadtentwicklung, Umweltschutz und Technologie' (see Umweltatlas Berlin, in prep.) for their excellent co-operation and to Dr. Uwe Starfinger, TU Berlin, for valuable advice.

References

Baader I, Blachnik-Göller T, Gödde M, Mittl S, Rodenheimer M, Wächter M (1991) Flächendeckende Erfassung der Vegetation als Instrument der räumlichen Planung, dargestellt am Beispiel einer Vegetationstypenkarte. Raumforsch u Raumordnung 1/1991: 20–26

Böcker R, Sukopp H (1985) Vegetation. In: Senator für Stadtentwicklung und Umweltschutz Berlin (ed) Umweltatlas, Bd 2, Karte 05.02. Senator für Stadtentwicklung und Umweltschutz Berlin, Berlin

Brzeziecki B, Kienast F, Wildi O (1993) A simulated map of the potential natural forest vegetation of Switzerland. J Veg Sci 4: 499–508

Feoli E, Zuccarello V (1994) Naiveté of fuzzy system spaces in vegetation dynamics? Coenoses 9: 25–32

Heinrich T, Rohner M-S (1992) Stadtbiotopkartierung Bremen: Stadtökologische Strukturkartierung. Gutachten i.A. Senator für Umweltschutz und Standentwicklung Bremen 1–258 np

Klotz S (1992) Probleme der Vegetationskartierung in Städten. Verh Ges Ökol 21: 201–206

Kopecky K (1992) Syntaxonomische Klassifizierung von Pflanzengesellschaften unter Anwendung der deduktiven Methode. Tuexenia 12: 13–24

Kowarik I (1995) Zur Gliederung anthropogener Gehölzbestände unter Beachtung urbanindustrieller Standorte. Verh Ges Ökol 24: 411–421

Scamoni A (1963) Forstgesellschaften. Biol Rundschau 1: 87–89

Schulte W, Sukopp H, Werner P (eds) (1993) Flächendeckende Biotopkartierung im besiedelten Bereich als Grundlage einer am Naturschutz orientierten Planung. Natur und Landschaft 68: 491–526

Seibert P (1968) Übersichtskarte der natürlichen Vegetationsgebiete von Bayern 1:500.000 mit Erläuterungen. Schriftenreihe Vegetationskde 3: 1–84

Seidling W (1993) Vegetationstypen der Wälder, Rieselfeldaufforstungen und Großsiedlungen für den Umweltatlas Berlin, Karte 05.02. Gutachten iA Senatsverwaltung für Stadtentwicklung und Umweltschutz Berlin 1–31 np

Seidling W (1994) Fortschreibung/Neubearbeitung der Karte 05.02 'Vegetationstypen' für den Umweltatlas Berlin. Gutachten iA Senatsverwaltung für Stadtentwicklung und Umweltschutz Berlin 1–69 np

SenSUT (Senatsverwaltung für Stadtentwicklung, Umweltschutz und Technologie) (in prep.) Umweltatlas Berlin, Karte 05.02 'Vegetationstypen'. Kulturbuch-Verlag Berlin

Sukopp H (1972) Wandel von Flora und Vegetation in Mitteleuropa unter dem Einfluß des Menschen. Ber Landw 50: 112–139

Sukopp H (ed) (1990) Stadtökologie: das Beispiel Berlin. R Reimer, Berlin

The Suburban National Park as a Form of Town Green Zone Territory Management

Mikhail I. Struk

One of the most important functions of a suburban area is its ability to guarantee essential town activities. Sanitary conditions in the city environment need to be improved; nature management plays a part in this and is directed towards satisfying the requirements of the town. Forms of nature management differ in high intensity utilisation of natural resources and in spatial structure in which high requirements for environmental quality are combined with combating environmental pollution. Areas managed for nature are, in particular, the recreation places for town inhabitants and sources of water for the town; sites where environmental pollution must be combated are agricultural and industrial enterprises and installation dumps.

This combination of both ecologically aggressive and vulnerable forms of nature management creates the danger of conflict situations. To prevent conflicts, or to reduce the resulting negative consequences, it is necessary to optimise the location of the different engineering buildings and structures, and of other sources influencing the natural environment and ecological stability. The latter must be managed accordingly for conflicts to be avoided; the basis of this organisation was the subject of the research reported in this chapter.

The work included two main tasks: (1) the creation of a scheme for the spatial management of the green zone surrounding the town; (2) the choice of the management form for realisation of the scheme. This necessitated carrying-out research in several directions:

1. Study of present suburban nature management development, taking into account the expected alterations in its structure.
2. Evaluation of the approach currently used for territory zoning.
3. Finding a basis for the optimal form of green zone management.

A suburban region of Minsk was chosen as the main area of research. The city itself has a population of about 1.7 million.

Institute for Problems of Use of Natural Resources & Ecology, Belorussian Academy of Sciences, 10 Staroborisovski trakt, 220114 Minsk, Belarus

J. Breuste, H. Feldmann, O. Uhlmann (Eds.)
Urban Ecology

1 Tendencies of Suburban Nature Management Development

The structure of a suburban nature management which complies with a city's requirements does not stay unchanged, but transforms in the course of time in accordance with the developments required. These may depend simply on the growth of the city and on changes in the social policy of the state. In the first case, the intensity of the resources used increases and thus the space required extends to satisfy the city's needs; in the second case, new forms of nature management are created.

These factors have influenced the development of nature management in the suburban region of Minsk. The city population has increased 1.8 times since 1970, which has entailed a twofold increase in the city's withdrawals of water from the suburban territory and an increase in building activity by 2.2 times.

The most significant changes in suburban nature management were connected with the alteration from a 6 to a 5 day working week at the end of the 1960's. This caused a critical rise in recreational demand of the urban population. The natural preconditions to satisfy this in Minsk and its surroundings were inadequate due to the shortage of large water bodies in the area. To compensate for this shortage, reservoirs were built on small, flowing rivers; these artificial water bodies, together with the adjoining forests, became the core for the formation of recreation zones.

Agricultural land provides the source for increased building and extraction of water on the suburban territory. During the research period, the area of agricultural land decreased from 55 to 51% of the total land; the area of forest in the region remained the same, constituting about 30%.

In the future we can expect that the main alteration in suburban nature management will be the increased colonisation of this territory. The growth in population in the Minsk region observed up to now was not due to settling of Minsk inhabitants in the suburban area. The basis for the evolution of this settling stage are now being created. They are expressed, particularly, in the state's housing policy of 1991, which aimed at a critical increase of single houses, mainly located in the suburban area.

The movement of part of the urban population to the suburbs, and the corresponding increase in local population, will lead to the area surrounding a city becoming the main centre of concentration of the population. This will cause a further increase on the environmental load, and at the same time will raise the requirements for improvement in environmental quality, thus increasing the role of the regulation of nature management.

2
The Current Approach to Territory Zoning

The basis of the system used at present to guarantee ecological stability in a suburban area and improvement in environmental health conditions in an urban environment necessitates forest conservation. For this purpose, green zones have been established. These are territories occupied by forests and forest parks, which fulfil the demands for environmental protection and improvement, while providing places for recreation.

The green zones are divided concentrically into internal forest parks and external economic forest depending on the level of their improvement. In Minsk, the first circle has a radius of 15–25 km and the second of 50–80 km.

Forests of the green zones belong to the first group (internal forest parks), and require management to guarantee their preservation. The destruction of these forests for economic purposes is forbidden. However, this does not completely exclude their being cut down when this is necessary for intensive building in the suburban zone. Moreover, some activities (road building, mining) cannot avoid using both forest and agricultural lands.

Forest or agricultural land can be used for the location of structures such as living and industrial buildings, recreation establishments etc. According to the regulations for land allocation for building, only forests and agricultural land of little value can be used for this purpose. The list of land where building is forbidden is defined by the local government. Such an administration expresses primarily the interests of inhabitants in these areas, but these do not coincide with the interests of the city. Under these conditions land suitable for the agriculture industry has the most protection, thus protecting the land's agricultural resources.

A different situation occurs with respect to forest protection. In spite of the fact that destruction of internal forest parks is forbidden in most cases, cutting down is permitted for the construction of recreational enterprises. Allocation of forest of little value for recreation is recommended, but this is ineffective as no unified criteria exist for defining the value of forests. The environmental and recreational functions of forests, which should be given priority in the suburban area, are fulfilled in different ways by different types of forest. For example, a swampy forest is unsuitable for recreation but has good water-regulating properties; according to the regulations, however, this type of land should be improved.

The legislative basis does not guarantee protection for suburban forests. The forests with conditions favourable for recreation, i.e. good transport access and proximity to water bodies, are the most vulnerable with respect to recreational building. They are most frequently exposed to tourists and, because the implemented measures (which aim to conserve the forest) are not always adequate, they may be misused. Sections of forest which have lost their ability for self-regeneration are defined during regular investigations carried out in the forest park belt of Minsk.

When forests are built on, or their environment is altered, their stabilising properties are reduced which may limit their ability to ameliorate the environment. This necessitates selective assimilation of forests surrounding the city; some should be kept in their natural condition and others transformed into forest parks.

The division of the green zone into forest park and economic forest sections does not fit in with the above proposal. Forests with the status of forest park belt are automatically available for improvement. Almost any part of the forest may be cut-down for recreational purposes or construction, or transformed according to the needs for recreation; this is ecologically not optimal.

3 Model-Drawing

For the selective assimilation of the green zone territory it is necessary to change from the concentric scheme of zoning to one based upon the mosaic interchange of forest-park cores and forest parts kept in their natural condition. The latter should be distributed to prevent completely splitting up the large tracts of forests; green wedges should be formed connecting the town's forested areas to the green zone. It is also necessary to prevent the breaching of forests and swamps which afford important natural protection, and also to maintain the most picturesque rivers and their surroundings in their natural condition.

For the practical realisation of this scheme, effective management is required for the regulation of area management. At the moment there is no such control, and nature management is developing spontaneously.

The most constructive way to overcome this management problem would be to give the green zone the status of a suburban national park. National park status is the most effective form of territory management, combining protection of nature and recreational functions. This combination must be provided in the suburban area as it is the principal recreation site for the town/city's inhabitants; it is necessary here to keep the separate natural complexes in their natural condition.

The suburban national park solution with its mosaic zoning would provide adequate management of the area to cope with increasing nature management. It would provide the opportunity to form an effective ecological scheme for the area as well as creating conditions favourable for recreation.

Landscape Ecological Criteria as a Basis for the Planning of a Suburban Park in Milan

Vittorio Ingegnoli

1 Introduction

It is necessary to integrate planning criteria with ecological principles, and the discipline of landscape ecology seems to provide the best method of achieving this (Naveh, Lieberman 1984, Haber 1990, Zonneveld 1995). An urbanised landscape often has to concentrate the ecological functions of the natural components in isolated patches which require connective corridors for linkage with the surrounding patches, sometimes for a large distance. Areas such as suburban parks are very sensitive to the changing of the ecological mosaic (Ingegnoli 1993). The urbanised landscape of Milan underwent alterations in the last century due to the increase from about 500,000 to 4 million inhabitants over an area of about 1600 km^2. The importance of parks has therefore increased enormously. Three main types of parks are found in the suburban landscape of Milan:

1. the historical parks (e.g. Monza);
2. the natural remnant parks (e.g. Groane);
3. the recently preserved agricultural parks (e.g. Parco Sud). Each of these parks has its own problems, some of which are impossible to solve without a landscape ecology analysis.

As in medicine, the environmental evaluation requires comparisons with "normal" patterns of behaviour of a complex (multi-dimensional) ecological mosaic or "ecotissue". Therefore the principal problem becomes how to recognise the normal state of an ecological system, and how to calculate the levels of alterations to that system. Thus particular ecological indices have been used to analyse the suburban parks of Milan. Examples of such indices are: the biological territorial capacity, or Btc (Ingegnoli 1991) which measures [$Mcal/m^2/yr$] the level of metastability of an ecological system, the SH, or

University of Milan, Faculty of Natural Sciences, Milan, Italy, Professional Studio: 45 via Senato, 20121 Milano, Italy

J. Breuste, H. Feldmann, O. Uhlmann (Eds.)
Urban Ecology

Social Habitat, (Ingegnoli 1993) which measures [m^2/inhab.], the ecological habitat per capita, and so on.

The results show quite a severe situation. The Park of Monza now has an ecological state lower than that at the time of its foundation in the 18th century. The Groane park today has a metastability lower than that of the entire Lombardy region. The Parco Sud is rural-agricultural landscape and is not able to incorporate the high disturbances caused by the urban structures.

2 The Parco Sud Landscape and the Lombardy Region

This case study regards the ecological criteria for planning the central zone of the Regional Parco Sud, around the abbey of Chiaravalle (12th–13th century). The area of the Chiaravalle district measures 2214.3 ha, three quarters of which falls within the border of Milan. First of all it was necessary to find the ecological unit of study (i.e. the local landscape) in this administrative/planning division.

The limits found are:

1. the barriers to the movement of species and humans;
2. the ecotonal belts between the town and its suburban area, and between the suburb and the countryside;
3. the main river corridors. We had to understand the principal functional relationships between the Chiaravalle district and its nearby ecotopes (as the source and sink areas for natural species), the historical linkages among old settlements, the coupling between the hydrologic web and paleo-riverbeds, the remnant corridors for a connection with the other zones of the Park, and, obviously, the types and the recurrence of the patches and corridors which form the local landscape pattern.

The local landscape (5916.2 ha, 7 municipalities) was studied in three periods of time: 1845, 1950, 1997. The difference between the situation in the last century and today is very impressive, although the landscape matrix remains agricultural. This landscape changed from being agricultural 1845 to rural in 1950, to the suburban/semi-urban landscape of today. It is necessary to compare this data with the regional scale if we want to calculate the best level of re-naturalisation. For this purpose we studied three periods of time close to the present: 1878, 1951 and 1993. The Btc (biological territorial capacity) media of the region remained almost constant, not withstanding the enormous changes in the last 115 years: 1878 = 2.04; 1951 = 2.01; 1993 = 1.96 which signifies the good incorporation of the ecological disturbances at that scale. At the same time the Btc of the Human Habitat (HH) changed from 1.79 in 1878 to 1.63 in 1993 and the Btc of the Natural Habitat (NH) from 2.75 in

Table 1. Landscape ecological criteria for the environmental planning of the Chiaravalle district of the Agricultural Park, South to Milan

Landscape indexes	1845	1950	1997	2040
HH-residential %	2.1	7.3	15.9	16.5
HH-subsidiary %	0.1	2.5	12.2	13.0
HH-productive %	87.5	83.3	66.6	53.0
NH-protective %	10.3	6.9	5.3	17.5
Btc (med) $Mcal/m^2/yr$	1.20	1.08	0.95	2.00
NH/HH	11.5	7.4	5.6	21.2

1878 to 2.78 in 1993 (HH = the human mosaic, NH = the natural mosaic of the same 'ecotissue').

These data differ sharply from those of the local landscape. This is normal, but if we want to transform the suburban Chiaravalle district into an agricultural park, we have to change its ecological function in the landscape, with a Btc greater than that of the region. In reference to the entire local landscape, it needs to increase in about 40 years from 5.3% to 17.5% of NH, as shown in Table 1. It is difficult to realise this forecasting data near a big town like Milan in only 40 years. The district of Chiaravalle presents today many disturbances and many symptoms of ecological degradation, but its medium Btc (1.30 $Mcal/m^2/yr$) is greater than that of its landscape (Btc = 0.95). The district is thus moving in the right direction, as we can see by the evaluation of its mosaic and the re-naturalisation which will be possible following the proposed plans.

3 Synthesis of the Main Mosaic

The evaluation of the "vertical" part of the "ecotissue" analysis, linking the ecological pattern of the region, the landscape and the district using landscape ecological indexes, permits a focus on operational quantified planning criteria. The "horizontal" part of the "ecotissue" analysis is equally necessary, and brings about the design of the ecological application. Diverse mosaics characterise the Chiaravalle district: geo-hydrological, vegetation, zoological, agricultural, architectural, technical, etc.

The vegetation mosaic has been mapped showing the remnant semi-natural corridors (mainly hedgerows) and the few patches in which the vegetation maintains some of the characteristics of the regional associations (*Querco-carpinetum, Salicetum-populetum albae)*, dominated by the English oak (*Quercus pedunculata*). Other larger patches represent sites in which the floristic richness is higher than in surrounding areas.

The zoological mosaic was based on the evaluation of the sensitivity (i.e. quality per vulnerability) of fauna evaluated in 6 districts, plus 8 marginal areas, in which patches of greater richness are plotted. The species with higher sensitivity are few, e.g. *Falco subbuteo, Lanius collurio, Picoides major, Asio otus, Martes foina, Myotis mystacinus, Plecotus austriacus.*

The results of the agricultural mosaic are clearly dependant on water pollution. Traditional barley and rice fields are substituted by maize in the eastern part of the district, where canals are polluted. Water meadows, locally called "marcita", are today rare in this mosaic.

The integration of the correlated mosaics of this 'ecotissue' into the main mosaic was not easy to synthesise. The findings of the above-mentioned mosaics of vegetation, fauna and agriculture, were very important, as were those derived from the inter-scale linkages. The distribution of local disturbances, the different roles of groups of animals, the presence of urban patches at the boundaries, the web of canals and ditches, the local points of connection with the surrounding parts of the landscape, the distribution of source and sink patches and corridors, and differences in the landscape matrix, helped to localise the ecotopes of this basic mosaic. The main mosaic of an 'ecotissue' must be a synthetic one, because it is formed by the integration of many other correlated mosaics. Generally it is the vegetation mosaic which facilitates this integration, but in a case like this, where the vegetation types have no connection and are widely invaded e.g. by North American species (*Robinia pseudoacacia, Populus x euroamericana, Phytolacca americana, Solidago canadensis, Erigeron annuus,* etc.), the synthesis is more difficult.

Our mosaic resulted in the formation of 14 ecotopes (Fig. 1) through which it is possible to make an ecological diagnosis and then to propose planning and design criteria. In achieving these objectives, it is absolutely necessary to remember some general principles of landscape ecology, in parallel with the characters and the ecological state of the main mosaic. The most important principles are:

1. "ecotissue" metastability;
2. aggregation-with-outliers;
3. indispensable patterns;
4. historical congruence of patterns (Ingegnoli 1993, Forman 1995).

After the development of a plan and/or a specific design, it is necessary to control the results, using the landscape ecological indexes and compare this information with the characters derived from other scales ("vertical ecotissue").

Figure 1 shows the map of the south Milanese suburbs (1 : 50,000), in which we can see the local landscape unit (thick line) circa 60 km^2, and the main ecological mosaic of the district of Chiaravalle (thin lines, numbered) circa 22 km^2, composed of 14 ecotopes. The ecotopes have been synthesised from

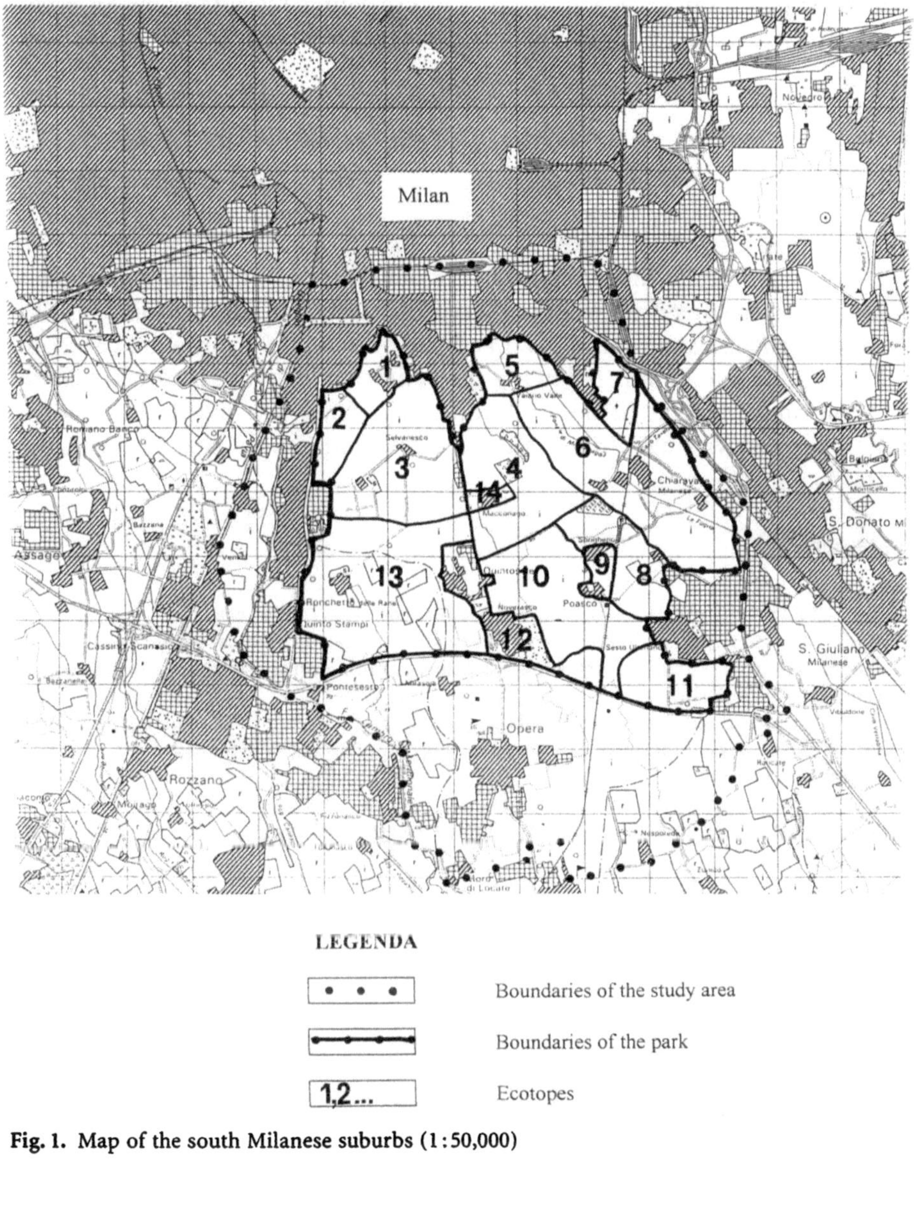

Fig. 1. Map of the south Milanese suburbs (1:50,000)

the following mosaics: vegetation, agriculture, fauna, hydrology, urban, technical infrastructure, boundary effects, with consideration of the inter-scales linkages.

4 Conclusions

The principal aim of the study described in this chapter is methodological. We tried to synthesise the necessity of utilising the most important principles of landscape ecology if we want to plan properly a complex environment like the suburban area south of Milan. Landscape ecological criteria are based on a double scale interaction, following the definition of "ecotissue". It is best to think of an ecological tissue, or "ecotissue", which is a multi-dimensional mosaic formed with a main mosaic and a hierarchic succession of correlated mosaics.

References

Forman RTT (1995) Land Mosaics. Cambridge University Press, New York

Ingegnoli V (1991) Human influences in landscape change: thresholds of metastability. In: Ravera O (ed) Terrestrial and aquatic ecosystems: perturbation and recovery. Ellis Horwood, Chichester pp 303–309

Ingegnoli V (1993) Fondamenti di ecologia del paesaggio. CittàStudi (Utet) Milano, pp 1–278

Haber W (1990) Basic concept of landscape ecology in planning and management. PhysEcol Japan: 131–146

Naveh Z, Lieberman A (1990) Landscape Ecology, Springer-Verlag, Berlin Heidelberg New York

Zonneveldt I (1995) Land Ecology. SPB Academic Pub, Amsterdam

The Influence of NGOs in Preserving Green Places in Buenos Aires City

Ana Maria Faggi

1 Summary

Buenos Aires, the capital of Argentina, with about 3 million inhabitants and an area of 200 square kilometres, has insufficient parks or green places. The deficit of these areas has been calculated as 2400 hectares. The non-conservation programmes related to failed development policies have caused a worrying depletion of such green spaces.

Private initiatives of NGOs which are attempting to enhance park areas are becoming very important. For three years over 30 associations have collaborated successfully to preserve these areas in Buenos Aires. Their work has already made authorities and citizens aware of the need to maintain and restore parks and reserves, not only for recreation but also because of their potential for nature conservation. Some of these NGOs are working to rescue our river coast, which is in poor shape due to uncontrolled settlement.

2 Introduction

Buenos Aires (200 km^2) is a huge city with about 3 million inhabitants. The daily population rises to over 7 million because most of the big companies like to have their head offices in the city. Fifteen thousand buses carry about 10 million passengers daily.

Like many other metropolitan areas, Buenos Aires has serious environmental problems. A subway network of only 43 km carries 700,000 passengers per day, which causes 81% of the air pollution by motor vehicles. Among the problems, the city is overloaded with traffic congestion, air pollution and water contamination, loss of public spaces and lack of green areas.

The World Health Organisation recommends between 10–15 m^2 of public green space per inhabitant for metropolitan areas. The situation in Buenos

Cefybo-Conicet, Serrano 669, 1414 Buenos Aires, Argentina

J. Breuste, H. Feldmann, O. Uhlmann (Eds.)
Urban Ecology

Aires has been deteriorating, especially since the sixties, with now less than 3 m^2/inhabitant. The deficit of green areas has been calculated to be 2400 ha (total size of public green area in Buenos Aires: 808 ha). There is no plan to create new areas as the government lacks interest in the protection of green places. The public spaces have been subjected to inappropriate use: the overwhelming majority of the society is unable to differentiate what is public and what is private; evidence for this is provided by the taking over of public spaces by people for different uses.

The distribution of green places is not homogeneous; most of them are concentrated near the river coast in the east, while many districts have no public green areas (Fig. 1).

85% of the woods near the coast have been built on, invaded, handed over as concessions, donated or sold as a loss since the beginning of this century (Guerrica Echeverria 1995).

The north riverbank was extended through landfills. This reclaimed land was given out as concessions, as a result of this, of the 40 km of riverbank, the public now has access to only 7 km.

In 1993 more than 25 neighbourhood associations, NGOs and professionals founded a forum to avoid this sort of development. Their efforts have made it possible to successfully recuperate, preserve and maintain existing green areas as well as creating new ones through the purchase or recuperation of confiscated land. Their efforts have blocked the construction of shopping centres or skyscrapers in recreational areas. They took part in environmental education in primary schools and in the outlining of the new Constitution, with reference to public property.

Presently they are working to stop some policies of the national and municipal government, which intend to:

- Develop the north railway station area of Buenos Aires giving priority to the building of skyscrapers. The citizens proposal is to turn this land into green areas.
- Construct an island in the La Plata river to build a new international airport. This would bring serious changes in the river sedimentation and cause problems with the water supply of the city.

Many NGOs collaborated to create and maintain a large reserve of about 350 ha, called "Costanera Sur". The location of this park close to the centre of Buenos Aires city could increase the ratio of green area (m^2/inhabitant) from about 3 to 5.15. Many ecosystems have developed on this artificial land enlargement at the riverside in only 10 years.

The reserve is devoted to bird watching, wildlife conservation, recreation and environmental education. The diversity of biotopes such as lagoons, wetlands and forests guarantees the presence of an interesting fauna.

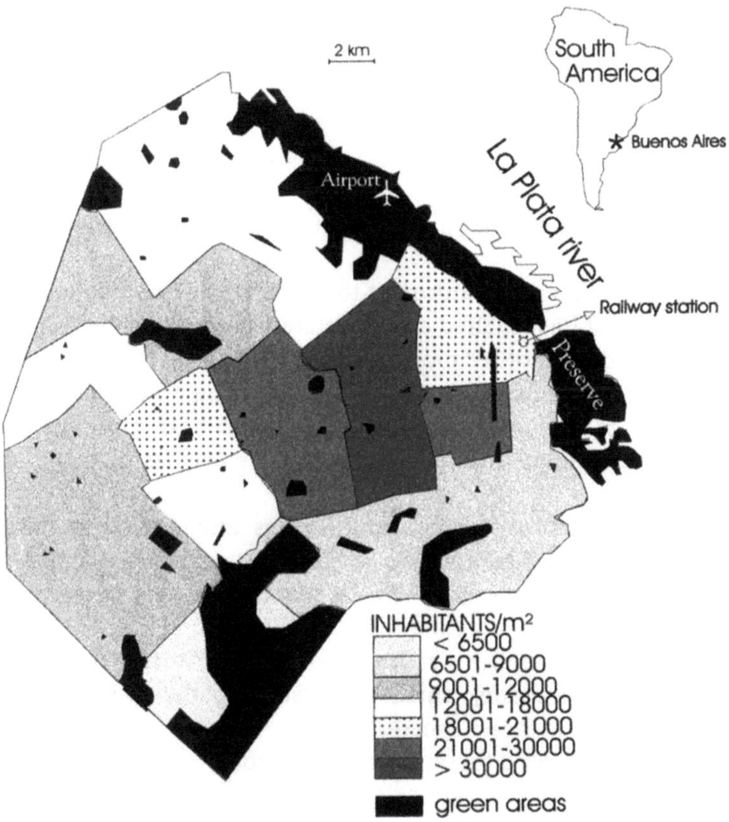

Fig. 1. Green areas distribution in Buenos Aires city

Vegetation monitoring has been carried out since the creation of the reserve and shows the need for restoration and rehabilitation of some ecosystems in order to reverse habitat degradation (Faggi and Cagnoni 1987, Faggi 1996, Cagnoni et al. 1996). Conservation values have been reduced by invasive exotic species such as *Ligustrum lucidum, Salix fragilis, Ailanthus altissima* and *Melia azedarach*. These species displace native trees like *Salix humboldtiana* and *Tessaria integrifolia,* altering the ecological dynamics of the area.

Restoration needs to be implemented effectively in order to return these degraded systems to some form of cover that is protective, valuable in a conservation sense and aesthetically pleasing.

A multi-disciplinary team was formed, composed of two botanists, a geomorphologist, a landscape architect and a conservationist. This team drew up a proposal for a solution to these problems in successive stages using a minimum of materials and labour.

The general principle of maintaining structural diversity should be kept in mind, because structure is closely correlated to organisms and ecological processes.

Through the proposal the following advice could be given to the administration of the reserve:

1. The water level in each lagoon should be brought to its original level by additional water supply.
2. It is not necessary to build an embankment to preserve the riverbank from surges because the coast has reached a stable state, shown in particular by the appearance of reeds and other pioneer plants.
3. Big concrete blocks from building demolition, now acting as waterbrakers in the south sector of the reserve, should be coated with concrete to improve their aesthetic quality and enable them to be used as solaria and observation sites;
4. Accelerating the propagation of reeds and rushes along the waterline by planting rhizomes of *Schoenoplectus californicus* and other kind of *Cyperaceae* as well as *Juncaceae*, will help to consolidate the substratum and filter the polluted water.
5. Re-creation of riverside shrubbery and restoration of riverine forest to increase tree density and shade, which will also help the spontaneous colonisation of hydrophytic forest species.
6. Development of a xeromorphic forest on squeletic soils.

3 Conclusion

During the military government (1976–1983) the participation of the citizens in solving problems related to green areas was very low; this situation could change with the restoration of democracy.

Nowadays there are new environmental problems to be considered as well as the lack of resolution of many previous ones because the city demands the strict fulfilment of the authorities' duties in conjunction with the committed participation of the citizens.

As regards green spaces, the decided joint action between citizens and urban planners can reverse years of misgovernment, usurpation and uneven distribution of privileges, giving the city the image its inhabitants deserve.

References

Cagnoni M, Faggi A, Otero, P (1996) Comparación de la vegetación en el Parque Natural Costanera Sur después de una década. Res. 25 Jornadas Argentinas de Botánica

Faggi A, Cagnoni M (1987) Parque Natural Costanera Sur. I. Las comunidades vegetales. Parodiana 5(1): 135–159
Faggi A (1996) El monitoreo de vegetación como base para la restauración ecológica. Actas II Cong. Universidad de Flores
Guerrica Echeverría O (1995) Palermo y después. Buenos Aires

Workshop: Concepts, Strategies, Management 5.3

Post-industrial Landscape

Wolfram Höfer

In this chapter I shall deal with concepts and strategies concerning nature in urban conditions in which the aesthetic perception of landscapes is of great importance. I would like to draw your attention to some aspects of the philosophical background of those strategies and aesthetic values because I believe an understanding of the ideological context of our action can improve professional practice. I will address some characteristics of utopian landscapes from a central European point of view and how those characteristics change with the movement of our society towards a post-industrial situation.

After the period of enlightenment with the beginning of the industrial revolution the concept of a natural agrarian landscape was emphasised as a counterweight to an industrial society. Landscape is beautiful harmony as opposed to ugly industry. I will develop the thought that some visual elements of this ideal of utopian landscapes change, rather than the structure of the ideal itself. My conclusion will be: post-industrial landscape is idealised nature *and* idealised industry (compare in more detail Höfer 1993). First, I will give some aspects of the philosophical roots of the idea of landscape.

1 Landscape in the Industrial Situation

In the middle ages nature as a topic did not exist. Man was part of God's creation; there was no aesthetic distance to nature. With the beginning of the period of enlightenment, man steps out of creation and becomes an independent thinking individual. Man is the opposite of nature.

The individual refers to its intellectual capabilities. Justification for action is not fulfilling God's will anymore, rather the independent will of the individual. With the emerging of the idea of nature, we see that this idea has an enlightening, liberal aspect.

In the tradition of landscape architecture this aspect is seen in the English landscape garden. In fact, the landscape garden has been a main step towards

Munich Technical University, Chair of Landscape Architecture and Planning, D-85350 Freising-Weihenstephan, Germany

J. Breuste, H. Feldmann, O. Uhlmann (Eds.)
Urban Ecology

recognising the surrounding space as an aesthetic landscape. Those gardens are constructed landscape paintings, in which the Arcadian utopia can be seen. There is no need to work, Gods and humans should enjoy themselves. It is important to keep in mind that those rural scenes had no relation to the real situation of the rural population. Landscape paintings were not a copy of reality but an expression of utopia. Therefore the *idea* of nature could become a refuge where present social constraints had no importance.

At the beginning of the eighteenth century paintings were built as landscape gardens. Clumps of trees, meadows and historic elements or monuments were composed as walkable pictures. Walking in those pictures made the surrounding space come into view as a landscape (v. Hollen 1991). The real landscape became the location of utopian escape. It is important to note that the supporters of this ideal were members of the upper class and were not involved in agrarian production.

I have talked about the liberal aspect of nature and landscape. On the other hand, with the process of enlightenment, ideas were also developed, in which lay the justification for activity not only in the individual, but also in nature itself. Nature may be seen as an original, higher law, a law which is not part of the rational discourse.

This position has been developed by the French philosopher Jean Jaques Rousseau. He criticised the French absolutist society as 'over-civilised'. From his point of view the original natural condition was good and the human being was good in this natural condition. Every step away from nature is bad (because nature itself is good). We see that a 'higher nature' becomes a justification for criticising existing social and political conditions. This cultural pessimism is an indication of a conservative position. The idea of nature is, therefore, a contradictory mix of conservatism and progress (Eisel 1982).

For Rousseau, reflection on nature could form a better society, but history itself has no determined goal. The German Philosopher Johann Gotfried Herder goes one step further. He sees a continuity of progress towards humanity as a goal of history. For him every chapter of history can be regarded as a cultural unit. Every individual, as part of a single unit, has the task to develop the unit successfully, so that the unit will make a positive step towards the goal of history. The success of each unit can be seen if the individual makes the best out of the environmental conditions through agriculture and craftsmanship. Looking into history gives Herder guidelines for a good life and a good future, history has a moral value (Eisel 1982).

Geography, as a field very close to our profession, transfers this moral value into the landscape itself. Each land is seen as an organic body which is part of the development of culture as a whole. One can 'read' in the landscape how farmers and craftsmen have made the best out of the natural conditions through concrete work (i.e. hand labour). A landscape gains its regional identity. It is important to note that in such a perspective, industrial abstract work does not contribute to the identity and harmony of a landscape. Therefore Geography, and later landscape planning, had a tendency against urban structures and industrial developments. A beautiful harmonic landscape is then defined as not industry and not city.

I summarise four main characteristics of utopian landscapes:

1. It requires an area that looks like a cultural landscape, which stimulates the mind
2. Landscape has a moral value
3. Landscape is a counterweight to the complex industrial society
4. This utopia makes it possible for the individual to feel part of a larger context; landscape has a regional identity.

2 Landscape in the Post-industrial Situation

A main aspect of regional identity is the definition of a local character against complex and unclear modernism. In a modern situation it is not possible to define one's position as a part of a clear unit. An urban industrial area looks like a symbol of the complex industrial society, a symbol of all those aspects which are the reason to seek refuge in a pastoral region. At the beginning I said, 'post-industrial landscape is idealised nature *and* idealised industry.' Now I want to show which aspects of the aesthetic landscape undergo changes and how it is possible. I will take the industrial area Ruhrgebiet (located in the west of Germany) as an example.

The sociologist Karl Rohe (Rohe 1984) describes a regional identity for the Ruhrgebiet which is connected to the conditions of industrial production of steel and coal. Working and living conditions formed a particular social structure. If we take the work of miners, this work was heavy and dangerous. The work was organised in small groups and it was essential for each member of a group to rely on his colleague producing a special solidarity. Today, the number of jobs in the coal and steel industry are in decline, and those working groups of comrades belong to the past. We might also look at living conditions. Settlements for workmen were not developed as additions to existing towns, rather they were located next to industrial plants and mines. The male population of an estate usually worked in the same factory. The neighbourhood was a place of particular solidarity, it was a clear and complete unit. Today, urban development destroys those structures, local units disappear through this 'urbanisation'. Obviously, changing economic conditions endanger regional identity, social structures and local identification disappear. However, this process opens up the possibility to see the Ruhrgebiet as a landscape, as a unit of land and people.

I have said that with the emergence of Geography, the anti-industrial aspect was developed to the full. Landscape was the counterweight to industry. Industry was abstract control of nature, distanced from nature, without any sense of meaning, not harmonic and not aesthetic. The emergence of regional identity in rural areas reflected these contrasts. Regional identity was a counter movement to the equalising tendencies of industrialisation and the individual tried to find refuge in those regional characteristics. In contrast, in the Ruhrgebiet regional identity does not refer to agrarian structures but to industrial production. Even if industrial production always has equalising tendencies, it is able to create characteristics of an everyday environment [Lebenswelt] of its social actors, which may take different regional shapes. In the Ruhrgebiet those social peculiarities were responsible for a local identity. The 'attack' on this identity is not forced by equalising tendencies of industrial production, as it is an essential aspect of it, but by equalising tendencies of a computerised service-oriented society. It was mentioned how economic changes alter local identity; industry can gain the qualities of cultural tradition and 'natural' conditions, which had been assigned to landscape up to now.

As a second aspect of utopian landscapes I mentioned its moral value. One can read in a landscape how concrete human activity has shaped a land into an harmonic unit, landscape as an example of a good life. In post-industrial conditions industrial work is seen as concrete work, no longer just farming and craftsmanship, but also coal and steel become a determinant for culture and a good life. With the transfiguration of industry, industrial work can be seen as concrete work as opposed to abstract computer work, therefore industry may naturally appear aesthetically anti-industrial. Concerning the development of an aesthetic appreciation of landscape I described the landscape garden as an important step to transfer the utopian landscape into real space. Supporters of this ideal were those sectors of society which were not directly involved in agrarian production. This is also true for the reception of industry as landscape. The social reality of the supporter of post-industrial landscapes is the service oriented society and no longer industrial production. This means that also in the post-industrial landscape the picture of idealised industry is conveyed as an historical element and a symbol for departure into the golden age of Arcadia. With reference to Rousseau, I described the aspect of nature to be a justification for an alternative draft of society. With regard to post-industrial landscapes, this means that the industrial society is considered as closer to the original, natural condition than the service oriented society. Thus an idealised industrial past can be taken as an alternative draft of society.

3 Summary

I described which structural elements of an idealised industry show similarities to idealised nature. Consequently the structure of the ideal of utopian landscapes does not change, but it does not refer anymore to agrarian Arcadian images *alone* because there is a new opposing equalising structure. The professional dealing with urban landscapes must realise that with the changing of our society into a post-industrial situation, utopian landscape as a symbol of alienation and as a counterweight to a complex society is still a sociological need. Also we should be aware that even in a post-industrial situation, dealing with landscape means dealing with a contradictory mix of conservatism and progress.

References

Eisel U (1982) Die schöne Landschaft als kritische Utopie oder als konservatives Relikt: Über die Kristallisation gegenerischer politischer Philosophien im Symbol "Landschaft". Soziale Welt 33(2): 157–168

Höfer W (1993) Postindustrielle Landschaft – Eine Diskussion am Beispiel der IBA Emscherpark. Diplomarbeit am FB 14 TU Berlin. Berlin

Hollen B v. (1991) Vom Bild zur Natur; Der Ursprung der landschaftlichen Wahrnehmung im Landschaftsgemälde. Diplomarbeit am FB 14 TU Berlin. Berlin

Rohe K (1984) Regionalkultur, regionale Identität und Regionalismus im Ruhrgebiet: Empirische Sachverhalte und theoretische Überlegungen. In: Lipp W (Hrsg) Industriegesellschaft und Regionalkultur: 123–153. München

Goals and Measures of Nature Conservation and Landscape Protection in Urban Cultural Landscapes of Central Europe – Examples from Leipzig

Jürgen Breuste and Sandra Wohlleber

1 Introduction

Over three quarters of the European population live in urban areas. This living space directly moulds the relationship of many people to their environment. Nature conservation and landscape protection in the urban cultural landscape are not only of great importance, but also require a modified approach which clearly differentiates itself from the protection of rural landscapes (Breuste 1995).

The main focus of nature conservation and landscape protection (that has been used up to now) must be examined and adapted to the city areas concerned.

The main aim of nature conservation and landscape protection in urban areas cannot be the conservation of the existing landscape and the reconstruction of pre-industrial landscape as this conservation approach does not fit in with the conditions of an ever changing society.

It is counterproductive to protect nature from people in urban areas thereby preventing their contact with it. Nature requires people's support which will depend on the benefits they receive, especially in city areas. Although intervention by people always causes changes, this must not always be seen as having a negative effect. In cities habitat conservation should not target at specific species or the provision of undisturbed areas for biotopes, rather it must be directed at people and their needs.

For more than 20 years the nature conservation laws of the Federal Republic of Germany (FRG) have encouraged nature conservation and landscape protection in order to ensure the sustainability of the basics of life of the people and the requirements for their recreation in nature and landscapes (Bundesnaturschutzgesetz § 1(1) (1976/1993). Those making the laws follow an anthropocentric approach to nature conservation whose effective transla-

UFZ Centre for Environmental Research Leipzig-Halle, Department Urban Landscapes, Permoserstr. 15, D-04318 Leipzig, Germany

J. Breuste, H. Feldmann, O. Uhlmann (Eds.)
Urban Ecology

tion through the states of the FRG and local authorities must emphasise different points (Bundesamt für Naturschutz 1996).

2
Nature Conservation and Landscape Protection in the Transformation Process of East German City Regions – Research in the Urban Cultural Landscape of Leipzig

The towns and cities especially (as urban areas and administrative planning and action representatives) are confronted with the implementation of this anthropogenic nature conservation approach in the new regions of the FRG. It is not only that the new legal rules must be applied but also that the urban areas themselves are undergoing enormous changes.

Research in the city region of Leipzig shows the far-reaching economic and landscape transformation process (Breuste 1994). The change-over here from a compact *old-industrial cultural landscape* with:

- high spatial concentration of built uses and traffic infrastructure (dominated by public transport and railways),
- high density of inhabitants in a small city centre and
- administration of the urban processes in the realm of a compact city centre, to the *dispersed post-industrial urban landscape* with:
- higher area dispersion of built-up uses and infrastructure facilities (mainly for private vehicles),
- fast expansion of urban areas and the city population away from the city centre, and
- spatial administrative responsibility lagging behind the suburbanisation process.

All of these changes express themselves in the demand and use of land. It is mainly agricultural areas which are affected by this and some scattered green areas which have not yet been built on- such as fallow land, gardens or scrub land.

In order to be able to influence the quantity and quality of land use, nature conservation and landscape protection in urban areas must portray a basis for specialised and general planning.

3 Instruments for Nature Conservation and Landscape Protection

3.1 General Planning and Models for Land Use

Through the creation of land use planning and planning for the Federal states and, thereby, the building up of regional planning, goals and measures for the development of the Federal states and their component regions were initialised in the FRG (ROG 1993, Schmidt-Eichstaedt 1993).

The *Landesentwicklungsplan* (Regional Development Plan) (Freistaat Sachsen 1994) as well as the *Regionalplan Westsachsen* (Regional Plan of West Saxony) give a structural and economic basis to serve as the goal and models for the urban area of Leipzig.

The *Landschaftsrahmenplan* (Regional Landscape Plan) which is included in the *Regionalplan* (Regional plan) of the planning unit of West Saxony, gives a stepwise improvement of the quality of the environment as a model for urban landscapes. Urban landscapes should be structured and revived through a network of green areas which will create a sufficient possibility for the people to experience nature and relax in their immediate living surroundings (Regionaler Planungsverband Westsachsen 1996). Regional green corridors should contribute to the connection of the inner-city green areas with the open landscape. Priority and reserved areas for nature and landscape portray a model for an 'Ecologically combined system'. The regeneration of landscape potential and the unspoilt development of flora and fauna are development goals about which all forms of use should be carefully considered.

3.2 Nature Conservation and Landscape Protection as Part of Local Plans

3.2.1 *Mapping of City Biotopes*

Representative biotope mapping (Workgroup 'Methods of biotope mapping in populated areas' 1993) was prepared on the basis of land-use in order to be used as a foundation for landscape planning area coverage. Its goal is the depiction of use of landscape in line with its potential (basis for recommendations) and the development of a basis for the minimisation of uses of land which are detrimental to nature (basis for regulation).

The differences between the procedures in the 'Inner-city' (predominantly urban-industrial nature and gardens) and the 'outer areas' (remaining natural landscape and agricultural land) can be seen clearly. These differentiated bases for observation lead to new points of emphasis (the protection and improvement of the provision of existing green areas, recreation uses, environmental education). The Leipzig biotope map is full of examples of these aims because the species and biotope protection do not alone give the value of the area and the importance of the protection efforts. What is especially important here is the contact of the city dweller to nature.

3.2.2
Landscape Plan (Landschaftsplan) and its Reference to the Use of the Area

Between 1992 and 1996 the design of Leipzig's landscape plan was worked upon. The plan aims to improve conditions for recreation; protect and recreate flood plains, expose built over river channels in the city; improve the climate of the city (through provision of open areas); protect soil and ground water and protect sufficient large areas for balancing and substitution measures for further intervention in the landscape (Stadt Leipzig 1996).

Due to the landscape planning having to ensue parallel to the development of the area land-use plans (Stadt Leipzig 1994) some of the results of work carried out are included in the land-use plans (FNP). Fundamental planning goals of the land-use plan are (as also in the landscape plan) security, protection and revitalisation of the natural landscape structure. Today's existing landscape outline plan includes some diverging statements on land-use from the FNP.

The fact that land use plans and also landscape plans can only be statements for the city's administrative area, clearly shows their limited effectiveness regarding the urban cultural landscape lying outside the city boundaries.

3.3
Co-operation and Supplementary Activities

3.3.1
Co-operation between Local Authorities

A co-operative approach in which the city and its surrounding area become integrated parts of the urban landscape can improve through overcoming local barriers to the goals of nature conservation and landscape protection.

The city of Leipzig is trying to carry out activities, together with the surrounding areas, to help the development and rehabilitation of the cultural landscape in the 'Grünen Ring' (green belt) around Leipzig through the 'City and Surrounding Countryside Conference'. The goal is to organise neighbouring communities/local authorities to work together in a larger landscape region, securing a network of important biotopes and attractive recreation places through concentration on key points which are of most use to the whole region of Leipzig (due to limited capacity).

A further example is the 'Zweckverband Partheaue'. The members of the association – towns, cities and communities- allowed common, sustainable and co-ordinated planning to be carried out in the the cultural landscape of the river Parthe and its flood plains (Stadt Leipzig 1995).

3.3.2
Co-operation between Local Authorities and Citizens

Before the reunification of Germany in 1990, the project 'Neue Ufer' ('New Riverbank') was founded. In this project architects, artists, town planners, landscape architects and water experts were engaged in order to re-open the Leipzig millrace. This project is now integrated into the planning and work of the Leipzig city authority, as well as being made known through publications and being supported through these (Stadt-Kultur Projekt 1995).

An important forum for the co-operation between politics, administration and citizens is Leipzig's Local Agenda 21, work on which began in early 1996 (Leipziger Agendareport 1/97).

The environmental quality targets form the first important step in the Local Agenda 21. Leipzig is one of the first cities in Germany to have concrete environmental quality targets in its approach to the city's development. The degree to which the objectives are fulfilled will be checked using recognised standards. The goals range from species and biotope protection to the support of the contact of the population to nature.

4
Aims and Measures for Nature Conservation and Landscape Protection in Urban Cultural Landscapes

Leipzig's example shows that there are needs as well as approaches in order to practice a new understanding of nature conservation and landscape protection. The practice not only conforms to the legal regulations but goes further in the choice of instruments which extend beyond those of nature conservation and planning and building laws.

Such modified actions are required in order to transplant nature conservation and landscape protection meaningfully into the cultural landscape.

General aims of nature conservation and landscape protection are:

- Protection of all living media which are particularly threatened,
- Structural diversity and extensive landscape differentiation,
- Refraining from all unnecessary operations in nature and landscape,

- Accessibility into various ecosystems and landscapes,
- Equal value inclusion of the landscape as an identification possibility in the process of surburbanisation,
- Equal value, equal rights and mutual functional supplements to different landscape types,
- Integration of voluntary nature conservation, from groups and societies to citizens, in area specific and regional activities in administrative planning and landscape development.

When dealing with measures for the development of the urban cultural landscape the following are of vital importance:

- Sponsorship for contact to and acceptance of nature,
- Public work as well as demand for the active shaping of nature and landscape through the population,
- Nature and landscape protection as general companions to users,
- Secondary protection for priority areas,
- Regionally differentiated focal points for nature conservation and landscape protection,
- Opening up of the possibility for spontaneous nature in building areas,
- Network of open spaces,
- Functional integration of building works in the ecosystem.

At the same time it is particularly important to maintain and support natural processes, continuation of historical uses of open spaces, the differences between places, different intensity of use of areas, a diversity of typical elements of urban landscape, and large inter-dependant open spaces (Breuste 1997).

References

Breuste J (1994) Ökologische Aspekte der Stadtentwicklung Leipzigs. Geogr Rundsch 46: 508–514

Breuste J (1995) Kulturlandschaft Stadt und Umland – Wandel und Perspektiven. In: ANL Vision Landschaft 2020, Laufener Seminarbeiträge 4/95. Laufen/Salzach

Breuste J (1997) Der suburbane Raum als neue Kulturlandschaft. In: Breuste J (ed) 2. Leipziger Symposium Stadtökologie: "Ökologische Aspekte der Suburbanisierung". UFZ Centre for Environmental Research Leipzig-Halle, Leipzig pp 3–16

Bundesamt für Naturschutz (1996) Landschaftsplanung am Scheideweg. Natur und Landschaft, Themenheft 11/96

Bundesnaturschutzgesetz (BNatSchG) vom 20.12.1976 (BGBl. I S. 3574), i.d.F. vom 12.03.1987 (BGBl. I S. 889), zuletzt geändert durch das Gesetz vom 22.04.1993 (BGBl. I S. 466)

Freistaat Sachsen, Sächsisches Staatsministerium für Umwelt und Landesentwicklung (1994) Landesentwicklungsplan Sachsen, Dresden

Stadt Leipzig, Dezernat für Planung und Bau (1994) Flächennutzungsplan Stadt Leipzig, Leipzig

Stadt Leipzig, Amt für Umweltschutz (1995) Umweltbericht 1995, Leipzig

Stadt Leipzig, Dezernat für Umwelt, Ordnung und Wohnen (1996) Landschaftsplan Leipzig, unveröff. Planungsunterl., Leipzig

Leipziger Agendareport (8.05.1997) (1997) Nr 1, Leipzig
ROG Raumordnungsgesetz (1993) in der Fassung vom 1.5.1993
Regionaler Planungsverband Westsachsen (1996) Regionalplan Westsachsen, Leipzig
Schmidt-Eichstaedt G (1993) Städtebaurecht. Kohlhammer, Stuttgart Berlin Köln
Stadt-Kultur-Projekt Leipzig (1995) Neue Ufer 4, Rahmenpläne zur Öffnung von Elster- und Pleißemühlgraben, Leipzig
Workgroup "Methods of biotope mapping in populated areas" (1993) Flächendeckende Biotopkartierung im besiedelten Bereich als Grundlage einer am Naturschutz orientierten Planung. Natur und Landschaft 10/93: 491–526

The Development of a Simple Geographic Information System for Use in Urban Ecological Landscape Analysis and Decision-Making

Andy Millard and Claire Freeman

1 Introduction

Urban areas are now recognised as important for the conservation of biodiversity. Apart from patches of relic, semi-natural habitats, a variety of uniquely urban habitats can occur. The latter, which may include areas like sports grounds, allotments and derelict industrial land, are not usually designated sites but can make a significant contribution to the biodiversity and amenity value of the city. Nonetheless, in the current economic climate many such sites are perceived as having valuable development potential and in these cases the lack of any designation frequently results in site development with no consideration for ecological or amenity value. Two factors contributing to this outcome are the paucity of ecological and amenity data on such sites and the effect this has on the subsequent decision-making process. Key reasons in the UK for the former are the time and resource constraints that prevent local authorities from using their ecologists, if they have any, to undertake surveys of all these sites.

There is thus a need for a methodology that can be used by non-ecologists, primarily planners, to identify a site's principal ecological and amenity characteristics and to enable them to recognise the type of site that warrants a more detailed ecological survey by a specialist. Data collected using such a methodology must also be stored and managed in a comprehensive, up-to-date, flexible and accessible database, a function for which a Geographical Information System (GIS) is admirably suited. A study was undertaken in Leeds, a large metropolitan city in the north of England, to develop such a methodology (Freeman 1997). This paper is concerned primarily with the development of the GIS component of the methodology.

Faculty of Health and Environment, Leeds Metropolitan University, Brunswick Terrace, Leeds LS2 8BU, Great Britain

J. Breuste, H. Feldmann, O. Uhlmann (Eds.)
Urban Ecology

2 GIS and Land Use Evaluation

GIS can be defined as a computer-based "system for capturing, storing, checking and manipulating data that are spatially referenced . . ." (Maguire 1991). Planning and land-use evaluation is an activity that is particularly suited to GIS. This has been recognised by local authorities in the UK, of which 47% have now installed a GIS (Ireland 1994).

In the environmental field the capacities of GISs have already been recognised and integrated into practice by a number of national and regional organisations involved in mapping and analysing natural vegetation and ecosystems. These have included data on national landcover for countries like the USA (Steffenson 1996) and UK (Department of the Environment 1993); more specific categories of land, e.g. UK common lands (Higgs et al. 1994), as well as regional databases like the Cornwall 'Life' Project (Cornwall Wildlife Trust 1994).

Most of these focus on coverage of relatively large areas of predominantly rural land, rather than on detailed, local, land-use cover and in the UK there is, as yet, no comprehensive land-use database available that can be used for the analysis of urban areas at a local scale.

3 Urban Greenspace in Leeds

Leeds is statistically one of the UK's greenest cities, containing within its boundaries 140 square miles (363 km^2) of greenspace, most of which is green belt land on the urban fringe (Leeds City Council 1993). However, as in most major cities, there is continuing pressure for development on open space in the city. Strategic Planning Guidance issued by central government (Department of the Environment 1989) directed that, the Leeds Unitary Development Plan "should provide land for the completion of 28,500 dwellings in the period between 1986 and 2001".

4 Site Survey and Evaluation

To test the methodology thirty sites in Leeds were selected for field study, all with naturally regenerating habitat and all threatened by development. They ranged from inner city derelict sites to some from the urban periphery.

Development information for each site was obtained from the Unitary Development Plan and a search of planning applications at Leeds City

Council. Field surveys comprised site mapping, the recording of easily quantifiable ecological attributes (e.g. area, age and identifiable habitat components, based on Wittig, Schreiber 1983) and evidence of amenity use. Criteria contributing to each of these three main data categories were then scored on a simple 5-point scale to arrive at overall ecological, amenity and development potential values for each site.

5
Producing the GIS Database

The GIS application selected was ArcView® because of its relative user-friendliness, superior mapping and paper copy production facilities and comparative economic price. In addition, it is the intended future standard GIS for Leeds City Council and is derived from the more substantial but less user-friendly ArcInfo, one of the most frequently used GISs in UK local authorities (Ireland 1994).

Baseplans were produced by scanning a paper map for the Leeds area and Planning Application maps for individual sites into Tiff image files. Initially a composite map of Leeds was produced. The relevant Tiff file was added as a theme to provide a baseplan and then a second theme was created as a shape file to mark the location of all the sites in Leeds. Numerical data on ecological, amenity, and development-potential values, together with key descriptive information on the land-use and type of proposed development for each site were then entered into the associated attribute table.

A number of sample sites were selected. For each, the appropriate Tiff file was added as a baseplan and then the digitising drawing tool used to draw the boundaries of the different habitat types on the site as a series of component features of a second theme. Additional data were added to the record of each feature in the attribute table as appropriate. An additional theme showing the extent and nature of any development plans was then added if appropriate. The composite map of Leeds and sample sites were each saved as separate Arcview project files.

6
Use of the GIS Database

Within an ArcView® map (known as a 'View'), themes effectively operate as overlays which can be individually selected and displayed in any order e.g. a theme mapping development proposals can overlay a habitat theme to illustrate the area of habitat that would be lost. Sites can be selected and/or sorted according to available criteria (e.g. ecological, amenity, development value etc.) and relevant features highlighted on individual site maps and values displayed on the composite map. There is also a facility for rapidly accessing different forms of site data held in other applications e.g. photographic images and detailed planning applications.

In this way the GIS provides an accessible, effective, site-level database which can be easily interrogated and updated by planners to assist with more effective decision-making on sites with ecological, amenity and development values. By easily integrating data from other sources within local government it can also facilitate the more interdisciplinary approach which is now recognised as essential for sustainable urban development in the future. These qualities also suit it to other functions outside the immediate concerns of planning e.g.: to serve as a foundation database for urban biodiversity and thereby assist in the production and monitoring of biodiversity species and habitat action plans; to facilitate landscape ecological studies (which, in the UK, have focused, up to now, chiefly on the rural environment) on the functioning of urban ecosystems.

7 Conclusions

The database as established at present is relatively simple and does not exploit fully the potential of ArcView®. Replacing Tiff file baseplans with Ordnance Survey vector data would allow habitat dimensions (e.g. length of hedgerow and area of scrub) to be incorporated, thereby providing additional criteria for comparison between sites. It would also enable distances between sites to be displayed and, say, the nearest example of a particular habitat type to a site to be identified. Addition of the ArcView® Spatial Analyst extension would also allow data in a cell-based, raster format (e.g. derived from aerial photographs of an urban area) to be incorporated and analysed.

Nonetheless, in its present state the GIS has made significant progress in realising its two principal aims. Firstly, it uses data that are easily collectable by, and comprehensible to, individuals without specialist training in ecology and yet also offers scope for more specialised data input at a later stage. Secondly, the choice of ArcView®, which operates on a Windows™or 'Windows™-like' platform with easily understandable graphics and mouse operations, means that users without extensive computing expertise can familiarise themselves with the system more rapidly than with previous GISs. When Leeds City Council have completed the installation of ArcView® it will then be possible to test the system in an operational planning department.

References

Cornwall Wildlife Trust (1994) Cornwall Life Project: Project Specification. Cornwall Wildlife Trust, Truro

Department of the Environment (1989) Strategic Guidance for West Yorkshire: Regional Planning Guidance 2. HMSO, London

Department of the Environment (1993) Countryside Survey 1990. Department of the Environment, London

Freeman C (1997) The development of a landuse decision-making model for use on sites with naturally regenerating habitat. Unpublished Ph.D. thesis, Leeds Metropolitan University

Higgs G, Aitchison J, Bracken I, Crosweller H, Jones P (1994) The national GIS demonstrator for common lands for England and Wales. J. of Env. Plan. & Manag. 37(1): 33–49

Ireland P (1994) Local Government GIS survey reveals all. Mapping Awareness and GIS in Europe. Nov.:14–15.

Leeds City Council (1993) Leeds Unitary Development Plan: Revised Draft. Department of Planning, Leeds City Council.

Maguire DJ (1991) An overview and definition of GIS. In: Goodchild MF, Rhind DW (eds) Geographic Information Systems. Longman, Essex

Steffenson J (1996) Uncle Sam map haven. Planning, July:14–17.

Wittig R, Schreiber, KF (1983) A quick method for assessing the importance of open spaces in towns for urban nature conservation. Biological Conservation, 26: 57–64

Regional Parks as Tools for Protecting the Landscape in the Suburban Development of Berlin

Manfred Kühn

1
The Aims, Strategies and Substance of Regional Parks

The joint planning of the Federal states Berlin and Brandenburg seeks to establish a polycentric settlement structure for minimising urban sprawl, land use and car dependence, which are criticised as being non-sustainable factors in the development of settlements. In the suburban space between the metropolis of Berlin and its surrounding countryside in Brandenburg good conditions for a sustainable development of settlements exist. As a result of the political divide between East and West -and contrary to most other European metropolitan regions- a sharp *contrast between city and countryside* has remained until today, with 'the Wall' acting as a barrier to development as in the Middle Ages. This has resulted in high-density housing developments being found right next to rural villages.

The aim of a polycentric settlement structure contradicts the actual process of *belated suburbanisation* ("nachholende Suburbanisierung") since the reunification of Germany and Berlin. Since 1989 large-scale industrial centres, shopping centres and residential areas have been built in the suburban area around Berlin. Most observers assume that the migration of private households and business will increase over the next few decades. Predictions for the year 2010 speak of 200 to 300 thousand people (7% to 9% of the total population today) leaving Berlin and primarily building detached houses on the cheaper land in Brandenburg.

The drafts of the "Joint State Development Programme" (Lepro), "Joint State Development Plan for the narrow Berlin region" (LEPeV) and initial Regional Plans are planning a chain of regional parks around Berlin as a *green belt between the city and its hinterland.* This, incidentally, is a very old planning idea from the 19th and early 20th century (e.g. Howard's garden-cities, Janssen-plan), which can be found in several other German and European city regions (e.g. Milan, Vienna, Frankfurt a.M., Leipzig) (Fig. 1).

Institute for Regional Development and Structural Planning (IRS), Flakenstr. 28–31, D-15537 Erkner, Germany

J. Breuste, H. Feldmann, O. Uhlmann (Eds.)
Urban Ecology

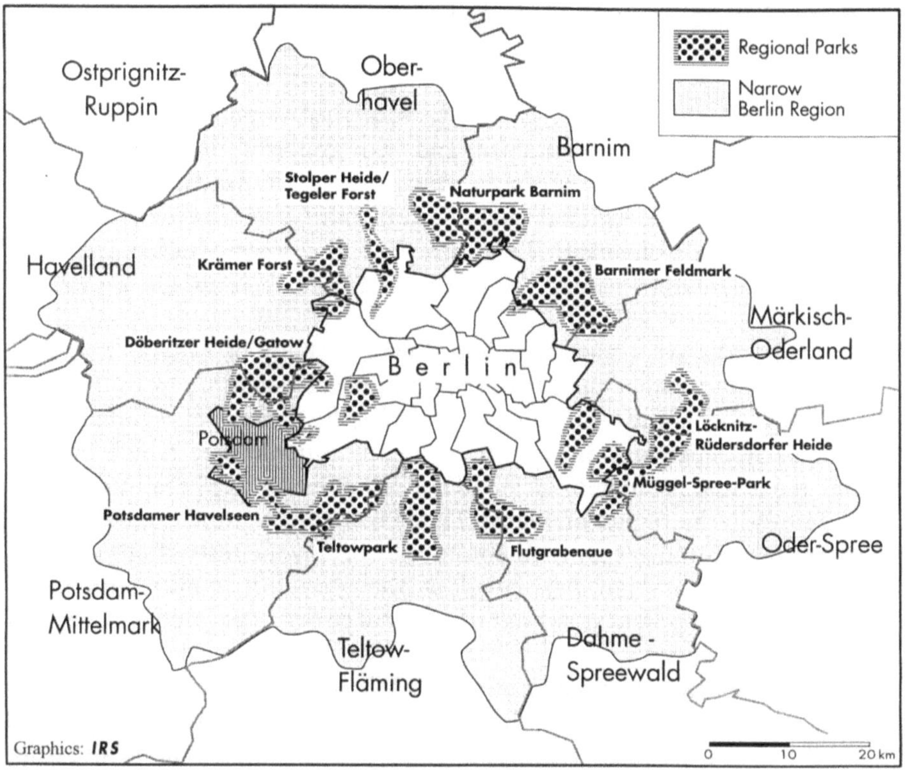

Fig. 1. The chain of regional parks in the narrow Berlin region

Taken as a new category of open space, regional parks can not be compared to either municipal gardens or parks, or to nature reserves (e.g. nature parks). Rather, regional parks are understood to be man-made landscapes which try to integrate land use with conservation of the suburban landscape. Due to their location close to towns, their most important intended use is recreation (bicycle-tourism, catering). The innovative and sustainable aspects of regional parks are seen to offer a strategy of protection and integration of ecology and economic land use.

So much for the theory, what about the practice?

2 Problems and Deficits in Implementation – the Case of the Teltowpark

The Teltowpark is located between the southern border of Berlin and the city of Ludwigsfelde and is currently being scientifically evaluated by the IRS. It will be used as an example to demonstrate the conflicts and problems of implementing this innovative planning concept, as described in the following points.

Regional parks have no restrictive effects against settlement development (no status of nature protection) and no concrete instruments and financial means to support land uses which protect the landscape; therefore there is a danger that regional parks *do not fulfil their functions either of protection or of development.* In contrast to the instrument of regional parks, the "orientation values" in LEPeV are able to limit the growth of communities.

Regional parks are a part of a "top down" strategy of Federal state and regional planning, rather than the idea and initiative coming from the local level ("bottom up"). The probability of realising the concept of regional parks is particularly dependent on *acceptance by the communities,* which have to become the actual promoters of the idea. This does not apply to most of the communities in the region of Teltow. The mayors and administrative leaders do not see the advantages for their communities, only the restrictions and disadvantages. They are not persuaded of the economic, cultural or social impulses for development as a result of regional parks, as promoted by Federal and regional planners.

Due to its central transport location the Teltowpark is one of the strongest areas of economic and structural growth in the suburban region of Berlin. This is the reason for many *conflicts between landscape protection and settlement development.* New business areas, new residential areas, new projects of transport infrastructure and the planning of the main airport Berlin-Schönefeld have a combined negative effect on protecting the landscape and its suitability for recreation.

Another difficulty of implementation is that the area of the plateau of Teltow is, like the area of Barnim in the north of Berlin, mostly characterised by large-scale agriculture and fallow land (former irrigation beds) and has had, and still does have, *little importance for recreation.* These areas have to compete with attractive landscapes with abundant water and forests (Wannsee, Müggelsee). Recreation and tourism have no or insufficient economic and social advocates who support the idea of a regional park in the region.

3 Approaches for Possible Solutions

Although it is possible that the development of other regional parks in the Berlin region are more successful than the case of the Teltowpark, it is necessary to improve the planning concept of regional parks. What are the ways of achieving a more successful implementation of regional parks?

To achieve a more successful planning strategy for regional parks the interests of the main planning addressees – the communities – has to be better considered and incorporated. It is important to see here that a polycentric settlement structure produces *winners and losers* amongst the communities. Whereas a few communities which are designed to become centres of settlements profit from this concept, most communities have to give up their plans of growth in the interest of protecting nature and the landscape. The future success and acceptance of strategies of spatial planning will depend on procedures to *compensate the costs of deferral in a regional context.*

An interesting approach for a regional balancing of interests is provided by the German regulation of intrusions into the environment ("naturschutzrechtliche Eingriffsregelung") which governs the avoidance, compensation and substitution of an intrusion into the balance of nature. In this context it is necessary to adapt the legislative framework to *compensate intrusions at an inter-communal level.* A pool of financial resources to realise and finance regional parks should be able to be obtained from large, building projects in the suburban area which take up much land area (e.g. the main airport Berlin-Schönefeld or logistic centres for freight).

Effects of Forest Fragmentation on Carabid Assemblages in the Urban Setting: Implications for Planning and Management

Jari Niemelä and Eero Halme

We compared carabid beetle assemblages (Coleoptera, Carabidae) between large (9.6–21.5 ha), medium-sized (4.2–8.2 ha), small forest fragments (0.5–3.0 ha), their surrounding agro-urban habitats and a nearby continuous forest in southern Finland. Carabid species richness was lowest in the continuous forest and highest in the surroundings. In a multivariate analysis of vegetation and carabid samples, sites from each fragment size class formed groups of their own and the small fragments were found to be similar to the surroundings. The proportion of open-habitat species increased with decreasing fragment size indicating that open habitat species easily invade the small fragments from the surroundings. Some specialised forest carabids were exclusively caught in the continuous forest. To preserve forest arthropod faunas in urban areas it is essential to leave large, continuous forest tracts untouched. Since this may be difficult to achieve, sizeable forest fragments must be set aside and ecological networks of forested habitats must be created.

1 Introduction

Urbanisation is one of the major causes of fragmentation of natural habitats. For instance, in our study area in Finland, the number of forest patches increased from 25 in the 1930's to 603 by the late 1970's due to urbanisation, while their size decreased dramatically (Wuorenrinne 1983). In this chapter we investigate the occurrence of carabids among different sized forest fragments and in the adjacent urban-agricultural matrix. We discuss the management implications of our findings.

2 Material and Methods

The study area is in Espoo, just west of Helsinki in southern Finland. Ten sites in a continuous coniferous forest, and five forest fragments in each of the fol-

Department of Ecology and Systematics, P.O. Box 17, FIN-00014 University of Helsinki, Finland

J. Breuste, H. Feldmann, O. Uhlmann (Eds.)
Urban Ecology

lowing size classes were studied: small (0.5–3.0 ha), medium-sized (4.2–8.2 ha) and large (9.6–21.5 ha). We also took samples in the cultivated fields, abandoned fields and lawns surrounding each fragment. The beetles were collected by 15 pitfall traps in each site (170 ml plastic jars, diameter 67 mm) during five days in June and five days in August, 1983. The field and ground layer vegetation was described in five randomly selected 1 m^2 squares at each site. For a more detailed description of the study area and methods see Halme and Niemelä (1993).

3
Results

Rarefaction estimate of carabid species richness was lowest in the continuous forest (12.8 species in a sample of 150 individuals) and in the medium-sized fragments (11.2), somewhat higher in large (16.8) and small fragments (17.6) and highest in the surroundings (26.4). However, species richness did not differ significantly between small and large fragments and continuous forest.

Three main patterns were distinguished in the DCA-ordination of the vegetation.

1. Vegetation in the ten sites in the continuous forest was very similar.
2. Variation in vegetation increased among the sites as the fragment size decreased and the small fragments (0.5–3 ha) were more similar (with respect to vegetation) to the surroundings than to the continuous forest.
3. There was much variation in the vegetation among the surroundings of the fragments.

A DCA-ordination based on the carabid samples was similar to that of the vegetation (Fig. 1). Two types of carabid assemblages can be distinguished based on the DCA-analysis: (1) assemblages in the surroundings and small fragments, and (2) assemblages in the larger fragments and in the continuous forest.

The proportion of carabid species of open habitat increased (habitat preferences obtained from literature) with decreasing size of fragments (proportion of open habitat species:forest species in continuous forest 0:81, large fragments 11:79, medium-sized fragments 7:73, small fragments 28:44, surroundings 43:7).

4
Discussion

In small fragments both forest species and open habitat species co-existed. This pattern is explained by increasing colonisation of species into the patches from the surroundings with decreasing patch-size. Similar results were

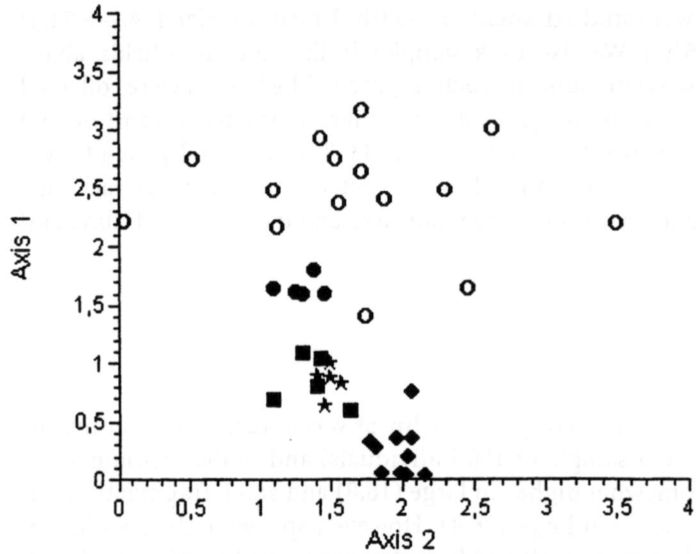

Fig. 1. Detrended Correspondence Analysis of the carabid samples of the sites. Eigenvalue of the first axis is 0.49 and the second axis 0.18. Symbols: ○ surroundings, ● small fragments, ★ medium-sized fragments, ■ large fragments, and ◆ continuous forest.

reported by Usher et al. (1993). In the small fragments, the dry microclimate and vegetation including many grass species tolerant to human disturbance favours carabids adapted to grasslands (Thiele and Weiss 1976).

Vegetation in the small forest fragments has changed not only due to human disturbance but also due to 'edge effects', e.g. light and wind conditions being different from in the forest interiors. Levenson (1981) estimated that the minimum fragment size for the survival of interior forest vegetation is 4 ha. Our small fragments were smaller and are thus actually edge habitat. However, whether the invasion of the small fragments by carabid species from the surroundings was due to habitat effects increasing the probability of survival in the fragment, or to the small size of the fragment leading to a high number of occasional invaders cannot be determined with the present material.

Two observations have implications for the management of urban landscapes:

1. The similarity of the carabid assemblages in the forest fragments with those in the strongly human-modified surrounding habitat increased with decreasing size of the forest fragment.
2. The strictly forest species occurred only in the continuous forest (e.g. *Carabus glabratus, C. violaceus*).

These forest specialists are flightless and probably reluctant to cross open habitat (Riecken and Raths 1996) which probably decreases their possibilities of moving between suitable forest patches. These observations imply that large forest fragments must be left undisturbed to maintain populations of forest species. However, as it may be difficult to protect large forest tracts near urban centres due to demands by other forms of land use, a compromise may be to set aside sizeable forest fragments as refuges for forest species and to create ecological networks of forested habitats. In order to monitor the success of such management, carabid beetles may function as a useful indicator taxon (Niemelä 1996).

References

Halme E, Niemelä J (1993) Carabid beetles in fragments of coniferous forest. Ann. Zool. Fennici 30: 17–30

Levenson J (1981) Woodlots as biogeographic islands in south-eastern Wisconsin. In: Burgess R, Sharpe D (eds) Forest island dynamics in man-dominated landscapes. Springer, Berlin Heidelberg New York

Niemelä J (ed) (1996) Population biology and conservation of carabid beetles. Ann. Zool. Fennici 33: 1–241

Riecken U, Raths U (1996) Use of radio telemetry for studying dispersal and habitat use of *Carabus coriaceus* L. Ann. Zool. Fennici 33: 109–116

Thiele H-U, Weiss H (1976) Carabids as bioindicators of microclimate changes caused by man in a meadow forest habitat. Sonderdruk aus der Schriftenreihe für Vegetationskunde Heft 10: 359–374

Usher M, Field J, Bedford S (1993) Biogeography and diversity of ground-dwelling arthropods in farm woodlands. Biodiversity Letters 1: 54–62

Wuorenrinne H (1983) Kaupunkilähiöiden metsistä. In: Kuusela S and Heikkilä T (eds) Uudenmaan metsäopas. Uudenmaan Ympäristönsuojelupiiri, Helsinki

Poster 5.4

Comparative Studies on the Size of Annual Rings, Leaf Growth and the Structure of Treetops of Urban Trees in Frankfurt/Main

Hans-Joachim Ballach, Jan Goevert, Sabine Kohlmann, and Rüdiger Wittig

1 Introduction

Trees growing along streets and roads are exposed to various detrimental factors in cities where motor vehicle traffic directly or indirectly plays an important role (such as in Frankfurt/Main). Such stress parameters comprise air pollution, compaction of soil, sealing of soil surface, mechanical injury, drought, flooding, as well as above normal temperatures and photoperiods. Many of these parameters, alone or in combination with others, can cause water stress. This is particularly important because it occurs frequently and affects so many processes (Kramer 1987). In this study, the growth of trees located in traffic-free areas (e.g. public green areas and parks) was compared with growth parameters of trees growing near streets and roads with heavy traffic (approximately 40,000 vehicles/24 h). Up to now, several studies have been published to describe the impairment of the vitality of urban trees near roads, but comparative studies using methods of demonstrating water stress on different tree levels (leaf, shoot, trunk) are quite rare. This chapter presents the results of such comparative measurements used for two tree species which are classified in the literature as being suitable for urban locations to a different degree.

2 Methods

The length of treetop shoots, the width of annual rings and the leaf area were measured. Five trees per species and location were selected. Trees were selected on the basis of their suitability for urban areas on the GALK scale (GALK 1992). *Platanus* × *hybrida* is recommended, whereas *Aesculus hip-*

Department of Geobotany and Plant Ecology, J. W. Goethe-University, Siesmayerstrasse 70, D-60323 Frankfurt/Main, Germany

J. Breuste, H. Feldmann, O. Uhlmann (Eds.)
Urban Ecology

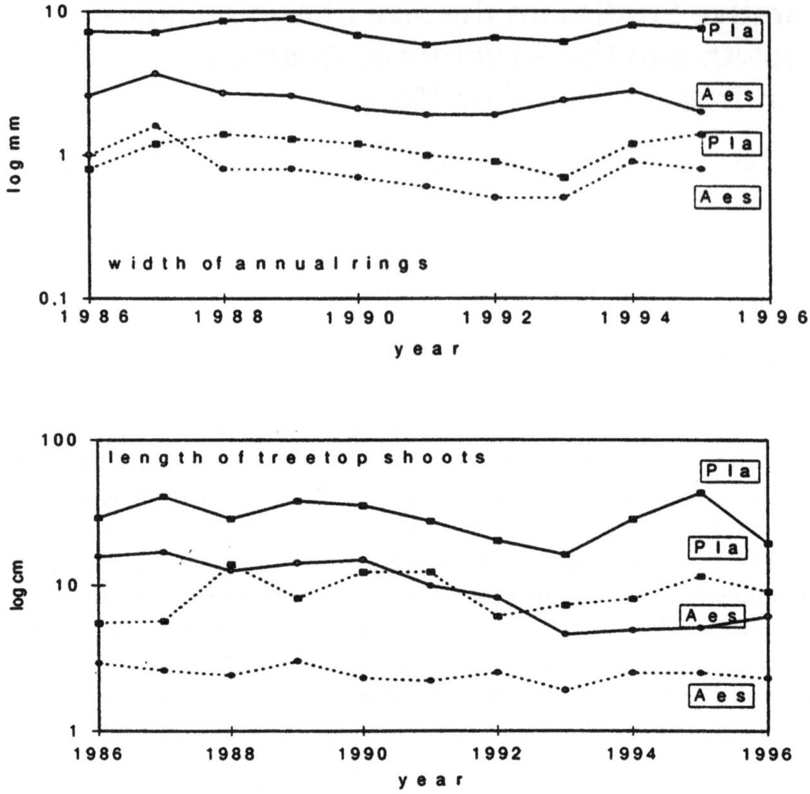

Fig. 1. Mean values (n = 5) of measurement of width of annual rings and of treetop shoots of *Platanus* × *hybrida* and *Aesculus hippocastanum* at traffic-free locations (solid curves) and at locations with heavy traffic (dotted lines) in Frankfurt/Main

pocastanum is classified as unsuitable. The age of the trees was between 40 and 100 years. As far as possible, care was taken to select trees of similar ages from the different study sites. For the analysis of annual rings, two drilled cores were taken from the trunk of each tree at a height of 1.30 m. The leaf area of 30 leaves of both tree species was analysed with an area meter (DIAS, Eijelkamp, the Netherlands); 3 branches per tree of the same 5 trees per species and location which were used for the other measurements.

3 Results and Discussion

Both analysis methods (measurement of width of annual rings and of the length of treetop shoots), revealed correspondingly that the two tree species under study suffered a continuous loss of growth at the locations with heavy traffic compared to the trees from traffic-free locations (Fig. 1). The more highly recommended species *Platanus* × *hybrida* exhibited without exception higher values of shoot and trunk growth than *Aesculus hippocastanum*. The latter revealed a higher rate of decrease of mean leaf area (211 cm^2: traffic-free location and 87 cm^2: heavy traffic location) than *Platanus* × *hybrida* (162 cm^2: traffic-free location and 113 cm^2: heavy traffic location).

Kramer (1987) reports that the most important direct effect of water stress is a decrease in cell enlargement, resulting in decreased leaf area, decreased width of annual rings, and decreased size of trees. Our studies show a good correlation between the decrease of width of annual rings, the growth rate of the branches of the treetop and the leaf growth rates caused by a complex of stress parameters at heavy traffic locations.

References

GALK (1992) Ständige Konferenz der Gartenamtsleiter beim deutschen Städtetag - Empfehlungen zur Schadstufenbestimmung für Bäume an Straßen und in der Stadt. Schriftenr. FLL, Troisdorf, 2S

Kramer PJ (1987) The role of water stress in tree growth. J. Arboriculture 13: pp. 33–38

Extensive Biotope Mapping in Cologne as a Basis for the Protection of Nature through Landscape Planning

Christian Hartwig and Betina Küchenhoff

1
Basics and Practical Application of Biotope Mapping

Due to the competition for land utilisation with regard to the remaining undeveloped areas in cities, nature conservation orientated planning can only be provided on the basis of competent inquiry of the ecological potential of the remaining areas.

The municipal administration of environmental affairs in Cologne was thus appointed by the municipal council to carry out an extensive biotope mapping project.

The project started in 1994 and some of the results can be used for planning urban development, especially in development plans and landscape planning.

Another aim of the extensive biotope mapping is to prevent additional environmental damage. Large coherent landscapes should be conserved, biotopes should be linked and the ecological value should be increased. The concept discussed should help to retain a great number of species in urban regions.

2
Method of Biotope Mapping and Result Transfer into Urban Development

Step one of the mapping is the evaluation of colour-infra-red (CIR) aerial photos on a scale of 1 : 5000.

Step two supports the photo interpretation by intensive mapping on the ground to supply specific details.

Part of the inquiry is a valuation of each biotope; this is necessary to calculate measures of compensation.

The mapping is guided by a special Cologne-code, which was based on the 'Instructions for biotope mapping in city areas' (Arbeitsgruppe 'Methodik der

City of Cologne, Department of Environment, Gürzenichstr. 6-16, D-50677 Cologne, Germany

J. Breuste, H. Feldmann, O. Uhlmann (Eds.)
Urban Ecology

Biotopkartierung im besiedelten Bereich' 1993) and the valuation system for biotopes established by Ludwig (1991). It was adapted further to suit the local conditions of Cologne.

Finally the results of the mapping are entered into the data base of a Geographical Information System (GIS) to guarantee a connection with other environmental data.

Faunal and floral surveys have been made in order to estimate the value of special biotope areas for single species, as a supplement to the biotope mapping. These surveys allow conclusions to be made about the biotic and abiotic conditions of the biotope areas. Based on this knowledge a concept for a habitat network can be developed to guarantee the survival of sensitive species.

Detailed investigations have been carried out at the gravel pit 'Meschenich' and a number of rare species have been detected. The inventory of species (*Oedipoda caerulescens*, *Bufo viridis*) concludes that this gravel pit represents an important substitute living space for species of the former floodplains of the Rhine. Moreover, other species (*Cicindela hybrida*, *Argyope bruennichi*) indicate that this area is favoured with regard to its temperature level.

Based on the described investigations, parts of the city areas will be selected for conservation because of their ecological importance, and a concept for the connection of biotopes can be developed.

These results can be linked to other environmental data (climate, soil, relief etc.) resulting in a so called "Ökologische Raumanalyse" (Ecological Area Analysis).

This analysis shows where urban development is possible (in accordance with the ecological results) and which areas should be conserved for nature.

This analysis includes no aspects of city planning to guarantee the optimum conditions with regard to ecology.

The weighing-up of environmental aspects in comparison with other aspects of urban development will follow in discussions with the municipal administration of planning affairs and the municipal council.

3
Example for an 'Ecological Area Analysis'

The region of 'Meschenich', south-west of the centre of Cologne, is shown as an example for a so called 'Ecological Area Analysis', which takes the place of an 'environmental impact assessment' for future city development.

This rural region is characterised by intensive farming and small villages and there are several large gravel pits in the area. The combination of environmental data described, together with data from the biotope mapping, resulted in the designation of 'taboo-areas' for future city development (Fig. 1). The ecological value of these areas should be increased by creating

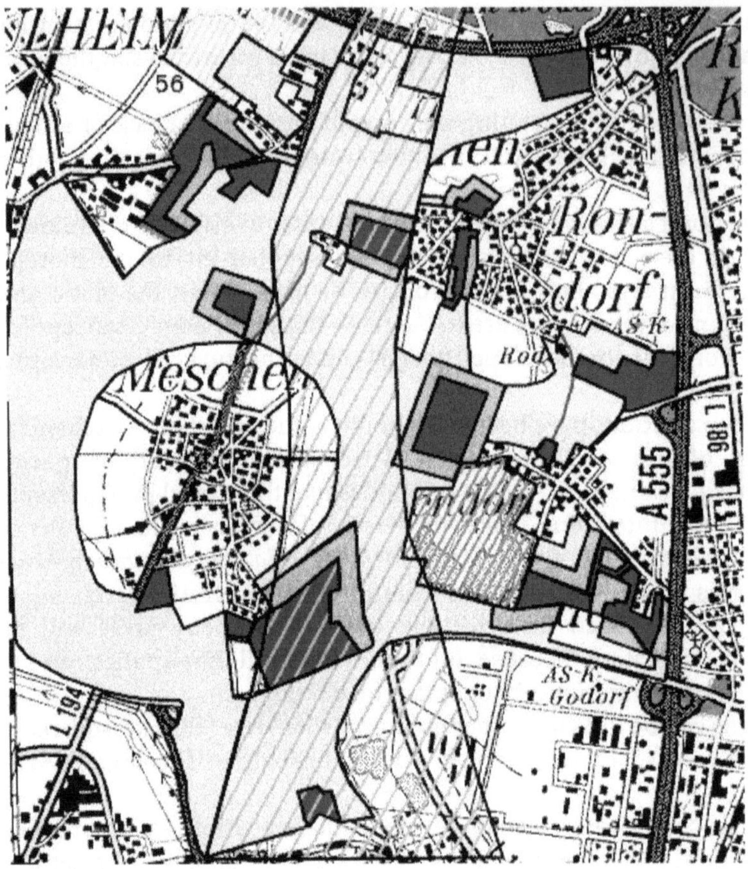

Legend
Taboo-Areas

- Important Biotopes
- Buffer Zones
- Areas for Habitat Network
- Regional Ventilation Corndor

Fig. 1. Results of the "Ökologische Raumanalyse" (Ecological Area Analysis) Köln - Meschenich, 1:35000

new biotopes and they should then become part of a habitat network. The areas should be developed according to the rural character of the landscape, which means that hedges, small wildwoods, extensive orchards etc. should be laid out.

The unmarked areas allow an enlargement of land set aside for building; the transition between countryside and urban settlement should be planned with care.

References

Arbeitsgruppe „Methodik der Biotopkartierung im besiedelten Bereich" (1993) Flächendeckende Biotopkartierung im besiedelten Bereich als Grundlage einer am Naturschutz orientierten Planung. Natur und Landschaft 68(10):491–526

Ludwig D (1991) Methode zur ökologischen Bewertung der Biotopfunktion von Biotoptypen. Froelich und Sporbeck, Bochum

Green Coverage for an Overcrowded City – an Evaluation Method for Urban Green Areas Using Landsat TM

M. Moriyama, H. Takebayashi, and B. Niimitsu

1 Introduction

Urban green areas should be planned and assessed based on their effects, such as mitigation of climate, purification of air, conservation of landscape, recreation areas for people and so on. The aim of the investigation described in this chapter is to produce an evaluation tool having the following functions:

1. Functional effective arrangement of green areas.
2. Assessment of environmental effects by the arrangement of green spaces.

Study areas were selected in Kobe City, Japan; Higashi-nada and Nagata districts. The areas suffered heavily from the Great Hanshin-Awaji earthquake of January 17, 1995, but are now going to be restored. This paper describes the process of producing the database for the planning of the green arrangement in the reconstruction plan.

2 Analysis of Land Cover Ratios

In this study maps of green coverage were made using Landsat TM data to evaluate in detail the spatial distribution of green coverage. The spatial scale of evaluation is assumed to be 100–1000 m and the minimum size of evaluation is assumed to be 30 m (this is a resolution of Landsat TM). The estimation of land cover ratios used the least-squares method, (Shimabukuro 1991).

Four categories were selected for a classification of urban surfaces, namely (1) green cover, (2) bare ground, (3) water surface and (4) pavement and buildings. 6 bands of Landsat TM, eliminating the thermal band (band 6) were used for the analysis.

Estimation accuracy was examined by means of the comparison with ground truth data. A sampling unit of the data is 5 × 5 pixels (about 150 × 150 m) and a total of 25 sites were sampled from each district. The data were

Research Centre for Urban Safety and Security, Kobe University, Nada, Kobe 657, Japan

J. Breuste, H. Feldmann, O. Uhlmann (Eds.)
Urban Ecology

produced using aerial photos and geographical maps. The standard deviation of errors of green coverage was 5% on average and was smallest in land cover categories.

3 Evaluation of Green Cover

3.1 Evaluation Method Using a Two Dimensional Weighted Moving Average

Three evaluation maps of green cover were produced from the green coverage map in the above mentioned land cover ratios. They were based on the evaluation by means of required green coverage, thermal environment and green sight distance.

The influence of the spread of green cover around the evaluation point or site should be considered to evaluate physical or psychological environmental effects of the green area. In this study an evaluation method of a two dimensional weighted moving average was applied.

$G(x, y) = \Sigma g(x', y')f(d)/\Sigma f(d)$: moving averaged value of green cover

$g(x, y)$: evaluation value of green cover

$f(d)$: weight by distance

$d = \left\{(x - x')^2 + (y - y')^2\right\}^{1/2}$: distance between (x, y) and (x', y')

3.2 Evaluation by Required Green Cover Degree (RGD)

Required green cover degree (RGD) is expressed as the deficiency of green cover compared with some standard. It is defined as follows.

$$RGD = 1 - GC/SRG$$

where, GC = green coverage; SRG = standard of required green coverage, when RGD is negative, then RGD = 0.

3.3 Evaluation by Thermal Environmental Effect of Green

In general, the effect of a drop in urban air temperature caused by green cover depends on the distance from green cover. This effect also depends on both the size of the green cover and the wind direction. The weight pattern of an exponential function reduction type is produced from the calculation results (Moriyama et al. 1997), assuming the influencing distance to be 250 m.

The wind direction data came from hourly data observed at 3 stations in each district.

3.4 Evaluation by Green Sight Distance

„Green Sight Distance" is defined as a distance at which people can 'feel' green cover by the visual sensation. Although it depends on individuals, the green sight zone was assumed here to be within a radius of about 250 m. The weight pattern of distance is assumed to be a fisheye lens type (cosine curve).

4 Results

1. RGD is defined as a deficiency from some standard of required green coverage (SRG). In this case, SRG is adopted as 12% according to the practical value for the objective areas. The higher values mean a great necessity to increase green cover.
2. Due to the drastic reduction in the weight distance, the patterns of „Thermal Environmental Effect of Green" appeared in irregular shapes at the boundary. The green cover of Mt. Rokko forest affects the results.
3. The values of the evaluation map of „Green Sight Distance" are higher than the values produced by „Thermal Environmental Effect of Green". The evaluation patterns are similar to the RGD.

References

Moriyama M, Takebayashi H, Fukumoto K (1997) Effects of green areas on urban air temperature by numerical solution. Mem. Grad. School Sci. & Technology, Kobe University 15-A:101–115

Shimabukuro YE, Smith JA (1991) The Least-Squares Mixing Models to generate fraction images derived from Remote Sensing Multi-spectral data. IEEE Trans. Geoscience and Remote Sensing 29(1):16–20

Effect of Urbanisation on Local Herpetofauna and its Possible Indicator Value, a Decade-Long Experience in Budapest

Miklós Puky

1 Introduction

Settlements have usually been built in areas where water was available in adequate quantities. As a consequence, they often include different water bodies and wetlands, which are, or have frequently been, valuable from a present nature conservation viewpoint, too. In spite of this the disappearance and isolation of these landscape elements is a common trend world-wide. Besides their obvious habitat function, they also serve as key areas in the maintenance of the corridor function of, for example, rivers flowing through cities. Settlements may bear conditions which can provide real (e.g. foraging) advantage (Ballasina 1984) attracting several species in different ways (Baker 1990).

The plant communities and birds of cities are usually quite well-known, while other groups are often neglected. In Hungary only limited information is available on amphibians (Puky, Kecskés 1991) though their bi-phasic life strategy, diverse ecological needs and relative abundance make them ideal organisms for the monitoring of environmental changes. The aim of this study was to investigate the herpetofauna of the Hungarian capital and to record its changes in the medium term.

2 Site and Methods

Budapest has 2.3 million inhabitants and covers 525.3 km^2. The altitude change is nearly 400 metres between the forest covered hills on the right bank and the lowland on the left side of the Danube. Diverse water bodies can be found within the border of the city. Mountain streams enter the Danube here (usually in a canalised form at their mouth) and the last remnants of the Danubian floodplain can also be found in low lying areas.

Hungarian Danube Research Station, Institute of Ecology and Botany, Hungarian Academy of Sciences, H-2131 Göd Jávorka S. u. 14., Hungary

J. Breuste, H. Feldmann, O. Uhlmann (Eds.)
Urban Ecology

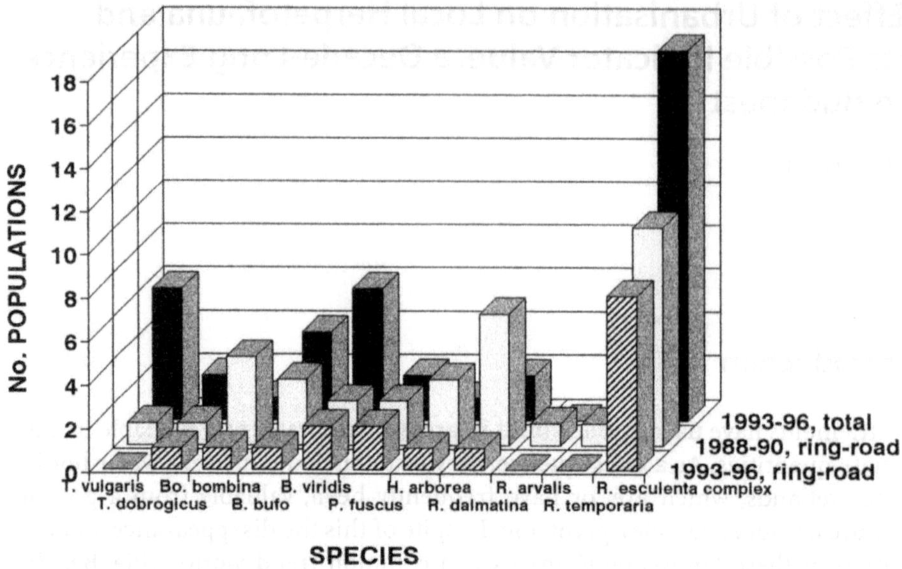

Fig. 1. **"Occurrence of amphibian species in Budapest"**

The building area of a 87 km ring-road was studied in 1988–90 and resurveyed for four years between 1993–96 to detect changes due to construction works and natural causes. Amphibians were selected for biomonitoring of water bodies; the chemical composition of the water together with a number of other parameters were also recorded. In the second phase (1993–96) another 21 sites, including all typical water types in the city from lowland streams to forest ponds, were added to the original list to obtain an overview of their general status.

Habitats were visited several times a year under different conditions and in different seasons. Where it was possible the population characteristics of the amphibian communities (species, sex ratios, length-weight distribution) were also described in detail to gain a broad overview of the ongoing processes.

3 Results and Discussion

A relatively high species number, altogether 10 species and a species group (68.7% of the total amphibian fauna of Hungary) were found at the end of the 1980's; this is similar to the situation in other large cities (see Langton 1991, Nabrowsky 1987). Unlike the situation in London, for example, no introduced species were observed (Puky, Kecskés 1991). *Rana esculenta complex* was the

Table 1. Chemical characteristics of small water bodies in Budapest between 1993–96

Locality	Chemical Parameters						
	NO_3^- [mg/l]	SO_4^{2-} [mg/l]	PO_4^{3-} [mg/l]	Cu^+ [mg/l]	Fe^{2-} [mg/l]	Conductivity [mS/cm]	TDS [g/l]
Csepel island	-	33	0,20	0,01	0,07	3,81	1,90
Lake Balázs	0,40	>75	0,02	0,01	0,03	0,9	0,45
Lake Jászberényi	-	38	1,21	0,19	0,16	-	-
Lake Naplás inflow	0,10	>36	0,04	0,08	-	0,90	0,43
	3,50	27	0,75	-	-	-	
Lake Temető	3,33	>75	0,29	0,71	-	1,02	0,52
Ördögárok	1,50	>75	2,09	1,54	-	1,19	0,60
Törökbálint, fish pond	-	>1300	0,04	1,38	0,09	-	-
Danubian floodplain	-	36	0,38	0,04	0,13	-	-
Lake Kavicsos	2,10	>75	0,01	-	-	1,80	0,84
Szilas stream	8,20	>75	0,10	-	-	-	-
Lake Istenszeme	0,50	>75	0,21	-	-	1,06	0,53
Lake Libegő	0,20	7	2,75!	-	-	2,06	1,04

commonest recorded taxon. The only lowland occurrence of *Rana temporaria* in the country was also detected during this survey (Stollmayerné et al. 1991). Several water bodies were inhabited by large and diverse communities (Fig. 1), others only by one or two species. By the mid-1990's two species disappeared and all but two (*Bufo viridis, Pelobates fuscus*) were present at fewer localities than before.

Amphibian decline is a well-known global phenomenon (Wake 1991), in this case the fast extinction rate was mainly in connection with a variety of human activities and drought. Isolation struck the investigated localities in two ways. Standing water bodies suffered the most from construction works. Running waters seemed to have barrier sections for some species. Pollution through dumping resulting in concentrations above the maximal acceptable limits was also shown to cause problems in the functioning of some aquatic systems (Table 1). Despite this, morphological deformities, which were reported in cases when urban pressure was high (Vershinin 1990), have not been recorded in Budapest so far. The most stable habitats were fish ponds, however, they did not contain the rarest species.

Urgent measures are needed to conserve the herpetofauna of the Hungarian capital. Besides regulations, a possible survival strategy has to include negotiations with the local councils and schools and the education of the residents on the value of their local waters in order to reverse some negative effects.

References

Baker J (1990) Toad Aggregations under Street Lamps. The British Herpetological Society Bulletin. 31:26–27

Ballasina D (1984) Amphibians of Europe: a colour field guide. David and Charles p 132

Langton L (1991) Distribution and Status of Reptiles and Amphibians in the London Area. The London Naturalist 70:97–124

Nabrowsky H (1987) Amphibien und Reptilien in Berlin. Naturschutzarbeit in Berlin und Brandenburg 23(2–3):48–51

Puky M, Kecskés F (1991) Herpetological investigation along a planned ring-road in and around Budapest – the M0 project. Proc Sixth Ord Gen Meet SEH, Budapest. pp 367–370

Stollmayerné Boncz E, Kecskés F, Ócsag A, Bognár A, Puky M, Bódi L (1991) Adatok a Naplás tó és környéke élôvilágához. Calandrella. V/1:65–84

Vershinin LV (1990) Features of amphibian populations of an industrial city. Proc Urban Ecological Studies in Central and Eastern Europe. Warsawa pp 112–119

Wake DB (1991) Declining amphibian populations – a global phenomenon? Alytes 9(2):33–42

Proposal for an Interdisciplinary Nature Conservation Assessment Approach Applied to an Exemplary Evaluation of Nunhead Cemetery, London, UK

Marcus Zisenis

Modern consciousness of nature conservation considers all demands of ecosystems and our human society, especially in urban areas. Nature conservation is understood as a spatial landscape planning process which needs general guidelines to allow assessments of comparable habitats.

Nunhead Cemetery in the London Borough of Southwark (UK) has been selected as an example to prove the practicability of an interdisciplinary nature conservation assessment approach, which takes into account both non-anthropocentric and anthropocentric reasons (Table 1).

Primary information was gathered on habitat structures, bird territories, butterflies, flora and vegetation, and on data on visitors. It was completed by obtaining secondary information for the research in 1993 (Zisenis, 1993).

Nunhead Cemetery is mainly covered by secondary Ash-Sycamore woodlands (*Fraxinus excelsior-Acer pseudoplatanus* woodlands). Ivy (*Hedera helix*) is dominant on the ground (Zisenis, 1996).

This is not only the typical and natural spontaneous vegetation of this succession stage on London clay, but also a highly endangered habitat on local, regional and national levels in the UK.

Historic gravestones and mature woodland were bulldozed during woodland management on Nunhead Cemetery on Nunhead beginning in the late 1970's. Sycamore was cut down and open glades were created for assumed nature conservation purposes to increase overall species diversity. This was carried out in accordance with recommendations by public nature conservation administrations, in particular the Joint Nature Conservation Committee (JNCC, which is responsible for nature conservation in Great Britain) and London Ecology Unit (LEU, which is in charge of most of the London Boroughs).

The result of this impact on Nunhead Cemetery's highly vulnerable woodlands has been a reduction of closed woodland species below their minimum viable populations, for example, woodland birds, beetles and such plants. In contrast, open land and wood-edge species were favoured by these actions e.g. scrub birds, butterflies and ruderal vegetation.

Eisenacher Straße 54, D-10823 Berlin, Germany

J. Breuste, H. Feldmann, O. Uhlmann (Eds.)
Urban Ecology

Table 1. Interdisciplinary nature conservation assessment approach

Nature conservation assessment			
Reasons			
Non-anthropocentric		Anthropocentric	
Ethical	Economic	Psychological	Social
	Educational	Culture-historical	Scientific
Criteria			
Rarity and endangerment	Naturalness/degree of human impact		Typicalness
Re-establishment ability	Vulnerability	Ecological function	Usability

The culture-historical jewel in the heart of London cannot be re-established because monuments and garden architecture of this Victorian cemetery represent a certain historical period. In addition, neither artistic nor financial resources are available for reconstruction. Furthermore, the concreted environment of this isolated green space prevents the recolonization of a similar species biodiversity.

Nunhead Cemetery is still used by plenty of visitors for recreation mainly from the surrounding area (up to 5 km away) as well as by school classes and scientists. Its ecological function as a habitat for birds, endangered beetles and other woodland species is outstanding for this area. The usability benefits are enormous, not only for psychological regeneration and social contacts of neighbours, but also for the economic value. This refers to the production of basic living resources such as clean air and ground water as well as to the climatic benefits of the vegetation.

Last but not least, it is an ethical duty to support the effort of each organism on earth to survive such as the woodland species on Nunhead Cemetery. The research on Nunhead Cemetery has shown that habitats fulfil many functions as complex as our human society. This can only be taken into account in a reformed interdisciplinary nature conservation approach. The applied evaluation guideline is a comprehensive attempt to fit in with this modern nature conservation attitude.

References

Zisenis M (1993) Nunhead Cemetery, London, United Kingdom: a case study of the nature conservation value of an urban woodland and associated habitats. In: Zisenis M, unpublished M.Sc. in Conservation dissertation. University College London, University of London, London, UK, p 294

Zisenis M (1996) Secondary woodland on Nunhead Cemetery, London, UK. In: Zerbe, S (ed.) Vegetationsökologie mitteleuropäischer Wälder. Kolloquium zum "Ökologietag" 1995 in Berlin. Landschaftsentwicklung und Umweltforschung. Schriftenreihe im Fachbereich Umwelt und Gesellschaft, Technical University Berlin, Berlin, Germany, 104: pp 73–79

MIX
Papier aus verantwortungsvollen Quellen
Paper from responsible sources
FSC® C105338

If you have any concerns about our products,
you can contact us on
ProductSafety@springernature.com

In case Publisher is established outside the EU,
the EU authorized representative is:
Springer Nature Customer Service Center GmbH
Europaplatz 3, 69115 Heidelberg, Germany

Printed by Libri Plureos GmbH
in Hamburg, Germany